AF661862

MANUFACTURING SCIENCE

ALSO FROM EAST-WEST PRESS

Clark, D S & W R Varney Physical Metallurgy for Engineers, *Second Edition*

Ghosh, A Rapid Prototyping: A Brief Introduction

Ghosh, A & A K Mallik Theory of Mechanisms and Machines, *Third Edition*

Iyengar, N G R & S K Gupta Structural Design Optimization

Mallik, A K Principles of Vibration Control

Mallik, A K & Shyamal Chatterjee Principles of Passive and Active Vibration Control

Ray, H S, R Sridhar & K P Abraham Extraction of Nonferrous Metals

Roy, D N Applied Fluid Mechanics

Srinath, L S Mechanical Reliability

Srinath, L S Reliability Engineering, *Fourth Edition*

Cover designed by Amrita Kar and Kuldeep Singh Singha

MANUFACTURING SCIENCE

Second Edition

Amitabha Ghosh

Formerly Professor of Mechanical Engineering
Indian Institute of Technology Kanpur

Senior Scientist
Indian National Science Academy, New Delhi

Honorary Distinguished Professor
Indian Institute of Engineering Science and Technology, Shibpur

Asok Kumar Mallik

Formerly Professor of Mechanical Engineering
Indian Institute of Technology Kanpur

Senior Scientist
Indian National Science Academy, New Delhi

Honorary Distinguished Professor
Indian Institute of Engineering Science and Technology, Shibpur

EWP

Affiliated East-West Press Private Limited

NEW DELHI

To
Meena & Kaberi

2022

© Affiliated East-West Press Private Limited

No reproduction in any form of this book, in whole or in part (except for brief quotations in critical articles or reviews), may be made without written permission of the publishers.

Published by Affiliated East-West Press Private Limited
New Delhi 110 002
e-mail: aewp.pub@gmail.com

ISBN 978-81-7671-063-3

Contents

Preface

This text was published first in 1985 by Affiliated East-West Press, New Delhi, and later, in 1986, by Ellis Horwood Ltd., Chichester (U.K.). Though it was thereafter reprinted a large number of times, no new edition was brought out. Considering its popularity both in India and abroad, we decided to produce a new edition. We have attempted this by adding a new chapter that provides the reader a reasonable idea about the major recent advances and future possibilities in manufacturing science and technology. We take this opportunity to record our appreciation of and sincere thanks to Dr. Rabibrata Mukherjee for his help in writing the section on "Self-Assembly". We also thank Chandranath Adak and Debasis Mukherjee for typing, and Pradipta Sarkar for drawing the initial figures of this new chapter.

We shall warmly welcome any suggestion for further improving the content of this study.

Amitabha Ghosh
Asok Kumar Mallik

Introduction

Since the beginning of civilization, man has been continuously engaged in converting the natural resources into useful products. The entire discipline of engineering is mainly concerned with this. Of all the subdisciplines of engineering, manufacturing is perhaps most important because it involves the conversion of a raw material into a final product.

The first industrial revolution was a major cornerstone in human history in that it set in motion the battle for abundance. The world today is totally dependent on science and technology and the nations are classified now according to their degree of industrialization. Since manufacturing plays a key role in extracting wealth from the mother Earth, it is quite easy to realize the importance of the subject which deals with this activity. Throughout the world market, there is a continuous struggle for cheaper production with better quality. This can be achieved only through optimal utilization of both material and human resources. Apart from this, with the advancement in technology, the manufacturing engineers are facing challenging problems requiring development of newer and better methods of production. The required rate of growth in manufacturing cannot be achieved through trial and error and a scientific approach is necessary. This has attracted the scientists and engineers and *manufacturing*, which was an art, has become a science.

The conversion of resources into raw materials is normally taken care of by two subdisciplines of engineering—mining and metallurgy. The real shaping starts from the stage a material is available in raw form. Changing a raw material into a final product generally involves various processes (*see figure*). The major

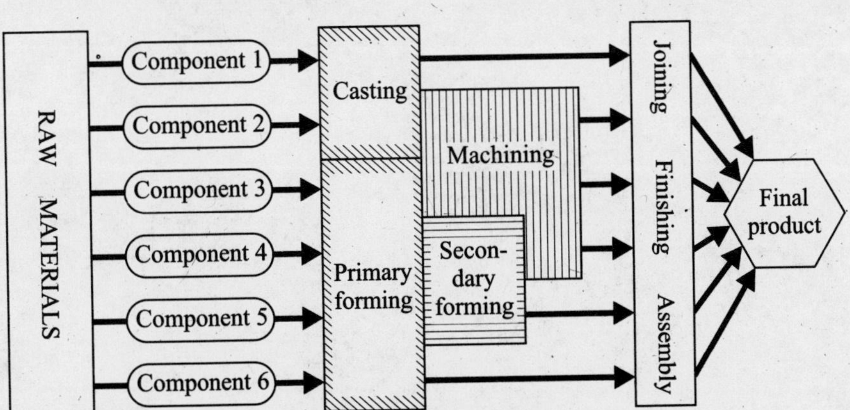

Manufacturing processes

processes a component undergoes before being converted into a final product are (i) casting, (ii) primary and secondary forming, (iii) machining, and (iv) joining, assembly, and finishing. A material is often subjected to primary forming for producing standard shapes such as those of rods, bars, and sheets. However, it may also undergo secondary forming, and this is performed when the objective is to obtain a shape other than the standard shapes.

The three engineering activities that require an understanding of the manufacturing processes are designing, production, and development of new techniques. At the design stage, manufacturing considerations have to be taken into account not only for producing the part in the most economic manner but also for imparting to the part some required properties such as strength and hardness. Further, in a production shop, a successful engineer must have a thorough understanding of the subject if he has to select and implement the right processes. The selection of the important process parameters is extremely important to achieve success in manufacturing. The third type of activity is development of new manufacturing processes and modification of the existing technology. It has already been mentioned that the production engineers are encountering more and more challenging problems, and often such problems cannot be solved with the help of existing conventional methods. Thus, a host of new processes has evolved; this group of manufacturing processes is generally known as *unconventional processes* or *new technology*. However, the recent emphasis on miniaturization has led to the development of micromanufacturing processes.

Unlike the fundamental subjects in science and engineering, the subject dealing with the manufacturing processes is interdisciplinary in nature. Obviously, therefore, understanding the science underlying the processes is not simple since, generally, the situations are not idealistic. However, many important generalized conclusions can be drawn with the help of simple models. The same information would otherwise require a large number of experiments or long experience.

1 Manufacturing Properties of Materials

1.1 STRUCTURE OF MATTER

The properties of a material are intimately connected to its basic molecular structure. Some knowledge of this structure is therefore essential for understanding the various macroscopic properties exhibited by the material. A general characteristic of all solids is their capability to retain definite shapes, and so we start from the mechanics of bonding between the molecules forming a solid.

1.1.1 BONDING OF SOLIDS

When two atoms are sufficiently close to each other, the outer electrons are shared by both the nuclei. This results in an attractive force between the two atoms. This force increases with the decrease in distance between the two atoms, as shown in Fig. 1.1. However, the two atoms do not collapse as a repulsive force is generated when the two nuclei come very close. This repulsive force increases rapidly with decreasing interatomic distance. The equilibrium interatomic distance d_e is that distance when the attractive and the repulsive forces are equal in magnitude (Fig. 1.1). The slope of the repulsive force curve is always more than that of the attractive force curve at the point of intersection *A* of the curves. Therefore, the equilibrium is of stable nature.

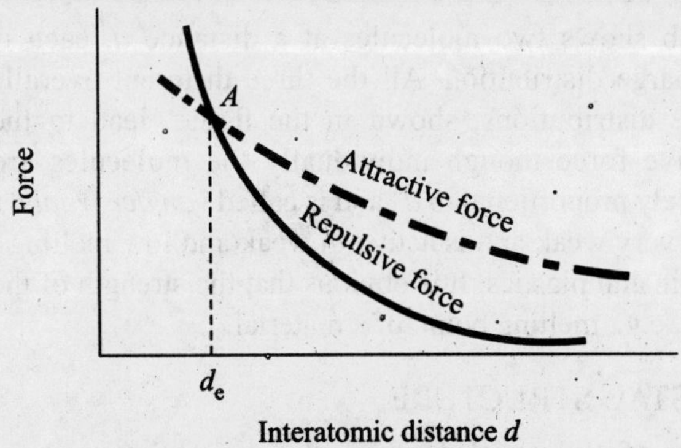

Fig. 1.1 Force between two atoms.

The mechanism we have discussed is one of the various possible interactions resulting in bonding between atoms, and is known as *covalent bonding*. In a given solid, one or more bonding mechanisms can be simultaneously operative.

The nature of a bonding mechanism depends on the electronic structure of the atoms involved. The bonding mechanisms predominant in solids include *metallic bonding* (in metals) and the *van der Waals bonding* (in molecular crystals). In a metal, a large number of free electrons are present, resulting in the formation of a common electron cloud; the rest of the system consists of positively-charged ions which are held together by the cloud (Fig. 1.2a). The mechanism of bonding in alloys is similar. Since the inert atoms do not possess free electrons, the metallic bonding mechanism cannot be operative. In such instances, however, a very weak short range attraction is generated due to the van der Waals force. The origin of this force is attributed to a rapidly-fluctuating dipole moment.

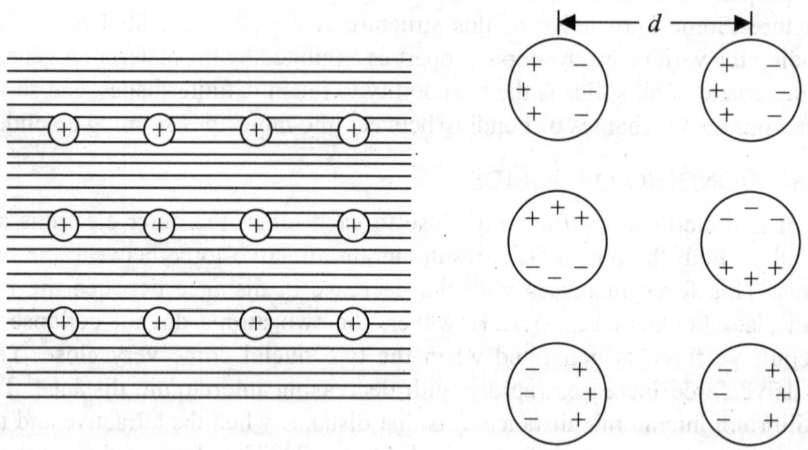

(a) Metallic bonding (b) Origin of van der Waals' force

Fig. 1.2 Bonding mechanisms.

Figure 1.2b shows two molecules at a distance d, each of which has a symmetric charge distribution. All the three different overall configurations of the charge distributions, shown in the figure, lead to the development of an attractive force though individually the molecules are neutral. This force is inversely proportional to d^4 and is called *van der Waals' force*. This type of bonding is very weak and is active in weak and low melting point materials such as paraffin and plastics. It is obvious that the strength of the bond controls the properties, e.g., melting point, of a material.

1.1.2 CRYSTAL STRUCTURE

The properties of a material depend not only on the bond strength but also on the arrangement of the atoms. In all metals and in many nonmetallic solids, the atoms are arranged in a well-ordered pattern. Such solids are commonly called *crystalline solids*. Of course, in a large number of situations, the whole solid is seldom composed of one single crystal. Instead, a very large number of small, randomly-oriented crystalline grains form the whole solid. Such materials are termed *polycrystalline*. Figures 1.3a and 1.3b show a single crystal and a polycrystalline solid, respectively.

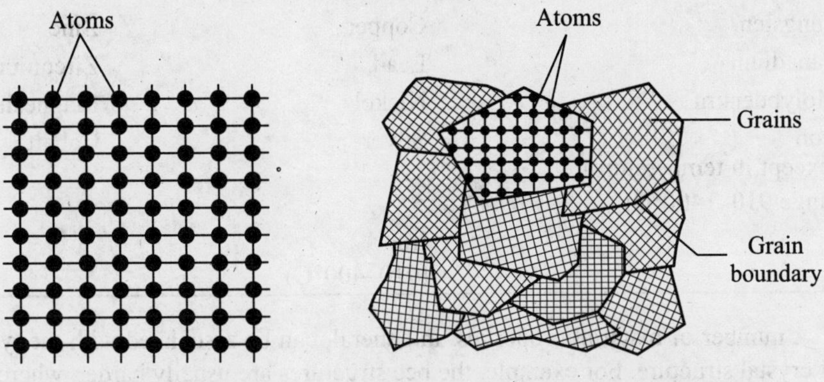

Fig. 1.3 Structure of crystals and polycrystalline solids.

In a crystal, we can identify the unit cell the repetition of which forms the whole crystal. The structure of a crystal is identified and described by this unit cell. The three commonly-observed crystal structures in metals are shown in Fig. 1.4. Of these three basic structures, the fcc and the cph crystals have

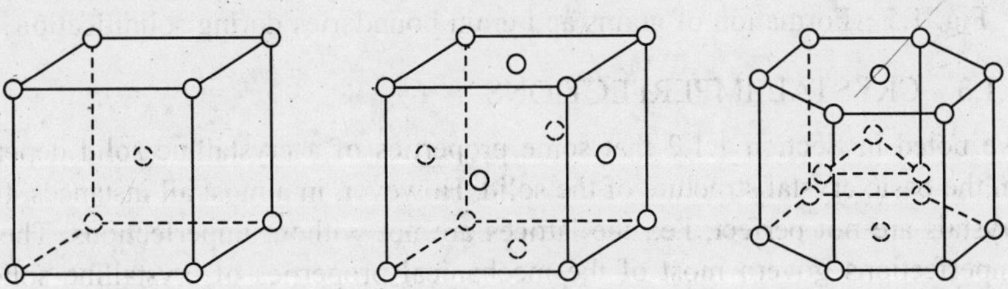

Fig. 1.4 Some common unit cells.

the most dense packing. The interatomic distance in such crystals is of the order of 10^{-7} mm. The crystal structures of some common metals are given in Table 1.1.

When a liquid metal solidifies by cooling, the atoms arrange themselves in regular space lattices, forming a crystal. The crystallization starts simultaneously at various places within the liquid mass. Figure 1.5 shows the growth of the crystal grains and the ultimate formation of the polycrystalline metal. Most metals have only one crystal structure. A few metals, however, can have more than one type of crystal structure. Such metals are called allotropic. Table 1.1 indicates that iron is an allotropic metal.

Table 1:1 . Crystal structure of some common metals

bcc	fcc	cph
Chromium	Aluminium	Titanium
Tungsten	Copper	Zinc
Vanadium	Lead	Zirconium
Molybdenum	Nickel	Magnesium
Iron (except in temperature) range 910–1400°C)	Silver	Cobalt
	Iron (910–400°C)	

A number of material properties, in general, can be associated with the type of crystal structure. For example, the bcc structures are usually harder, whereas the fcc structures are more ductile. In cph structures, the ductility is low.

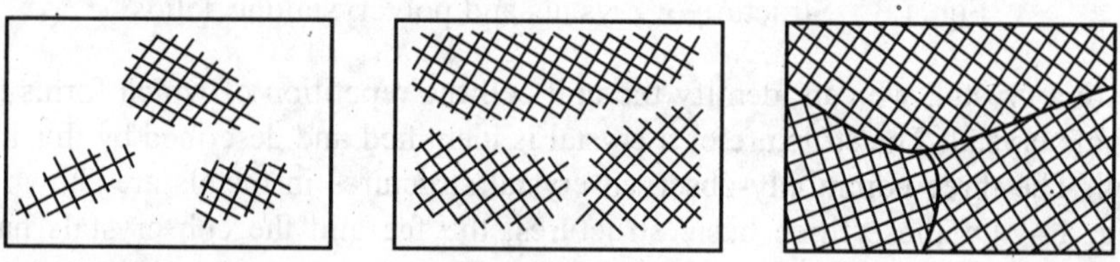

Fig. 1.5 Formation of grains and grain boundaries during solidification.

1.1.3 CRYSTAL IMPERFECTIONS

We noted in Section 1.1.2 that some properties of a crystalline solid depend on the basic crystal structure of the solid. However, in almost all instances, the crystals are not perfect, i.e., the lattices are not without imperfections. These imperfections govern most of the mechanical properties of crystalline solids (see Table 1.2). The study of the crystal imperfections and their effects on the properties of a material is a subject by itself. In our discussion therefore, we shall give only those concepts that will be required for an understanding of the various phenomena associated with different processes, e.g., plastic deformation and heat treatment.

The imperfections in a crystal lattice structure can be classified as follows:

(i) If an imperfection is restricted to the neighbourhood of a lattice point, the imperfection is referred to as a *point defect*. This type of imperfection can be due to various reasons. Figure 1.6 illustrates the three different types of point defects. In Fig. 1.6a, one lattice atom is missing, creating a *vacancy*. Since an atom vibrates about its lattice position, the tendency of the atom to jump out of its regular position creating a vacancy increases rapidly with its energy, i.e., the temperature. For example, the usual order of vacancy at 500°C is one in 10^{10} which increases to one in 300 at 2000°C. It is possible to increase the vacancy

Table 1.2 Properties dependent on crystal structure and defect

Property dependent on basic crystal structure	Property dependent on crystal imperfection
Density	Electrical conductivity
Specific heat	Yield stress
Coefficient of thermal expansion	Creep
Melting point	Fracture strength
Elastic constants	Semiconductivity
Hardness and ductility	Work hardening
	Fatigue strength

density at a given temperature by rapid cooling or extensive plastic deformation. In Fig. 1.6b, an atom is occupying an abnormal position. Such an atom is called an *interstitial impurity atom*. An interstitial impurity can be caused when an atom possesses large enough thermal energy or when its energy is increased by nuclear bombardment. In Fig. 1.6c, a regular lattice position is occupied by an atom of a different material.

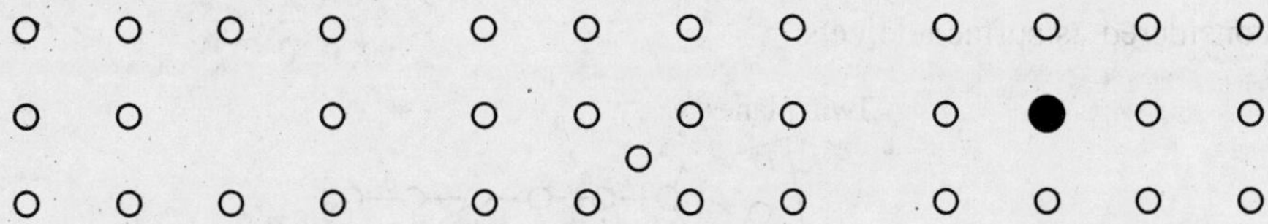

(a) Vacancy (b) Interstitial impurity (c) Substitutional impurity

Fig. 1.6 Types of point defects.

(ii) If an imperfection extending along a line has a length much larger than the lattice spacing, the imperfection is called a *line defect* or, commonly, a *dislocation*. Two common, simple types of dislocations are illustrated in Fig. 1.7. When an extra half-plane of atoms is accommodated by distorting the regular lattice arrangement (as done with the *XX′* half-plane in Fig. 1.7a), the resulting defect is termed as *edge dislocation*. Another type of distortion, resulting from a movement of the lattice atoms from their regular ideal positions, is shown in Fig. 1.7b. Such a defect is called a *screw dislocation*. The line separating the deformed and the undeformed regions is normally called the *dislocation line*. In Figs. 1.7a and 1.7b, the dislocation lines are *XY*. The dislocation density is defined as the total length of all the dislocation lines per unit volume. In a single crystal, the minimum attainable dislocation density is of the order of 100–1000 per cm^2, whereas the density in a normal polycrystalline solid is as high as 10^7–10^8 per cm^2. In Section 1.3.1, we shall show that plastic deformation takes place mainly through a movement of dislocations.

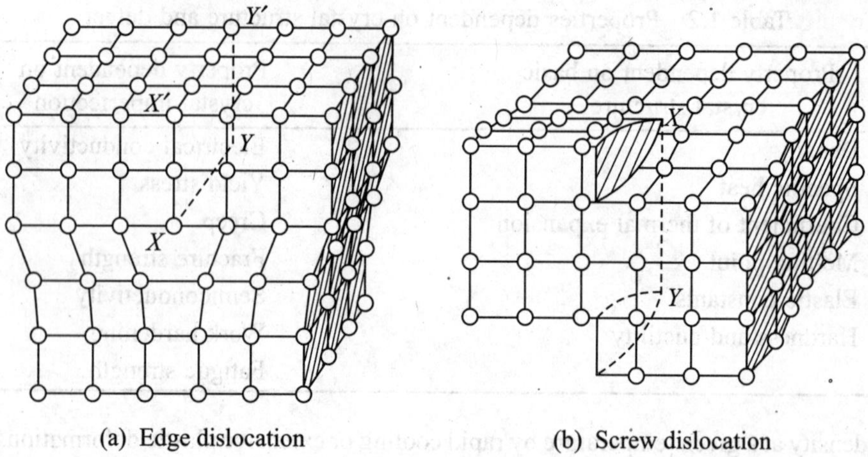

Fig. 1.7 Edge and screw dislocations.

(iii) When an imperfection extends over a surface, the imperfection is known as surface defect. Figure 1.8 shows a common type of surface defect known as twins. Twins are normally produced when a metal is stressed at a low temperature. The grain boundaries in a polycrystalline solid can also be considered as surface defects.

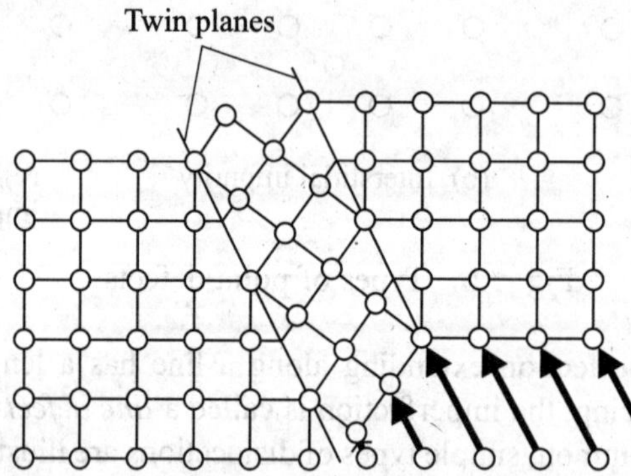

Fig. 1.8 Twinning mechanism.

There are other types of crystal imperfections and the real situation, in general, is quite complex.

1.2 METALS AND ALLOYS

The various types of materials used in engineering practice include, among others, metals, alloys, ceramics, and polymers. Of these, metals and alloys are commonly used. Metals are rarely used in pure form, and the desired properties are normally obtained by suitably alloying different metals. Alloys, unlike most pure metals, do not have a fixed melting point. Also, certain conditions have to be satisfied to make an alloy of two or more materials. In this section, we shall study the formation, the composition, and the control of properties of alloys.

1.2.1 SOLID SOLUTIONS

An alloy can be defined as a mixture of two or more materials, of which at least one must be a metal. The material having the largest percentage composition in the mixture is called the solvent and the remaining are called the solutes. Such a mixture is called a solid solution. In the solid state, the solute atoms can be present in the solvent in two different ways. When the size of the solute atoms is small enough so that they can occupy the interstitial spaces of the solvent matrix (Fig. 1.9a), the solid solution is of the interstitial type. For normal metals, the only useful material which can be accommodated in the interstitial spaces is carbon. The other type of solid solution is formed when the solute atoms occupy the regular matrix position by replacing some solvent atoms (see Fig. 1.9b). Such a solution is normally termed as a substitutional solid solution. In some solid solutions with two components, there is no restriction on the percentage composition.

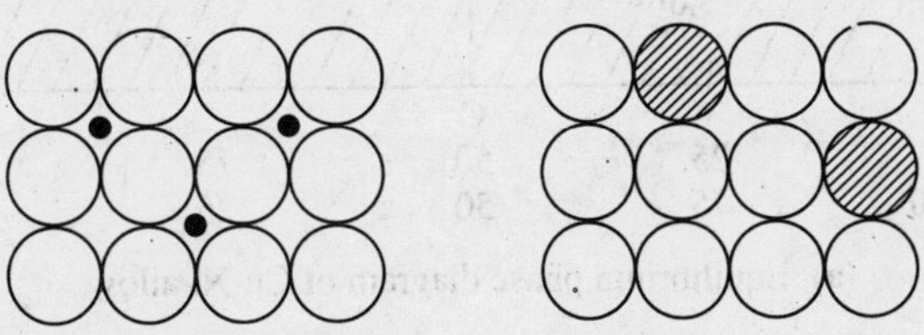

(a) Interstitial solid solution

(b) Substitutional solid solution

Fig. 1.9 Interstitial and substitutional solid solutions.

1.2.2 EQUILIBRIUM PHASE DIAGRAMS

The important metallurgical changes that take place when a mixture of different metals and/or materials is gradually cooled from a liquid state are best described with the help of an equilibrium phase diagram. Phases are characterized by the boundaries across which discontinuities exist in the physical properties. Even a pure material can be in different phases, namely, solid, liquid, or vapour. Moreover, even within the solid state, there can be different phases, each characterized by a different crystal structure. When cooling is sufficiently slow, we can assume that all the phases involved in the transformation process at a given temperature are in equilibrium with one another. Though, in general, the transformation of phases is governed by temperature composition and pressure, the latter plays an insignificant role in the processes that we shall consider.

A convenient way of describing the phase transformations is a diagram where the phases at different combinations of temperatures and compositions are indicated. Such a diagram is called an equilibrium phase diagram. For example, let us take the simple case of the Cu-Ni alloy which forms a solid solution without any restriction on the percentage composition. The equilibrium phase diagram for this alloy is shown in Fig. 1.10a. It has been obtained by considering the cooling curves for various compositions of the alloy. Figure

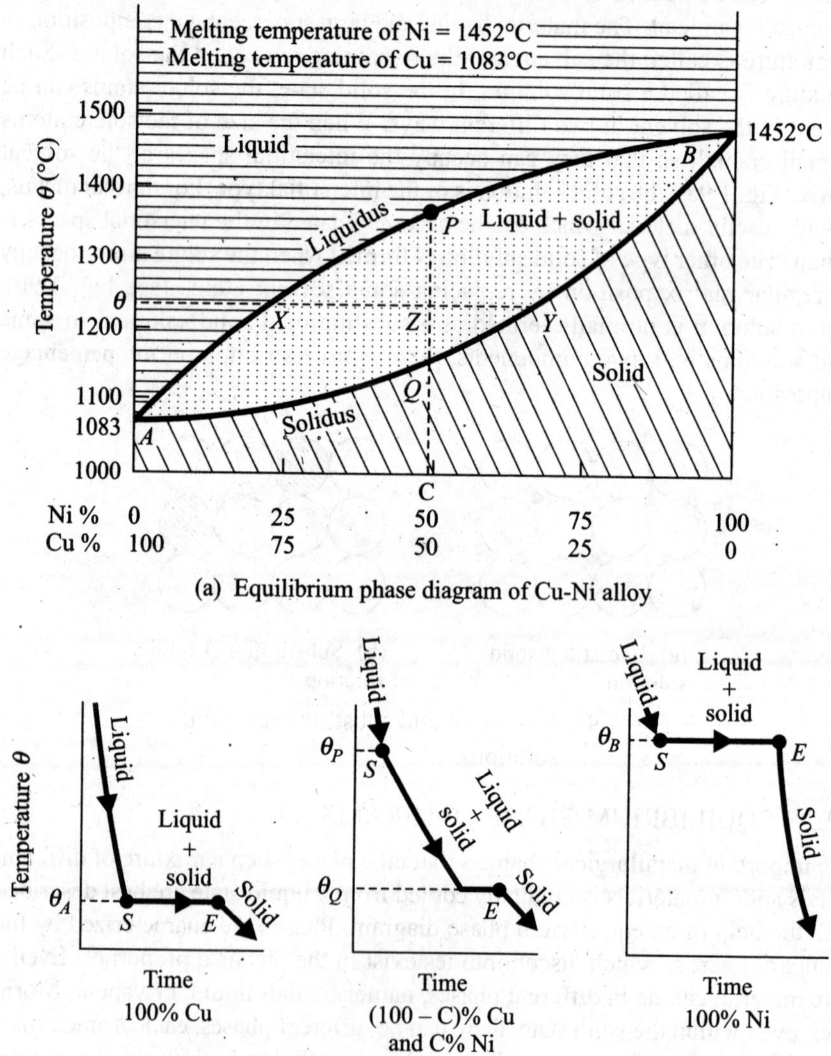

(a) Equilibrium phase diagram of Cu-Ni alloy

(b) Cooling and solidification of Cu-Ni alloy for various compositions

Fig. 1.10 Characteristics of Cu-Ni alloy

1.10b shows three typical cooling curves; here, the start and the end of the solidification process are indicated by *S* and *E*. Figure 1.11 shows the physical state of the mixture. It is evident that there are three distinct regions in this phase diagram. In the region at the top, the entire material is in the liquid state, whereas in the bottom region, the entire material is in the solid state. Further, in the intermediate region, the material is in the form of a mixture of solid and

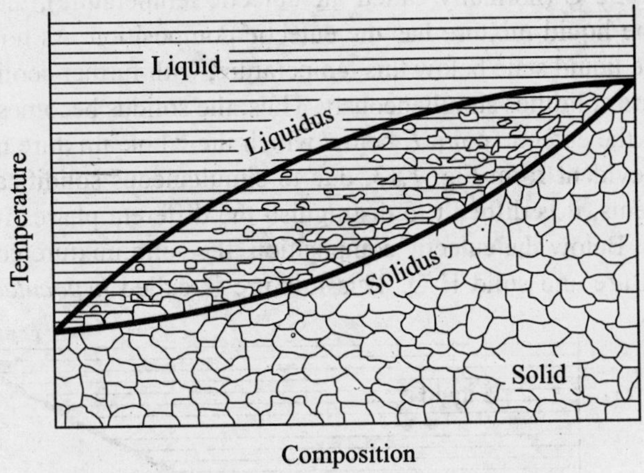

Fig. 1.11 Physical state of various phases.

liquid. The boundary *APB* (Fig. 1.10a) is called the *liquidus* and indicates the temperature beyond which the composition will be entirely in the liquid state. Similarly, the boundary *AQB*, indicating the temperature below which the composition will be totally solid, is called the *solidus*.

Using Fig. 1.10a, we can determine the relative proportion of liquid and solid. The ratio of solid and liquid for a composition C at a temperature θ is *XZ*/*ZY*. Moreover, the composition of the solid portion at the temperature θ is given by the abscissa of the point *Y*. Similarly, the composition of the liquid portion at this temperature is given by the abscissa of the point *X*. It may be noted from Fig. 1.11 that when there is no restriction on solid solubility, the solid state of an alloy looks like a pure metal. Such an alloy is called a single-phase alloy where nothing but the grain boundaries are distinguishable.

The example we have considered is quite simple as there is no restriction on solid solubility. However, the situation becomes complex when two metals, having no restriction on solubility in the liquid state, are only partially soluble in the solid state. Complete insolubility in the solid state is another extreme case. Let us consider the equilibrium phase diagram of this extreme situation. A solution of NaCl and H_2O can be taken up as an example in this category. Figure 1.12 shows the corresponding equilibrium phase diagram. From the nature of this diagram, it is obvious that at a particular composition (i.e., 23.5% NaCl), the mixture, like a pure material, has a specific freezing point (i.e., –22°C). This signifies simultaneous solidification of both NaCl and H_2O. Such a mechanical mixture of two solids is referred to as the eutectic. This eutectic composition is seen to have the minimum melting (freezing) temperature, and hence the name *eutectic* (which, in Greek, means *easy melting*). As before, let us consider a mixture of composition C at a temperature θ. The solid portion (Fig. 1.12) is seen to be pure ice (H_2O), whereas the liquid has a composition given by the point X. On further cooling, the point X shifts towards the eutectic point E. At –22°C (normally called the eutectic temperature), the entire solid is ice and the liquid mixture has the eutectic composition. As no mixture can remain in the liquid state below this temperature, with further cooling the entire liquid mixture solidifies simultaneously. Thus, the solidus becomes a horizontal line through the eutectic point E below which the whole mixture is in the solid state. As is evident from Fig. 1.13, due to simultaneous solidification of both the components, it is difficult to distinguish the different phases in an eutectic composition. Below the eutectic composition, the solid mixture consists of the eutectic mixture and solid H_2O. This mixture is called *hypoeutectic*. Beyond

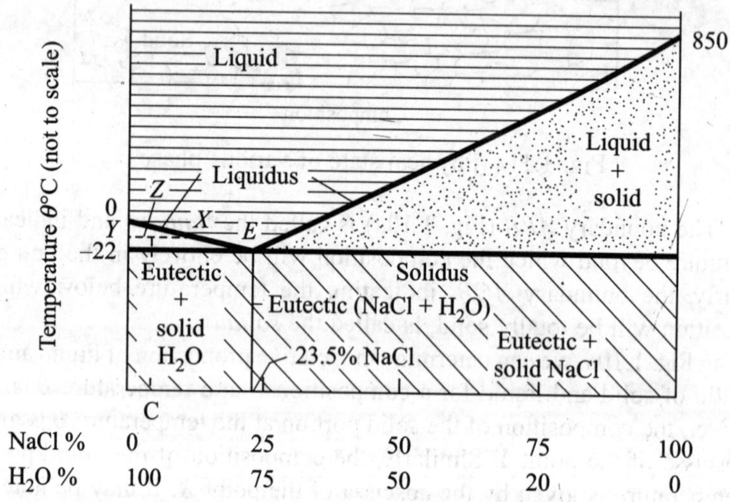

Fig. 1.12 Equilibrium phase diagram of NaCl + H_2O.

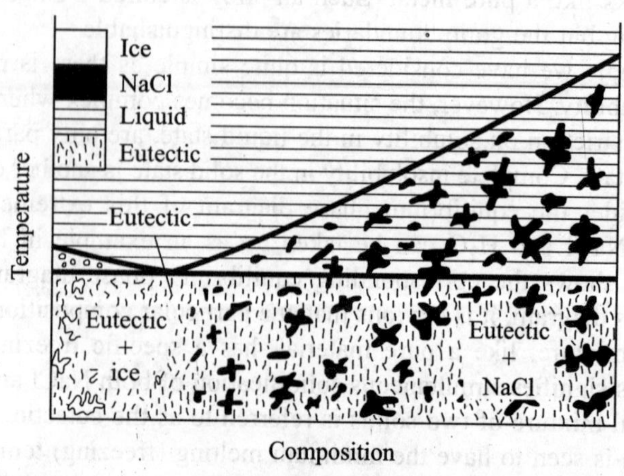

Fig. 1.13 Equilibrium phase diagram of eutectic.

the eutectic composition, the solid mixture has the eutectic mixture and solid NaCl; this region is normally referred to as *hypereutectic*.

The situation is slightly more complicated when the materials are partially soluble in the solid state. Partial solubility means that one component can form a solid solution with the other only up to a maximum concentration. This is normally the case with metallic alloys. A typical equilibrium phase diagram is shown in Fig. 1.14. First of all, let us assume that the maximum solid solubility of A in B is independent of temperature and is given by the percentage

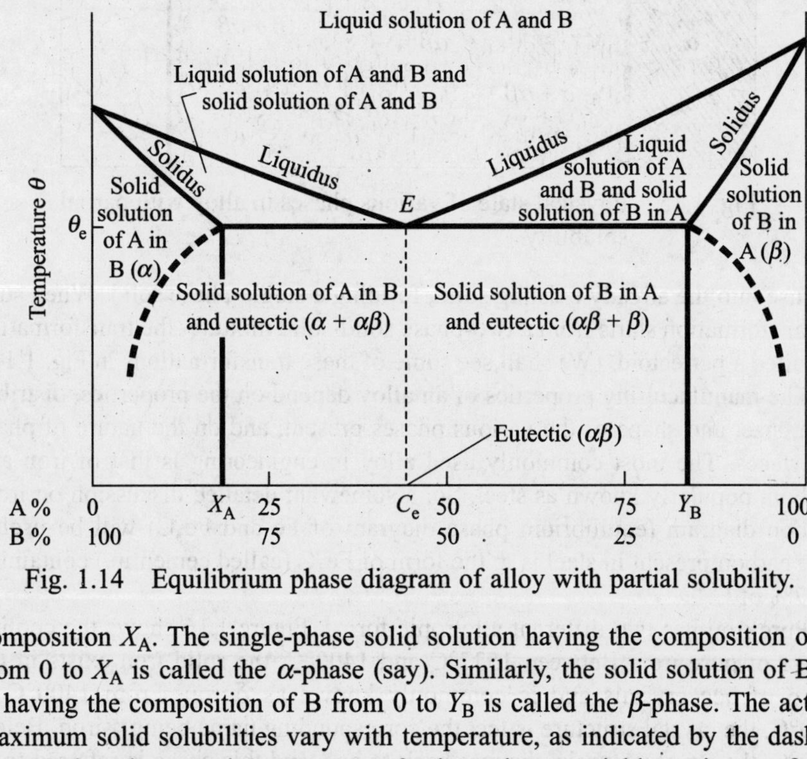

Fig. 1.14 Equilibrium phase diagram of alloy with partial solubility.

composition X_A. The single-phase solid solution having the composition of A from 0 to X_A is called the α-phase (say). Similarly, the solid solution of B in A having the composition of B from 0 to Y_B is called the β-phase. The actual maximum solid solubilities vary with temperature, as indicated by the dashed lines. Below the eutectic temperature θ_e, the entire material is a mixture of two solid solutions, namely, α and β. However, at the eutectic composition, it is difficult to distinguish between the α- and the β-phase and this composition is therefore represented as $\alpha\beta$ instead of $\alpha + \beta$. Figure 1.15 shows the nature of the alloy at various states.

There are other types of phase transformations similar to the eutectic transformation we have just described. We have observed that during an eutectic transformation, a single-phase liquid changes to a two-phase solid. A similar transformation from a single-phase solid is called an eutectoid transformation. Another transformation, also taking place at a constant temperature, is known as peritectic. In this case, however, above the peritectic temperature, the system exists in the form of a two-phase liquid-solid mixture. When the system is cooled to the peritectic temperature, all the liquid solidifies and the atoms

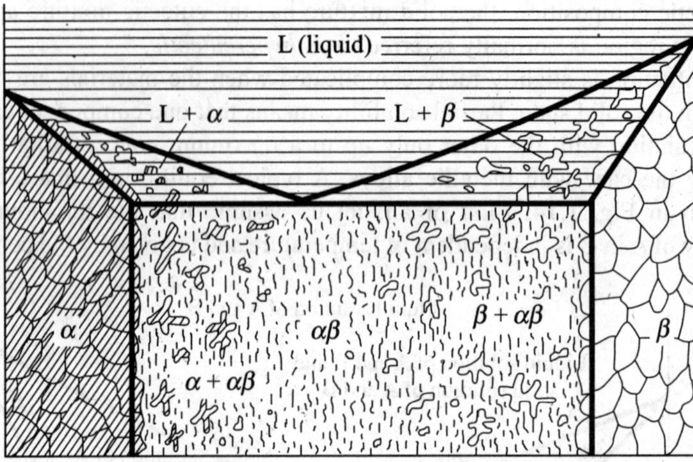

Fig. 1.15 Physical state of various phases in alloy with partial solubility.

diffuse into the already existing solid, forming a single-phase solid. When such a transformation starts from a two-phase solid-solid mixture, the transformation is called a pertectoid. (We shall see some of these transformations in Fig. 1.17.)

The manufacturing properties of an alloy depend on the properties, distribution, size, and shape of the various phases present, and on the nature of phase interfaces. The most commonly used alloy in engineering is that of iron and carbon, popularly known as steel. So, a somewhat detailed discussion on iron-carbon diagram (equilibrium phase diagram of Fe and Fe_3C) will be useful. The carbon present in steel is in the form of Fe_3C (called cementite) containing 6.67% of C.

Pure iron has two different allotropic forms. Figure 1.16 shows the cooling curve of pure iron. Between 1537°C and 1400°C, the solid iron exists in the form of bcc crystals and is commonly known as δ-iron. From 1400°C to 910°C, the crystal structure is fcc, the corresponding name being γ-iron. Below 910°C, the structure again changes back to bcc, and this phase is referred to as α-iron (however, there is no basic structural difference between the α- and the δ-phase). Figure 1.17 shows the iron-carbon equilibrium diagram. In this figure, the portion involving the δ-phase is not of much interest so far as the normal manufacturing processes are concerned because the temperature is very high. For casting processes, the liquid-solid transformation at 1125°C is significant, whereas for heat treatment of steels, the transformations around 723°C play an important role.

At 1125°C, the solubility of cementite in γ-iron is limited to 2% as indicated by the point *A* in Fig. 1.17. This solid solution of γ-iron and Fe_3C is commonly termed as austenite. In the bcc phases (i.e., α- and δ-phase) of iron, the solubility of Fe_3C is much smaller (around 0.33% in the α-phase and 0.1% in the δ-phase, as indicated by the points *C* and *B* in Fig. 1.17). The solid solution of Fe_3C in

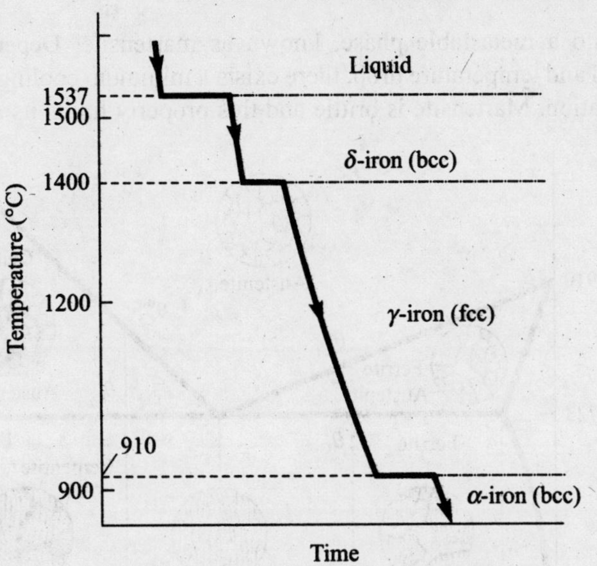

Fig. 1.16 Cooling curve of pure iron.

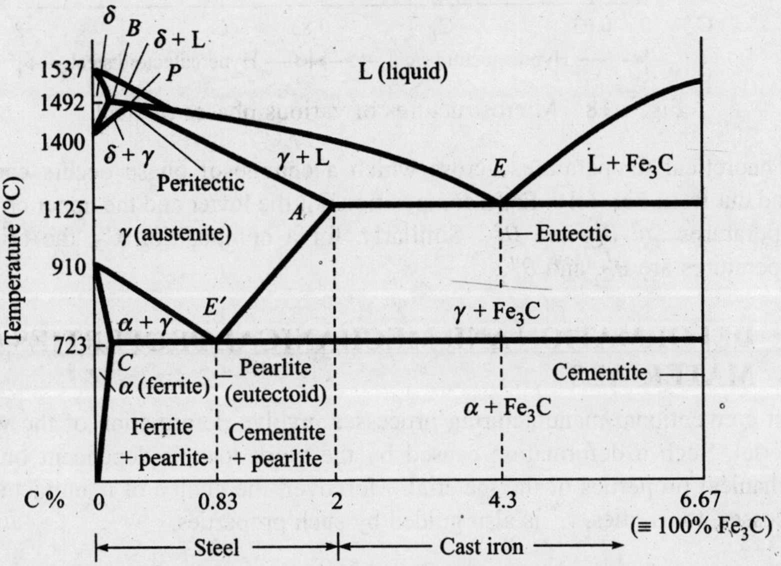

Fig. 1.17 Iron-carbon equilibrium diagram.

α-iron is called ferrite. The eutectoid (E') composition of ferrite and cementite is referred to as pearlite which consists of alternate thin laminates of cementite and ferrite. The different structures for the various phases of steel are indicated in Fig. 1.18. As can be noticed, the structure of ferrite is thick and rounded, whereas that of cementite tends to be thin and needle-like. Ferrite is soft and cementite is very hard. The transformation of austenite into ferrite and cementite is achieved only when the cooling is slow. A rapid cooling rate transforms austenite into a metastable phase, known as martensite. Depending on the composition and temperature drop, there exists a minimum cooling rate for such a transformation. Martensite is brittle and this property limits its applicability.

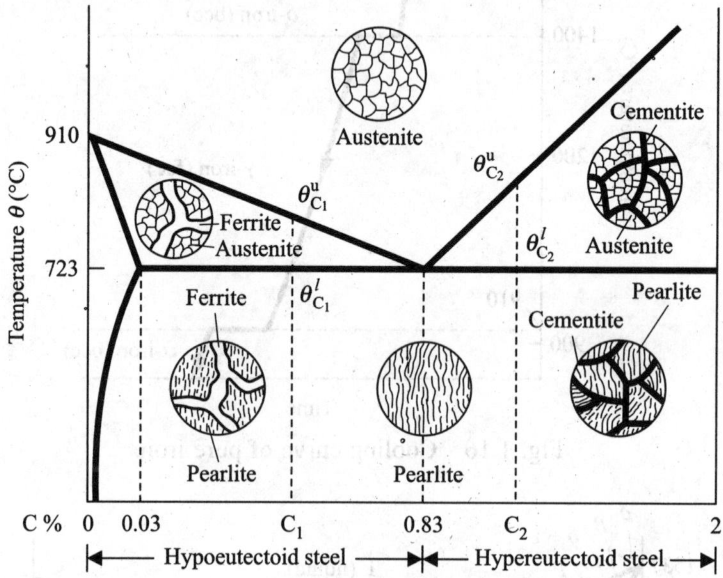

Fig. 1.18 Microstructures of various phases of steel.

Theoretical temperatures across which a change of phase occurs can be found out from Fig. 1.18. For a composition C_1, the lower and the upper critical temperatures are $\theta^l_{C_1}$ and $\theta^u_{C_1}$. Similarly, for a composition C_2, the critical temperatures are $\theta^l_{C_2}$ and $\theta^u_{C_2}$.

1.3 DEFORMATION AND MECHANICAL PROPERTIES OF MATERIALS

Most conventional manufacturing processes involve deformation of the work material. Such a deformation caused by the work load is dependent on the mechanical properties of the material. Moreover, the choice of manufacturing processes, tools, dies, ... is also guided by such properties.

1.3.1 ELASTIC AND PLASTIC DEFORMATION

In the absence of any external force, the distance between a pair of atoms is d_e (see Fig. 1.1). The net interatomic force varies with the atomic spacing in a manner shown in Fig. 1.19. Under the application of an external tensile force, the interatomic distance increases beyond d_e to maintain the equilibrium. If the external tensile force is of magnitude P, then the interatomic distance should be d_A so that the net interatomic force is an attractive force of the same magnitude. If d_A is not very much different (of the order of 5%) from d_e, then, upon removal of the external force, the atoms attain their original positions. A similar behaviour is also observed with an external compressive force (when $d_A < d_e$). This behaviour is called the elastic behaviour and the associated deformation is termed as elastic deformation.

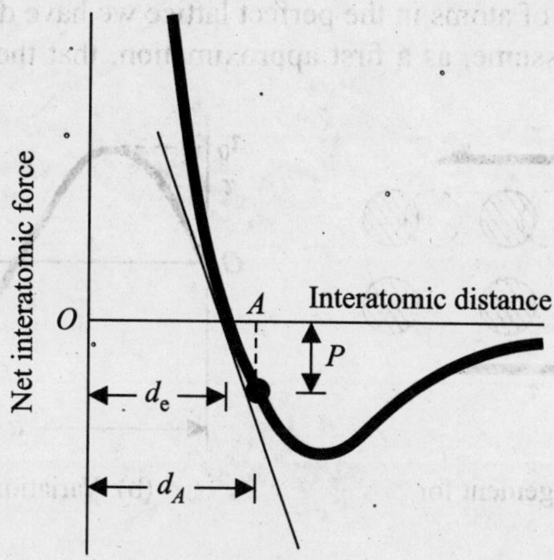

Fig. 1.19 Variation of net interatomic force with interatomic distance.

The phenomenon we have described for a pair of atoms is true also for normal solids even on a macroscopic scale. It may be noted from Fig. 1.19 that the tangent to the curve at the point d_e coincides with the curve over a small range on either side of the point d_e. Thus, the external force is proportional to the change in the interatomic distance. Hence, within the elastic behaviour, most solids follow a linear force deformation rule, and are therefore called linear elastic solids.

Now, let us consider a crystal lattice with regularly spaced atoms, as shown in Fig. 1.20a. Under the externally applied shear force (indicated in this figure), the upper layers of atoms move to the right and the lower layers move to the left. When the applied force reaches a sufficiently high value, the crystal lattice looks as in Fig. 1.20b. Here, all the atoms are again in equilibrium and will remain thus if the external force is removed. Thus, a permanent deformation is produced

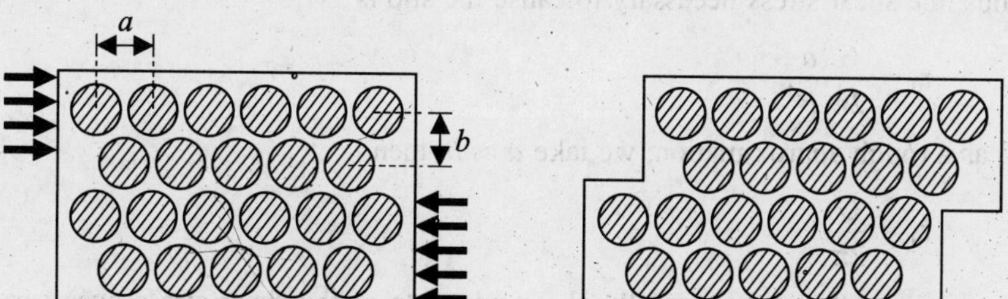

(a) Original position (b) After slipping through distance a

Fig. 1.20 Scheme of plastic deformation in perfect crystal.

in the crystal lattice. This permanent set is termed as plastic deformation and cannot be recovered upon withdrawal of the external load.

Next, let us find the amount of shear stress necessary to effect the slip between two layers of atoms in the perfect lattice we have described. Referring to Fig. 1.21a, we assume, as a first approximation, that the shear stress τ and

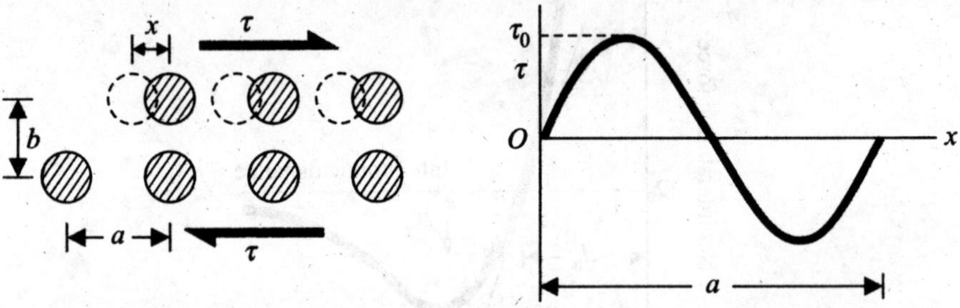

(a) Atomic arrangement for slip through x

(b) Variation of τ with slip

Fig. 1.21 Mechanics of slip in perfect crystal.

the amount of movement of the layer, x, are related as

$$\tau = \tau_0 \sin \frac{2\pi x}{a}.$$

This variation is shown in Fig. 1.21b. For small values of x/a, the foregoing relation can be rewritten approximately as

$$\tau \approx \tau_0 \frac{2\pi x}{a}.$$

Now, the shear strain produced is given by x/b. Using linear stress-strain relationship, we can write

$$\tau = \tau_0 \frac{2\pi x}{a} = G\frac{x}{b},$$

where the constant G is known as the shear modulus or modulus of rigidity. Thus, the shear stress necessary to cause the slip is

$$\tau_0 = \frac{G}{2\pi}\frac{a}{b}.$$

If, as a rough approximation, we take $a \approx b$, then

$$\tau_0 = \frac{G}{2\pi}.$$

Surprisingly, the experimentally-observed value of the shear stress necessary to produce a slip is found to be much less (of the order of 100 times) than the value we have calculated. This discrepancy can be explained only by bringing in the concept of imperfections which are always present in an actual lattice structure. Figure 1.22 explains how the movement of an edge dislocation, rather than the bodily movement of a whole plane of atoms, can cause a slip. This necessitates a much smaller value of the shear stress as can be appreciated from the following analogy.

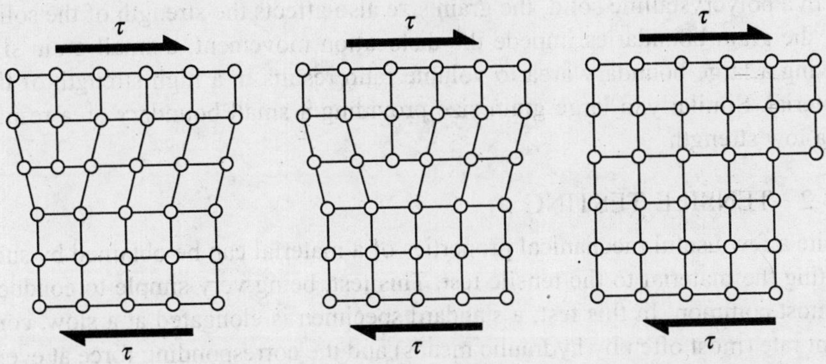

Fig. 1.22 Crystal deformation through movement of edge dislocation.

Let us consider a thick, heavy carpet, lying on a floor, which has to be moved through a distance δ (Fig. 1.23). We can immediately visualize that a very large force will be needed if the whole carpet has to be bodily moved over the distance δ. The same effect, however, can be realized very easily if first a fold is made at one end of the carpet to move that end by the distance δ and then this fold is moved across the entire carpet length.

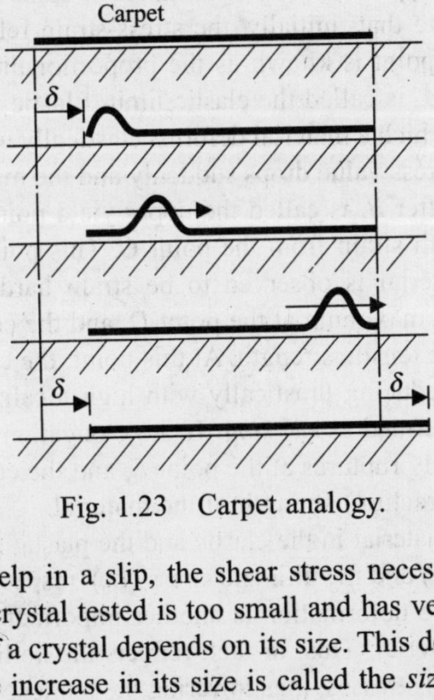

Fig. 1.23 Carpet analogy.

As dislocations help in a slip, the shear stress necessary to cause the slip will increase if the crystal tested is too small and has very few imperfections. Thus, the strength of a crystal depends on its size. This decrease in the strength of a crystal with the increase in its size is called the *size effect.* Further, with increasing deformation, more and more dislocations move and start interacting with one another. These interactions impede the mobility of various dislocations and, as a result, the strength of the material increases. This increase in the strength due to the immobility of dislocations is known as strain hardening; it implies an increase in the strength of a material due to its deformation.

In a polycrystalline solid, the grain size also affects the strength of the solid. As the grain boundaries impede the dislocation movement, a small grain size having a large boundary area to volume ratio results in a high strength of the material. Similarly, a large grain size providing a small boundary area results in a low strength.

1.3.2 TENSILE TESTING

Quite a few useful mechanical properties of a material can be obtained by subjecting the material to the tensile test. This test, being very simple to conduct, is most common. In this test, a standard specimen is elongated at a slow, constant rate (most often by hydraulic means) and the corresponding force at every instant is recorded. The force deformation relation is often expressed in terms of an engineering stress-strain curve. The engineering stress (σ) is defined as the applied load (P) divided by the original cross-sectional area (A_0) of the specimen. The engineering strain (e) is defined as the ratio of the increment in length (Δl) over the original length (l_0). l_0 is sometimes called the gauge length and Δl represents the change in this length when the applied load is P. The stress-strain description renders the load deformation relation independent of the specimen geometry and can be taken as the description of the material properties.

Figure 1.24 shows typical stress-strain curves of some engineering materials. In Fig. 1.24a, we note that, initially, the stress-strain relationship is linear up to the point A. This point is known as the proportionality limit. The point B, just above the point A, is called the elastic limit. Elastic limit is defined as the greatest stress up to which a material deforms elastically without any permanent set. Beyond e_B, the stress value drops suddenly and the material is said to yield. This point (Y), just after B, is called the upper yield point. The stress value is seen to rise again with strain from the point C. This point is called the lower yield point. The material is observed to be strain hardened beyond C. The stress value reaches a maximum at the point D, and the corresponding stress is known as the ultimate tensile strength. At this point, the cross-sectional area of the specimen starts reducing drastically with high localized deformation, and this phenomenon is termed as *necking*. If the elongation is continued further, the specimen ultimately ruptures at the point E, and the corresponding stress is called the breaking (fracture) strength of the material.

The strength of a material in the elastic and the plastic regions is represented by the yield stress (σ_Y) and the ultimate stress (σ_u), respectively. The capability of withstanding plastic deformation is another important mechanical property of an engineering material. This, in fact, reflects the ability of the material to distribute the localized stresses, thus lowering the tendency of crack formation. This property is commonly referred to as the ductility of the material. Ductility

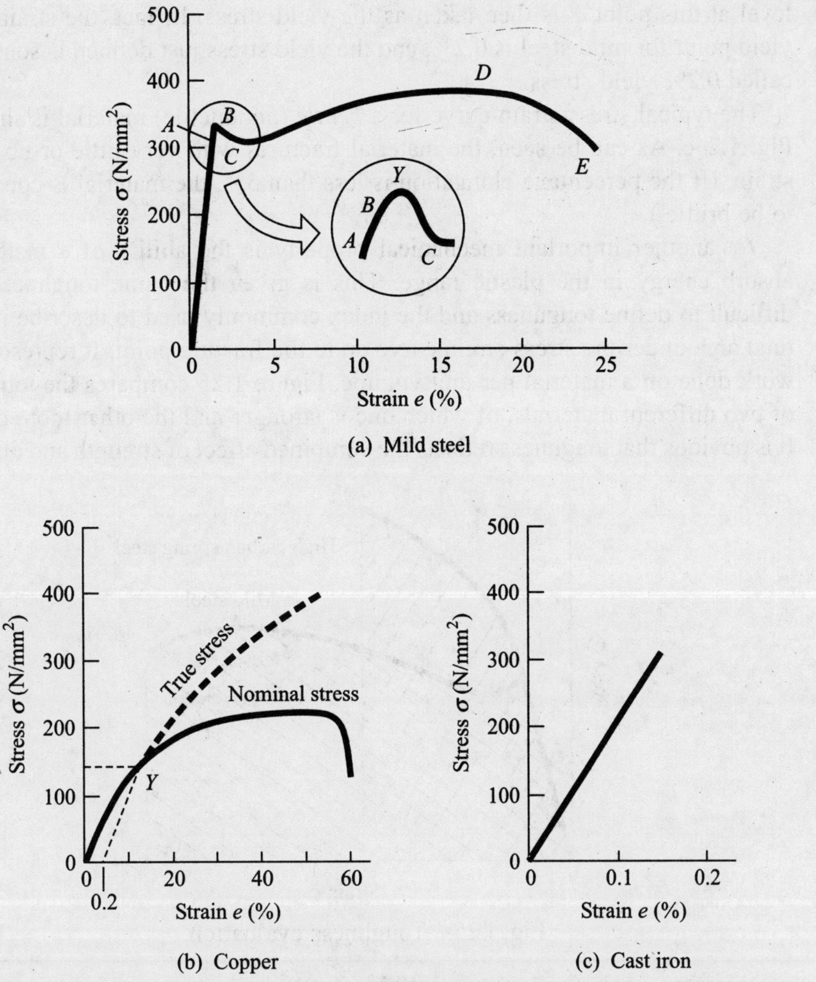

Fig. 1.24 Stress-strain relations for engineering materials.

is expressed by the percentage elongation, i.e., the percentage strain at the fracture point. Thus, larger percentage elongation means higher ductility. Moreover, the strain (see Fig. 1.24a) at the point *D* (where necking starts) represents the amount of plastic strain a material can withstand without localized deformation. This is used as an index of the formability of the material and is useful in some sheet metal forming operations. In the linear elastic behaviour, the constant of proportionality between the engineering stress and the engineering strain is known as Young's modulus or the modulus of elasticity.

It should be noted at this stage that most of the materials, unlike mild steel, do not show any precise proportional limit, elastic limit, or yield point (Fig. 1.24b). In such instances, the yield stress is defined as follows. A line parallel to the tangent of the stress-strain curve at the origin is drawn from the point representing 0.2% strain. This line intersects the stress-strain curve at the point *Y*. The stress level at this point Y is then taken as the yield stress. In fact, the strain at the yield point for mild steel is 0.2%, and the yield stress just defined is sometimes called 0.2% yield stress.

The typical stress-strain curve for a *brittle* (nonductile) material is shown in Fig. 1.24c. As can be seen, the material fractures with very little or no plastic strain. (If the percentage elongation is less than 5%, the material is considered to be brittle.)

Yet another important mechanical property is the ability of a material to absorb energy in the plastic range. This is given the name toughness. It is difficult to define toughness and the index commonly used to describe it is the total area under the stress-strain curve up to the fracture point. It represents the work done on a material per unit volume. Figure 1.25 compares the toughness of two different materials, of which one is stronger and the other more ductile. It is obvious that toughness reflects the combined effect of strength and ductility.

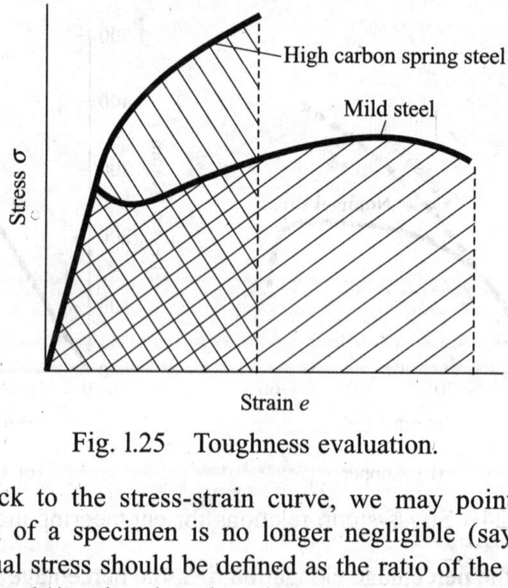

Fig. 1.25 Toughness evaluation.

Referring back to the stress-strain curve, we may point out that when the deformation of a specimen is no longer negligible (say, when necking begins), the actual stress should be defined as the ratio of the load (P) and the instantaneous area (A_i) rather than the original area. This ratio P/A_i is called the true stress. Similarly, the true strain should also be defined on the basis of the instantaneous deformation. True strain (ε), when the length is l_i, is defined as

$$\varepsilon = \int_{l_0}^{l_i} \frac{dl}{l} = \ln\left(\frac{l_i}{l_0}\right) = \ln(1 + e),$$

where e is the engineering strain $= (l_i - l_0)/l_0$. It is obvious that, for $e \ll 1$, $\varepsilon \approx e$. The true stress versus true strain curve is indicated by the dashed line in Fig. 1.24b. This curve shows a continuous strain hardening of the material up to the fracture point, a phenomenon not revealed by a stress-strain curve.

The stress-strain curve of a material is too complicated to be represented by a simple mathematical relationship amenable to analysis. As such, for purposes of analysis, the curve is idealized in various ways, keeping only those basic features that are important for a given problem. Figure 1.26 shows some idealized curves.

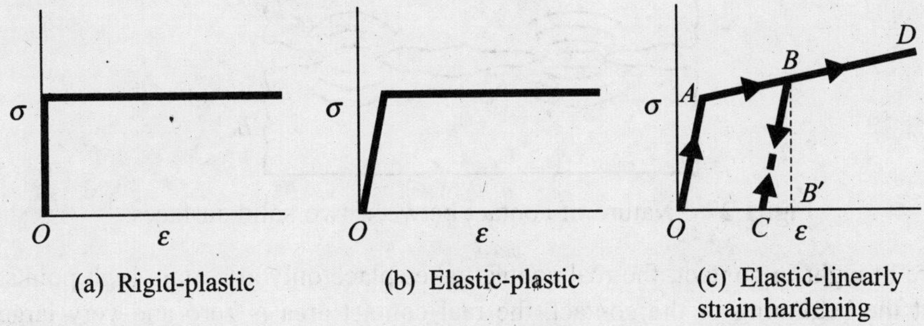

Fig. 1.26 Various idealizations of material.

Let us now see what happens if a material, loaded beyond the yield point, is unloaded completely and reloaded. If the material is unloaded from the point B (see Fig. 1.26c), the unloading curve BC is then parallel to the initial elastic curve given by the line OA. The permanent strain is given by OC, whereas the amount of strain recovered is given by CB'. Thus, the total strain at B is thought of as consisting of two parts, namely, the elastic part CB' and the plastic part OC. The reloading curve follows the lines CB and BD.

Another important mechanical property, which is not obtained from tensile testing but is relevant in the context of manufacturing processes, is *hardness*. Hardness is a very ill-defined term and is normally used to indicate the resistance of a material to plastic deformation. However, the nature of plastic deformation needs to be specified, and hardness refers to the plastic deformation caused by indentation. There are various standard tests which designate the hardness of different materials by using numbers on prescribed scales. The most commonly used numbers are Vickers, Brinell, and Rockwell. These empirical numbers are calculated on the basis of indentation tests with a known applied load and the area or depth of the resulting impression.

1.3.3 FRICTION AND WEAR

It is well-known that whenever a solid surface slides over another, a resisting force, commonly referred to as the friction force, develops. The friction phenomenon was first scientifically studied by Amonton and Coulomb. Since this phenomenon is extremely complicated, we shall restrict our discussion on it to a very elementary level. Then, with the help of such an oversimplified and elementary model, certain important and fundamental aspects of friction can be easily understood.

Let us consider two solid surfaces in contact, as in Fig. 1.27. Though a surface may appear smooth and plain, in reality no solid surface is perfectly smooth. Asperities are always present in a solid surface, and when two bodies

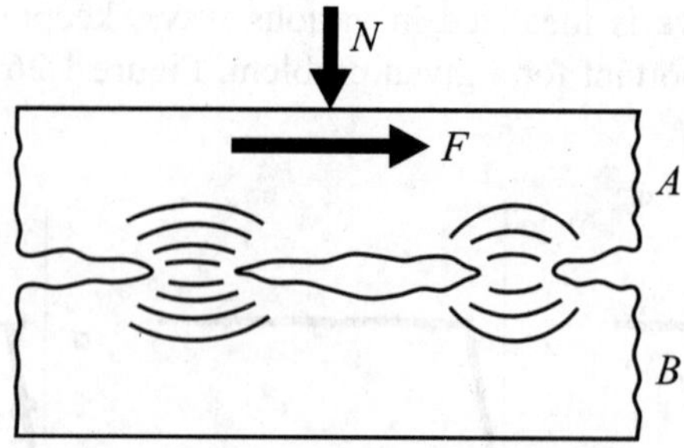

Fig. 1.27 Nature of contact between two solid surfaces.

are brought in contact, the real contact takes place only at certain high points. At the beginning of the contact, the real contact area is zero and very large localized stresses develop, causing plastic deformation of the contact regions. Thus, the area of real contact continues to increase till it is large enough so that the corresponding stresses do not cause any further plastic deformation. If we assume a rigid plastic model of the materials and if σ is the yield stress in compression (of the weaker material), then the real area of contact all over the mating surfaces can be expressed as $A_{real} \approx N/\sigma$, where N represents the applied load. Large stresses and plastic deformations cause the upper contamination layers, which are always present, to tear off and the real materials come in contact. This results in a welding of the asperity junctions, and the sliding of one body above the other will be possible only after these welded asperity junctions are sheared. If τ is the yield shear stress of the weaker material, then the force required to shear off a junction with a total area of A_{real} is $F \approx A_{real}\tau$. This shearing force is the friction force. The ratio of the friction force and the normal force is found out as

$$\frac{F}{N} = \frac{A_{real}\,\tau}{A_{real}\,\sigma} = \frac{\tau}{\sigma} = \mu.$$

It is obvious from this equation that the ratio μ, normally termed as the coefficient of friction, depends only on the materials in contact. For better results, τ and σ should be taken for the alloy formed (at the junctions) due to heavy cold work and welding.

The foregoing model, yielding $\mu = \tau/\sigma$, is valid only when the real area of contact is much smaller than the apparent area of contact. If the normal load N is gradually increased, A_{real} increases and approaches the apparent area of contact A. Once A_{real} reaches a value equal to A, the shear force F (i.e., the friction force) will not increase even if N is increased. This is illustrated in Fig. 1.28. Under such a situation, mechanisms other than the welding of asperity junctions become active, making the friction phenomenon quite complex. One such important mechanism is the locking of asperities. The coefficient of friction then varies and tends to increase with increasing load.

When a solid surface slides over another, both the surfaces are subjected to a gradual loss of material. A fraction of the material lost from one surface may be

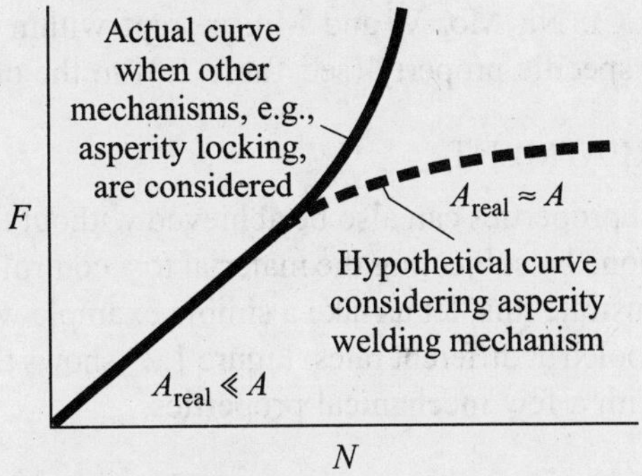

Fig. 1.28 Variation of friction force with normal load.

transferred to the other body, whereas the rest gets removed in the form of small (wear) particles. This process of gradual loss or transfer of material from a body (in contact with another) is known as wear. The three major mechanisms of wear, relevant to the manufacturing processes, are abrasion, adhesion, and diffusion.

If one of the surfaces contains very hard particles, then these, during the process of sliding, may dislodge material from the other surface by the ploughing action. This mechanism is called abrasion.

When the bodies in contact are of a similar nature, the asperities on the contacting surfaces tend to get welded. Sliding causes fracture of these welded junctions and material is lost from both the surfaces. The wear due to this mechanism is referred to as the adhesion wear.

Atoms in a metallic crystal lattice always move from a region of high concentration to that of low concentration. This process is known as diffusion. The rate of diffusion depends on the concentration gradient and the existing temperature. Since the rate increases exponentially with temperature, the diffusion mechanism plays a predominant part at high temperatures. When two dissimilar bodies slide over each other, the atoms of various constituent elements diffuse across the junction, leading to wear on both the bodies; this process is known as diffusion wear.

1.4 CONTROL OF MATERIAL PROPERTIES

Metals and alloys may not possess all the properties required in a finished product. The material properties can, however, be controlled by various methods to make the material suitable for a given application. These methods include (i) alloying, (ii) heat treatment, and (iii) mechanical working and recrystallization. We shall briefly discuss each of these processes with reference to the most commonly used engineering material, namely, steel.

1.4.1 ALLOYING

Alloy steels are broadly classified into two categories, viz., (i) low alloy steels and (ii) high alloy steels. In low alloy steels, the total content of the alloying elements, such as Cr, Ni, Mo, V, and Mn, is kept within 5%. Each alloying element imparts a specific property (see Table 1.3) to the original material.

1.4.2 HEAT TREATMENT

Control of material properties can also be achieved without the addition of other elements. This is done by subjecting the material to a controlled cycle of heating and cooling. To illustrate this, let us take a simple example where austenite steel (above 723°C) is cooled at different rates. Figure 1.29 shows the various resulting structures along with a few mechanical properties.

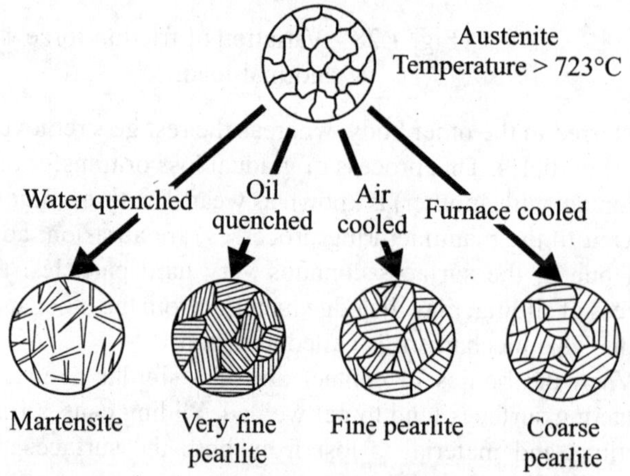

	Martensite	Very fine pearlite	Fine pearlite	Coarse pearlite
Tensile stress (N/mm^2)	175×10^4	112×10^4	88×10^4	52×10^4
Yield stress (N/mm^2)		56×10^4	28×10^4	14×10^4
Reduction in area (%)	Low	22	27	30
Rockwell hardness (C)	65	35	25	15

Fig. 1.29 Effect of cooling rate on microstructure and properties.

It is thus obvious that by changing only the rate of cooling, different phases can be achieved. The information on the change of phase with the cooling rate can be conveniently displayed with the help of a time-temperature-transformation diagram (commonly known as the TTT diagram). In such a diagram, the temperature is plotted along the vertical axis (using a linear scale), whereas the abscissa represents the time on a logarithmic scale. The TTT diagram for carbon steel is given in Fig. 1.30. When austenite is brought to a temperature θ_1 from θ_0 (in essentially zero time) and thereafter held at θ_1, the transformation to pearlite begins after a lapse of time t_1, as shown by the point A in the figure. Such a transformation, taking place at a constant temperature, is known as an isothermal

Table 1.3 Effects of alloying elements on steel

Element	Effect(s)	Remark(s)
Al	Promotes deoxidization Promotes nitriding Restricts grain growth	Typically low percentage
B	Increases hardenability	Typical percentage 0.001–0.003 Used in steels with carbon content less than 0.6%
Co	Impairs impact strength slightly Contributes to red hardness Sustains hardness during tempering	
Cr	Increases hardenability Increases resistance to corrosion, abrasion, and wear Increases high temperature strength	Typical percentage 0.5–2 to increase hardenability Typical percentage 4–8 to increase corrosion and wear resistance
Cu	Increases corrosion resistance Counteracts brittleness from S	Typical percentage 0.1–0.4 Typical percentage 0.25–0.40 to counteract brittleness from S
Mn	Increases hardenability Increases resistance Reduces ductility and weldability	Typical percentage >1 to increase hardenability

Table 1.3 Effects of alloying elements on steel (cont.)

Element	Effect(s)	Remark(s)
Mo	Increases hardenability significantly	Typical percentage 0.2–5
	Increases strength, toughness, red hardness, and hot strength when used with Cr, Mn, and V	Used for HSS (high speed steel) cutting tools, forged crankshafts, turbine rotors, high pressure cylinders and boiler plates, and gears
	Enhances corrosion and abrasion resistance	
Ni	Increases toughness and impact strength	Typical percentage 2–5 to increase toughness and strength
	Improves corrosion resistance	Typical percentage 12–20 to increase corrosion resistance
		Used in case hardened parts such as high speed gears and bearings
P	Increases hardenability	
	Improves machinability	
	Increases strength in low carbon steel	
	Improves corrosion resistance	
Si	Strengthens low alloy steels	Typical percentage 0.2–0.9 to increase strength
	Increases hardenability	Typical percentage 2 Used in spring steels
	Acts as deoxidizer	
	Improves magnetic properties when present in large percentage	Typically high percentage to improve magnetic properties

Table 1.3 Effects of alloying elements on steel (cont.)

Element	Effect(s)	Remark(s)
S	Improves machinability of very low carbon steels	Typical percentage 0.08–0.15 Normally considered an impurity
Ti	Increases austenitic hardenability Reduces martensitic hardness in Cr steels	Fixes carbon in inert particles, resulting in remarkable carbide forming effect
V	Increases strength while retaining ductility Produces fine grain size Increases hardenability	Typical percentage 0.15 Forms stable carbides that persist at quite high temperature Normally used in combination with chromium
W	Imparts hardness and wear resistance Significantly improves red hardness Imparts strength at high temperature	Typical percentage 4 to impart wear resistance Typical percentage 18 to improve red hardness Used in high speed tool materials

transformation. The point B indicates a time t_2, after which the transformation is complete. In Fig. 1.30, the transformations corresponding to other temperatures, viz., θ_2 and θ_3, are also shown. At about 600°C, the transformation starts after a minimum lapse of time, and this part of the diagram is called the nose. Below this temperature, austenite transforms into bainite which is an intimate mixture of ferrite and cementite (cementite exists in the form of tiny spheroids). Bainite cannot be produced by continuous cooling.

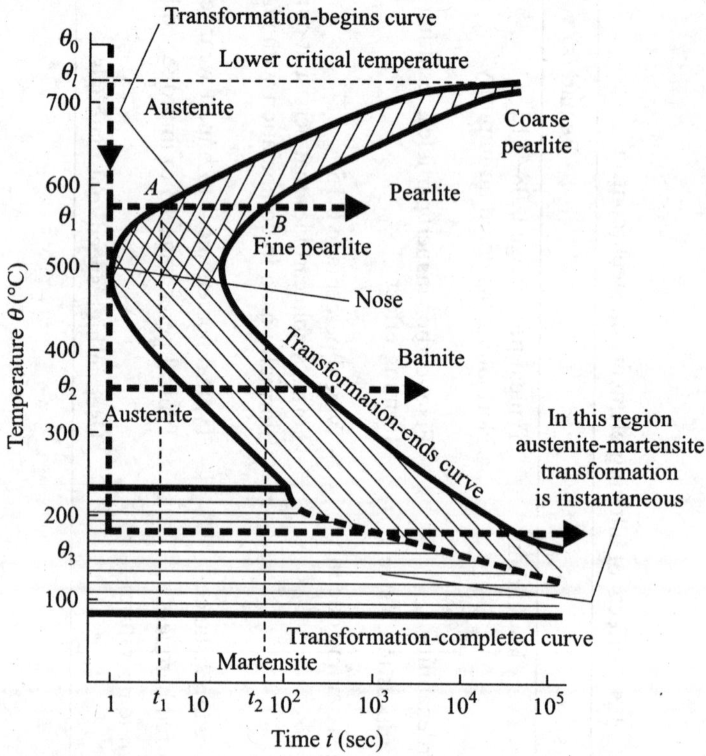

Fig. 1.30 TTT diagram for carbon steel.

When the temperature of isothermal transformation is decreased (above 600°C), the time required for the transformation reduces. This results in a finer grain structure as less time is available for the growth of new nuclei.

It may be noted that the curve, indicating the beginning of the transformation, does not exist below about 220°C. Below this temperature, austenite instantaneously starts transforming into martensite.

A TTT diagram is quantitatively valid only when the transformations are isothermal. In practical situations of heat treatment where a continuous cooling is involved, a modified TTT diagram (see Fig. 1.31) is used. The use of such a diagram can be explained as follows. If the cooling rate is very high (as shown by cooling curve 1), the entire austenite is transformed into martensite because the cooling curve does not enter the pearlite region. With a moderate cooling rate (depicted by cooling curve 2), a portion of austenite is transformed into pearlite and the rest of it into martensite. The percentage of pearlite depends on the point of intersection of the cooling curve and the line *AB* (Fig. 1.31). When this point is nearer *A*, lesser pearlite is produced. So, when the desired structure is prescribed, steel should be heated beyond 723°C (where it is austenitic) and then cooled in a manner dictated by the TTT diagram.

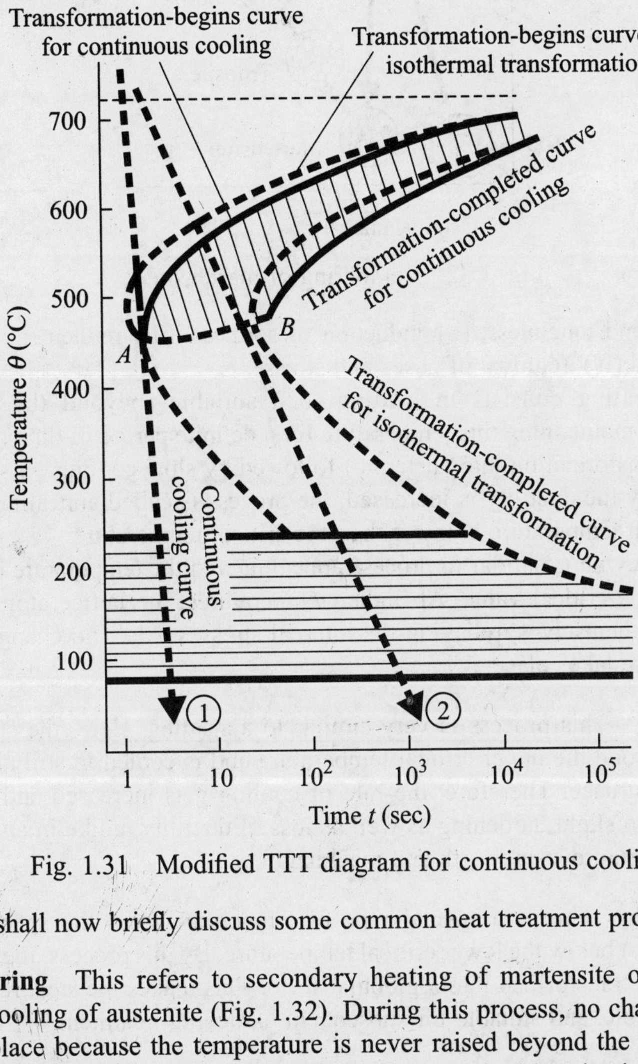

Fig. 1.31 Modified TTT diagram for continuous cooling.

We shall now briefly discuss some common heat treatment processes.

Tempering This refers to secondary heating of martensite obtained by a rapid cooling of austenite (Fig. 1.32). During this process, no change of phase takes place because the temperature is never raised beyond the lower critical temperature (723°C). This process hardens the steel with reduction in strength; also, it adds to the toughness and ductility. The different structures, indicated in Fig. 1.32, result from the dispersion of carbides.

Annealing It includes, among others, full annealing, stress relieving, and process annealing. The general purposes served by annealing are (i) alteration

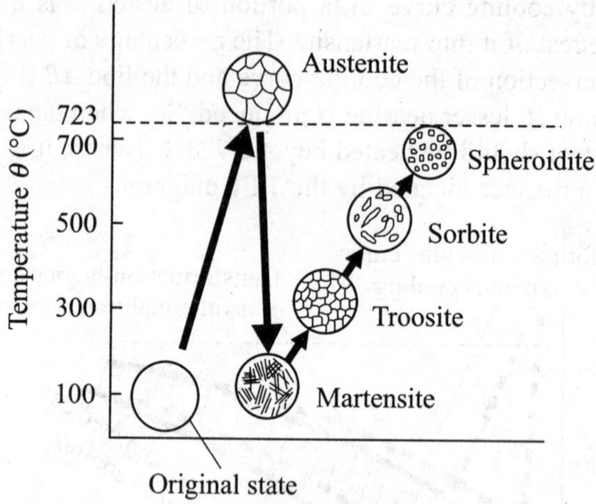

Fig. 1.32 Tempering of martensite.

of ductility and toughness, (ii) induction of softness, (iii) refinement of grain structure, and (iv) removal of gases and stresses.

Full annealing consists in heating to a suitable (beyond the critical) temperature, maintaining this temperature for a definite period of time (to allow complete transformation into austenite) followed by slow cooling.

When only the ductility is increased, the process is called annealing, where the maximum temperature is lower than the critical temperature.

Stress relieving is similar to process annealing and the temperature does not go beyond the critical value. At higher temperatures, the lattice atoms move to rearrange themselves, relieving the internal stresses. Also, no change in the microstructure takes place.

Normalizing This process is very similar to annealing. Here, the specimen is heated beyond the upper critical temperature and is cooled in still air rather than in the furnace. Therefore, the rate of cooling gets increased and this, in turn, results in slight hardening as well as loss of ductility unlike in annealing. This process improves strength and machinability.

Spheroidizing This is a special type of tempering where the specimen is reheated to just below the lower critical temperature. By this process, the carbide in the steel is transformed into a globular form. This makes the steel relatively soft, machinable, and suitable for subsequent hardening treatment.

Case hardening For low carbon steels, the nose of the TTT diagram suggests that no practical cooling rate can achieve direct transformation of austenite into martensite. One way to solve this problem is to shift the nose to the right through alloying. However, as hardness is normally required only at the surface of the specimen, alloying of the whole specimen is not necessary. There are various treatments to impart surface hardness, such as (i) carburizing, (ii) nitriding, and (iii) cyanide hardening. A process using any of these treatments is called case hardening.

(i) In carburizing, the specimen is heated beyond the upper critical temperature in a sealed container having the atmosphere of carbon. The heating is continued for 4–10 hours depending on the depth of penetration required. As a result, carbon diffuses into the surface layer, making the specimen harder.

(ii) Nitrogen, in place of carbon, can also be used as a hardening agent. Here, the ammonia atmosphere is used. Also, the temperature required is about 1/2–1/3 of that in carburizing, but the heating is continued for a period almost twice that in carburizing. Normally, alloy steels containing chromium, vanadium, molybdenum are subjected to nitriding.

(iii) Using the sodium cyanide atmosphere, both carbon and nitrogen can be effectively employed for imparting hardness. During heating, the nitrides are formed, whereas during subsequent quenching, the carbides are formed. Since heat is dissipated through the surface of a specimen, it is possible to achieve a high cooling rate if the heat content of the specimen is confined to the surface layer. As already mentioned, higher hardness is normally required at the surface. So, if the surface layer is quickly heated to a suitable high temperature (keeping the core temperature unaffected) and then rapidly quenched, the austenite at the surface gets hardened. The two methods generally used for doing this are (a) flame heating and (b) induction heating. Such surface hardening treatments are normally done on steels containing 0.35% or more carbon.

(a) *Flame hardening* In this, an oxyacetylene flame is moved over the specimen followed by a quenching spray (Fig. 1.33). The velocity of the flame and the material property determine the depth of the layer being hardened. Flame hardening does not produce any sharp boundary between the hardened layer and the core; as a result, there is no danger of the surface layer chipping out. The flame should be kept at a sufficient distance from the sharp corners to avoid overheating.

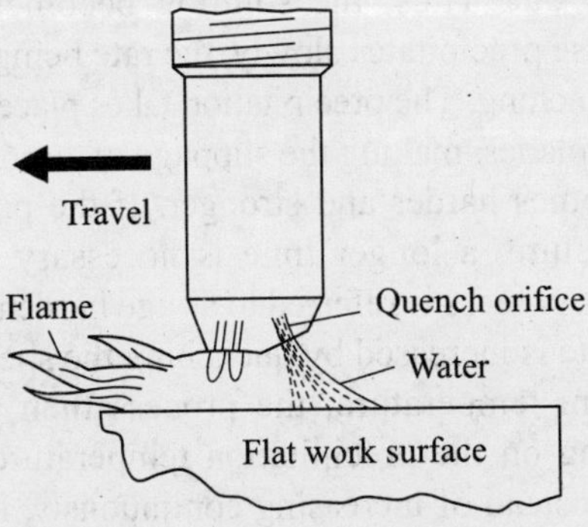

Fig. 1.33 Flame hardening.

(b) *Induction hardening* Here, the heating is done by placing the specimen in a high frequency magnetic field. The depth of penetration decreases as the frequency increases. So, the surface hardening of thin-walled sections requires high frequency.

Age hardening and precipitation hardening This process of hardening is applicable only for those alloys that exist as a two-phase material at the room temperature and can be heated up to a single phase. The phase diagram of one such alloy is shown in Fig. 1.34. Assuming that the composition is 3% Cu and 97% Al, the alloy exists as a two-phase material ($\alpha + \beta$) below a temperature

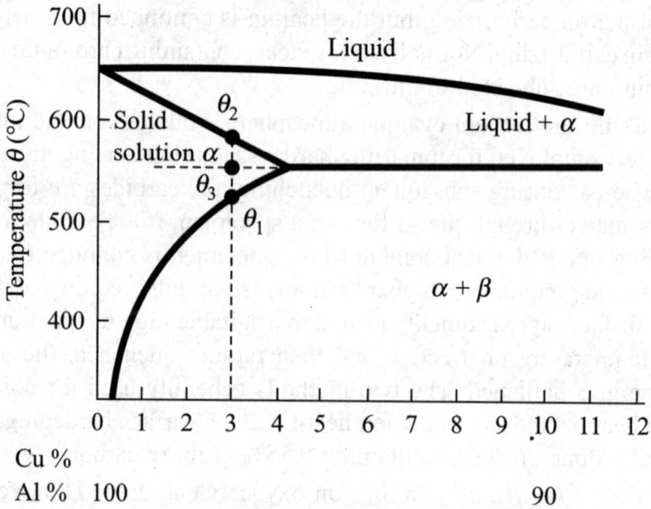

Fig. 1.34 Principle of precipitation hardening.

θ_1. However, between θ_1 and θ_2, it exists as a single-phase solid solution (α). A solution heat treatment process consists in heating the alloy to a temperature between θ_1 and θ_3. Also, a sufficient time is given at this temperature for the material to homogenize. A subsequent rapid quenching does not allow all the β-phase to separate out. Thus, the solution becomes supersaturated. This supersaturated β-phase precipitates slowly, the rate being dependent on the final temperature after quenching. The precipitation takes place at the grain boundaries and crystallographic planes, making the slippage of atomic layers more difficult. Thus, the alloy becomes harder and stronger. If the precipitation takes place at the room temperature, a longer time is necessary for the completion of precipitation, and this process is referred to as age hardening. On the other hand, if the precipitation rate is increased by quenching the specimen to a temperature higher than the room temperature, the process then is called precipitation hardening. Depending on the precipitation temperature, the hardness (at the room temperature), instead of increasing continuously, may attain a maximum before it starts decreasing. The optimal properties are very sensitive to both the temperature (at which the precipitation takes place) and the time elapsed after quenching. The nature of the variation of hardness with these two parameters is shown in Fig. 1.35.

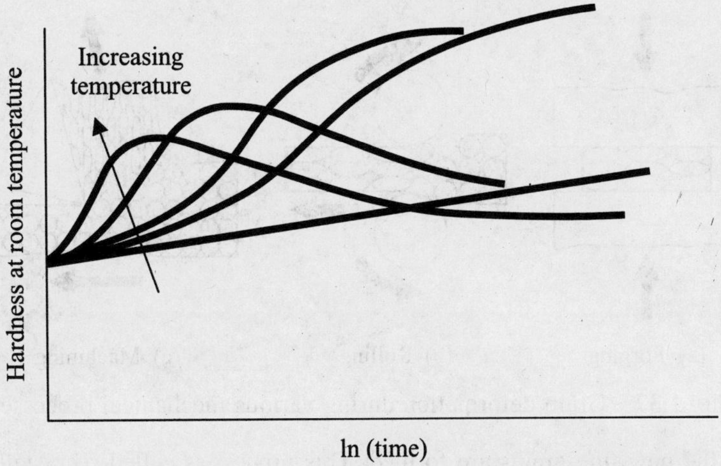

Fig. 1.35 Variation of hardness with precipitation temperature and time.

1.4.3 MECHANICAL WORKING AND RECRYSTALLIZATION

As already mentioned, the mechanical properties, e.g., strength and hardness, of a polycrystalline material are also governed by the grain size of the material. Figure 1.36 shows the nature of variation of two mechanical properties with the average grain size. The grain size can be controlled by mechanical

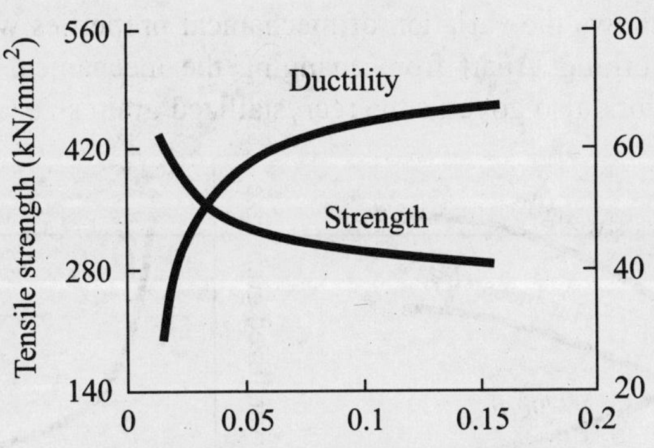

Fig. 1.36 Effect of grain size on strength and ductility of annealed brass.

working and recrystallization. As we have already noted, the effective size of a grain is decided by the volume to surface area ratio. Figure 1.37 shows how the grains are deformed by various mechanical working processes. Here, we can easily see that the effective grain size is reduced because the surface area of each grain increases, whereas the volume remains the same. It is possible to restore the original grain geometry by heating the material up to a temperature

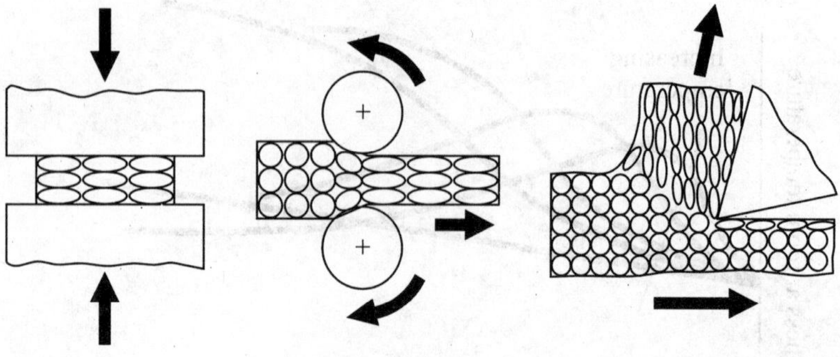

(a) Forging (b) Rolling (c) Machining

Fig. 1.37 Grain deformation during various mechanical processes.

where the new fine grains are formed. This process is called recrystallization and the temperature needed to achieve this is known as the recrystallization temperature. If the material is kept for a considerable period of time at this temperature, the newly formed fine grains grow in size. On the other hand, if the material is cooled quickly, the grain size remains small. It should be noted that in the mechanical working processes, the grains are deformed, whereas in recrystallization, new grains are formed and, what is more, their size can be controlled. Normally, the recrystallization temperature of a metal is about 40% of its melting temperature in the absolute scale but this temperature depends also on the amount of prior cold work.

Figure 1.38a shows the variation of mechanical properties with the amount of mechanical working. Apart from changing the mechanical properties, the amount of cold work also governs the recrystallized grain size (see Fig. 1.38b).

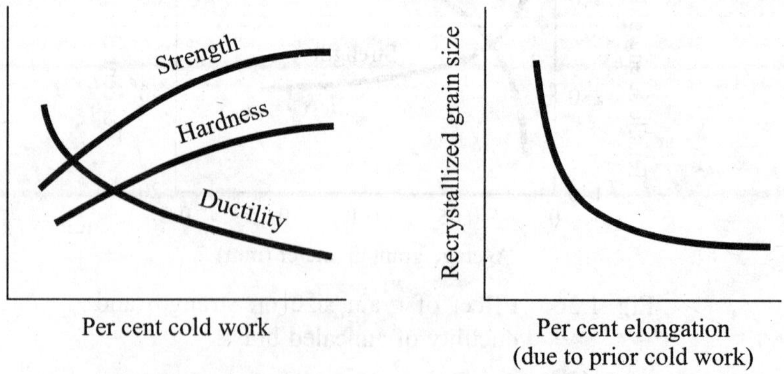

(a) Variation of mechanical properties with amount of mechanical work

(b) Dependence of recrystallized grain size on prior cold work

Fig. 1.38 Effects of mechanical working.

2 Casting Processes

2.1 INTRODUCTION

Casting is one of the oldest manufacturing processes, and even today is the first step in manufacturing most products. In this process, the material is first liquefied by properly heating it in a suitable furnace. Then, the liquid is poured into a previously prepared mould cavity where it is allowed to solidify. Subsequently, the product is taken out of the mould cavity, trimmed, and cleaned to shape.

It is clear from the definition of the process that a successful casting operation needs a knowledge in the following areas:

(i) Preparation of moulds and patterns (used to make the mould).

(ii) Melting and pouring of the liquefied metal.

(iii) Solidification and further cooling to room temperature.

(iv) Defects and inspection.

There are various types of casting processes depending, among others, on the material, the type of patterns and moulds, and the pouring technique. Before going into the details of these processes, we shall discuss the basic common features among the various casting processes in the context of the four areas we have just mentioned.

The suitability of the casting operation for a given material depends on

(i) the melting temperature of the job and the mould materials,

(ii) the solubility of and the chemical reaction between the job and the mould materials,

(iii) the solubility of the atmosphere in the material at different temperatures to be encountered in the casting operation,

(iv) the thermal properties such as conductivity and coefficient of linear expansion of both the mould and job materials.

2.2 PATTERN AND MOULD

A pattern is the replica of the part to be cast and is used to prepare the mould cavity. Patterns are made of either wood or metal. A mould is an assembly of two or more metal blocks, or bonded refractory particles (sand) consisting of a primary cavity. The mould cavity holds the liquid material and essentially acts as a negative of the desired product. The mould also contains secondary cavities for pouring and channelling the liquid material into the primary cavity and to act as a reservoir, if necessary.

A four-sided frame in which a sand mould is made is referred to as a *flask*. If the mould is made in more than one part, the top portion is called the *cope* and the bottom one is termed as the *drag*. For producing hollow sections, the entry of the liquid metal is prevented by having a *core* in the corresponding portion of the mould cavity. The projections on the pattern for locating the core in the mould are called *core prints*. There are diverse types of patterns and moulds depending on the material, the job, and the number of castings required.

2.2.1 PATTERN ALLOWANCES

A pattern is always made somewhat larger than the final job to be produced. This excess in dimensions is referred to as the pattern allowance. There are two categories of pattern allowances, namely, the *shrinkage allowance* and the *machining allowance*.

The shrinkage allowance is provided to take care of the contractions of a casting. The total contraction of a casting takes place in three stages, and consists of

(i) the contraction of the liquid from the pouring temperature to the freezing temperature,

(ii) the contraction associated with the change of phase from liquid to solid,

(iii) the contraction of the solid casting from the freezing temperature to the room temperature.

It must be noted, however, that it is only the *last* stage of the contraction which is taken care of by the shrinkage allowance. (The other two categories of contraction will be discussed in Section 2.5.) Obviously, the amount of shrinkage allowance depends on the linear coefficient of thermal expansion α_l of the material. The higher the value of this coefficient, the more the value of shrinkage allowance. For a dimension l of a casting, the shrinkage allowance is given by the product $\alpha_l l(\theta_f - \theta_0)$, where θ_f is the freezing point of the material and θ_0 is the room temperature. This is normally expressed per unit length for a given material. Table 2.1 gives some quantitative idea about the shrinkage allowance for casting different materials.

Usually, a cast surface is too rough to be used in the same way as the surface of the final product. As a result, machining operations are required to produce the finished surface. The excess in the dimensions of the casting (and consequently in the dimensions of the pattern) over those of the final job to take care of the machining is called the machining allowance. The total machining allowance also depends on the material and the overall dimension of the job, though not linearly as the shrinkage allowance. Table 2.1 gives also an idea of the machining allowance for various materials. For internal surfaces, the allowances provided should obviously be negative, and normally the machining allowances are 1 mm more than those listed in the table.

There is another deviation from the original job dimensions and is intentionally provided in the pattern; this is called *draft*. It refers to a taper put on

Table 2.1 Machining allowance for various metals

Material	Shrinkage allowance	Machining allowance for dimensions	
		0–30 cm	30–60 cm
Cast iron	1/96	2.5 mm	4.0 mm
Cast steel (low carbon)	1/48	3 mm	4.5 mm
Aluminium	3/192	1.5 mm	3 mm
Bronze	3/192	1.5 mm	3 mm
Brass	1/48	1.5 mm	3 mm

the surface parallel to the direction of withdrawal of the pattern from the mould cavity. A draft facilitates easy withdrawal of the pattern. The average value of the draft is between $\frac{1}{2}^\circ$ and 2°.

2.2.2 TYPES OF PATTERNS

The commonly-used patterns are classified as follows:

(i) *Loose pattern* It is made in one piece, usually from wood, and is used for castings numbering up to 100.

(ii) *Gated pattern* This is simply one or more than one loose pattern with attached gates and runners and provides a channel through which the molten metal can flow from the pouring sprue to the mould cavity. This pattern is frequently set on a *follow board* conforming to the *parting surface* of the mould. The follow board helps in an easy removal of the pattern after the mould has been prepared.

(iii) *Match plate pattern* This pattern is made in two halves mounted on both sides of a *match plate* (of wood or metal) conforming to the contour of the parting surface. The match plate is accurately placed between the cope and the drag flasks by means of locating pins. For small castings, several patterns can be mounted on the same match plate.

(iv) *Cope and drag pattern* Here, the cope and drag halves of a split pattern (Fig. 2.1) are separately mounted on two match plates. Thus, the cope and the drag flasks are made separately and brought together (with accurate relative location) to produce the complete mould.

(v) *Sweep pattern* Normally made of wood, it is used to generate surfaces of revolution in large castings, and to prepare moulds out of a paste-like material. Here, "sweep" refers to the section that rotates about an edge to yield circular sections.

(vi) *Skeleton pattern* This consists of a simple wooden frame outlining the shape of the casting. It is used to guide the moulder for hand-shaping the mould and for large castings having simple geometrical shapes.

While designing a pattern, the parting line should be chosen so as to have the smallest portion of the pattern in the cope. As the moulding sand has greater strength in compression than in tension, the heavier sections of the pattern should be included in the drag. The possible defects due to loose sand in the mould are more frequent in the cope half. For this reason, the most critical surface should also be included in the drag. Figure 2.1 shows a typical split pattern (with allowances) for a cast iron wheel. The reader is advised to carefully note all the allowances, positive and negative.

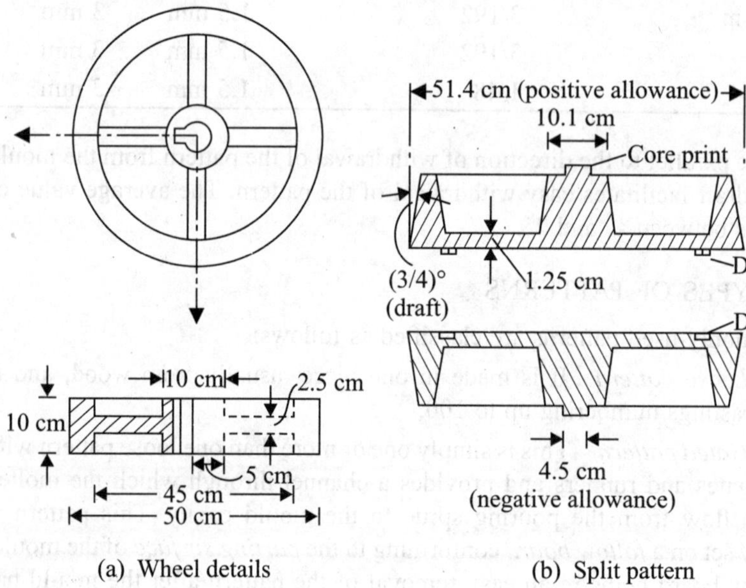

Fig. 2.1 Cast iron wheel and its split pattern (shown to a different scale).

2.2.3 TYPES OF MOULDS

Moulds can be classified on the basis of either the material, i.e., green sand mould, plastic mould, metal mould, or on the method of making them, e.g., shell mould and investment mould. Metal moulds are permanent in the sense that a large number of castings can be made from a single mould; on the other hand, moulds of refractory materials can be used only once. Generally, the green sand moulds are used; in what follows, we shall consider some of their important characteristics. (For a discussion on the other types of moulds, see Section 2.7.)

2.2.4 GREEN SAND MOULD

The material for a green sand mould is a mixture of sand, clay, water, and some organic additives, e.g., wood flour, dextrin, and sea coal. The percentage of these ingredients on weight basis is approximately 70–85% sand, 10–20% clay, 3–6% water, and 1–6% additives. This ratio may vary slightly depending on whether the casting is ferrous or nonferrous.

Sand is an inexpensive refractory material, but natural sand may not have all the desirable qualities of a moulding material. For example, it normally has higher clay content than desired. The sand used as a moulding material should have a specified clay, water, and additive content; in addition, it must have a specific grain size distribution. The importance of the grain size distribution would be clear from the discussion that follows.

Both the shape and the size of sand grains vary over a wide range. The grains may be smooth and round in shape or may have sharp angular corners. The bulk density of a sand-mix is very low if the grains are of almost equal size with smooth round shape. Such grains result in an increased void and a higher permeability. Higher permeability permits an easy outflow of the gases (produced during the casting operation) which may otherwise be entrapped within the casting. The situation gets reversed if the grains are of various sizes and have sharp corners. To study the grain size distribution, the screening test is performed. This is done by taking a fixed sample weight of sand and screening it through standard sieves. The screening is accomplished by shaking the sieves. The amount of sand that collects in the different sieves is then plotted. Finally, from this plot, the distribution of grain size and the average grain size are computed.

Clay, together with water, acts as a bonding agent and imparts tensile and shear strength to the moulding sand. The organic additives burn out at high temperatures and make room for the moulding sand to expand, and thus save the mould from crumbling.

The success of a casting process depends greatly on the properties of the moulding sand. These include (i) strength, (ii) permeability, (iii) deformation, (iv) flowability, and (v) refractoriness. (Standard specimens and tests are recommended for an evaluation of these properties[1].) Strength refers to the compressive strength and deformation indicates the change in length of a standard specimen at the point of failure. Permeability is expressed as the gas flow rate through the specimen under a specified pressure difference across it. Flowability refers to the ability of the sand to flow around and over the pattern when the mould is rammed. Refractoriness measures the ability of the sand to remain solid as a function of temperature. For a given sand-clay ratio, the nature of variation of these properties with water content is as shown in Fig. 2.2. It is obvious, both from strength and permeability considerations, that there is an optimum water content. At a low water content, dry clay powder, being finer than sand grains, fills up the void between the sand particles, and thus reduces the permeability. With higher water content, moist clay forms a coating over the sand particles keeping them further away, thus enhancing the permeability. Beyond the optimum water content, water itself fills up the void and reduces the permeability.

2.2.5 PREPARATION OF MOULD

Moulds are made by hand if the number of moulds to be prepared is small. If a large number of simple moulds are required, moulding machines are then used.

[1]Datsko, J., Material Properties and Manufacturing Processes, Wiley, New York, 1966.

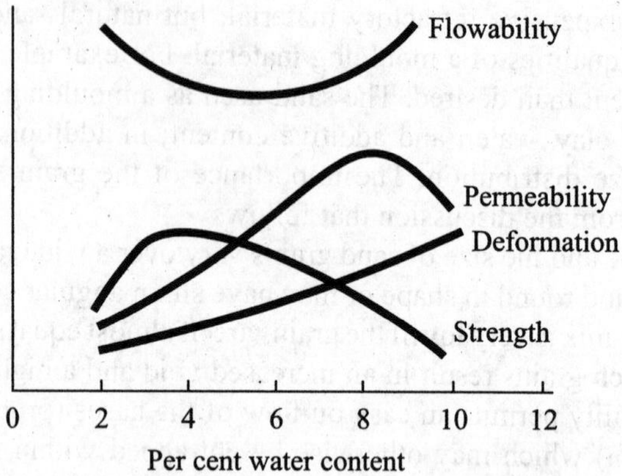

Fig. 2.2 Effect of water content on moulding sand properties.

In this section, we shall briefly discuss some important features of mould making; also, some typical moulding machines will be outlined.

To facilitate an easy removal of the pattern, a parting compound, e.g., non-wetting talc, is dusted on the pattern. Fine grain facing sand is used to obtain a good surface on the casting. Normally, a dead weight is placed on the cope flask to prevent the cope flask from floating due to hydrodynamic forces of the liquid metal. For a large mould, care should be taken to prevent the sand from falling off the cope flask when it is lifted to remove the pattern. This can be done by providing extra supports, called *gaggers*, within the cope flask. For a casting with re-entrant surfaces, e.g., a wheel with a groove at the rim, the mould can be made in three parts (Fig. 2.3). The part between the cope and the drag is termed as the *cheek*. For an easy escape of the gases, vent holes are provided in the cope flask.

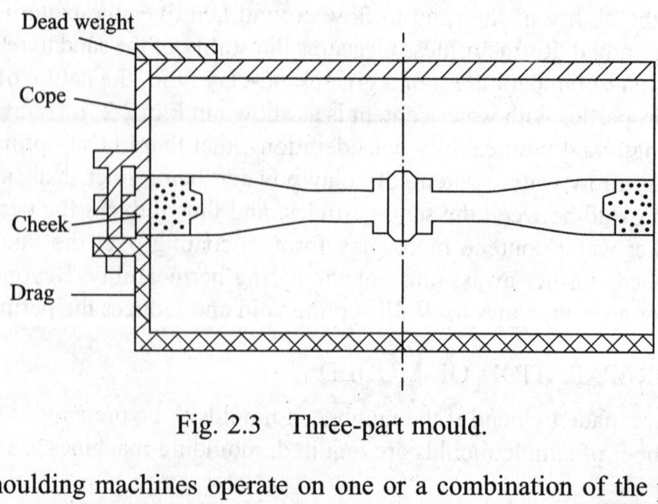

Fig. 2.3 Three-part mould.

The moulding machines operate on one or a combination of the principles explained in Fig. 2.4. In jolt ramming, the mould is lifted through a height of

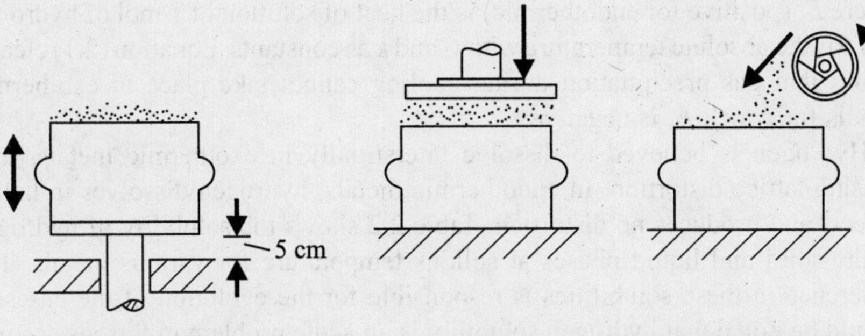

Fig. 2.4 Principles of machine moulding operation.

about 5 cm and dropped 50–100 times at a rate of 200 times per minute. This causes somewhat uneven ramming, but is quite suitable for horizontal surfaces. On the other hand, squeezing is found satisfactory for shallow flasks. The sand slinging operation is also very fast and results in uniform ramming. This, however, incurs high initial cost.

2.3 MELTING

A proper care during melting is essential for a good, defect-free casting. The factors to be considered during melting include gases in metals, selection and control of scrap, flux, furnace, and temperature. We shall now give a short discussion on these.

2.3.1 GASES IN METALS

The gases in metals normally lead to faulty castings. However, the presence of a controlled amount of specific gases can be beneficial in imparting certain desirable qualities to the castings. In metal castings, the gases

(i) may be mechanically trapped (in such situations, proper venting arrangements in the mould prevent their occurrence),

(ii) may be generated due to the variation in their solubility at different temperatures and phases, and

(iii) may be produced due to chemical reactions.

The gases most commonly present are hydrogen and nitrogen. Metals are divided into two groups so far as the solubility of hydrogen is concerned. One group is called endothermic; this includes common metals such as aluminium, magnesium, copper, iron, and nickel. The other group, called exothermic, includes, amongst others, titanium and zirconium. Endothermic metals absorb less hydrogen than exothermic metals. Further, in endothermic metals, the solubility of hydrogen increases with temperature. The reverse is true for exothermic metals. In both cases, the solubility (S) can be expressed as

$$S = C \exp [-E_s/(k\theta)], \tag{2.1}$$

where E_s (positive for endothermic) is the heat of solution of 1 mol of hydrogen and θ is the absolute temperature with C and k as constants. Equation (2.1) clearly shows that gas precipitation during cooling cannot take place in exothermic metals for which E_s is negative.

Hydrogen is believed to dissolve interstitially in exothermic metals, thus causing lattice distortion. In endothermic metals, hydrogen dissolves in lattice defects and produces no distortion. Table 2.2 shows the solubility of hydrogen in the solid and liquid phases at solidus temperature for various metals. The difference in these solubilities is responsible for the evolution of the gases. It should be noted that hydrogen solubility is an acute problem in ferrous casting. Here, although the amount of hydrogen by weight appears negligible, the volume evolved during solidification is quite large. Sievert's law states that the amount of hydrogen dissolved in a melt varies as

$$\% \text{ hydrogen present} = K\sqrt{p_{H_2}}, \tag{2.2}$$

where p_{H_2} is the partial pressure of hydrogen in the atmosphere over the melt, and the constant K can be evaluated from Table 2.2.

Table 2.2 Solubility of hydrogen in various metals

	Pressure = 1 atm	
Metal	Liquid solubility (cc/kg)	Solid solubility (cc/kg)
Iron	270	70
Magnesium	260	180
Copper	55	20
Aluminium	7	0.4

The primary sources of hydrogen in a melt are furnace dampness, air, oil, and grease. There is no simple dehydrogenating addition to eliminate hydrogen in the form of slag. So, care should be taken to maintain the hydrogen level to a minimum.

Most hydrogen removal techniques are based on equation (2.2), i.e., reducing the partial pressure of hydrogen by bubbling some other dry insoluble gas through the melt. For nonferrous metals, chlorine, nitrogen, helium, or argon is used. Nitrogen cannot be used for ferrous and nickel based alloys since it is soluble in these, and also it may form nitrides which affect the grain size; therefore, in ferrous alloys in particular, an accurate control of the nitrogen is necessary. In such situations, carbon monoxide bubbles are used. This removes not only hydrogen but also nitrogen; the carbon content is controlled by subsequent oxidation and recarburization. For ferrous metals, a marked decrease in the solubility of nitrogen during the change of phase may give rise to porosity in the casting. The re-entry of nitrogen from the air is prevented by the impermeable slag at the top of the melt.

Currently, vacuum melting is increasingly being used for preventing the solution of gases in metals and the combination of reactive elements in the melt. Additions in the ladle, rather than in the melt, have been found to be more effective for controlling the gases and chemical compositions.

2.3.2 FURNACES

The furnaces used for melting metals differ widely from one another. The selection of a furnace depends mainly on the metal chemistry, the maximum temperature required, and the metal delivery rate and mode. The other important factors in making a selection are the size and shape of the available raw materials.

The metal chemistry decides not only the control of standard elements but also some important mechanical properties, e.g., machinability.

The optimum temperature after melting is decided by a property, called *fluidity*, of the metal. Fluidity refers to the relative ability of the liquid metal to fill in the mould at a given temperature. Normally, the lower the viscosity, the higher the fluidity. The fluidity of a metal can be checked as follows. A spiral of standard dimensions is poured with the liquid metal at various temperatures. The length of the spiral which can be fed in this way before the solidification starts gives the measure of fluidity. If we examine the temperature-fluidity curves for various metals, we find that the higher the fluidity of a metal, the lower the difference needed between the pouring temperature (furnace temperature) and the melting temperature. For completely filling up the intricate, thin sections of the mould, this difference should be a minimum. A large difference implies higher cost and more gas solubility.

The rate and mode of liquid metal delivery are largely decided by the process—batch or continuous melting—used.

Figure 2.5 gives the sketches of various furnaces normally used in foundries; the maximum obtainable temperatures are also shown.

2.4 POURING (GATING DESIGN)

After melting, the metal is poured or injected into the mould cavity. We shall now discuss the difficulties faced in doing this and explain how these can be overcome by using an appropriate gating design. A good gating design ensures distribution of the metal in the mould cavity at a proper rate without excessive temperature loss, turbulence, and entrapping gases and slags.

If the liquid metal is poured very slowly, then the time taken to fill up the mould is rather long and the solidification may start even before the mould has been completely filled up. This can be avoided by using too much superheat, but then gas solubility may cause a problem. On the other hand, if the liquid metal impinges on the mould cavity with too high a velocity, the mould surface may be eroded. Thus, a compromise has to be made in arriving at an optimum velocity.

The design of a gating system depends on both the metal and mould compositions. For example, an elaborate gating design is needed to avoid dross (e.g., oxides) in easily oxidized metals of low melting point such as aluminium.

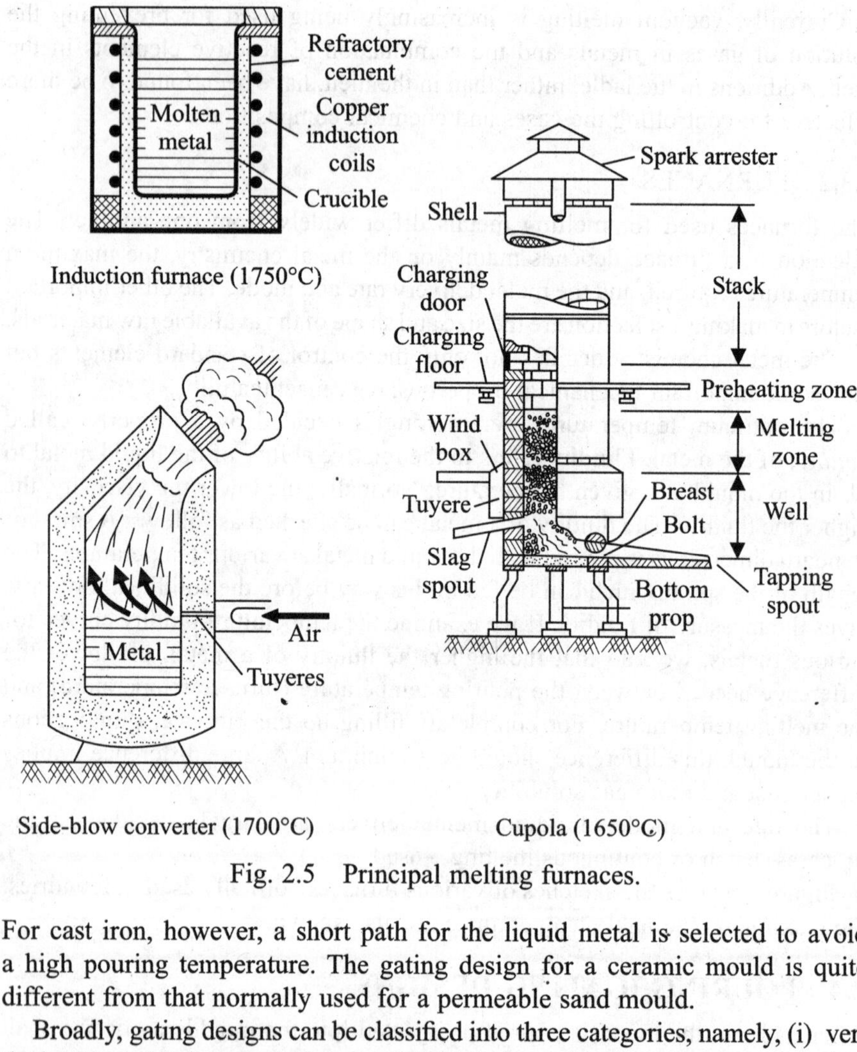

Fig. 2.5 Principal melting furnaces.

For cast iron, however, a short path for the liquid metal is selected to avoid a high pouring temperature. The gating design for a ceramic mould is quite different from that normally used for a permeable sand mould.

Broadly, gating designs can be classified into three categories, namely, (i) vertical gating, (ii) bottom gating, and (iii) horizontal gating. In vertical gating, the liquid metal is poured vertically to fill the mould with atmospheric pressure at the base. In bottom gating, on the other hand, the liquid metal is filled in the mould from bottom to top, thus avoiding the splashing and oxidation associated with vertical gating. Figure 2.6 shows a simple vertical gating and a bottom gating design. In the horizontal gating system, additional horizontal portions are introduced for better distribution of the liquid metal with minimum turbulence.

Simple calculations based on principles of fluid flow can lead to an estimate of the time taken to fill up a mould. We shall illustrate this for the two designs in Fig. 2.6. The integrated energy balance equation on the basis of per unit mass flow, more commonly known as Bernoulli's equation, will be used. For example, in Fig. 2.6a, it is assumed that the pressure at points 1 and 3 is equal (i.e., $p_1 = p_3$) and that level 1 is maintained constant. Thus, the velocity at station

1 (v_1) is zero. Moreover, the frictional losses are neglected. Then, the energy balance equation between points 1 and 3 gives

$$gh_t = v_3^2/2$$

or

$$v_3 = \sqrt{2gh_t}, \tag{2.3}$$

where g is the acceleration due to gravity and v_3 is the velocity of the liquid metal at the gate, subsequently referred to as v_g. So, the time taken to fill up the mould (t_f) is obtained as

$$t_f = \frac{V}{A_g v_3}, \tag{2.4}$$

where A_g and V are the cross-sectional area of the gate and the volume of the mould, respectively.

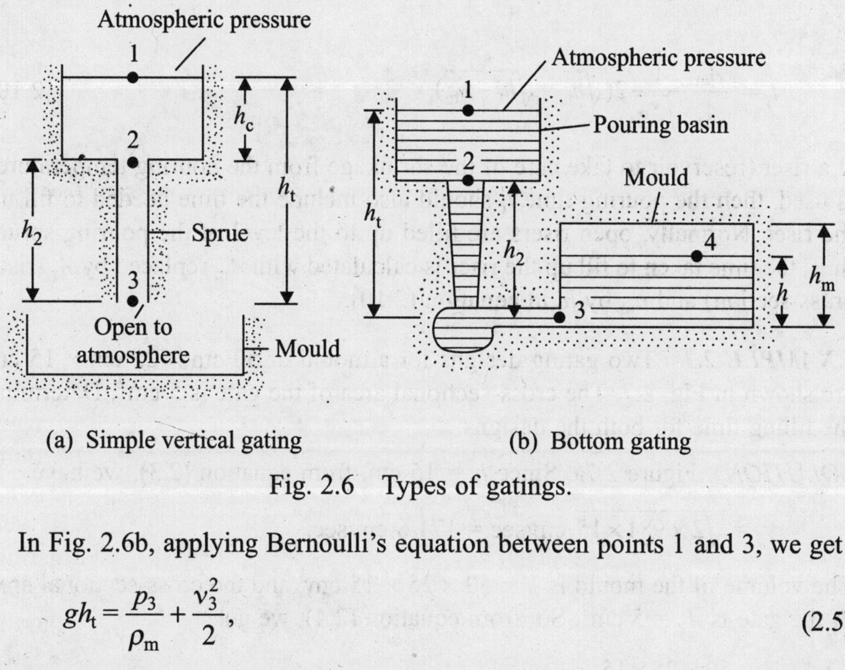

Fig. 2.6 Types of gatings.

In Fig. 2.6b, applying Bernoulli's equation between points 1 and 3, we get

$$gh_t = \frac{p_3}{\rho_m} + \frac{v_3^2}{2}, \tag{2.5}$$

where ρ_m is the density of the liquid metal, p_3 is the gauge pressure at station 3, and h_t is again assumed to be constant. Further, applying Bernoulli's equation between points 3 and 4, with the assumptions that v_4 is very small and all the kinetic energy at station 3 is lost after the liquid metal enters the mould, we can write

$$p_3/\rho_m = gh. \tag{2.6}$$

From equations (2.5) and (2.6), the velocity of the liquid metal at the gate we obtain is

$$v_g = v_3 = \sqrt{2g(h_t - h)}. \tag{2.7}$$

Equation (2.7) gives the velocity of a jet discharging against a static head h, making the effective head as $(h_t - h)$. Now, for the instant shown, let the metal level in the mould move up through a height dh in a time interval dt; A_m and A_g are the cross-sectional areas of the mould and the gate, respectively. Then,

$$A_m\, dh = A_g v_g\, dt. \tag{2.8}$$

Using equations (2.7) and (2.8), we get

$$\frac{1}{\sqrt{2g}} \frac{dh}{\sqrt{h_t - h}} = \frac{A_g}{A_m} dt. \tag{2.9}$$

At $t = 0$, $h = 0$ and at $t = t_f$ (filling time), $h = h_m$. Integrating equation (2.9) between these limits, we have

$$\frac{1}{\sqrt{2g}} \int_0^{h_m} \frac{dh}{\sqrt{h_t - h}} = \frac{A_g}{A_m} \int_0^{t_f} dt$$

or

$$t_f = \frac{A_m}{A_g} \frac{1}{\sqrt{2g}} 2(\sqrt{h_t} - \sqrt{h_t - h_m}). \tag{2.10}$$

If a riser (reservoir to take care of the shrinkage from the pouring temperature) is used, then the pouring time t_f should also include the time needed to fill up the riser. Normally, open risers are filled up to the level of the pouring sprue; thus, the time taken to fill up the riser is calculated with A_m replaced by A_r (riser cross-section) and h_m by h_t in equation (2.10).

EXAMPLE 2.1 Two gating designs for a mould of 50 cm × 25 cm × 15 cm are shown in Fig. 2.7. The cross-sectional area of the gate is 5 cm^2. Determine the filling time for both the designs.

SOLUTION Figure 2.7a. Since $h_t = 15$ cm, from equation (2.3), we have

$$v_3 = \sqrt{2 \times 981 \times 15} \text{ cm/sec} = 171.6 \text{ cm/sec}.$$

The volume of the mould is $V = 50 \times 25 \times 15$ cm^3 and the cross-sectional area of the gate is $A_g = 5$ cm^2. So, from equation (2.4), we get

$$t_f = \frac{50 \times 25 \times 15}{5 \times 171.6} \text{ sec} = 21.86 \text{ sec}.$$

Figure 2.7b. Here, $h_t = 15$ cm, $h_m = 15$ cm, $A_m = 50 \times 25$ cm^2, and $A_g = 5$ cm^2. Using equation (2.10), we have

$$t_f = \frac{50 \times 25}{5} \frac{\sqrt{2}}{\sqrt{981}} \sqrt{15} \text{ sec} = 43.71 \text{ sec}.$$

It should be noted that in Fig. 2.7b the time taken is double of that in Fig. 2.7a. We can easily verify that this will always be so if $h_m = h_t$.

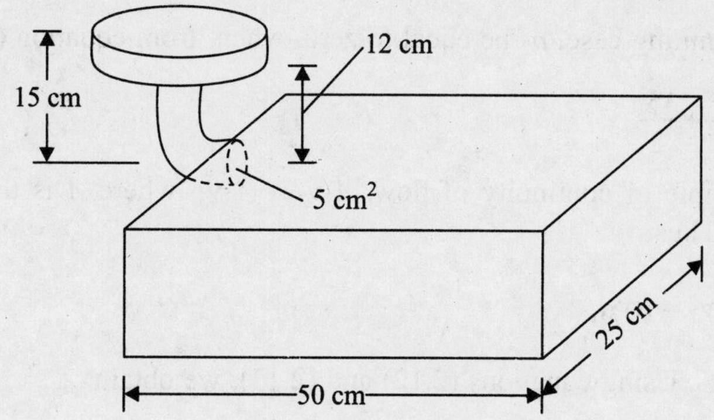

(a) Top gating

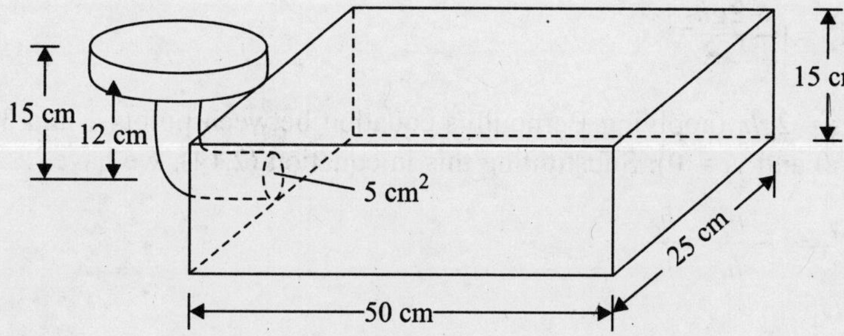

(b) Bottom gating

Fig. 2.7 Top and bottom gating designs.

2.4.1 ASPIRATION EFFECT

For a mould made of a permeable material (e.g., sand), care should be taken to ensure that the pressure anywhere in the liquid metal stream does not fall below the atmospheric pressure. Otherwise, the gases originating from baking of the organic compounds in the mould will enter the molten metal stream, producing porous castings. This is known as the aspiration effect.

Referring to Fig. 2.6a and applying Bernoulli's equation between points 2 and 3, we obtain

$$gh_2 + \frac{p_2}{\rho_m} + \frac{v_2^2}{2} = \frac{p_3}{\rho_m} + \frac{v_3^2}{2}, \tag{2.11}$$

where p and v refer to the pressure and velocity, respectively, of the liquid metal at stations 2 and 3. If the pressure at point 3 is atmospheric, i.e., $p_3 = 0$, then $p_2 = -\rho_m g h_2$ as $v_2 = v_3$. Hence, the design in Fig. 2.6a is not acceptable. To avoid negative pressure at point 2 (to ensure positive pressure anywhere in the liquid column), the sprue should be tapered, the ideal shape of which can be determined as follows.

Let, in the limiting case, p_2 be equal to zero, when, from equation (2.11),

$$\frac{v_3^2}{2} = gh_2 + \frac{v_2^2}{2}. \tag{2.12}$$

From the principle of continuity of flow, $A_2v_2 = A_3v_3$, where A is the cross-sectional area. Thus,

$$v_2 = \frac{A_3}{A_2}v_3 = Rv_3, \tag{2.13}$$

where $R = A_3/A_2$. Using equations (2.12) and (2.13), we obtain

$$\frac{v_3^2}{2g} = h_2 + \frac{R^2v_3^2}{2g}$$

or

$$R^2 = 1 - \frac{2gh_2}{v_3^2}. \tag{2.14}$$

Again, $v_3^2 = 2gh_t$ (applying Bernoulli's equation between points 1 and 3, with $p_1 = p_3 = 0$ and $v_1 = 0$). Substituting this in equation (2.14), we have

$$R^2 = 1 - \frac{h_2}{h_t} = \frac{h_c}{h_t}$$

or

$$R = \frac{A_3}{A_2} = \sqrt{\frac{h_c}{h_t}}. \tag{2.15}$$

[This can easily be seen to be the shape of a freely falling stream when $v_2 = \sqrt{2gh_c}$ and $v_3 = \sqrt{2gh_t}$.] Thus, ideally, the sprue profile should be as shown by the solid lines in Fig. 2.8 when the pressure throughout the stream is just atmospheric. However, a straight tapered sprue (shown by the dashed lines) is safer (pressure everywhere, except at points 2 and 3, is above atmospheric) and easier to construct. The sprue design in Fig. 2.6b is better than that in Fig. 2.6a.

Another situation where aspiration effect comes into the picture is associated with a sudden change in the flow direction. As shown in Fig. 2.9a, the liquid metal stream contracts around a sharp corner due to the momentum effect. In vertical gating, this has got nothing to do with acceleration due to gravity. The constricted region shown at station 2 in Fig. 2.9a is known as *vena contracta.* To avoid the creation of vacuum around station 2, the mould is made to fit the vena contracta, as done in Fig. 2.9b. In other words, a sharp change in the flow direction is avoided. If the runner diameter is d and the diameter at the entrance is d', then, normally, d'/d is maintained at a value approximately equal to 1.3[1]. This means $r \approx 0.15d$.

[1]Flinn, R.A., Fundamentals of Metal Casting, Addison-Wesley, Reading, Massachusetts, 1963.

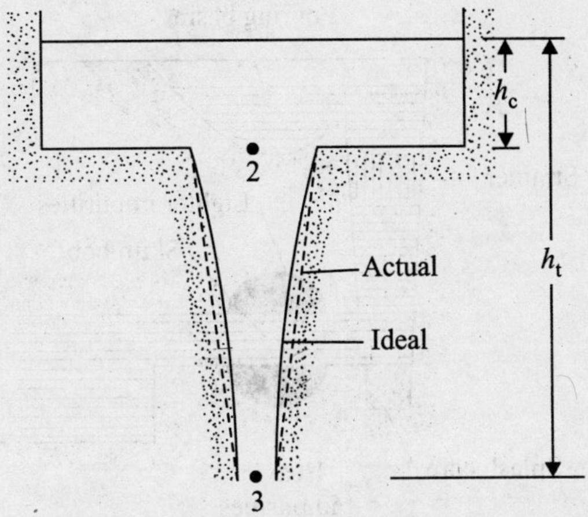

Fig. 2.8 Ideal and actual shapes of sprue.

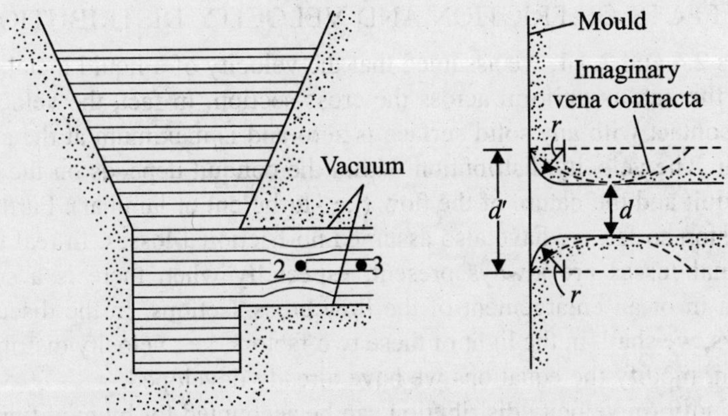

(a) Mechanism of vacuum generation

(b) Outlet dimensions to prevent vacuum generation

Fig. 2.9 Principle of avoiding vacuum generation.

The common items employed in a gating design to prevent impurities in the casting are as follows (see also Fig. 2.10).

(i) *Pouring basin* This reduces the eroding force of the liquid metal stream coming directly from the furnace. A constant pouring head can also be maintained by using a pouring basin.

(ii) *Strainer* A ceramic strainer in the sprue removes dross.

(iii) *Splash core* A ceramic splash core placed at the end of the sprue also reduces the eroding force of the liquid metal stream.

(iv) *Skim bob* It is a trap placed in a horizontal gate to prevent heavier and lighter impurities from entering the mould.

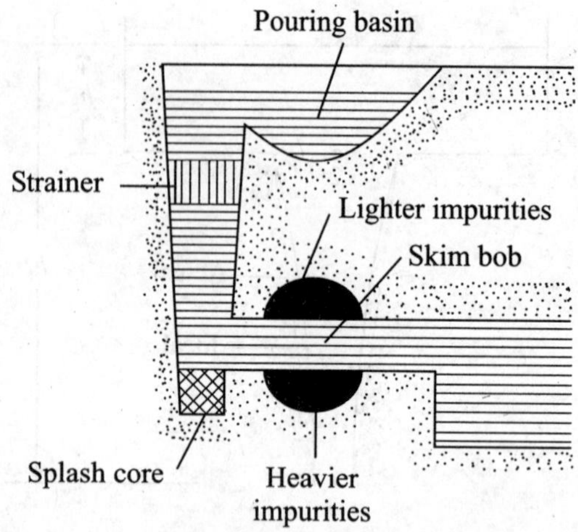

Fig. 2.10 Gating design to prevent impurities.

2.4.2 EFFECTS OF FRICTION AND VELOCITY DISTRIBUTION

In Sections 2.4 and 2.4.1, we assumed that the velocity of a liquid metal in the sprue and the gate is uniform across the cross-section. In fact, the velocity of a fluid in contact with any solid surface is zero and is maximum at the axis of the conduit. The velocity distribution within the conduit depends on the shape of the conduit and the nature of the flow (i.e., turbulent or laminar). Further, in our discussion so far, we have also assumed no frictional losses. In real fluids, the frictional losses are always present, especially when there is a sudden contraction in or an enlargement of the flow cross-sections. In the discussion that follows, we shall, in the light of these two factors, i.e., velocity distribution and friction, modify the equations we have already developed.

The nonuniform velocity distribution can be accounted for by modifying the kinetic energy term in the integrated energy balance equation by replacing the (velocity)^2 term by $\overline{v}^2/\beta$, where $\overline{v}$ is the average velocity and β is a constant. For a circular conduit, the value of β is equal to 0.5 for laminar flow and approximately equal to 1 for turbulent flow.

The energy loss due to friction in a circular conduit (on the basis of per unit mass) is given by

$$E_{f_1} = 4f\frac{L}{D}\frac{\overline{v}^2}{2}, \tag{2.16}$$

where

$\overline{v}$ = average velocity,

D, L = diameter and length, respectively, of the conduit, and

f = friction factor.

The value of f depends on the roughness of the conduit and the nature of the flow. Thus, while using the integrated energy equation between two points, say, 1 and 2, in that order in the direction of flow, E_{f_1} should be added to the energy of station 2.

For a smooth conduit, the value of f is given by the equations

$$f = 16/Re \quad \text{for laminar flow} \qquad (Re < 2000), \tag{2.17}$$

$$\frac{1}{\sqrt{f}} = 4\log_{10}(Re\sqrt{f}) - 0.4 \quad \text{for turbulent flow} \qquad (Re > 2000), \tag{2.18}$$

where Re is Reynolds number. For the range $2100 < Re < 10^5$, equation (2.18) can be simplified to the form

$$f = 0.0791(Re)^{-1/4}. \tag{2.19}$$

Frictional losses also occur due to a gradual change in the flow direction, e.g., in a 90° bend and other similar fittings. In such a situation, the associated frictional losses are accounted for by using equation (2.16) with an equivalent (L/D) factor for the bend.

The frictional loss (per unit mass) associated with a sudden enlargement or contraction of a flow area is expressed as

$$E_{f_2} = \tfrac{1}{2}e_f\overline{v}^2, \tag{2.20}$$

where $\overline{v}$ is the average velocity of the fluid in the smaller cross-section, and e_f is the friction loss factor and it depends on the ratio of the flow area and Reynolds number of the flow. For a laminar flow, the length and diameter of the smaller flow cross-section have also to be taken into account. The value of e_f depends on whether the flow area is enlarging or contracting in the direction of flow. The values of e_f for a sharp change in the flow cross-section are shown in Figs. 2.11a and 2.11b for sudden expansion and contraction, respectively. In these figures, the Reynolds numbers of the flow have been calculated on the basis of the average flow velocity in the smaller cross-sections, and l and d are the length and diameter, respectively, of the smaller flow cross-sections. The values of e_f for some other types of changes in flow geometry are listed in Fig. 2.12; here, e_{f_s} refers to the values corresponding to a sharp change of geometry with identical initial and final dimensions (i.e., the values of e_{f_s} can be obtained from Figs. 2.11a and 2.11b). Now, going back to Fig. 2.6a, let us modify the analysis presented in Section 2.4 in the light of the effects of friction and velocity distribution.

Using the integrated energy balance equation between points 1 and 3, after accounting for the loss due to a sudden contraction at 2, we get

$$\frac{p_1}{\rho_m} + 0 + gh_t = \frac{p_3}{\rho_m} + \frac{\overline{v}_3^2}{2\beta} + E_{f_1} + E_{f_2}, \tag{2.21}$$

where $\overline{v}_3$ is the average fluid velocity in the sprue. With $p_1 = p_3$ and using equations (2.16) and (2.20), we get

$$2gh_t = \overline{v}_3^2\left(\frac{1}{\beta} + e_f + 4f\frac{l}{d}\right), \tag{2.22}$$

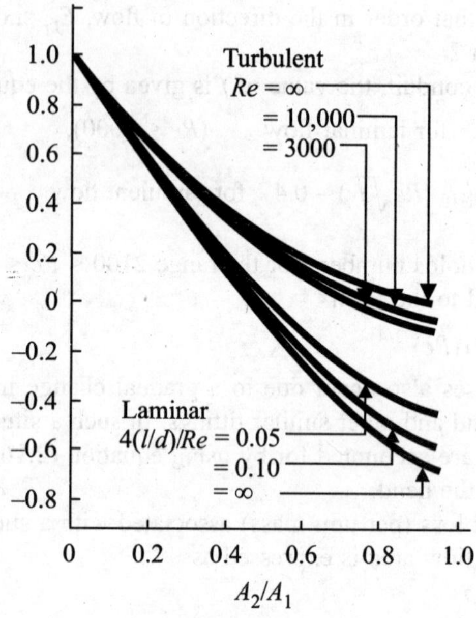

(a) For sudden expansion

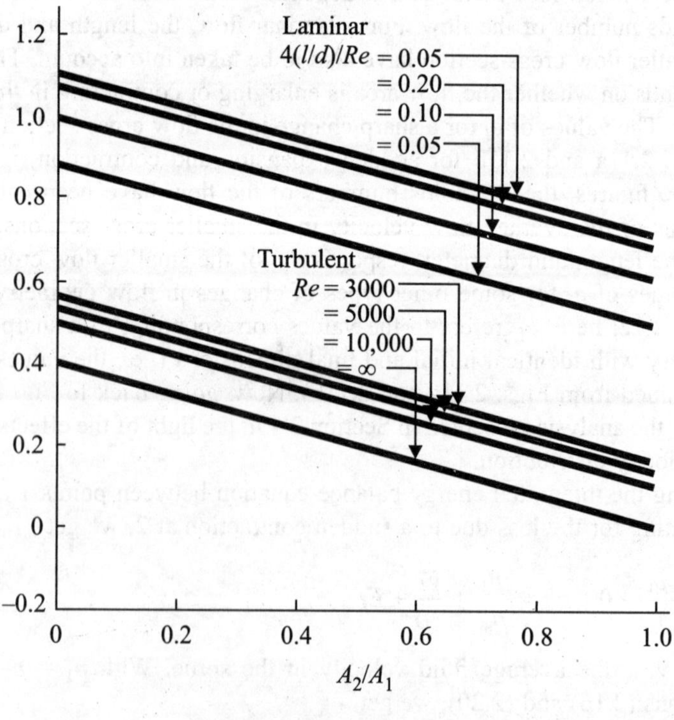

(b) For sudden contraction

Fig. 2.11 Friction loss factor.

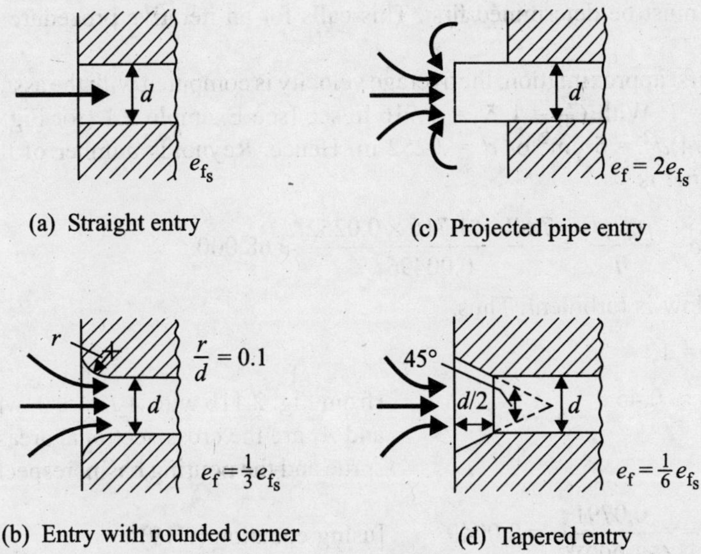

Fig. 2.12 Values of e_f for some types of changes in flow geometry.

where d is the diameter of the sprue and l is the length of the sprue (=h_2 in Fig. 2.6a). Thus, equation (2.3) can be modified as

$$\overline{v}_3 = C_D\sqrt{2gh_t}, \tag{2.23}$$

where the discharge coefficient

$$C_D = (\frac{1}{\beta} + e_f + 4f\frac{l}{d})^{-1/2}. \tag{2.24}$$

In this analysis, we have neglected the fluid velocity (and hence the loss) between points 1 and 2. If the sprue also has a bend or fitting, then E_{f_1} is modified as

$$E_{f_1} = 4f\overline{v}_3^2[\frac{l}{d} + (\frac{L}{D})_{eq}].$$

[The values of $(L/D)_{eq}$ for various types of fittings are listed in standard tables.] In such a case, the discharge coefficient C_D is, finally, given by

$$C_D = [\frac{1}{\beta} + e_f + 4f\{\frac{l}{d} + (\frac{L}{D})_{eq}\}]^{-1/2}. \tag{2.25}$$

EXAMPLE 2.2 Solve Example 2.1 (for Fig. 2.7a) by including the friction and velocity distribution effects. The liquid being poured is molten Fe with the properties $\rho_m = 7800$ kg/m^3, kinetic viscosity $\eta = .00496$ kg/m-sec. For the 90° turn at the end of the sprue, $(L/D)_{eq} = 25$.

SOLUTION It should be noted that to obtain β, f, and e_f (before calculating C_D), we should determine whether the flow is laminar or turbulent. So, the velocity must be determined first. This calls for an iterative procedure as now outlined.

As a first approximation, the average velocity is computed with the assumption that $C_D = 1$. With $C_D = 1$, $\overline{v}_3 = 1.716$ m/sec [see Example 2.1 (for Fig. 2.7a)]. Now, $(\pi/4)d^2 = 5$ cm^2 or $d = .0252$ m. Hence, Reynolds number of the flow in the sprue is

$$Re = \frac{\rho_m \overline{v}_3 d}{\eta} = \frac{7800 \times 1.716 \times 0.0252}{0.00496} \approx 68{,}000.$$

So, the flow is turbulent. Thus,

$\beta = 1,$

$e_f \approx 0.45$ [from Fig. 2.11b with $A_2/A_1 \approx 0$, where A_2 and A_1 are the cross-sectional areas of the sprue and the pouring basin, respectively],

$$f = \frac{0.0791}{(68{,}000)^{1/4}} = 0.0049 \quad \text{[using equation (2.19)].}$$

Now, using equation (2.25) with $l = 0.12$ m, we have

$$C_D = [1 + 0.45 + 4 \times 0.0049(\frac{0.12}{0.0252} + 25)]^{-1/2} = 0.7.$$

From equation (2.23),

$$\overline{v}_3 = C_D\sqrt{2gh_t} = 0.7 \times 1.716 \text{ m/sec} = 1.20 \text{ m/sec.}$$

We recalculate C_D on the basis of this value of $\overline{v}_3$ until it converges.

For the second approximation,

$$Re = \frac{1.2}{1.716} \times 68{,}000 = 47{,}500$$

when

$\beta = 1,$

$e_f = 0.46,$

$$f = \frac{0.0791}{(47{,}500)^{1/4}} = 0.0054,$$

$$C_D = [1 + 0.46 + 4 \times 0.0054(\frac{0.12}{0.0252} + 25)]^{-1/2}$$

$$= 0.69 \quad \text{(starting value of } C_D = 0.7\text{).}$$

So,

$$\overline{v}_3 = 0.69 \times 1.716 \text{ m/sec} = 1.18 \text{ m/sec.}$$

Thus, the value of $\overline{v}_3$ has converged with sufficient accuracy. Now, using equation (2.4), we find the time taken to fill up the mould is

$$t_f = \frac{V}{A_g \overline{v}_3}$$

$$= \frac{50 \times 25 \times 15}{5 \times 118} \text{ sec} = 31.7 \text{ sec.}$$

This time is approximately 43% higher than that obtained in Example 2.1 (Fig. 2.7a) (neglecting all the losses).

EXAMPLE 2.3 Figure 2.13 shows a ladle having an internal diameter of 1 m with a capacity height of 1.2 m. It has a 45°-tapered nozzle to a 75-mm exit diameter.

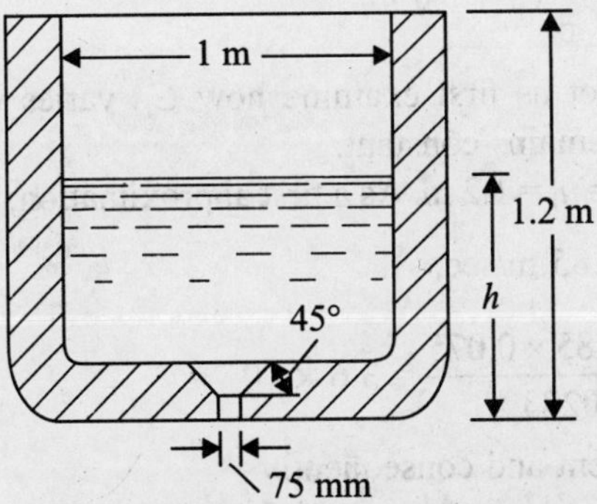

Fig. 2.13 Description of ladle.

(i) Calculate the time required to empty the ladle if it is filled with an Al-Si alloy at 704°C.

(ii) Estimate the discharge rate in kg/sec (a) initially and (b) when the ladle is 75% empty.

Given

$$\rho_m = 2700 \text{ kg/m}^3,$$

$$\eta = 0.00273 \text{ kg/m-sec.}$$

SOLUTION Let at any instant the level of liquid metal be h when the average velocity of the liquid metal through the nozzle is

$$\overline{v}_n = C_D \sqrt{2gh}.$$

The mass flow discharge rate is

$$\dot{m} = \rho_m A_n \overline{v}_n = \rho_m A_n C_D \sqrt{2gh}, \tag{a}$$

where A_n is the nozzle area. Again, $\dot{m} = -\rho_m A_l \, dh/dt$, where A_l is the ladle cross-sectional area. Thus,

$$-\frac{dh}{dt} = \frac{A_n}{A_l} C_D \sqrt{2gh}. \tag{b}$$

This is, of course, under the assumption that C_D remains almost constant for all values of h. Now, the time taken (t_f) for the level to come down from h_i to h_f can be obtained by integrating equation (b) as

$$-\int_{h_i}^{h_f} \frac{dh}{\sqrt{h}} = \frac{A_n}{A} C_D \sqrt{2g} \int_0^{t_f} dt$$

or

$$2(\sqrt{h_i} - \sqrt{h_f}) = \frac{A_n}{A_l} C_D \sqrt{2g} t_f$$

or

$$t_f = \sqrt{\frac{2}{g}} \frac{A_l}{A_n} \cdot \frac{1}{C_D} (\sqrt{h_i} - \sqrt{h_f}). \qquad \text{(c)}$$

For the given data, let us first examine how C_D varies with h to justify the assumption that C_D remains constant.

First, let us take $h = h_i = 1.2$ m. As a first approximation, we take $C_D = 1$ when

$$\overline{v}_n = \sqrt{2gh} = 4.85 \text{ m/sec},$$

$$Re = \frac{2700 \times 4.85 \times 0.075}{0.00273} = 3.6 \times 10^5.$$

So, the flow is turbulent and consequently

$$\beta = 1,$$

$$e_f = \tfrac{1}{6} e_{f_s} = \tfrac{1}{6} \cdot 0.45 \qquad \text{(from Figs. 2.12 and 2.11b with } A_2/A_1 \approx 0)$$
$$= 0.075.$$

The frictional losses within the ladle are negligible. Hence,

$$C_D = (1 + 0.075)^{-1/2} = 0.96.$$

As this value of C_D is very near unity, there is no need to iterate further (as done in Example 2.2).

Second, let us consider the case when $h \approx 0$, $\overline{v}_n \approx 0$, $Re \to 0$, i.e., the flow is laminar. So, $\beta = \frac{1}{2}$ and $e_f \approx 0.82$. Thus,

$$C_D = (2 + 0.82)^{-1/2} \approx 0.6.$$

Third, let $h = 5$ cm $= 0.05$ m and assume that $C_D = 1$. Then,

$$\overline{v}_n = \sqrt{2 \times 9.81 \times 0.5} \text{ m/sec} \approx 1 \text{ m/sec}.$$

Thus,

$$Re = \frac{2700 \times 1 \times 0.075}{0.00273} = 74175 \qquad \text{(i.e., the flow is turbulent).}$$

So,

$$\beta \approx 1,$$

$$e_f = \tfrac{1}{6} e_{f_s} \approx \tfrac{1}{6} \times 0.45 \approx 0.075 \qquad \text{(again from Figs. 2.12 and 2.11b with } A_2/A_1 \approx 0\text{)},$$

i.e.,

$$C_D \approx 0.96.$$

So, C_D remains equal to 0.96 for almost the entire period (except when $h \to 0$), justifying our assumption. We are now in a position to calculate the time and discharge rate.

(i) Here,

$$A_l = \frac{\pi}{4} \times 1 \text{ m}^2 = 0.785 \text{ m}^2,$$

$$A_n = \frac{\pi}{4} \times (0.075)^2 \text{ m}^2 = 4.4 \times 10^{-3} \text{ m}^2 \qquad [\text{with } h_f = 0 \text{ from equation (c)}],$$

$$t_f = \sqrt{\frac{2}{9.81}} \times \frac{0.785}{4.4 \times 10^{-3}} \cdot \frac{1}{0.96} \sqrt{1.2} \text{ sec}$$

$$= 91.9 \text{ sec.}$$

(ii) The initial discharge rate, from equation (a) with h = 1.2 m, is given as

$$\dot{m} = 2700 \times 4.4 \times 10^{-3} \times 0.96 \sqrt{2 \times 9.81 \times 1.2} \text{ kg/sec}$$

$$= 55.34 \text{ kg/sec.}$$

The discharge rate when the ladle is 75% empty, i.e., with $h = 0.25 \times 1.2$ m, is obtained from equation (a) as $\dot{m} = 27.67$ kg/sec.

2.5 COOLING AND SOLIDIFICATION

A clear understanding of the mechanism of solidification and cooling of liquid metals and alloys is essential for the production of successful castings. During solidification, many important characteristics such as crystal structure and alloy composition at different parts of the casting are decided. Moreover, unless a proper care is taken, other defects, e.g., shrinkage cavity, cold shut, misrun, and hot tear, also occur.

2.5.1 MECHANISM OF SOLIDIFICATION

Pure Metals

Liquids need to be cooled below their freezing points before the solidification begins. This is because energy is required to create surfaces for new crystals. The degree of supercooling necessary is reduced by the presence of other surfaces (particles) which serve as the initial nuclei for crystal growth. When a liquid metal is poured into a mould, initially (at time t_0 in Fig. 2.14) the temperature everywhere is θ_0. The mould face itself acts as the nucleus for crystal growth,

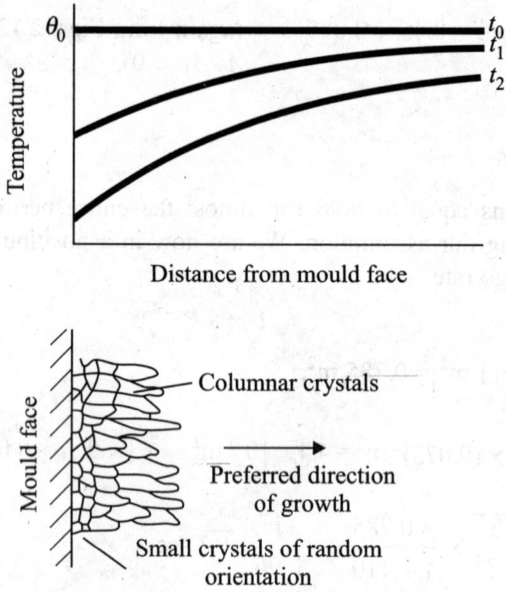

Fig. 2.14 Development of columnar crystals.

and if the conductivity of the mould is high, randomly-oriented small crystals grow near the mould face. Subsequently, a temperature gradient results within the casting, as indicated in Fig. 2.14 for t_1 and t_2. As the solidification progresses gradually inwards, long columnar crystals, with their axes perpendicular to the mould face, grow. This orientation of crystal growth is desirable from the point of view of strength of the casting.

Alloys

As we have already noted in Chapter 1, an alloy, unlike a pure metal, does not have a sharply defined freezing temperature. The solidification of an alloy takes place over a range of temperature. During this process, the solids separating out at different temperatures possess varying compositions. Due to all these facts, the direction of crystal growth in an alloy depends on various factors, such as

(i) the composition gradient within the casting,

(ii) the variation of solidus temperature with composition, and

(iii) the thermal gradient within the mould.

We shall discuss each of these factors by considering the example of a solid solution alloy whose phase diagram is shown in Fig. 2.15.

Let the liquid alloy have the composition C_0 (of B in A). Also, let θ_f be the freezing point of pure metal A, and θ_0 and θ_0', respectively, be the liquidus and the solidus temperatures of the alloy of composition C_0. As the liquid alloy is cooled down to the temperature θ_0, solids start to separate out. The concentration of B in these solids is only C_1 ($<C_0$) as is evident from Fig. 2.15. As a result, the concentration of B in the liquid, near the solid-liquid interface, increases to a

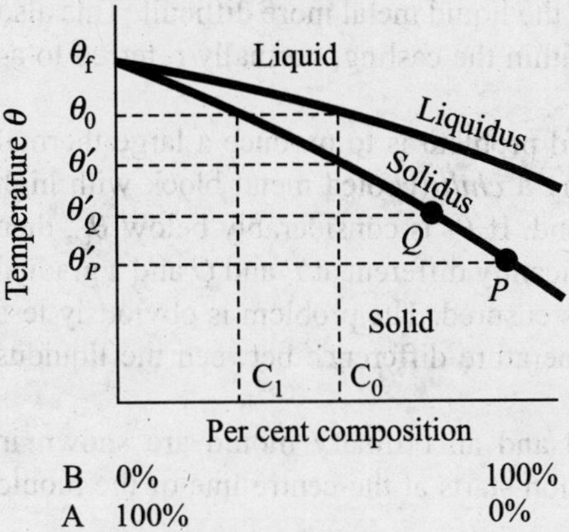

Fig. 2.15 Equilibrium phase diagram of A-B alloy.

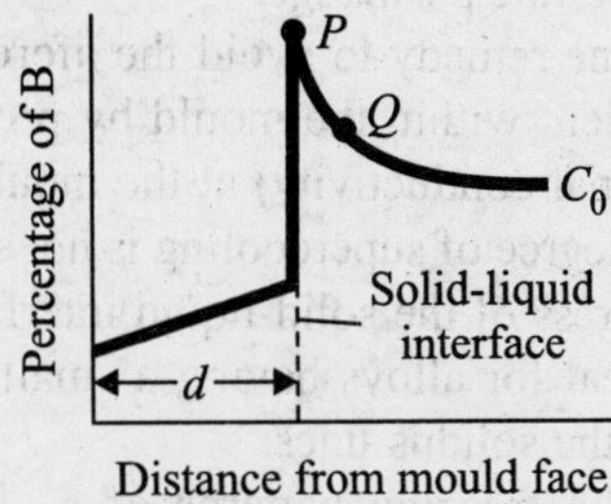

Fig. 2.16 Variation of concentration of B with distance from mould face.

value more than C_0. Figure 2.16 shows this for the situation where solidification front has progressed up to some distance *d* from the mould face.

Now, let us consider two points *P* and *Q* within the liquid alloy, *P* being just beyond the solid-liquid interface, as indicated in Fig. 2.16. The solidus temperatures corresponding to the compositions at *P* and *Q* are θ'_P and θ'_Q, respectively (see Fig. 2.15). Let θ_P and θ_Q be the actual temperatures at the points *P* and *Q*, respectively. θ_Q is greater than θ_P due to the thermal gradient within the casting (see Fig. 2.14). If both θ_Q and θ_P lie in the range θ'_P to θ'_Q, then the liquid at *Q* is supercooled, whereas that at *P* is not. This implies that the crystallization starts at *Q* sooner than at *P*. If this difference is very prominent, then the columnar growth of crystals starting from the mould surface is hampered. The crystal growth in such a situation may appear as in Fig. 2.17. Thus, a dendritic

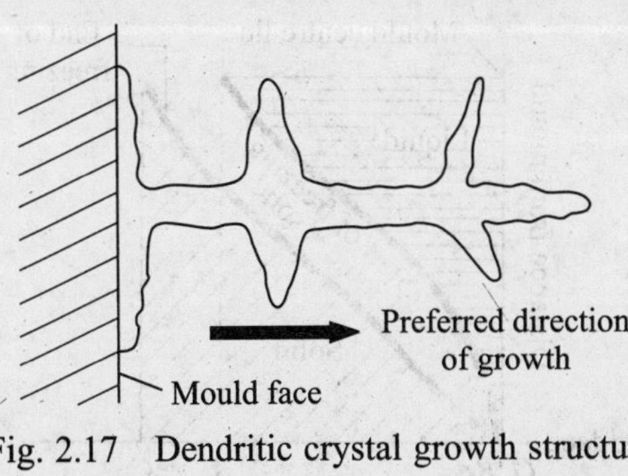

Fig. 2.17 Dendritic crystal growth structure.

structure is produced. If the crystallization at *Q* gets completed before it starts at *P* (due to a very small thermal gradient, with a very high concentration difference and a very slopy solidus line), then randomly-orientated crystals may appear inside the casting. Moreover, the presence of solid crystals ahead of the solid-liquid interface makes feeding of the liquid metal more difficult. This also implies greater risk of having voids within the casting, normally referred to as *centre-line shrinkage*.

One remedy to avoid the aforestated problem is to produce a large thermal gradient within the mould by providing a *chill* (cooled metal block with high thermal conductivity) at the mould's end. If θ_P is considerably below θ_Q, then the degree of supercooling is not significantly different at P and Q and a gradual progress of the solid-liquid interface is ensured. The problem is obviously less critical for alloys having a small temperature difference between the liquidus and the solidus lines.

The freezing patterns of a chilled and an ordinary mould are shown in Fig. 2.18. In Fig. 2.18a, the solidification starts at the centre line of the mould

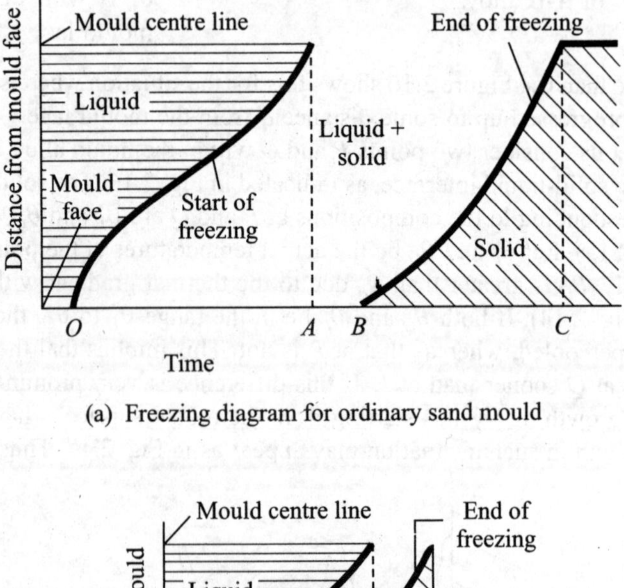

(a) Freezing diagram for ordinary sand mould

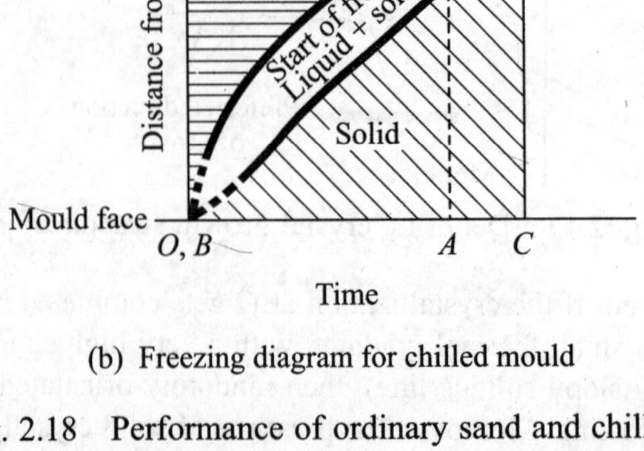

(b) Freezing diagram for chilled mould

Fig. 2.18 Performance of ordinary sand and chilled moulds.

before the solidification is complete even at the mould face. In the chilled mould (Fig. 2.18b), on the other hand, due to rapid heat extraction, a narrow liquid-solid zone quickly sweeps across the molten metal.

The difficulty of feeding a given alloy in a mould is expressed by a quantity, called *centre-line feeding resistance* (CFR). It is defined as

$$\text{CFR} = \frac{\text{time interval between start and end of freezing at centre line}}{\text{total solidification time of casting}} \times 100\%.$$

Referring to Fig. 2.18, we find that

$$\text{CFR} = \frac{AC}{OC} \times 100\%. \tag{2.26}$$

Normally, feeding is considered to be difficult if CFR > 70%.

2.5.2 RATE OF SOLIDIFICATION

A reservoir of liquid metal, called riser, is used to compensate for the shrinkage that takes place from the pouring temperature up to solidification (i.e., during the first two phases mentioned in Section 2.2.1). In this respect, grey cast iron is an interesting exception where solidification occurs in two stages. The shrinkage associated with the first stage may well be compensated by the *expansion* that takes place during the second stage, and as such, a riser may not be necessary. To ensure that the riser does not solidify before the casting, we should have an idea of the time taken by the casting to solidify. Moreover, the placement (location) of the riser can be judiciously chosen if an estimate of the time taken by the casting to solidify up to a certain distance from the mould face is available.

The heat rejected by the liquid metal is dissipated through the mould wall. The heat, released as a result of cooling and solidification of the liquid metal, passes through different layers. The temperature distribution in these layers, at any instant, is schematically shown in Fig. 2.19. The thermal resistances which

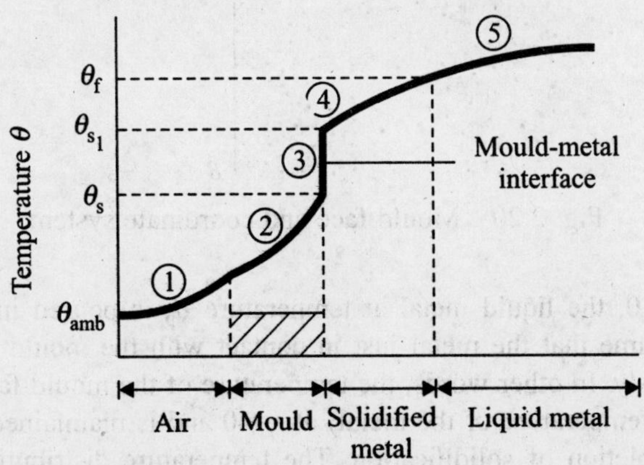

Fig. 2.19 Temperature distribution in different layers.

govern the entire solidification process are those of the liquid, the solidified metal, the metal-mould interface, the mould, and the ambient air. These five different regions are indicated by the numbers 1 to 5 in Fig. 2.19. The solidification process is quite complicated especially when complex geometry, freezing of alloys, or temperature dependence of thermal properties is considered. In what follows, we shall discuss the solidification of pure metals in some cases of practical interest. In doing so, we shall, depending on the situation, make simplifying assumptions to neglect the thermal resistance of one or more of the regions shown in Fig. 2.19.

2.5.3 SOLIDIFICATION OF A LARGE CASTING IN AN INSULATING MOULD

During the solidification of a large casting in an insulating mould, like the one used in the sand or investment casting (see Section 2.7), almost the entire thermal resistance is offered by the mould. Hence, the analysis we give computes the freezing time by considering only the thermal resistance of region 2 (Fig. 2.19).

Consider a mould face *AB* shown in Fig. 2.20. The large mould, initially at a temperature θ_0, is assumed to be extended up to infinity in the *x*-direction.

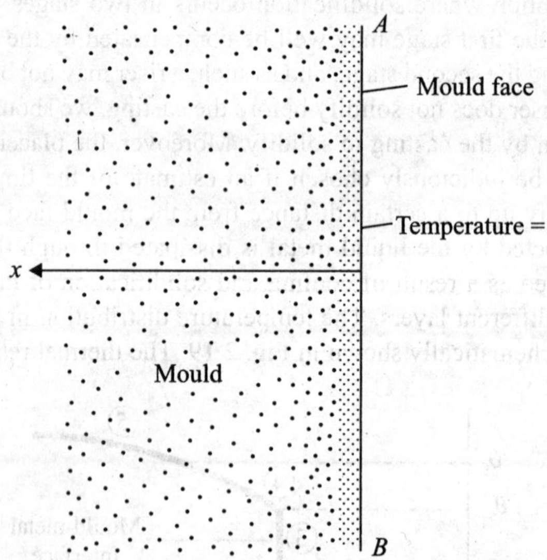

Fig. 2.20 Mould face and coordinate system.

At time $t = 0$, the liquid metal at temperature θ_p is poured into the mould. We also assume that the metal just in contact with the mould face solidifies instantaneously. In other words, the temperature of the mould face is raised to θ_f (freezing temperature of the metal) at $t = 0$ and is maintained at that value till the completion of solidification. The temperature distribution within the mould at a subsequent time t (assuming one-dimensional heat conduction in the x-direction) for such a case is given by[1]

$$\theta_x(t) = \theta_0 + (\theta_f - \theta_0)[1 - \operatorname{erf} \frac{x}{2\sqrt{\alpha t}}]. \tag{2.27}$$

---

[1]Carslaw, H.S. and Jaeger, J.C., Conduction of Heat in Solids, Clarendon Press, Oxford, 1959.

erf (z) is called the error function. This function is very useful for heat conduction problems in semi-infinite mediums. It represents the convergent series

$$\operatorname{erf}(z) = \frac{2}{\sqrt{\pi}}(z - \frac{z^3}{3 \cdot 1!} + \frac{z^5}{5 \cdot 2!} - \frac{z^7}{7 \cdot 3!} + \ldots) \tag{2.28}$$

or the integral

$$\operatorname{erf}(z) = \frac{2}{\sqrt{\pi}} \int_0^z e^{-x^2} dx. \tag{2.29}$$

The values of this function for various values of its argument z are given in mathematical tables and handbooks. We shall use the result

$$\frac{d}{dz} \operatorname{erf}(z) = \frac{2}{\sqrt{\pi}} e^{-z^2} \tag{2.30}$$

in this chapter.

In (2.27), $\theta_x(t)$ is the temperature at a distance x from the mould face at an instant t and α is thermal diffusivity of the mould material and is equal to $k/(\rho c)$ with k = conductivity, ρ = density, and c = specific heat of the mould material. Now, the rate of heat flow through the mould face at any instant t is given by

$$\dot{Q} = -kA \frac{\partial \theta_x}{\partial x}\bigg|_{x=0},$$

where A is the cross-sectional area of the mould-metal interface (approximately the surface area of the casting). Using equations (2.27) and (2.30), we obtain

$$\dot{Q} = \frac{kA(\theta_f - \theta_0)}{\sqrt{\pi \alpha t}}. \tag{2.31}$$

Thus, the total quantity of heat flow across the mould face up to a certain time t_0 is

$$Q_{t_0} = \int_0^{t_0} \dot{Q}\, dt = \frac{2Ak(\theta_f - \theta_0)}{\sqrt{\pi \alpha}} \sqrt{t_0}. \tag{2.32}$$

Next, let us calculate the heat the liquid metal rejects in order to solidify. If the liquid metal has a latent heat L, a specific heat c_m, and a density ρ_m, then the heat rejected is given by

$$Q_R = \rho_m V[L + c_m(\theta_p - \theta_f)], \tag{2.33}$$

where V is the total volume of the casting. Hence, the solidification time t_s, obtained by equating (2.32) and (2.33) at t_s, is

$$\frac{2Ak(\theta_f - \theta_0)}{\sqrt{\pi\alpha}}\sqrt{t_s} = \rho_m V[L + c_m(\theta_p - \theta_f)]$$

or

$$t_s = \gamma(\frac{V}{A})^2, \tag{2.34}$$

where the constant γ is given by

$$\gamma = \{\frac{\rho_m \sqrt{\pi\alpha}[L + c_m(\theta_p - \theta_f)]}{2k(\theta_f - \theta_0)}\}^2. \tag{2.35}$$

It should be noted that the foregoing analysis assumes a plane metal-mould interface AB, not usually encountered in engineering practice. Often, we are required to find out the freezing time of complex contours. For such contours, all we need to do is observe (without any precise calculations) the following basic features to know whether the analysis we have given underestimates or overestimates the actual freezing time. To observe these features, we consider three types of metal-mould interfaces (see Fig. 2.21), namely, (i) convex, (ii) plane (used in our analysis), and (iii) concave.

In Fig. 2.21a, the heat flow is more divergent, and consequently the rate is somewhat more than that in Fig. 2.21b. Thus, the freezing time in such a case is overestimated by the foregoing analysis. Similarly, in Fig. 2.21c, the heat flow is more convergent, and consequently the rate is somewhat less than that in Fig. 2.21b. So, the freezing time in such a case is underestimated by the analysis we have given.

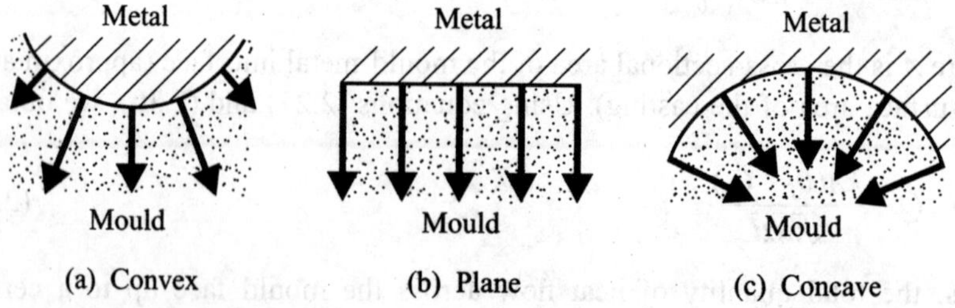

Fig. 2.21 Heat flow across various types of metal-mould interfaces.

The quantitative results of the effect of the mould-casting interface on the freezing time can be obtained for some basic shapes. Before we give these results, we define two nondimensional parameters, namely,

$$\beta = \frac{V/A}{\sqrt{\alpha t_s}}, \tag{2.36a}$$

$$\lambda = \frac{\theta_f - \theta_0}{\rho_m L'} \rho c \tag{2.36b}$$

with

$$L' = L + c_m(\theta_p - \theta_f). \tag{2.36c}$$

In terms of these parameters, equation (2.34) can be rewritten as

$$\beta = \lambda \frac{2}{\sqrt{\pi}} \qquad \text{(for an infinite plane).} \tag{2.37}$$

A similar result for an infinitely long cylinder is

$$\beta = \lambda\left(\frac{2}{\sqrt{\pi}} + \frac{1}{4\beta}\right) \tag{2.38}$$

and for a sphere is

$$\beta = \lambda\left(\frac{2}{\sqrt{\pi}} + \frac{1}{3\beta}\right). \tag{2.39}$$

EXAMPLE 2.4 A large plate of cross-sectional area A is being cast. Establish a relationship between the time after pouring and the distance of the solidification front from the mould face (Fig. 2.22), assuming no superheat.

SOLUTION During the solidification of this plate-shaped casting, most of the heat is rejected through two side faces, each having the cross-sectional area A. The heat rejected through the other four faces (having very small area as compared with A) is negligible. So, the solidification fronts move from the two sides, as indicated in Fig. 2.22.

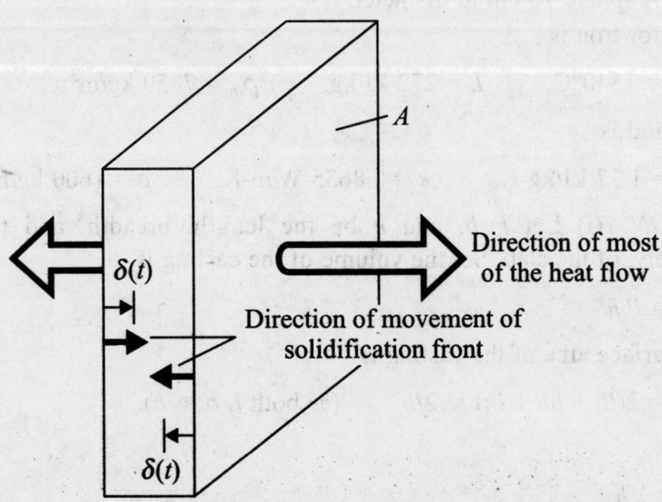

Fig. 2.22 Movement of solidification fronts.

If the solidification fronts move through a distance $\delta(t)$ at any instant t from the respective mould-metal interfaces, then the heat rejected by each solidified half is

$$Q_R = \rho_m A \delta L. \tag{a}$$

Now, the time taken to reject this much of heat through the mould face of the area A is t. Then, from equation (2.32),

$$Q_R = \frac{2Ak(\theta_f - \theta_0)}{\sqrt{\pi\alpha}}\sqrt{t}. \tag{b}$$

Equating the right-hand sides of (a) and (b), we get

$$t = \gamma\delta^2$$

or

$$\delta(t) = \frac{1}{\sqrt{\gamma}}\sqrt{t}, \tag{c}$$

where γ is given by equation (2.35).

A smooth progress of the solidification fronts, as envisaged in the foregoing problem, is not observed in real life, especially in the case of alloys. In fact, all our foregoing analyses should be used only to provide an estimate and not the precise answer.

EXAMPLE 2.5 Determine the solidification time of the following two iron castings when both are poured, with no superheats, into sand moulds at the initial temperature 28°C:

(i) A slab-shaped casting 10 cm thick.

(ii) A sphere 10 cm in diameter.

The data for iron is

$$\theta_f = 1540°\text{C}, \qquad L = 272 \text{ kJ/kg}, \qquad \rho_m = 7850 \text{ kg/m}^3;$$

and for sand is

$$c = 1.17 \text{ kJ/kg-K}, \qquad k = 0.8655 \text{ W/m-K}, \qquad \rho = 1600 \text{ kg/m}^3.$$

SOLUTION (i) Let l, b, and h be the length, breadth, and thickness, respectively, of the slab. So, the volume of the casting is

$$V = lbh$$

and the surface area of the casting is

$$A = 2(lb + bh + lh) \approx 2lb \qquad (\text{as both } l, b \gg h).$$

Hence,

$$\frac{V}{A} \approx \frac{h}{2} = 5 \times 10^{-2} \text{ m}.$$

As superheat is zero, $\theta_p = \theta_f$, $L' = L$, and $\theta_0 = 28°C$. Using equation (2.36b), we find that

$$\lambda = \frac{(1540-28)\times 1600\times 1.17\times 10^3)}{7850\times 272\times 10^3} = 1.3256.$$

So, from equation (2.37),

$$\beta = \frac{2}{\sqrt{\pi}}\times 1.3256 = 1.4957.$$

Now,

$$\alpha = \frac{k}{\rho c} = \frac{0.8655}{1600\times 1.17\times 10^3}\ \text{m}^2/\text{sec} = 0.46\times 10^{-6}\ \text{m}^2/\text{sec}.$$

Using equation (2.36a), we have

$$\alpha t_s = \frac{(V/A)^2}{\beta^2} = \frac{25\times 10^{-4}}{2.24}\ \text{m}^2.$$

So, the solidification time

$$t_s = \frac{25\times 10^{-4}}{2.24\times 0.46\times 10^{-6}}\ \text{sec} = 2430\ \text{sec} = 0.675\ \text{hr}.$$

(ii) Using equation (2.39) with $\lambda = 1.3256$, we obtain

$$\beta = 1.4957 + \frac{0.4419}{\beta} \quad \text{or} \quad \beta^2 - 1.4957\beta - 0.4419 = 0.$$

If we take only the positive real root, then $\beta = 1.75$. Now, for a sphere of radius $R(=5\ \text{cm})$,

$$\frac{V}{A} = \frac{(4/3)\pi R^3}{4\pi R^2} = \frac{R}{3} = \frac{5}{3}\times 10^{-2}\ \text{m}.$$

Again, using equation (2.36a), we find

$$\alpha t_s = \frac{(V/A)^2}{\beta^2} = \frac{25}{9}\times\frac{10^{-4}}{3.05}\ \text{m}^2$$

or

$$t_s = \frac{25}{9}\times\frac{10^{-4}}{3.05}\times\frac{10^6}{0.46}\ \text{sec} = 198\ \text{sec} = 0.055\ \text{hr}.$$

2.5.4 SOLIDIFICATION WITH PREDOMINANT INTERFACE RESISTANCE

In some common casting processes, the heat flow is controlled significantly by the thermal resistance of the mould-metal interface (indicated as region 3 in Fig. 2.19). These processes include permanent mould casting and die casting. The condition of no contact resistance exists only when the mould-metal contact is so intimate that a perfect wetting occurs, i.e., the casting gets soldered to the mould face. In this section, we shall consider the solidification process assuming that the thermal resistance at the interface is of overriding importance. In such a case, the temperature distribution, assuming no superheat, is as shown in Fig. 2.23. We are considering again a problem of one-dimensional heat flow.

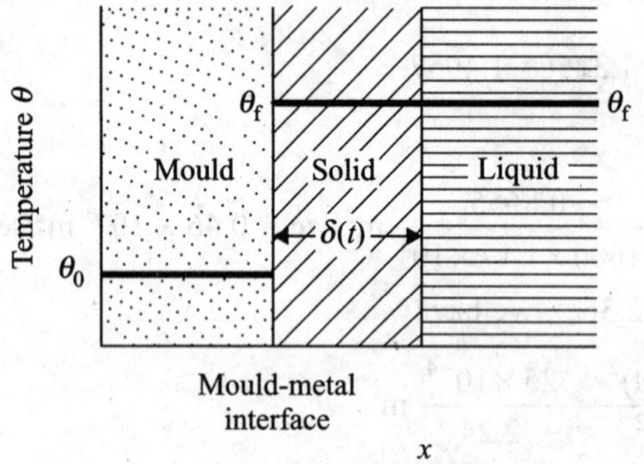

Fig. 2.23 Temperature distribution in mould, and solidified and liquid metal.

Here, the rate of heat flow through the interface is

$$\dot{Q} = h_f(\theta_f - \theta_0)A, \tag{2.40}$$

where h_f is the film heat transfer coefficient of the interface and A is the surface area of the interface. If the solidification front at this instant is at a distance δ from the mould face, then, on solidification,

$$\text{rate of heat released} = \rho_m AL \frac{d\delta}{dt}. \tag{2.41}$$

Therefore, equating the right-hand sides of (2.40) and (2.41), we have

$$\frac{d\delta}{dt} = \frac{h_f(\theta_f - \theta_0)}{\rho_m L}. \tag{2.42}$$

Integrating this equation with $\delta = 0$ at $t = 0$, we get

$$\delta(t) = \frac{h_f(\theta_f - \theta_0)}{\rho_m L} t. \tag{2.43}$$

Comparing equation (2.43) with equation (c) of Example 2.4, we see that, in the former, the depth of solidification increases linearly with time, whereas in the latter, it is proportional to the square root of time. The total heat rejected by the casting is given by equation (2.33) with $\theta_p = \theta_f$. The heat flow through the interface during the period of solidification t_s is $h_f(\theta_f - \theta_0)At_s$. So, t_s is obtained by equating this expression with equation (2.33), after substituting $\theta_p = \theta_f$, as

$$t_s = \frac{\rho_m L}{h_f(\theta_f - \theta_0)} \cdot \frac{V}{A}. \tag{2.44}$$

Equation (2.44) is helpful in estimating the solidification time of small, thin-section parts cast in a heavy metal mould as used in a die or permanent mould casting.

It may be noted at this stage that over and above the interface resistance we have discussed, there are significant differences between the solidification process in a sand mould and that in a chill or metal mould. We give here two important ways in which the latter differs from the former:

(i) The thermal conductivity of the solidified metal may provide considerable thermal resistance, as shown by region 4 of Fig. 2.19. Because of this, the surface temperature of the casting (θ_s), as can be seen, becomes much lower than the freezing temperature θ_f.

(ii) Because of the subcooled solidified metal, more total heat than that considered in the earlier sections has to be removed. Thus, the heat capacity of the solidifying metal also plays an important role in the rate of solidification.

We shall see more of these two aspects in Sections 2.5.5 and 2.5.6 where additional examples of the solidification process will be considered.

2.5.5 SOLIDIFICATION WITH CONSTANT CASTING SURFACE TEMPERATURE

If a large, slab-shaped casting (say, of steel) is produced in a thin, water cooled mould made out of a metal (say, of copper) having a much higher conductivity than the solidified casting, then the thermal resistance provided by the solidifying metal itself is significant. In such a case, the predominant thermal resistance is offered by region 4 (see Fig. 2.19). Neglecting the thermal resistances of all the other regions, the temperature distribution at any instant takes the shape shown in Fig. 2.24. Here, the mould-metal interface (or the casting surface) temperature θ_s can be assumed to remain constant at its initial value θ_0, and θ_f indicates the freezing temperature of the metal and this is also taken as the pouring temperature. At any instant t, $\delta(t)$ indicates the depth of solidification. The process can be idealized, without much error, as a one-dimensional one. Hence, the solidification time t_s is obtained from $\delta(t_s) = h/2$, where h is the thickness of the slab being cast. The temperature profile within the range $0 < x < \delta(t)$ is given by[1]

$$\frac{\theta - \theta_s}{\theta_\infty - \theta_s} = \text{erf}\,\left(\frac{x}{2\sqrt{\alpha_s t}}\right), \tag{2.45}$$

where α_s is the thermal diffusivity of the solidified metal and θ_∞ is a constant of integration. At $x = \delta(t)$, $\theta = \theta_f$; substituting these in equation (2.45), we get

$$\text{erf}\,\left[\frac{\delta(t)}{2\sqrt{\alpha_s t}}\right] = \frac{\theta_f - \theta_s}{\theta_\infty - \theta_s} = \text{constant} = \lambda\,(\text{say}). \tag{2.46}$$

[1]Geiger, G.H. and Poirier, D.R., Transport Phenomena in Metallurgy, Addison-Wesley, Reading, Massachusetts, 1973.

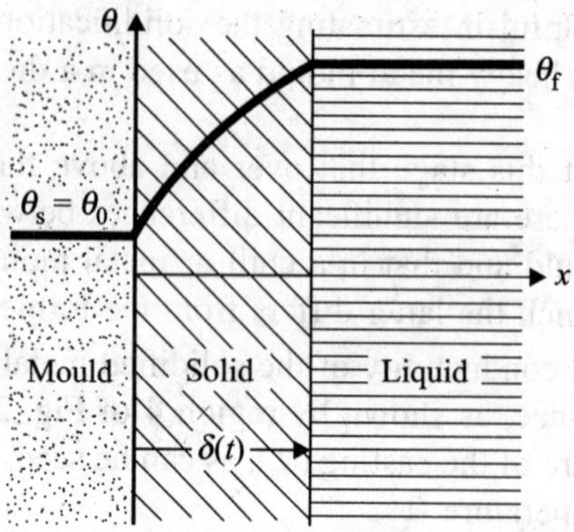

Fig. 2.24 Temperature distribution with constant casting surface temperature.

Hence,

$$\delta(t) = 2\zeta\sqrt{\alpha_s t}, \tag{2.47}$$

where erf $(\zeta) = \lambda$. Once again, we find the depth of solidification varies as the square root of time. Now, the constant ζ can be determined as follows. Considering the rate of energy flow balance at the solid-liquid interface, we have

$$k_s \left.\frac{\partial \theta}{\partial x}\right|_{x=\delta} = \rho_m L \frac{d\delta}{dt}, \tag{2.48}$$

where

k_s = conductivity of the solidified metal = $\alpha_s \rho_m c_s$,

ρ_m = density of the metal (same in the solid and the liquid states),

c_s = specific heat of the solidified metal, and

L = latent heat of fusion of the metal.

From equation (2.45),

$$\frac{\partial \theta}{\partial x} = (\theta_\infty - \theta_s)\frac{d}{dx}[\text{erf}\,(\frac{x}{2\sqrt{\alpha_s t}})]$$

$$= (\theta_\infty - \theta_s)\frac{1}{2\sqrt{\alpha_s t}}\frac{2}{\sqrt{\pi}}$$

$$\times \exp\,[-(\frac{x}{2\sqrt{\alpha_s t}})^2] \qquad \text{[using equation (2.30)].} \tag{2.49}$$

Again, using equation (2.49) in equation (2.48), we obtain

$$\frac{k_s}{\sqrt{\alpha_s}}(\theta_\infty - \theta_s)\frac{1}{\sqrt{\pi t}}\exp\,[-(\frac{\delta}{2\sqrt{\alpha_s t}})^2] = \rho_m L \frac{d\delta}{dt}. \tag{2.50}$$

Now, substituting from equations (2.46) and (2.47) for $(\theta_\infty - \theta_s)$ and δ, respectively, and simplifying, we can rewrite equation (2.50) as

$$\sqrt{k_s \rho_m c_s} \frac{(\theta_f - \theta_s)}{\text{erf}(\zeta)} \frac{1}{\sqrt{\pi t}} e^{-\zeta^2} = \rho_m L \zeta \sqrt{\alpha_s} \frac{1}{\sqrt{t}}$$

or

$$\zeta e^{\zeta^2} \text{erf}(\zeta) = \frac{(\theta_f - \theta_s)}{\sqrt{\pi}} \frac{c_s}{L}. \tag{2.51}$$

The solution of equation (2.51) for ζ can be obtained by trial and error. A simpler procedure is to first plot a graph of ζe^{ζ^2} erf (ζ) for various values of ζ. Then, from this graph, the value of ζ corresponding to the value of the right-hand side of equation (2.51) can be obtained. Thus, once ζ is known, the solidification time t_s can be obtained, from equation (2.47) with $\delta(t_s) = h/2$, as

$$2\zeta\sqrt{\alpha_s t_s} = \frac{h}{2} \quad \text{or} \quad t_s = \frac{h^2}{16\zeta^2 \alpha_s}. \tag{2.52}$$

This analysis is valid only after the initial solidification stage (0.5–1 cm) is over. Similar results for the solidification time of the other shapes can be found from the available literature.

2.5.6 SOLIDIFICATION WITH PREDOMINANT RESISTANCE IN MOULD AND SOLIDIFIED METAL

Let us now consider the problem discussed in Section 2.5.5 under a little different situation. We assume that the copper mould is quite thick and is not water cooled. Then, the mould-metal interface temperature θ_s can no longer be assumed to remain at its initial value θ_0. The value of θ_s, still assumed to be constant, is decided by the thermal properties of the mould and the solidified metal. Moreover, after the initial stage of solidification, the interface resistance also becomes negligible. Thus, the only significant thermal resistance is offered by regions 2 and 4 (Fig. 2.19) and the resulting temperature distribution at any instant is as shown in Fig. 2.25. Assuming the mould to be a semi-infinite medium in the negative x-direction, the temperature distribution in the mould is[1]

$$\frac{\theta - \theta_s}{\theta_0 - \theta_s} = \text{erf}\left(-\frac{x}{2\sqrt{\alpha t}}\right) \quad \text{for } x < 0 \tag{2.53}$$

and that in the solidified metal is

$$\frac{\theta - \theta_s}{\theta_\infty - \theta_s} = \text{erf}\left(\frac{x}{2\sqrt{\alpha_s t}}\right) \quad \text{for } x > 0, \tag{2.54}$$

where θ_0 is the initial temperature of the mould, θ_s is the mould-metal interface temperature ($\neq\theta_0$), and θ_∞ is a constant of integration. The other conditions to be satisfied are

(i) the balance of heat flux at the mould-metal interface, i.e.,

[1]Geiger, G.H. and Poirier, D.R., Transport Phenomena in Metallurgy, Addison-Wesley, Reading, Massachusetts, 1973.

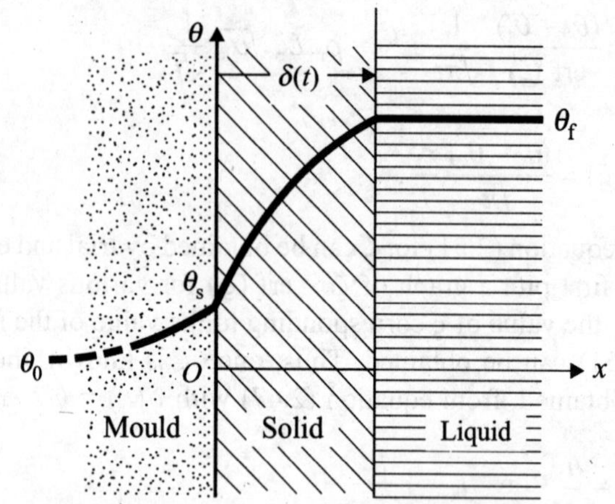

Fig. 2.25 Temperature distribution.

$$k_s \left.\frac{\partial \theta}{\partial x}\right|_{x=0^+} = k \left.\frac{\partial \theta}{\partial x}\right|_{x=0^-}, \tag{2.55}$$

(ii) the rate of energy flow balance at the solid-liquid interface [see equation (2.48)], i.e.,

$$k_s \left.\frac{\partial \theta}{\partial x}\right|_{x=\delta} = \rho_m L \frac{d\delta}{dt}, \tag{2.56}$$

(iii)

$$\theta(\delta, t) = \theta_f. \tag{2.57}$$

Equations (2.53)–(2.57) can be combined to give the three equations

$$\frac{(\theta_f - \theta_s)c_s}{L\sqrt{\pi}} = \zeta e^{\zeta^2} \text{erf}\,(\zeta), \tag{2.58}$$

$$\frac{(\theta_\infty - \theta_s)c_s}{L\sqrt{\pi}} = \zeta e^{\zeta^2}, \tag{2.59}$$

$$\frac{\theta_s - \theta_0}{\theta_\infty - \theta_s} = \left(\frac{k_s \rho_m c_s}{k\rho c}\right)^{1/2} = \phi \text{ (constant)}, \tag{2.60}$$

where ζ is the same as in Section 2.5.5. The three unknowns, namely, θ_s, θ_∞, and ζ, can be solved from equations (2.58)–(2.60) as follows. Using equations (2.59) and (2.60), we can write

$$(\theta_s - \theta_0)\frac{c_s}{L\sqrt{\pi}} = \zeta e^{\zeta^2} \phi. \tag{2.61}$$

Then, adding equations (2.58) and (2.61), we get

$$\frac{(\theta_f - \theta_0)c_s}{L\sqrt{\pi}} = \zeta e^{\zeta^2}[\text{erf}(\zeta) + \phi]. \tag{2.62}$$

Now, the left-hand side and ϕ in equation (2.62) are known; so, ζ can be determined either graphically or by trial and error, as mentioned in Section 2.5.5. In the former approach, a graph of ζe^{ζ^2} [erf (ζ) + ϕ] versus ζ should be drawn for the given value of ϕ, and ζ can then be solved for with the known value of the left-hand side of equation (2.62). Once ζ is known, the depth of solidification can be computed from equation (2.47) and the solidification time from equation (2.52). For such a casting to be feasible, it should be ensured that θ_s works out to be less than the melting point of the mould metal.

EXAMPLE 2.6 Determine the solidification time of the slab-shaped casting considered in Example 2.5 when

(i) the casting is done in a water cooled copper mould,

(ii) the casting is made in a very thick copper mould.

In both cases, assume no resistance at the mould-metal interface and use the following data:

For iron	*For copper*
$c_s = 0.67$ kJ/kg-K	$c = 0.376$ kJ/kg-K
$k_s = 83$ W/m-K	$\rho = 8960$ kg/m^3
$\theta_f = 1540$°C	$k = 398$ W/m-K
$L = 272$ kJ/kg	
$\rho_m = 7850$ kg/m^3	

SOLUTION (i) In this case, $\theta_s = 28$°C and $h = 0.1$ m. So, from equation (2.51),

$$\zeta e^{\zeta^2} \text{ erf}(\zeta) = \frac{(1540 - 28) \times 0.67}{272\sqrt{\pi}} = 2.1$$

or

$$\zeta = 0.98.$$

Now,

$$\alpha_s = \frac{k_s}{\rho_m c_s} = \frac{83}{7850 \times 0.67 \times 10^3} \text{ m}^2/\text{sec} = 15.8 \times 10^{-6} \text{ m}^2/\text{sec}.$$

The solidification time t_s is given by equation (2.52) as

$$t_s = \frac{(0.1)^2}{16 \times (0.98)^2 \times 15.8 \times 10^{-6}} \text{ sec} = 0.0115 \text{ hr}.$$

(ii) From the given data, using equation (2.60), we have

$$\phi = \left(\frac{83 \times 7850 \times 0.67}{398 \times 8960 \times 0.376}\right)^{1/2} = 0.57.$$

From equation (2.62) with $\theta_0 = 28°C$,

$$\zeta e^{\zeta^2}[\text{erf}(\zeta) + 0.57] = \frac{(1540 - 28) \times 0.67}{272\sqrt{\pi}} = 2.1$$

or

$$\zeta = 0.815.$$

The solidification time, again from equation (2.52), is obtained as

$$t_s = \frac{(0.1)^2}{16 \times (0.815)^2 \times 15.8 \times 10^{-6}} \text{ sec} = 0.0165 \text{ hr.}$$

The surface temperature of the mould (θ_s) can be found from equation (2.58). Thus,

$$\frac{(1540 - \theta_s) \times 0.67}{272\sqrt{\pi}} = 0.815 e^{0.664} \text{ erf } (0.815)$$

$$= 0.815 \times 1.943 \times 0.75 = 1.188$$

or

$$\theta_s = (1540 - 854)°C = 685.4°C.$$

Since the melting point of copper is 1080°C, the mould will not melt away.

Comparing the solidification times obtained in Example 2.5 with those in Example 2.6, we find that the solidification time in the thick copper mould is almost twice that in the water cooled copper mould and both are of an order of magnitude less than that in the sand mould. Thus, the production rate in a metal mould is much higher than that in a sand mould.

In the foregoing analysis, the thermal resistance at the interface was totally neglected. However, in real practice, the mould wall expands due to heating and the casting surface shrinks due to cooling, resulting in a gap. Moreover, the wetting of the mould face by the liquid metal is never complete. All these facts suggest that a thermal resistance at the interface is inevitable. In what follows, we attempt to approximately account for the thermal resistance of layers 2, 3, 4 (Fig. 2.19). The temperature profile under such circumstances is as shown in Fig. 2.26. Here, two separate surfaces are imagined on either side of the interface, at temperatures θ_{s_1} (on the casting side) and θ_{s_2} (on the mould side). The film heat transfer coefficient h_f is apportioned on these two sides according to the equations

$$h_2 = \left(1 + \sqrt{\frac{k\rho c}{k_s \rho_m c_s}}\right) h_f, \tag{2.63a}$$

$$h_1 = \left(1 + \sqrt{\frac{k_s \rho_m c_s}{k\rho c}}\right) h_f. \tag{2.63b}$$

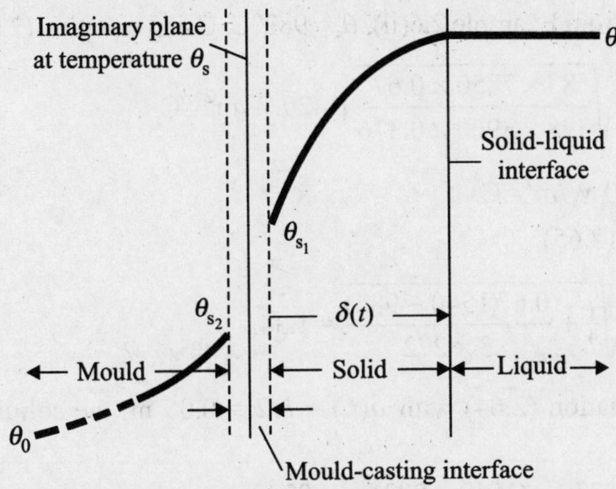

Fig. 2.26 Methodology for taking thermal resistance at mould-casting interface into consideration.

The temperature θ_s at the imaginary interface is calculated by neglecting the interface resistance (as done in Example 2.6). The solidification depth $\delta(t)$ can then be obtained as[1]

$$\delta(t) = \frac{h_1(\theta_f - \theta_s)}{\rho_m L a} t - \frac{h_1}{2k_s}\delta^2(t), \tag{2.64}$$

where the factor a is given by the relation

$$a = \tfrac{1}{2} + \sqrt{\tfrac{1}{4} + \frac{c_s(\theta_f - \theta_s)}{3L}}. \tag{2.65}$$

Further, the solidification time t_s can be obtained from equation (2.64) with $\delta(t_s) = h/2$, h being the thickness of the slab. The mould surface temperature θ_{s_2}, when the casting has completely solidified, is given by[1]

$$\frac{\theta_{s_2} - \theta_0}{\theta_s - \theta_0} = 1 - e^{\gamma}[1 - \text{erf}\,(\sqrt{\gamma})], \tag{2.66}$$

where

$$\gamma = \frac{h_2^2}{k^2}\alpha t_s \tag{2.67}$$

and θ_0 is the initial temperature of the mould.

EXAMPLE 2.7 Estimate the solidification time and the mould surface temperature for the casting considered in Example 2.6(ii) with the total heat transfer coefficient across the casting-mould interface given as 1420 W/m^2-°C.

[1]Geiger, G.H. and Poirier, D.R., Transport Phenomena in Metallurgy, Addison-Wesley, Reading, Massachusetts, 1973.

SOLUTION From Example 2.6(ii), $\theta_s = 982°C$. Using equation (2.63b), we have

$$h_1 = [1 + \sqrt{\frac{83 \times 7850 \times 0.67}{398 \times 8960 \times 0.376}}]1420 \text{ W/m}^2\text{-°C}$$

$$= 2230 \text{ W/m}^2\text{-°C}.$$

From equation (2.65),

$$a = \tfrac{1}{2} + \sqrt{\tfrac{1}{4} + \frac{0.67(1540 - 982)}{3 \times 272}} = 1.34.$$

Now, using equation (2.64) with $\delta(t_s) = h/2 = 0.05$ m, the solidification time t_s we obtain is

$$0.05 = \frac{2.230 \times (1540 - 982)}{7850 \times 272 \times 1.34} t_s - \frac{2230}{2 \times 83}(0.05)^2 \text{ sec}$$

or

$$t_s = 192.2 \text{ sec} \equiv 0.0533 \text{ hr}$$

which is significantly more than that estimated in Example 2.6(ii). To compute the mould surface temperature, we first calculate h_2, using equation (2.63a), as

$$h_2 = (1 + \sqrt{\frac{398 \times 8960 \times 0.376}{83 \times 7850 \times 0.67}}) \times 1420 \text{ W/m}^2\text{-°C} = 3909 \text{ W/m}^2\text{-°C}.$$

Now, from equation (2.67),

$$\gamma = \frac{3909 \times 3909}{398 \times 398} \frac{0.398 \times 192.2}{8960 \times 0.376} = 2.19.$$

So, using equation (2.66), the mould surface temperature θ_{s_2}, when the entire casting has solidified, we obtain is (with $\theta_0 = 28°C$ and using the error function tables)

$$\theta_{s_2} = 28 + (982 - 28)[1 - 8.935(1 - 0.962)]°C = 658.1°C$$

which, again, is much less than the value of θ_s estimated in Example 2.6(ii).

2.5.7 CONTINUOUS CASTING PROCESS

Another situation where estimation of the solidification rate and heat transfer calculations are important is the continuous casting process. Here, no riser is used, and the casting is withdrawn continuously from one end of a water cooled mould while the liquid metal is fed continuously at the other end. Figure 2.27 schematically shows the continuous casting process. The heat transfer calculation in such a case is most conveniently represented by means of three nondimensional groups, namely[1],

[1]Geiger, G.H. and Poirier, D.R., Transport Phenomena in Metallurgy, Addison-Wesley, Reading, Massachusetts, 1973.

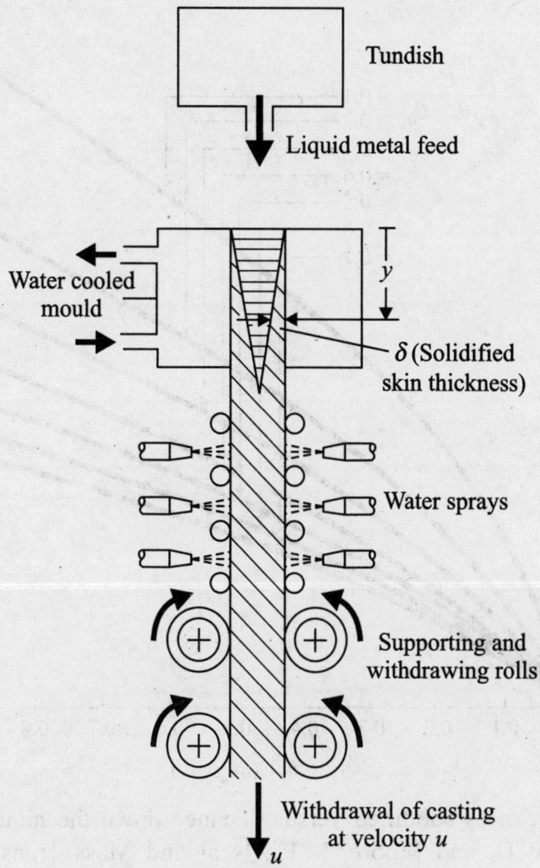

Fig. 2.27 Continuous casting process.

$$\frac{h_f^2 y}{u k_s \rho_m c_s}, \quad \frac{L'}{c_s(\theta_f - \theta_0)}, \quad \frac{h_f \delta}{k_s},$$

where h_f is the film heat transfer coefficient at the mould-casting interface, y is the distance along the mould, u is the velocity of withdrawal of the casting, θ_0 is the temperature of the water cooled mould, δ is the solidified skin thickness,

$$L' = L + c_m(\theta_p - \theta_f),$$

and all the other symbols are the same as used in the earlier sections. The results are presented in Figs. 2.28–2.30 in terms of these nondimensional parameters. These figures can be used for an estimation of important quantities such as the minimum mould length for a desired withdrawal rate, the skin depth of the casting, and the minimum cooling water flow rate for a maximum allowable temperature rise of the cooling water (see also the example that follows).

EXAMPLE 2.8 Determine (i) the mould length, and (ii) the cooling water requirement to produce a continuous casting of steel slab with a cross-section

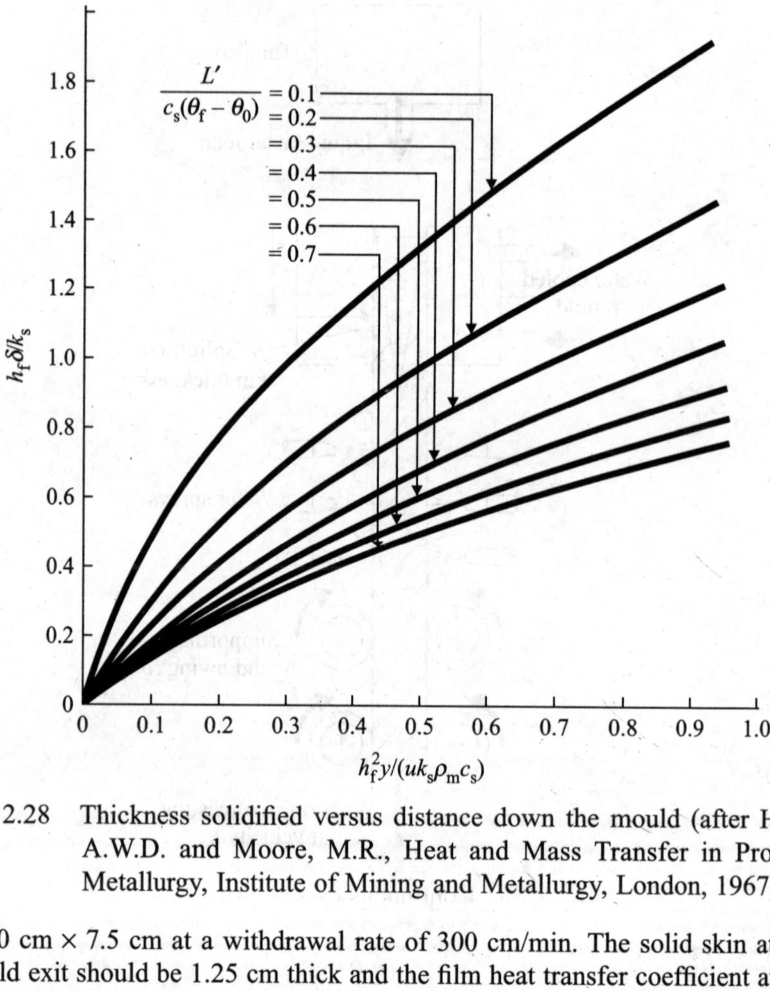

Fig. 2.28 Thickness solidified versus distance down the mould (after Hills, A.W.D. and Moore, M.R., Heat and Mass Transfer in Process Metallurgy, Institute of Mining and Metallurgy, London, 1967).

of 60 cm × 7.5 cm at a withdrawal rate of 300 cm/min. The solid skin at the mould exit should be 1.25 cm thick and the film heat transfer coefficient at the mould-casting interface is 1420 W/m^2-°C. The temperature of the cooling water should not rise by more than 10°C. The mould is maintained at 20°C. The data for steel is

θ_p = 1550°C, θ_f = 1500°C, L = 268 kJ/kg,

ρ_m = 7680 kg/m^3, c_s = 0.67 kJ/kg-K, k_s = 76 W/m-K,

c_m = 0.755 kJ/kg-K.

SOLUTION From the given data, we note that with $y = l_m$, the value of the skin thickness δ = 0.0125 m, u = 0.05 m/sec, θ_0 = 20°C, and

$$L' = 268 + 0.755 \times (1550 - 1500) \text{ kJ/kg} = 305 \text{ kJ/kg}.$$

So,

$$\frac{L'}{c_s(\theta_f - \theta_0)} = \frac{305}{0.67 \times 1480} = 0.308,$$

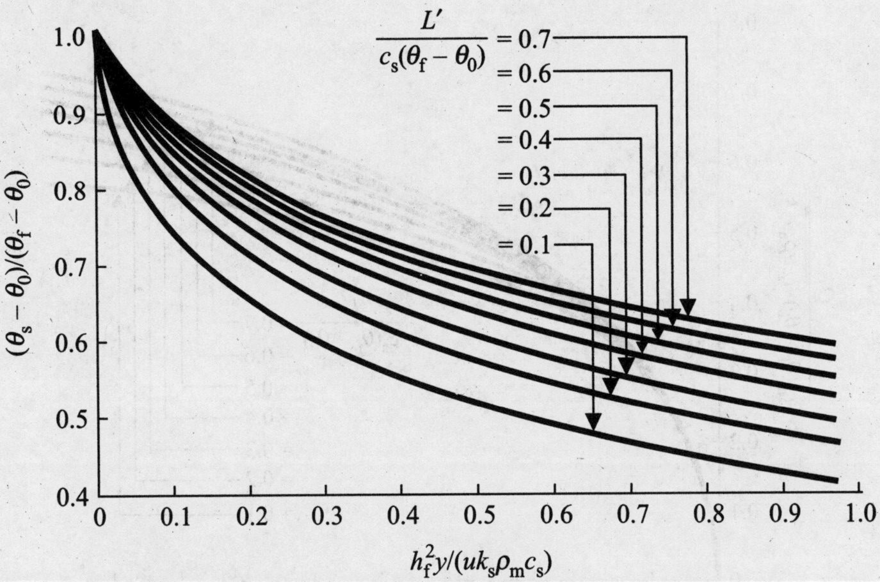

Fig. 2.29 Surface temperature versus distance down the mould (after Hills, A.W.D. and Moore, M.R., Heat and Mass Transfer in Process Metallurgy, Institute of Mining and Metallurgy, London, 1967).

$$\frac{h_f \delta}{k_s} = \frac{1420 \times 0.0125}{76} = 0.234.$$

From Fig. 2.28, for these values of the parameters (with $y = l_m$), we get

$$\frac{h_f^2 l_m}{uk_s \rho_m c_s} = 0.11.$$

Hence,

$$l_m = \frac{0.11 \times 0.05 \times 76 \times 7680 \times 670}{(1420)^2} = 1.07 \text{ m}.$$

Again, from Fig. 2.30,

$$\frac{Q}{l_m(\theta_f - \theta_0)\sqrt{l_m u \rho_m c_s k_s}} = 0.28 \quad \text{or} \quad Q = 2.12 \text{ MW}.$$

Hence, the cooling water requirement $\dot{m}$ is found out from the relation

$$\dot{m} c_w \Delta\theta = 2.12 \text{ MW} \qquad (\text{with } c_w = 4.2 \text{ kJ/kg}, \Delta\theta = 10°\text{C})$$

or

$$\dot{m} = \frac{2.12 \times 10^6}{4.2 \times 10^3 \times 10} \text{ kg/sec} = 5.05 \text{ kg/sec}.$$

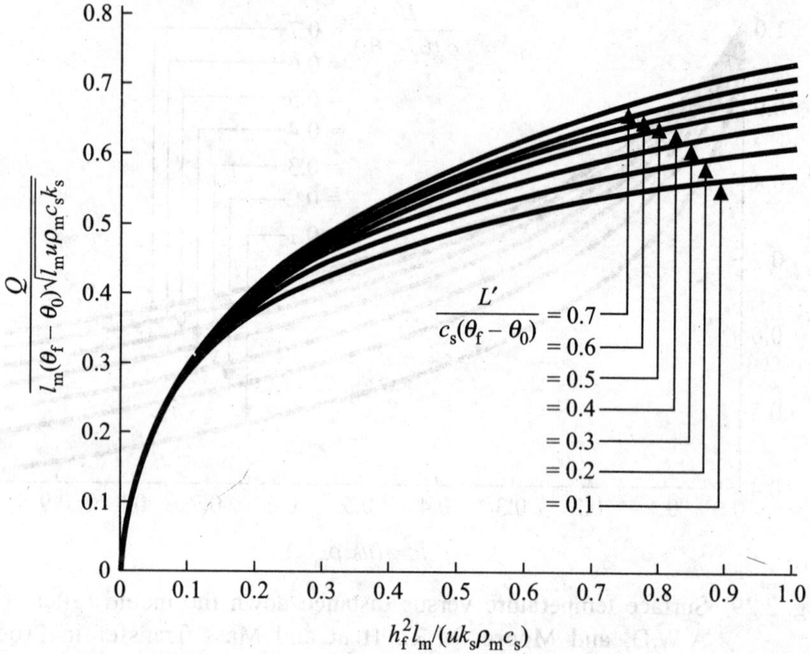

Fig. 2.30 Rate of heat removal by mould cooling water Q versus mould length l_m (after Hills, A.W.D. and Moore, M.R., Heat and Mass Transfer in Process Metallurgy, Institute of Mining and Metallurgy, London, 1967).

2.5.8 RISER DESIGN AND PLACEMENT

From the heat transfer analysis presented in Sections 2.5.5 and 2.5.6 we see that the solidification time depends primarily on the ratio V/A, where V is the volume of the casting and A is the surface area of heat dissipation (i.e., of the casting). This is also to be expected intuitively since the amount of heat content is proportional to volume and the rate of heat dissipation depends on the surface area. This information is utilized when designing a riser to ensure that the riser solidifies after the casting. However, the information on the amount of liquid metal needed from the riser is used only to compensate for the shrinkage that takes place from the pouring temperature till solidification. Depending on the metal, the percentage of this shrinkage varies from 2.5 to 7.5. Thus, the use of a large riser volume (to ensure large solidification time) is uneconomical. So, a riser should be designed with the minimum possible volume while maintaining a cooling rate slower than that of the casting.

It may be noted that a casting with a high surface area/volume ratio requires a riser larger than that determined by considering only the cooling rate. This is shown clearly by the example that follows.

Let us consider a steel plate of dimensions 25 cm × 25 cm × 0.25 cm. The casting then has the A/V ratio as

$$\left(\frac{A}{V}\right)_c = \frac{2 \times 625 + 4(25 \times 0.25)}{25 \times 25 \times 0.25} \text{ cm}^{-1} = 8.16 \text{ cm}^{-1}.$$

A cubical riser with sides 1.25 cm has the A/V ratio as

$$\left(\frac{A}{V}\right)_r = \frac{6 \times 1.25 \times 1.25}{1.25 \times 1.25 \times 1.25} = 4.8 \text{ cm}^{-1}.$$

Thus, the riser is assured to have a much slower cooling rate (more solidification time) than that of the casting. The volume shrinkage of steel during solidification is 3%. So, the minimum volume of the riser necessary is

$$0.03 \times \tfrac{625}{4} \text{ cm}^3 = 4.69 \text{ cm}^3.$$

The riser we have considered has the volume 1.95 cm^3 only. Therefore, a much larger riser is required.

For a given shape of the riser, the dimensions of the riser should, however, be chosen so as to give a minimum A/V ratio, and the minimum volume should be ensured from the shrinkage consideration. It must be remembered that a liquid metal flows from the riser into the mould only during the early part of the solidification process. This necessitates the minimum volume of the riser to be approximately three times that dictated by the shrinkage consideration alone.

EXAMPLE 2.9 Determine the dimensions of a cylindrical riser to be used for casting as aluminium cube of sides 15 cm. The volume shrinkage of aluminium during solidification is 6.5%.

SOLUTION First of all, let us determine the diameter/height ratio of the most compact cylinder so that, for a given volume, the surface area is minimum. With the diameter and the height of the cylinder as d and h, respectively, the surface area of the cylinder is

$$A = \pi dh + 2\frac{\pi}{4}d^2$$

and the volume of the cylinder is

$$V = \frac{\pi}{4}d^2 h \quad \text{or} \quad h = \frac{4V}{\pi d^2}.$$

Hence,

$$A = \pi d \frac{4V}{\pi d^2} + 2\frac{\pi}{4}d^2 = \frac{4V}{d} + \frac{\pi}{2}d^2.$$

For A to be minimum,

$$\frac{\partial A}{\partial d} = 0 \quad \text{or} \quad -\frac{4V}{d^2} + \pi d = 0 \quad \text{or} \quad d^3 = \frac{4V}{\pi}.$$

Again,

$$\frac{4V}{\pi} = d^2 h = d^3$$

or

$$h = d^1 \qquad (\text{when } \frac{A}{V} = \frac{6}{d}).$$

Now, the minimum volume necessary for the riser is $V_r = 3 \times 0.065 V_c$, where V_c is the volume of the casting and is equal to 3375 cm³. So, $V_r = 658.2$ cm³. Thus, the diameter of the riser (d), which is also equal to the height of the riser, can be written as

$$h = (\frac{4V_r}{\pi})^{1/3} = 9.43 \text{ cm}.$$

Now,

$$(\frac{A}{V})_r = \frac{6}{d} = \frac{6}{h} = \frac{6}{9.43} \text{ cm}^{-1} = 0.636 \text{ cm}^{-1},$$

$$(\frac{A}{V})_c = \frac{6 \times 15 \times 15}{15^3} \text{ cm}^{-1} = 0.4 \text{ cm}^{-1} \qquad [\text{this is less than } (\frac{A}{V})_r].$$

So, the riser will not have a longer solidification time. The dimensions of the riser can be recalculated as follows. For

$$(\frac{A}{V})_r \leqslant (\frac{A}{V})_c,$$

we need

$$\frac{6}{d} \leqslant 0.4 \text{ cm}^{-1} \quad \text{or} \quad d \geqslant 15 \text{ cm}.$$

With the minimum value of d, $V_r = (\pi/4)d^2h = (\pi/4)d^3 = 2650$ cm³. This volume is much more than the minimum V_r necessary. Let us now consider the top riser when the optimum cylindrical shape is obtained with $h = d/2$ and again $(A/V)_r = 6/d$. However, with a large top riser, the cube loses its top surface for the purpose of heat dissipation. Hence,

$$(\frac{A}{V})_c = \frac{5 \times 15 \times 15}{15^3} \text{ cm}^{-1} = \tfrac{1}{3} \text{ cm}^{-1}.$$

To have

$$(\frac{A}{V})_r \leqslant (\frac{A}{V})_c \quad \text{or} \quad \frac{6}{d} \leqslant \tfrac{1}{3} \text{ cm}^{-1},$$

d should be greater than or equal to 18 cm. So, the riser volume with minimum diameter is given as

[1]This optimum ratio $h/d = 1$ for a cylindrical riser is true only if the riser is attached to the side of the casting. For a riser attached to the top of a casting, the surface area $A = \pi dh + (\pi/4)d^2$ when the optimum ratio h/d turns out to be equal to 1/2. Sometimes, the dimensions to yield the minimum value of A for a given value of V are determined by using the Lagrange multiplier technique for constrained optimization (see Exercise 2.14).

$$V_r = \frac{\pi}{4} d^2 h = \frac{\pi}{4} 18^2 \times 9 \text{ cm}^3 = 2289 \text{ cm}^3$$

which is greater than the minimum V_r necessary. Though we see that with a top riser there is a little saving of material as compared with the side riser, we have to use, however, a deeper mould with the top riser. Thus, in this case, the side riser may be chosen.

To check the adequacy of the riser size for a steel casting, Caine's relationship is normally used. Equation (2.34) shows that the solidification time is proportional to the square of the ratio volume/surface area. Caine's relationship, however, is based on the assumption that the cooling rate is linearly proportional to the ratio surface area/volume. A typical risering curve is depicted in Fig. 2.31. Here, the ordinate of a point on the curve shows the volume ratio and the abscissa the freezing ratio; also, the subscripts c and r refer to the casting and the riser, respectively. For a given casting-riser combination, if the point in Fig. 2.31 falls to the right of the curve, the adequacy of the riser is ensured. The equation for a risering curve is of the form

$$x = \frac{a}{y - b} + c, \tag{2.68}$$

where a is the freezing constant for the metal, b is the contraction ratio from liquid to solid, and c is a constant depending on the different media around the riser and the casting. The value of c is unity if the mould material around the casting and the riser is the same. For steel, the typical values are $a = 0.1$ and $b = 0.03$.

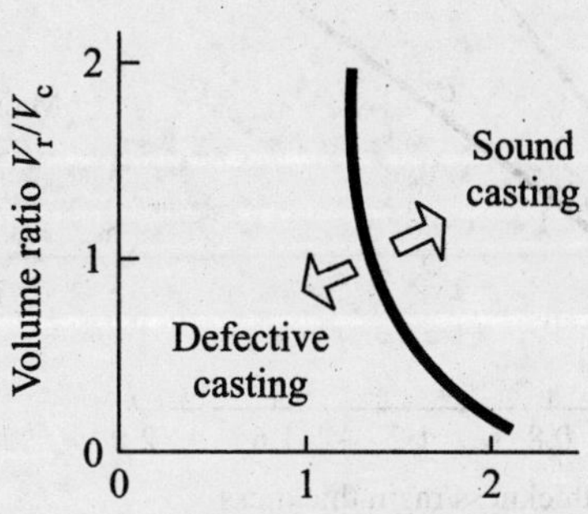

Fig. 2.31 Proper combinations of volume and freezing ratios.

The tedious calculation of $(A/V)_c$ for a complex casting has given rise to another method where a risering curve of the type shown in Fig. 2.32 is used. In this method, the shape factor $(l + w)/h$, instead of $(A/V)_c$, is plotted along the x-axis, where l, w, and h denote, respectively, the maximum length, the maximum width, and the maximum thickness of the casting. This method and Caine's relationship give almost identical results for a casting of simple shape. If the appendages to the main body (of a simple, regular shape) of a casting are thin, then the solidification time does not alter significantly. As a result, a

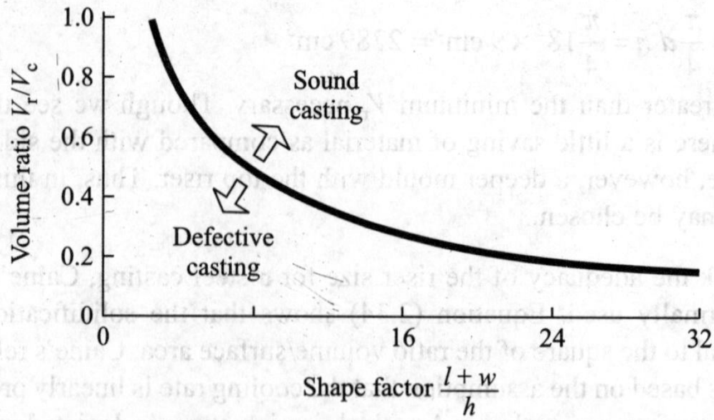

Fig. 2.32 Proper combinations of volume ratio and shape factor.

marginal increase in the calculated volume (on the basis of the main body) of the riser performs the job satisfactorily. As the appendages become heavier, the riser volume required is calculated on the basis of a modified total volume of the casting. The total volume of the casting is taken as the volume of the main section plus the effective percentage of the appendage volume, called the parasitic volume. The effective percentage is estimated from curves of the type shown in Fig. 2.33. A shape is called *plate-like* or *bar-like* depending on whether the width of the cross-section is more or less than three times the depth.

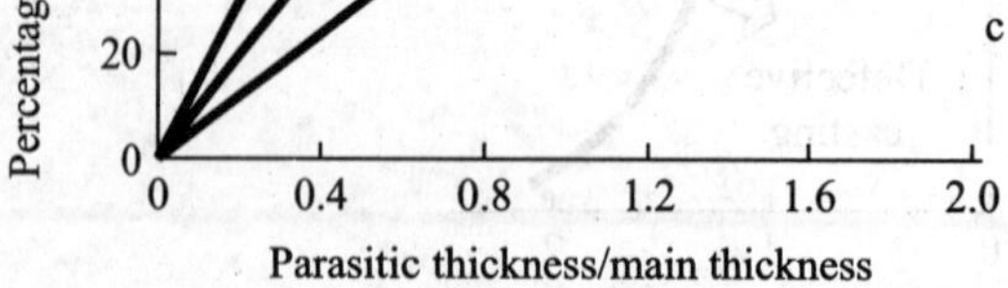

	Shape	
	Main casting	Appendage
a	Bar	Plate
b	Bar plate	Bar plate
c	Plate	Bar

Fig. 2.33 Effective parasitic volume.

In the foregoing discussion, we assumed no special means of controlling the cooling rate (and hence the solidification time) of either the casting or of the riser. In practice, however, chill blocks (see Section 2.5.1) or thin fins are used on the casting to increase its cooling rate. Chilling is less effective for a metal having a thermal conductivity higher than that of the chill. Similarly, to increase the solidification time of the riser, some exothermic compounds are added in the riser to keep it molten for a longer period.

So far, we have restricted our discussion to the adequacy of the riser size from the points of view of shrinkage and cooling rate. Another important aspect of risering is to ensure that the available liquid metal in the riser can be fed to the desired locations within the casting. For this, we should recall the concept of centre-line feeding resistance explained in Section 2.5.1.

In fact, the thermal gradient, within the casting, during the last stage of cooling is the most important factor. The minimum allowable gradient depends on the shape and size of the cross-section. Normally, for a casting with low (A/V) ratio (e.g., cube and sphere), one central riser is able to feed the entire casting. On the other hand, for a casting with high (A/V) ratio (e.g., for a bar and a plate), usually more than one riser is necessary. In such a case, a proper location of the riser has to be decided.

For a steel plate of up to 100 mm thickness, one central riser is satisfactory if the maximum feeding distance is less than 4.5 times the plate thickness. The feeding distance should be measured from the edge of the riser, as explained in Fig. 2.34a. It should be noted that, of the total distance $4.5t$, the riser gradient prevails up to a distance $2t$, whereas the end-wall gradient prevails in the remaining distance $2.5t$. Thus, the maximum distance between the edges of two consecutive risers is $4t$ and not $9t$ (see Fig. 2.34b).

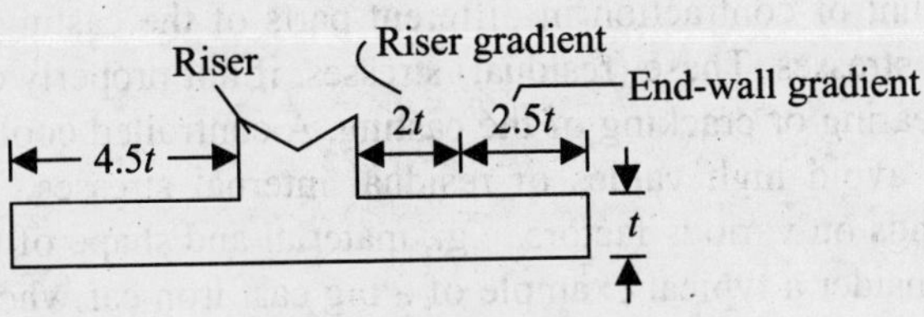

(a) Plate with one central riser

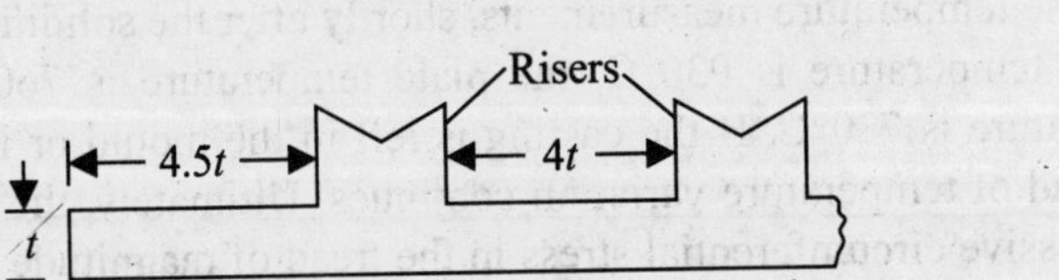

(b) Maximum distance between two consecutive risers

Fig. 2.34 Placement of risers.

A bar of square cross-section with sides measuring 50–200 mm can be fed satisfactorily from a single riser, up to a maximum distance of $30\sqrt{s}$, where s is the side of the square expressed in mm. The maximum distance between the edges of two consecutive risers is found to be $1.2s$ (and not $60\sqrt{s}$).

As mentioned in Section 2.5.1, the presence of a chill in the mould increases the feeding distance of the riser. This is achieved by providing a sharp thermal gradient with consequent decrease in the feeding resistance. It is obvious that the chill should be placed at the ends if a single riser is used. For more than one riser, the chill should be placed midway between the two risers. Figure 2.35

schematically explains the proper placement of risers and chills. The maximum permissible distances for various cases are also indicated in this figure.

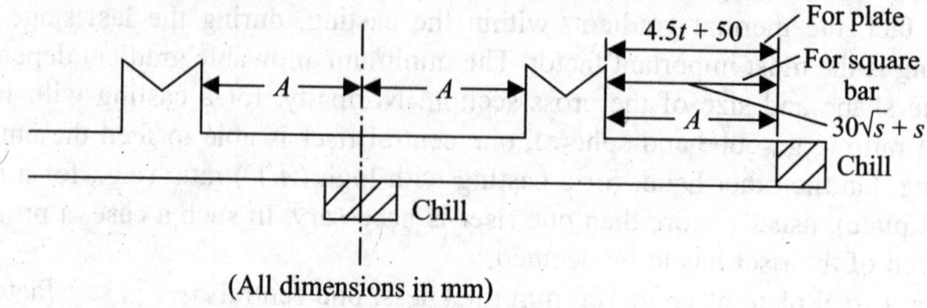

Fig. 2.35 Placement of risers and chill blocks.

2.5.9 RESIDUAL STRESS

A sharp temperature gradient exists within a hot solidified casting of non-uniform cross-section. This is caused by the variation of heat content and the cross-sectional area available for its dissipation. This, in turn, results in varying amount of contraction in different parts of the casting, giving rise to high internal stresses. These (residual) stresses, if not properly controlled, may even cause tearing or cracking of the casting. A controlled cooling is therefore necessary to avoid high values of residual internal stresses. The method of control depends on various factors, e.g., material and shape of the casting. We shall now consider a typical example of a big cast iron car wheel.

Figure 2.36 shows the cross-section of the casting we are considering. The cooling rate of the heavier rim is improved considerably by providing a chill. As a result, the temperature measurements, shortly after the solidification, show that the hub temperature is 930°C, the plate temperature is 760°C, and the tread temperature is 730°C. If the casting is left in the mould or is air cooled, the same trend of temperature variation continues. Ultimately, the final casting has a compressive circumferential stress in the tread of magnitude 70.7 N/mm^2 and a radial tensile stress at the front hub fillet measuring 222 N/mm^2. Such huge internal stresses can be avoided by taking out the casting at an average temperature of 760°C and putting it in an insulated pit when the cooling rate is brought down to 5.5°C/hr. The entire cooling takes about three days. At such

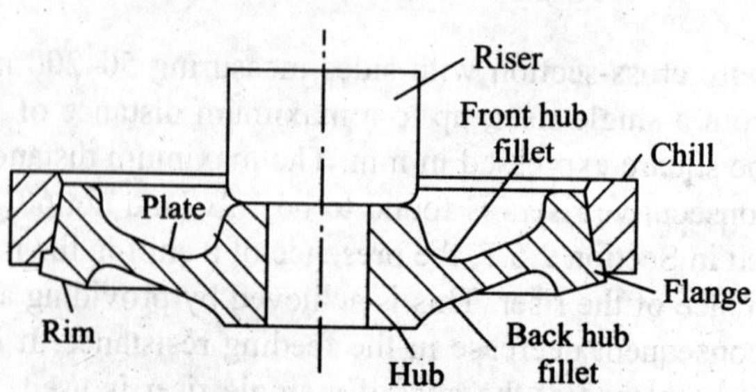

Fig. 2.36 Cross-section of wheel.

a low cooling rate, temperature equalization takes place above 540°C. Any temperature gradient existing within the casting above this temperature does not give rise to elastic strain. In the temperature range exceeding 540°C, the elastic strain is relaxed to the plastic strain due to the high rate of creep. This avoids the development of high internal residual stresses. The stresses are found to be compressive, both in the tread (circumferential 8 N/mm^2) and in the fillet (radial 4 N/mm^2). If high internal residual stresses do exist in a casting, these should be taken care of either by subsequent heat treatment or by the other methods of stress relieving.

2.6 DEFECTS IN CASTINGS

In this section, we shall discuss the different types of defects in castings, and their origin and remedies. The treatment is restricted essentially to the sand mould castings. The defects in a casting may arise due to the defects in one or more of the following:

(i) Design of casting and pattern.

(ii) Moulding sand and design of mould and core.

(iii) Metal composition.

(iv) Melting and pouring.

(v) Gating and risering.

The following defects are most commonly encountered in the sand mould castings (Fig. 2.37):

(i) *Blow* It is a fairly large, well-rounded cavity produced by the gases which displace the molten metal at the cope surface of a casting. Blows usually occur on a convex casting surface and can be avoided by having a proper venting and an adequate permeability. A controlled content of moisture and volatile constituents in the sand-mix also helps in avoiding the blow holes.

(ii) *Scar* A shallow blow, usually found on a flat casting surface, is referred to as a scar.

(iii) *Blister* This is a scar covered by the thin layers of a metal.

(iv) *Gas holes* These refer to the entrapped gas bubbles having a nearly spherical shape, and occur when an excessive amount of gases is dissolved in the liquid metal.

(v) *Pin holes* These are nothing but tiny blow holes, and occur either at or just below the casting surface. Normally, these are found in large numbers and are almost uniformly distributed in the entire casting surface.

(vi) *Porosity* This indicates very small holes uniformly dispersed throughout a casting. It arises when there is a decrease in gas solubility during solidification.

(vii) *Drop* An irregularly-shaped projection on the cope surface of a casting is called a drop. This is caused by dropping of sand from the cope or other overhanging projections into the mould. An adequate strength of the sand and the use of gaggers can help in avoiding the drops.

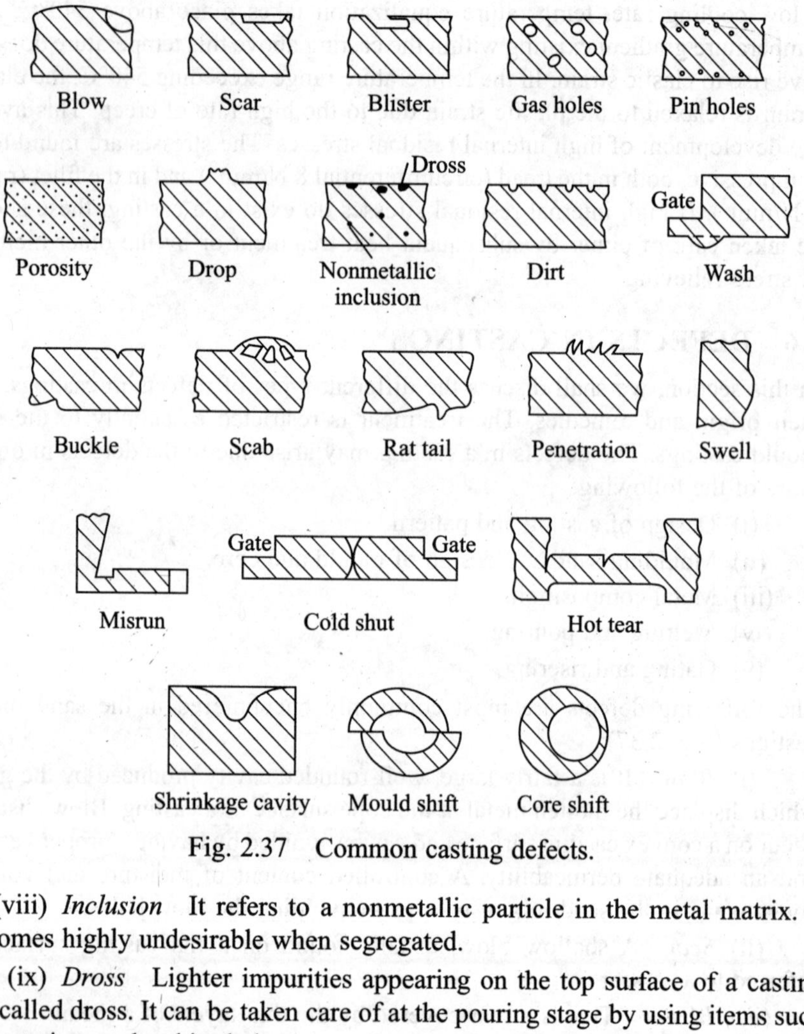

Fig. 2.37 Common casting defects.

(viii) *Inclusion* It refers to a nonmetallic particle in the metal matrix. It becomes highly undesirable when segregated.

(ix) *Dross* Lighter impurities appearing on the top surface of a casting are called dross. It can be taken care of at the pouring stage by using items such as a strainer and a skim bob.

(x) *Dirt* Sometimes sand particles dropping out of the cope get embedded on the top surface of a casting. When removed, these leave small, angular holes, known as dirts. Defects such as drop and dirt suggest that a well-designed pattern should have as little a part as possible in the cope. Also, the most critical surface should be placed in the drag.

(xi) *Wash* A low projection on the drag surface of a casting commencing near the gate is called a wash. This is caused by the erosion of sand due to the high velocity jet of liquid metal in bottom gating.

(xii) *Buckle* This refers to a long, fairly shallow, broad, vee-shaped depression occurring in the surface of a flat casting of a high temperature metal. At this high temperature, an expansion of the thin layer of sand at the mould face takes place before the liquid metal at the mould face solidifies. As this expansion is obstructed by the flask, the mould face tends to bulge out, forming the vee shape. A proper amount of volatile additives in the sand-mix is therefore essential to make room for this expansion and to avoid the buckles.

(xiii) *Scab* This refers to the rough, thin layer of a metal, protruding above the casting surface, on top of a thin layer of sand. The layer is held on to the casting by a metal stringer through the sand. A scab results when the upheaved sand is separated from the mould surface and the liquid metal flows into the space between the mould and the displaced sand.

(xiv) *Rat tail* It is a long, shallow, angular depression normally found in a thin casting. The reason for its formation is the same as that for a buckle. Here, instead of the expanding sand upheaving, the compressed layer fails by one layer, gliding over the other.

(xv) *Penetration* If the mould surface is too soft and porous, the liquid metal may flow between the sand particles up to a distance, into the mould. This causes rough, porous projections and this defect is called penetration. The fusion of sand on a casting surface produces a rough, glossy appearance.

(xvi) *Swell* This defect is found on the vertical surfaces of a casting if the moulding sand is deformed by the hydrostatic pressure caused by the high moisture content in the sand.

(xvii) *Misrun* Many a time, the liquid metal may, due to insufficient superheat, start freezing before reaching the farthest point of the mould cavity. The defect that thus results is termed as a misrun.

(xviii) *Cold shut* For a casting with gates at its two sides, the misrun may show up at the centre of the casting. When this happens, the defect is called a cold shut.

(xix) *Hot tear* A crack that develops in a casting due to high residual stresses is called a hot tear.

(xx) *Shrinkage cavity* An improper riser may give rise to a defect called shrinkage cavity, as already detailed.

(xxi) *Shift* A misalignment between two halves of a mould or of a core may give rise to a defective casting, as shown in Fig. 2.37. Accordingly, this defect is called a mould shift or a core shift.

2.7 MISCELLANEOUS CASTING PROCESSES

We have so far discussed the basic features of the casting processes mainly with reference to the most common type of green sand mould casting. In this section, we shall briefly consider the other types of casting processes.

2.7.1 DRY SAND MOULD CASTING

The dry sand mould casting uses expendable moulds, i.e., each mould is used only once. A dry sand mould is basically a green sand mould baked in an oven at 100–250°C for several hours. The sand-mix contains 1–2% of pitch. The oxidation and polymerization of pitch increases the hot strength of the mould. As the water is driven out from the sand-mix by heating, the defects caused by the generation of steam, e.g., blows and porosity, are less frequent in dry sand mould casting.

2.7.2 SHELL MOULD CASTING

The shell mould casting is a semiprecise method for producing small castings repetitively in large numbers. The mould material contains phenolic resin mixed with fine, dry silica. These are mixed either dry or in the presence of alcohol; no water is used. Normally, a machined pattern of gray iron, aluminium, or brass is used in this process. First, the pattern is heated to 230–260°C, and then the sand-resin mixture is either dumped or blown over its surface. This way, the heated pattern melts and hardens the resin which, in turn, bonds the sand grains closely together. After a dwell time of 20–30 sec, the pattern and sand are inverted. When this happens, a layer of sand adheres to the pattern in the form of a shell of about 6 mm thickness. The rest of the sand is cleaned off. The thickness of the shell can be accurately controlled by controlling the dwell time. The thickness of the shell is so decided that the shell has the required strength and rigidity to hold the weight of the liquid metal to be poured into the mould. Then the mould is heated in an oven (at 300°C) for 15–60 sec. This curing makes the shell rigid when it can be stripped off by means of ejector pins mounted on the pattern. The shell thus formed constitutes one-half of the mould. Two such halves, placed one over the other, make the complete mould. While pouring the liquid metal, the two halves are clamped down together by clamps or springs.

It should be noted that in this process, the smoothness of the mould wall is independent of the moulder's skill. This contributes to a better dimensional accuracy and consistency when compared with green sand moulding. Smooth mould walls also offer less resistance to the flow of liquid metal in the mould. This is why smaller gates can be used. Moreover, thin sections, sharp corners, small projections, which are not possible in green sand moulds, can be accommodated. Further, subsequent machining operations are also reduced. Often, only grinding can produce the finished product. The increased cost of the metal pattern (as compared with the wooden pattern used in green sand moulding), however, can be justified only if the casting is produced in large enough numbers.

2.7.3 INVESTMENT CASTING

The process of investment casting is suitable for casting a wide range of shapes and contours in small-size parts, especially those that are made of hard-to-machine materials. The process produces excellent surface finish for the casting. Here, the mould is made in a single piece, and consequently there is no parting line to leave out fins. This also adds to the dimensional accuracy of the casting. As will be apparent from the description of the process, no complication arises when withdrawing a pattern from the mould. Though the process is elaborate and expensive, it has been found very suitable for casting turbine and jet engine parts made of high temperature and high strength alloys. We now describe the steps to be followed in this process.

A rather accurately dimensioned metal pattern is used. The dimensions of the pattern are calculated to compensate for the several size adjustments which take place in the process—in the die, in the wax, in the investment material, and, finally, in the casting material. The determination of the pattern dimensions is a tedious task and requires considerable experimentation. This makes the pattern in an investment casting very costly.

This pattern is used to make a die out of a soft material, e.g., aluminium. Thereafter, wax or plastic is injected into the die to form an expendable pattern. The expendable pattern is rinsed in alcohol to remove grease and dirt. After drying, the pattern is dipped in a slurry composed of silica flour, water, and some bonding agent. Then, the pattern is taken out of the slurry and rotated to produce a uniform coating, to fill inside corners and to drain out the excess slurry. Sometimes, a number of expendable patterns are assembled as a 'tree' for economy. Finally, fine-grain silica sand is sprinkled onto the wet slurry surface. The coating thus produced on the expendable pattern after drying is called a precoat.

The pattern with the precoat is then placed on a steel base and is covered by an open-ended steel can. Both the pattern and the can are secured to the base by molten wax. Then, the can is filled with a slurry of heavy, self-hardening refractory concentrate. The concentrate sets in after a lapse of 24 hours when the can is placed in an oven. Thus, most of the wax or plastic melts and flows out of the mould, leaving a cavity with the shape of the intended casting. The residual wax is removed by firing the mould in a furnace for about 24 hours.

The liquid metal is poured into this mould while it is still hot. This saves the liquid metal from acquiring the moisture and avoids high thermal gradient between the liquid metal and the mould. In critical cases, the pouring is conducted in a vacuum chamber or in a protective inert atmosphere (such as argon). Frequently, the mould is clamped to a special type of furnace which is then inverted for pouring directly from the furnace into the mould. After cooling, the can is removed and the hard refractory investment is knocked off by a hammer or other vibratory means. Finally, the adhered investment material is removed from the casting surface by sand-blasting or a tumbling operation.

2.7.4 GRAVITY DIE CASTING

In gravity die casting, a permanent mould is used. The liquid metal is poured into a non-expendable mould under the force of gravity. The process is normally used for cast iron and, occasionally, for a nonferrous casting. The mould is made of heat-resistant cast iron, and is provided with fins on its outer surface for efficient air-cooling. The inner surface of the mould cavity is sprayed with an oil-carbon-silica mixture before each pouring.

2.7.5 DIE CASTING

In the die casting process, unlike in gravity casting, the liquid metal is forced into the mould cavity under high pressure. The process is used for casting a low melting temperature material, e.g., aluminium and zinc alloys. The mould, normally called a die, is made in two halves (see Figs. 2.38–2.40), of which one is fixed and the other moving. Medium carbon, low alloy tool steel is tbe most common die material. The die is cooled by water for an efficient cooling of the casting. This also increases the die life. The process is referred to as a hot chamber (Fig. 2.39) or a cold chamber (Fig. 2.38) process, depending on whether or not the melting furnace is an integral part of the mould. Since the liquid metal is forced into the die with high external pressure, much thinner sections can be cast by this process. The process, when applied to a plastic casting, is called injection moulding (Fig. 2.40).

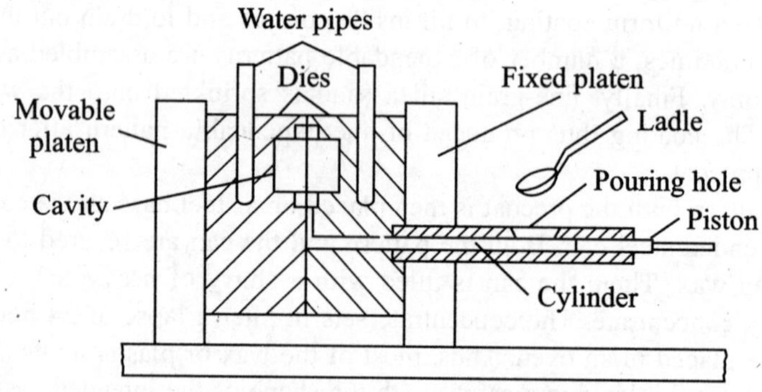

Fig. 2.38 Cold chamber.

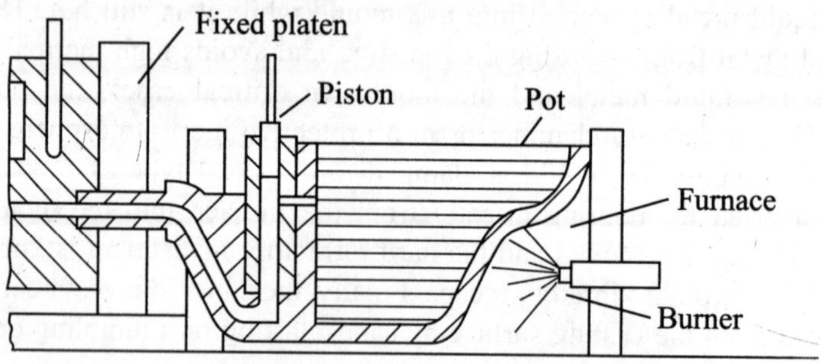

Fig. 2.39 Hot chamber.

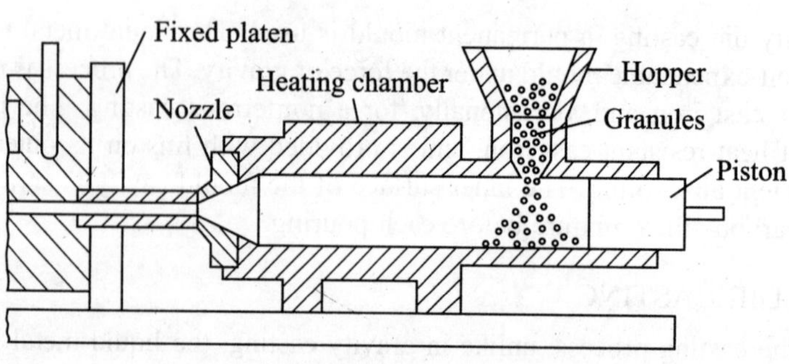

Fig. 2.40 Injection moulding.

2.7.6 CENTRIFUGAL CASTING

The centrifugal casting process is normally carried out in a permanent mould which is rotated during the solidification of a casting (Fig. 2.41). For producing a hollow part, the axis of rotation is placed at the centre of the desired casting. The speed of rotation is maintained high so as to produce a centripetal acceleration of the order of 60*g* to 75*g*. The centrifuge action segregates the less dense nonmetallic inclusions near the centre of rotation. It should be noted that the casting of hollow parts needs no core in this process. Solid parts can also be cast by this process by placing the entire mould cavity on one side of the axis of rotation. The castings, produced by this method, are obviously very dense. By having several mould cavities, more than one casting can be made simultaneously.

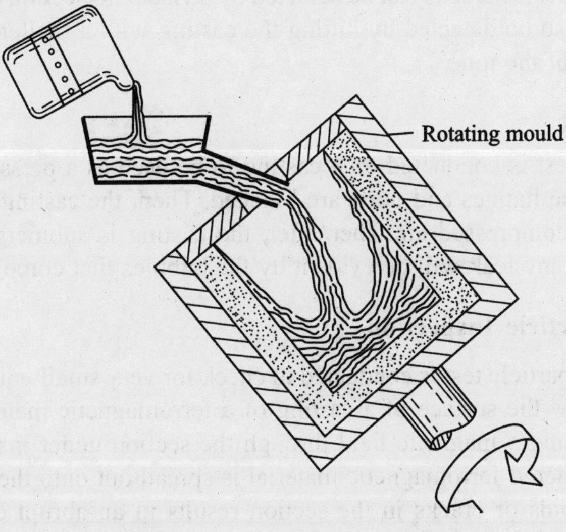

Fig. 2.41 Centrifugal casting.

2.7.7 SLUSH CASTING

A slush casting is produced by pouring the liquid material into an open-top permanent mould and inverting the mould after a thin layer of the liquid has solidified on its surfaces. Thus, the liquid at the centre of the mould is drained out. The process results in shell-like castings which are widely used for ornamental objects, e.g., lamp shades and toys. Normally, low temperature materials such as tin, lead, and zinc are used in this process.

2.7.8 CO_2 PROCESS

The CO_2 process is essentially a sand moulding process where the sand-mix does not contain any oil, resin, or clay as the bonding agent. This eliminates the use of driers and the heating cycle. Instead, the sand-mix contains 2–6% of sodium silicate solution. This sand-mix has a very high flowability to fill up corners and intricate contours. The sand is hardened by passing CO_2 for about one minute. The CO_2 gas forms a weak acid that hydrolyzes the sodium silicate (Na_2O, SiO_2) solution to form amorphous silica which acts as the bond. Sodium silicate itself also provides some bonding action.

2.8 INSPECTION OF CASTINGS

Nondestructive inspection techniques are essential for creating a confidence when using a cast product. In this section, we shall briefly outline some of these techniques for testing the various kinds of defects.

Visual Inspection

Common defects such as rough surfaces (fused sand), obvious shifts, omission of cores, and surface cracks can be detected by a visual inspection of the casting. Cracks may also be detected by hitting the casting with a mallet and listening to the quality of the tone.

Pressure Test

The pressure test is conducted on a casting to be used as a pressure vessel. In this, first all the flanges and ports are blocked. Then, the casting is filled with water, oil, or compressed air. Thereafter, the casting is submerged in a soap solution when any leak will be evident by the bubbles that come out.

Magnetic Particle Inspection

The magnetic particle test is conducted to check for very small voids and cracks at or just below the surface of a casting of a ferromagnetic material. The test involves inducing a magnetic field through the section under inspection. This done, the powdered ferromagnetic material is spread out onto the surface. The presence of voids or cracks in the section results in an abrupt change in the permeability of the surface; this, in turn, causes a leakage in the magnetic field. The powdered particles offer a low resistance path to the leakage. Thus, the particles accumulate on the disrupted magnetic field, outlining the boundary of a discontinuity.

Dye-Penetrant Inspection

The dye-penetrant method is used to detect invisible surface defects in a nonmagnetic casting. The casting is brushed with, sprayed with, or dipped into a dye containing a fluorescent material. The surface to be inspected is then wiped, dried, and viewed in darkness. The discontinuities in the surface will then be readily discernible.

Radiographic Examination

The radiographic method is expensive and is used only for subsurface exploration. In this, both X- and γ-rays are used. With γ-rays, more than one film can be exposed simultaneously; however, X-ray pictures are more distinct. Various defects, e.g., voids, nonmetallic inclusions, porosity, cracks, and tears, can be detected by this method. On the exposed film, the defects, being less dense, appear darker in contrast to the surrounding.

Ultrasonic Inspection

In the ultrasonic method, an oscillator is used to send an ultrasonic signal through the casting. Such a signal is readily transmitted through a homogeneous medium. However, on encountering a discontinuity, the signal is reflected back. This reflected signal is then detected by an ultrasonic detector. The time interval between sending the signal and receiving its reflection determines the location of the discontinuity. The method is not very suitable for a material with a high damping capacity (e.g., cast iron) because in such a case the signal gets considerably weakened over some distance.

2.9 EXERCISE PROBLEMS

2.1 Sketch the patterns with allowances for casting the following articles: (i) cast iron bearing block (Fig. 2.42a), (ii) aluminium bracket (Fig. 2.42b).

2.2 The grain size distribution of two samples of green sand for moulding is shown in Fig. 2.43. Which mould will have the higher permeability? Justify your answer.

2.3 In order to remove hydrogen from a liquid melt of iron of mass 100 kg, carbon monoxide bubbles are used so that the partial pressure of hydrogen falls to 0.1 atm. Determine the total volume of hydrogen in the liquid.

2.4 Figure 2.44 shows a mould along with the riser for casting a plate 20 cm × 20 cm × 5 cm. Determine the area A_g such that the mould and the riser get filled up within 10 sec after the downsprue has been filled. It should

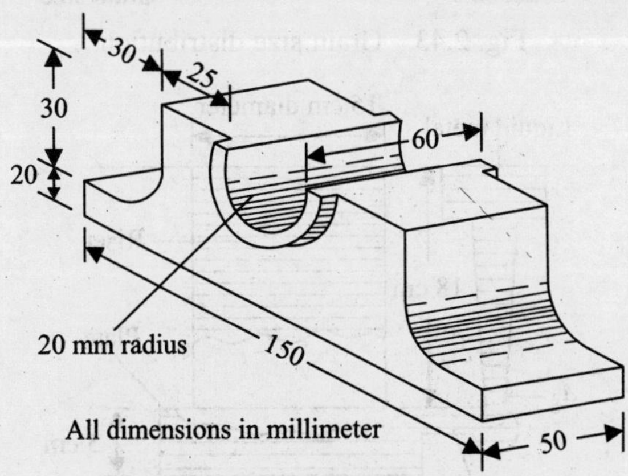

(a) Cast iron bearing block

Fig. 2.42 Jobs of Problem 2.1 (cont.).

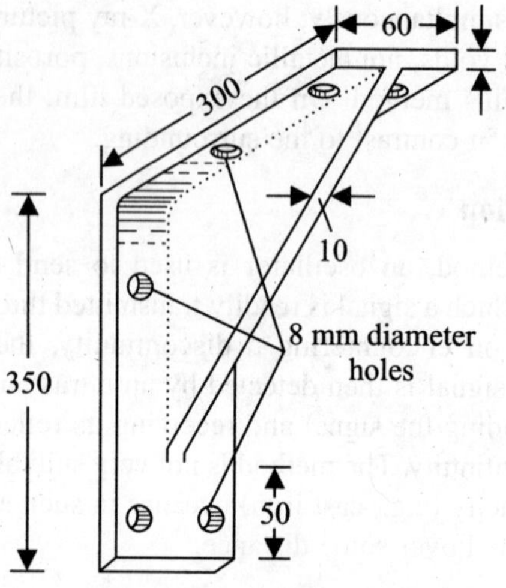

(b) Aluminium bracket

Fig. 2.42 Jobs of Problem 2.1.

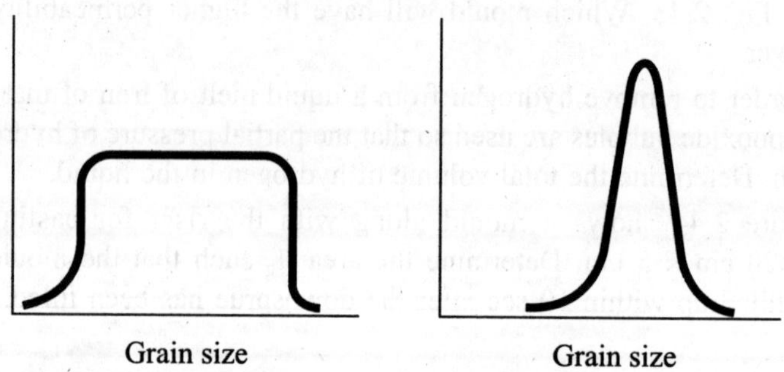

Fig. 2.43 Grain size distribution.

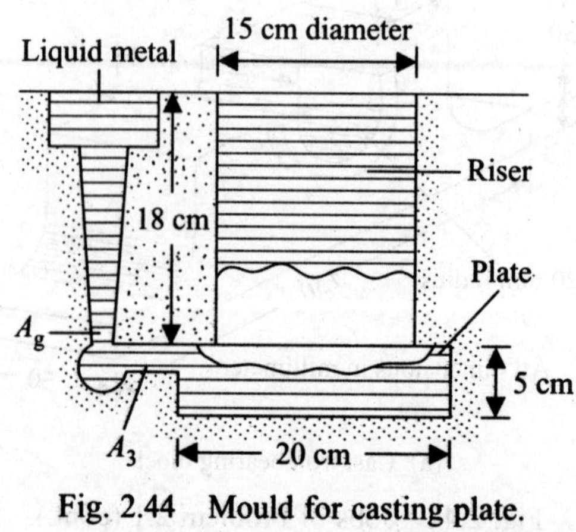

Fig. 2.44 Mould for casting plate.

be noted that $A_3 \gg A_g$ since below the downsprue a flat gate is attached to the casting. Neglect the frictional and orifice effects.

2.5 Design the downsprue, avoiding aspiration, shown in Fig. 2.45 to deliver liquid cast iron ($\rho_m = 7800$ kg/m^3) at a rate of 10 kg/sec against no head at the base of the sprue. Neglect the frictional and orifice effects.

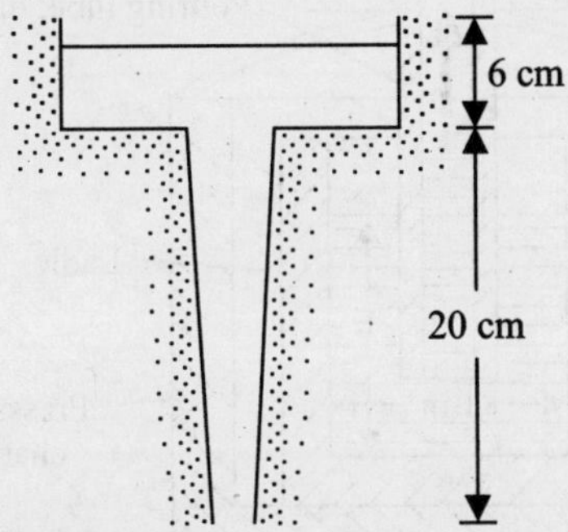

Fig. 2.45 Downsprue for delivering liquid cast iron.

2.6 Figure 2.46 shows the casting of a seamless metal pipe in a cooled mould. A pressure pouring technique is used to fill the mould quickly, as depicted in the figure. Determine the expression for estimating the time taken to fill up only the mould. Assume that the level h_1 in the ladle does not change. You may neglect the frictional losses but not the entrance losses.

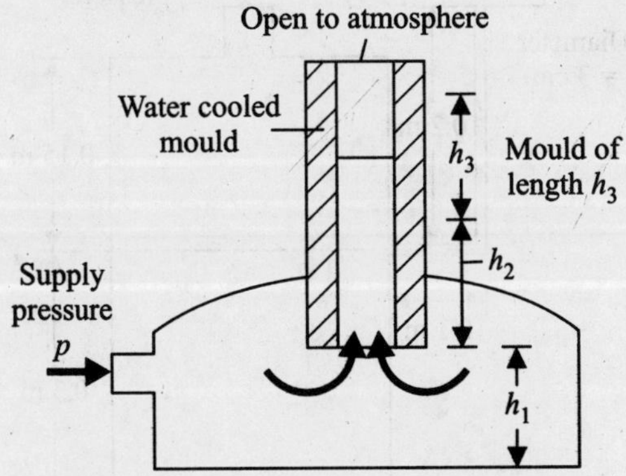

Fig. 2.46 Pressure pouring technique.

2.7 A ladle containing molten steel is placed in a pressurized chamber (Fig. 2.47). The air pressure in the chamber is increased to 0.3 N/mm^2 in order to force the molten metal into the mould through the pouring tube. Estimate the time taken to fill up the mould. Also, account for the frictional losses in the tube and the losses due to sudden contraction and expansion. Initially, half of the tube length is submerged into the melt. For steel, $\rho_m = 7600$ kg/m^3, and $\eta = 0.00595$ kg/m-sec.

2.8 Estimate the time to fill up the mould shown in Fig. 2.48. Assume that the

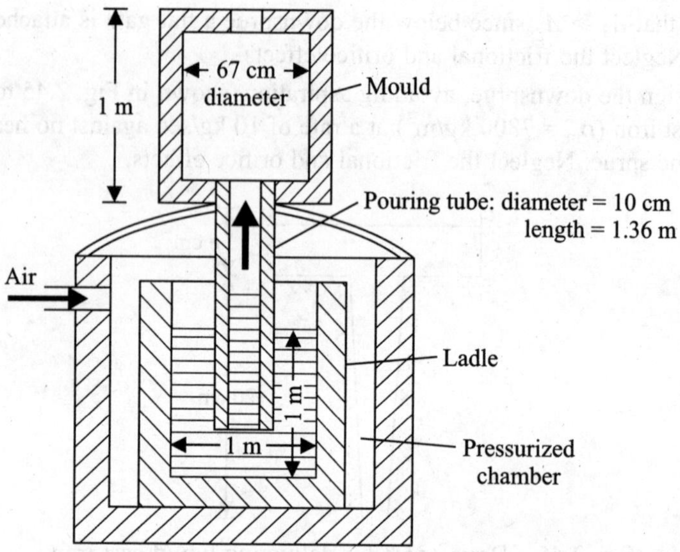

Fig. 2.47 Filling up mould using pressurized chamber.

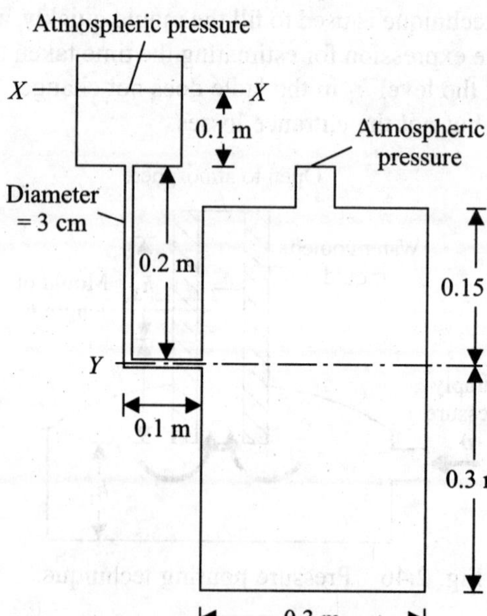

Fig. 2.48 System of Problem 2.8.

liquid metal level at *X-X* is maintained constant and the time to fill the runner is negligible. The following data can all be treated as constants:

$$\rho_m = 6000 \text{ kg/m}^3, \qquad \eta = 0.00165 \text{ kg/m-sec}, \qquad f(\text{runner}) = 0.0025,$$

$$e_f\,(\text{contraction}) = 0.1, \qquad \beta = 1.0, \qquad (L/D)_{eq} \text{ for } 90° \text{ turn} = 25,$$

e_f (enlargement for levels below Y-Y) = 0,

e_f (enlargement for levels above Y-Y) = 1.0.

2.9 A continuous casting operation is illustrated in Fig. 2.49. The liquid level in the tundish and the degasser is maintained at the height indicated in the figure. Determine the tundish and degasser nozzle sizes for a production rate of 25,000 kg/hr per strand. The given data are

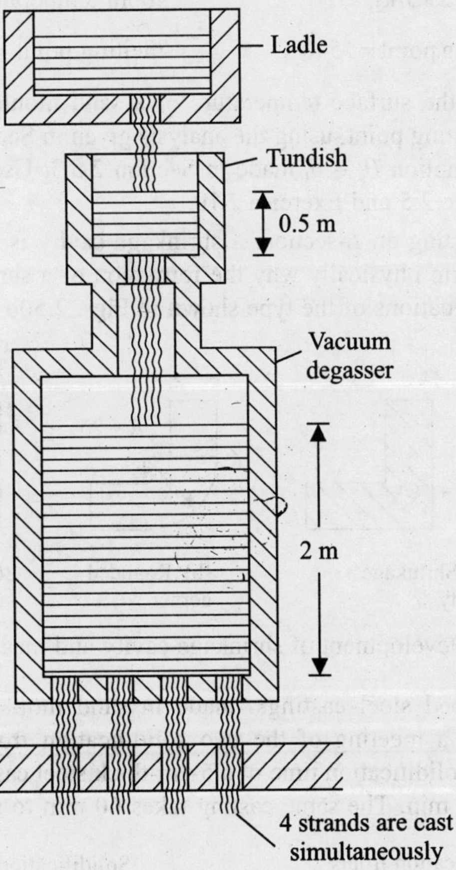

Fig. 2.49 Continuous casting with degasser.

$\rho_m = 7600$ kg/m^3,

discharge coefficient for each nozzle = 0.8,

vacuum pressure = 10^{-3} atm.

2.10 Is it possible to cast an iron slab, poured at its melting point, in a very thick aluminium mould? Justify your answer quantitatively. Assume no thermal resistance at the mould-metal interface. Use the following data:

For iron	*For aluminium*
$c = 0.67$ kJ/kg-K	$c = 1.12$ kJ/kg-K
$k = 83$ W/m-K	$k = 498$ W/m-K
$\rho = 7850$ kg/m^3	$\rho = 2748$ kg/m^3
$L = 272$ kJ/kg	room temperature = 28°C
melting point = 1540°C	melting point = 660°C

2.11 Estimate the surface temperature of a sand mould while casting iron, poured at its melting point, using the analysis given in Section 2.5.6, and hence justify the assumption $\theta_s = \theta_f$ made in Section 2.5.3. Use the property values given in Example 2.5 and Exercise 2.10.

2.12 While casting an *L*-section, a shrinkage cavity is formed, as shown in Fig. 2.50a. Explain physically why the formation of a shrinkage cavity should be avoided in situations of the type shown in Figs. 2.50b and 2.50c.

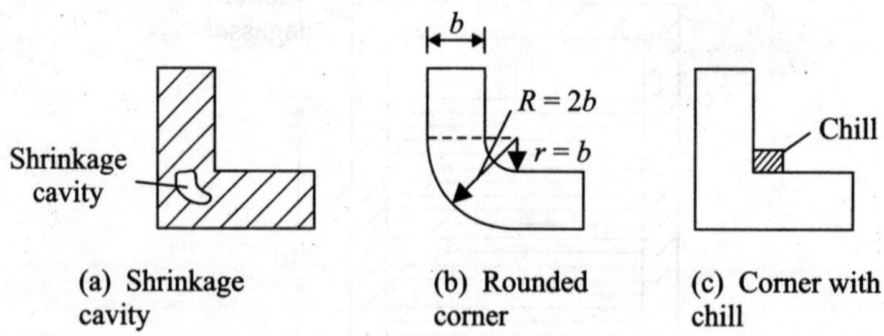

Fig. 2.50 Development of shrinkage cavity and means of avoiding it.

2.13 Slab-shaped steel castings, made in sand moulds, show centre-line porosity due to a meeting of the two solidification fronts, as explained in Fig. 2.51a. The solidification time of a 5-cm-thick steel casting in a sand mould is found to be 6 min. The same casting takes 60 min to solidify when placed

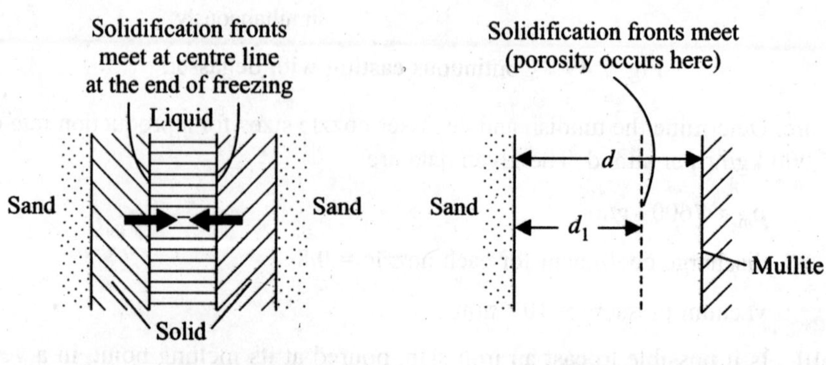

Fig. 2.51 Solidification front in symmetrical and composite moulds.

in an insulating mullite mould. If the casting is made in a composite mould, as sketched in Fig. 2.51b, determine the thickness *d* to be cast such that, after machining, a 47-mm-thick, sound casting is obtained.
[*Hint* The machining allowance for cast steel is 3 mm (see Table 2.1) for the dimension given. Hence, $d_1 = 47 + 3 = 50$ mm.]

2.14 Compare the solidification time of two *optimum* risers of the same volume when one has a cylindrical shape and the other is of the form of a rectangular parallelopiped.

2.15 Determine the dimensions of an optimum cylindrical riser attached to the side of a steel plate casting having the dimensions 25 cm × 12.5 cm × 5 cm by

(i) using Caine's relationship, i.e., equation (2.68),

(ii) using Fig. 2.32,

(iii) assuming that the volume shrinkage on solidification is 3% for steel and that the volume of the riser is three times that dictated by the shrinkage consideration alone.

2.16 Estimate the riser volume necessary if a bar of cross-section 2.5 cm×2.5 cm and 10 cm long is added to the plate considered in Exercise 2.15.

2.17 Design an economic risering system for the steel casting shown in Fig. 2.52, taking into account the problem of feeding distance.

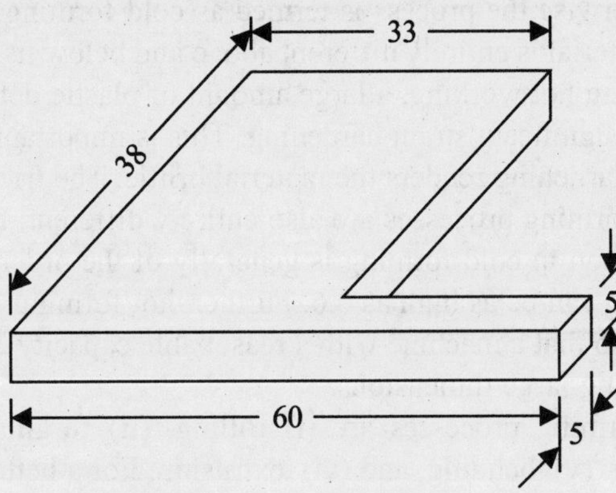

All dimensions in millimeter

Fig. 2.52 Steel casting.

3 Forming Processes

3.1 INTRODUCTION

Forming can be defined as a process in which the desired size and shape are obtained through the plastic deformation of a material. The stresses induced during the process are greater than the yield strength, but less than the fracture strength, of the material. The type of loading may be tensile, compressive, bending, or shearing, or a combination of these. This is a very economical process as the desired shape, size, and finish can be obtained without any significant loss of material. Moreover, a part of the input energy is fruitfully utilized in improving the strength of the product through strain hardening.

The forming processes can be grouped under two broad categories, namely, (i) cold forming, and (ii) hot forming. If the working temperature is higher than the recrystallization temperature of the material, then the process is called hot forming. Otherwise the process is termed as cold forming. The flow stress behaviour of a material is entirely different above and below its recrystallization temperature. During hot working, a large amount of plastic deformation can be imparted without significant strain hardening. This is important because a large amount of strain hardening renders the material brittle. The frictional characteristics of the two forming processes are also entirely different. For example, the coefficient of friction in cold forming is generally of the order of 0.1, whereas that in hot forming can be as high as 0.6. Further, hot forming lowers down the material strength so that a machine with a reasonable capacity can be used even for a product having large dimensions.

The typical forming processes are (i) rolling, (ii) forging, (iii) drawing, (iv) deep drawing, (v) bending, and (vi) extrusion. For a better understanding of the mechanics of various forming operations, we shall briefly discuss each of these processes.

Two other common production processes, namely, punching and blanking, though not classified under the forming processes, will also be briefly considered because of their similarity to the forming processes.

(i) *Rolling* In this process, the job is drawn by means of friction through a regulated opening between two power-driven rolls (Fig. 3.1). The shape and size of the product are decided by the gap between the rolls and their contours. This is a very useful process for the production of sheet metal and various common sections, e.g., rail, channel, angle, and round.

(ii) *Forging* In forging, the material is squeezed between two or more

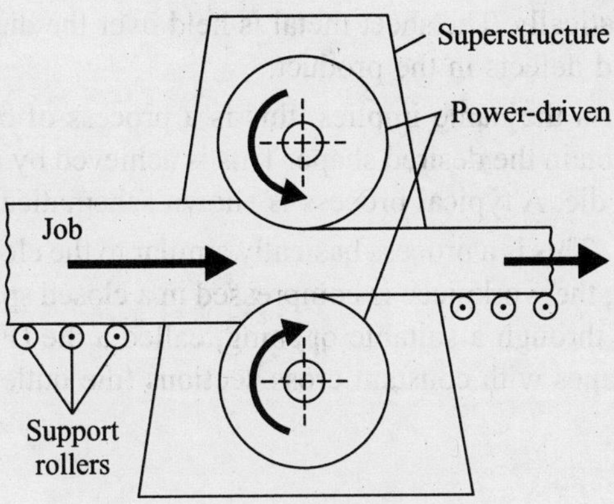

Fig. 3.1 Rolling operation.

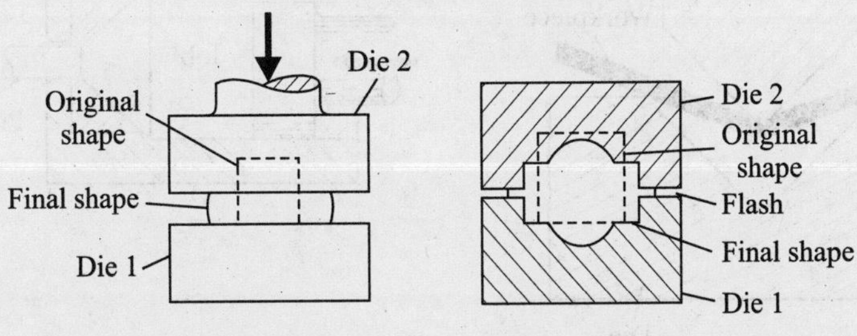

(a) Open die forging (b) Closed die forging

Fig. 3.2 Forging operation.

dies to alter its shape and size. Depending on the situation, the dies may be open (Fig. 3.2a) or closed (Fig. 3.2b).

(iii) *Drawing* In this process, the cross-section of a wire or that of a bar or tube is reduced by pulling the workpiece through the conical orifice of a die. Figure 3.3 represents the operation schematically. When high reduction is required, it may be necessary to perform the operation in several passes.

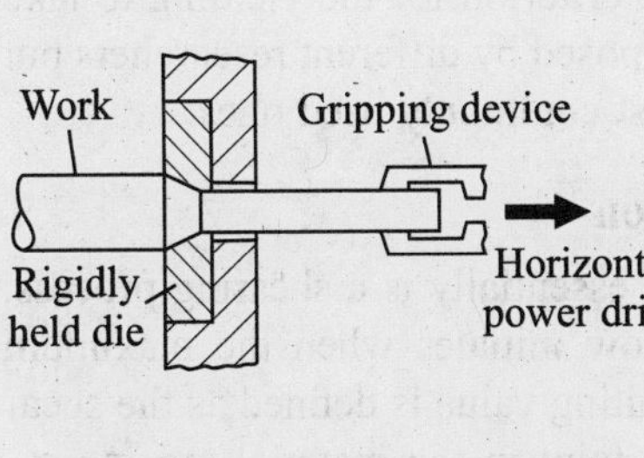

Fig. 3.3 Drawing operation.

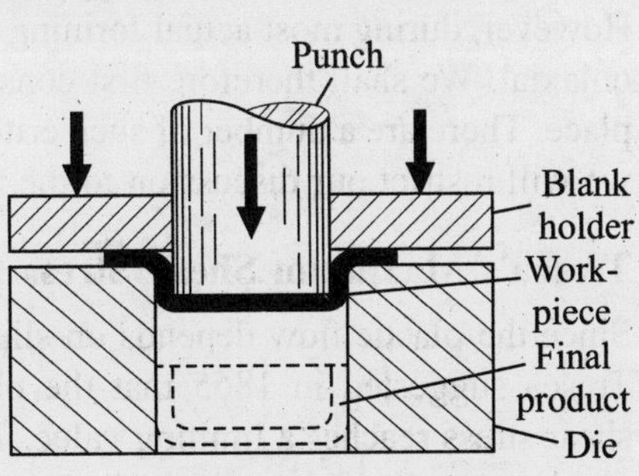

Fig. 3.4 Deep drawing.

(iv) *Deep drawing* In deep drawing, a cup-shaped product is obtained from a flat sheet metal with the help of a punch and a die. Figure 3.4 shows the operation schematically. The sheet metal is held over the die by means of a blank holder to avoid defects in the product.

(v) *Bending* As the name implies, this is a process of bending a metal sheet plastically to obtain the desired shape. This is achieved by a set of suitably designed punch and die. A typical process is shown schematically in Fig. 3.5.

(vi) *Extrusion* This is a process basically similar to the closed die forging. But in this operation, the workpiece is compressed in a closed space, forcing the material to flow out through a suitable opening, called a die (Fig. 3.6). In this process, only the shapes with constant cross-sections (die outlet cross-section) can be produced.

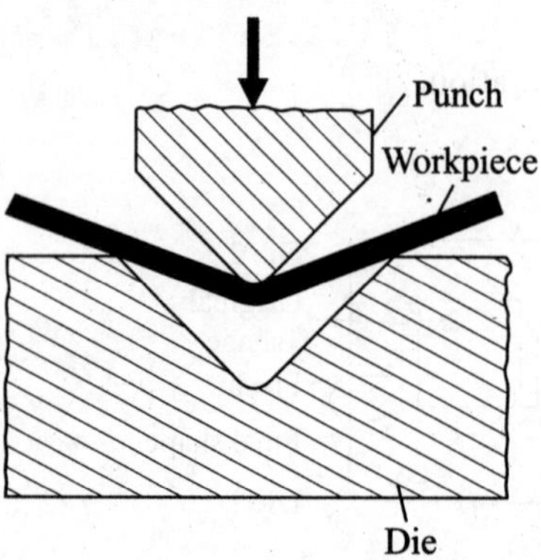

Fig. 3.5 Bending.

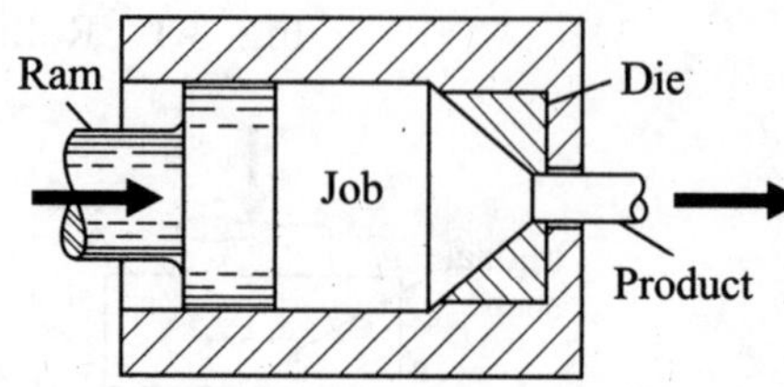

Fig. 3.6 Extrusion.

Apart from the foregoing processes, there are various other forming operations which we shall take up in Section 3.5.

3.2 PLASTIC DEFORMATION AND YIELD CRITERIA

In Chapter 1, we mentioned that plastic deformation takes place when the applied stress level exceeds a certain limit defined as yield stress. It should be recalled that we discussed this behaviour only with respect to uniaxial loading. However, during most actual forming operations, the loading conditions are not uniaxial. We shall therefore first consider the criterion for the yielding to take place. There are a number of such criteria proposed by different researchers but we will restrict our discussion to the two most commonly used ones.

Tresca's Maximum Shear Stress Criterion

Since the plastic flow depends on slip which essentially is a shearing process, Tresca suggested in 1865 that the plastic flow initiates when the maximum shear stress reaches a limiting value. This limiting value is defined as the shear yield stress *K*. If the principal stresses at a point in the material are σ_1, σ_2, σ_3 ($\sigma_1 \geqslant \sigma_2 \geqslant \sigma_3$), then the maximum shear stress τ_{max} is given by $\tau_{max} = \frac{1}{2}(\sigma_1 - \sigma_3)$. Plastic deformation occurs when τ_{max} is equal to *K*. So, Tresca's criterion becomes

$$\tfrac{1}{2}(\sigma_1 - \sigma_3) = K. \tag{3.1}$$

It is evident from equation (3.1) that the yielding is independent of the intermediate principal stress σ_2.

von Mises' Maximum Distortion Energy Criterion

In 1913, von Mises proposed that the plastic flow occurs when the shear strain energy reaches a critical value. The shear strain energy per unit volume (ε) can be expressed in terms of the three principal stresses as

$$\varepsilon = \frac{1}{6G}[(\sigma_1 - \sigma_2)^2 + (\sigma_2 - \sigma_3)^2 + (\sigma_3 - \sigma_1)^2], \tag{3.2}$$

where G is the shear modulus of the material. Hence, according to this criterion, the plastic flow initiates when the right-hand side of equation (3.2) reaches a particular value, say, A. Finally, the von Mises criterion takes the form

$$(\sigma_1 - \sigma_2)^2 + (\sigma_2 - \sigma_3)^2 + (\sigma_3 - \sigma_1)^2 = 6GA = C \text{ (constant)}. \tag{3.3}$$

It should be noted that, according to this criterion, the initiation of plastic flow depends on all the principal stresses.

3.3 RELATIONSHIP BETWEEN TENSILE AND SHEAR YIELD STRESSES

To apply the foregoing yield criteria, it is necessary to know the right-hand sides of equations (3.1) and (3.3) for a given material. In most cases, the material properties are determined purely from uniaxial tensile tests. Such tests give the value of the tensile yield stress σ_Y which can be used to determine the shear yield stress K, as now explained. When yielding occurs under uniaxial tensile loading, $\sigma_1 = \sigma_Y$, $\sigma_2 = \sigma_3 = 0$. Hence, the constant in equation (3.3) can be written as

$$C = (\sigma_Y - 0)^2 + 0 + (0 - \sigma_Y)^2 = 2\sigma_Y^2. \tag{3.4}$$

Now, let us consider yielding under pure torsion. The state of stress in a material, for a two-dimensional situation, is shown with the help of Mohr's circle in Fig. 3.7. It is clear from this figure that $\sigma_1 = K$, $\sigma_3 = -K$, $\sigma_2 = 0$. Substituting these values in equation (3.3), we get

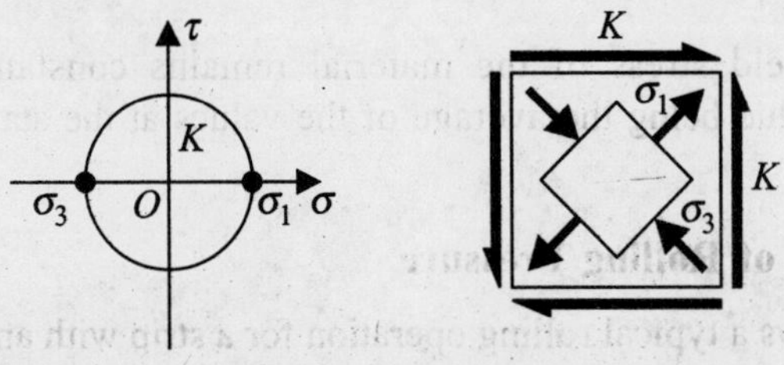

Fig. 3.7 Mohr's circle representation of two-dimensional state of stress.

$$C = (K - 0)^2 + (0 + K)^2 + (-K - K)^2 = 6K^2. \tag{3.5}$$

Since the magnitude of C in the von Mises criterion is independent of the type of loading, the relationship between K and σ_Y, obtained by equating the right-hand sides of equations (3.4) and (3.5), is

$$2\sigma_Y^2 = 6K^2$$

or

$$K = \sigma_Y/\sqrt{3}. \tag{3.6}$$

Applying Tresca's yield criterion to these two different, pure loading situations, we can obtain K in terms of σ_Y as

$$K = \sigma_Y/2. \tag{3.7}$$

The von Mises criterion being more realistic than Tresca's criterion, equation (3.6) is normally used to relate K with σ_Y.

3.4 MECHANICS OF FORMING PROCESSES

In this section, we shall give an elementary analysis of the various basic metal forming processes. In doing so, we shall show how (i) the work load can be estimated from the knowledge of the material properties and the working conditions, and (ii) certain other aspects of the processes can be better understood.

When studying the mechanics of forming processes, the acceleration, and consequently the inertia forces, of the flowing materials are negligible.

Hence, the equations for static equilibrium can be applied in all cases.

3.4.1 ROLLING

The basic objectives of the analysis we give here are to determine (i) the roll separating forces, (ii) the torque and power required to drive the rolls, and (iii) the power loss in bearings. An analysis considering all the factors in a real situation is beyond the scope of this text, and therefore the following simplifying assumptions will be made:

(i) The rolls are straight and rigid cylinders.

(ii) The width of the strip is much larger than its thickness and no significant widening takes place, i.e., the problem is of plane strain type.

(iii) The coefficient of friction μ is low and constant over the entire roll-job interface.

(iv) The yield stress of the material remains constant for the entire operation, its value being the average of the values at the start and at the end of rolling.

Determination of Rolling Pressure

Figure 3.8a shows a typical rolling operation for a strip with an initial thickness t_i which is being rolled down to a final thickness t_f. Both the rolls are of equal radius R and rotate with the same circumferential velocity V. The origin of the coordinate system xy is taken at the midpoint of the line joining the centres O_1 and O_2. (The operation is two-dimensional, and so the position of O along the axis mutually perpendicular to Ox and Oy is of no significance. In our analysis, we shall assume that the width of the strip is unity.) The entry and exit velocities of the strip are V_i and V_f, respectively. In actual practice, $V_f > V > V_i$. Therefore, at a particular point in the working zone, the velocity of the strip will be equal to V, and this point will hereafter be referred to as the neutral point.

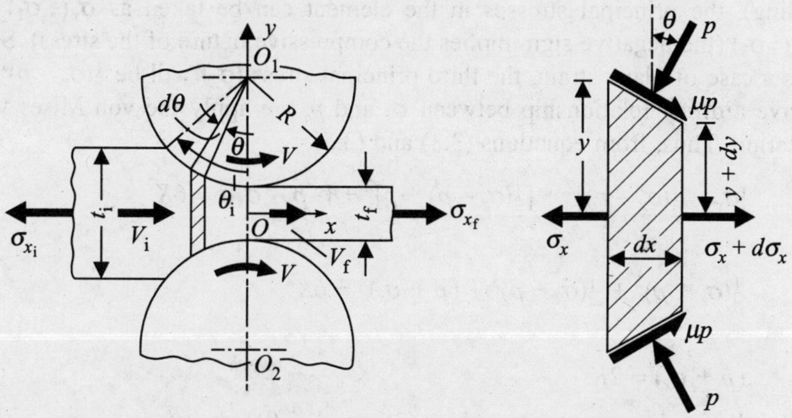

(a) Details of rolling operation (b) Stresses on element

Fig. 3.8 Forces and stresses during rolling.

Considering a general case, we assume that the stresses σ_{x_i} and σ_{x_f} are acting on the entry and the exit sides (Fig. 3.8a). However, depending on the situation, either one or both of these stresses may be absent.

Now, let us consider the forces (shown in Fig. 3.8a) acting on the element of length dx. The element and the various stresses acting on it are indicated clearly in Fig. 3.8b. The direction of the friction force (per unit area) μp (p being the pressure exerted by the rolls on the job) depends on the relative velocity of the job and the roll periphery. Thus, it is obvious that the direction, shown in Fig. 3.8b, is valid before the element reaches the neutral point. After the element crosses the neutral point, the friction force changes its direction. Another important fact to be remembered is that the total angle θ_i, subtended by the work zone (Fig. 3.8a), is quite small in all actual operations. Considering the equilibrium of the forces acting on the element in the x-direction, we have

$$2(y + dy)(\sigma_x + d\sigma_x) - 2y\sigma_x - 2R\, d\theta\, \mu p \cos\theta + 2R\, d\theta\, p \sin\theta = 0.$$

Since θ is always small,

$$2(y + dy)(\sigma_x + d\sigma_x) - 2y\sigma_x - 2R\, d\theta\, \mu p + 2R\, d\theta\, p\theta = 0.$$

Neglecting the higher order terms, we get

$$2y d\sigma_x + 2\sigma_x\, dy - 2R\, d\theta\, \mu p + 2R\, d\theta\, p\theta = 0$$

or

$$d(y\sigma_x) - Rp(\mu - \theta)\,d\theta = 0.$$

So, rearranging the terms, the differential equation we obtain is

$$\frac{d}{d\theta}(\sigma_x y) - (\mu - \theta)Rp = 0. \tag{3.8}$$

Because the friction force is assumed to be small (especially true for cold rolling), the principal stresses in the element can be taken as $\sigma_x(=\sigma_1)$ and $-p(=\sigma_3)$ (the negative sign implies the compressive nature of the stress). Since it is a case of plane strain, the third principal stress (σ_2) will be $\frac{1}{2}(\sigma_x - p)$[1]. To derive a direct relationship between σ_x and p, we apply the von Mises yield criterion. Thus, from equations (3.3) and (3.5),

$$[\sigma_x - \tfrac{1}{2}(\sigma_x - p)]^2 + [\tfrac{1}{2}(\sigma_x - p) + p]^2 + [-p - \sigma_x]^2 = 6K^2$$

or

$$\tfrac{1}{4}(\sigma_x + p)^2 + \tfrac{1}{4}(\sigma_x + p)^2 + (p + \sigma_x)^2 = 6K^2$$

or

$$(p + \sigma_x) = 2K. \tag{3.9}$$

Now, eliminating σ_x from equations (3.8) and (3.9), we get

$$\frac{d}{d\theta}[(2K - p)y] - (\mu - \theta)Rp = 0. \tag{3.10}$$

As already mentioned, the change in the direction of the friction force before and after the neutral point can be incorporated in equation (3.10) as

$$\frac{d}{d\theta}[(2K - p)y] - (\pm\mu - \theta)Rp = 0. \tag{3.11}$$

The positive sign applies for the region before the neutral point and the negative sign after that. As the material being rolled undergoes strain hardening, the shear yield stress K increases (though, for simplicity, K has been considered to be constant in the rest of the analysis) as the rolling progresses. On the other hand, it is clear from Fig. 3.8a that y decreases during rolling. As a consequence, it may be assumed for simplification (without resulting in much error) that the product Ky remains constant. With this simplification, equation (3.11) can be rewritten as

[1]From Hooke's law (for plane strain), the strain

$$\varepsilon_2 = \frac{1}{E}[\sigma_2 - \nu(\sigma_1 + \sigma_3)] = 0$$ (where ν is the Poisson ratio and E is Young's modulus of the material)

or

$$\sigma_2 = \tfrac{1}{2}(\sigma_1 + \sigma_3)$$ (since $\nu = \frac{1}{2}$ in plastic deformation).

$$2Ky\frac{d}{d\theta}(1 - \frac{p}{2K}) + (\theta \mp \mu)Rp = 0. \tag{3.12}$$

As θ is small, y can be expressed in the form

$$y = \frac{t_f}{2} + \frac{R\theta^2}{2}.$$

Expanding equation (3.12) and substituting y from this relation, we obtain

$$-(t_f + R\theta^2)\frac{d}{d\theta}(\frac{p}{2K}) + 2(\theta \mp \mu)R(\frac{p}{2K}) = 0$$

or

$$\frac{d(\frac{p}{2K})}{(\frac{p}{2K})} = \frac{2R(\theta \mp \mu)\, d\theta}{(t_f + R\theta^2)}.$$

Integrating, we get

$$\int \frac{d(\frac{p}{2K})}{(\frac{p}{2K})} = \int \frac{2R\theta\, d\theta}{t_f + R\theta^2} \mp \int \frac{2R\mu\, d\theta}{t_f + R\theta^2} + C_1$$ (with C_1 being the constant of integration)

or

$$\ln(\frac{p}{2K}) = \ln(t_f + R\theta^2) \mp 2\mu\sqrt{R}\frac{1}{\sqrt{t_f}}\tan^{-1}\sqrt{\frac{R}{t_f}}\theta$$

$$+ \ln(\frac{C}{2R})$$ (with C being another constant)

or

$$\frac{p}{2K} = C\frac{y}{R}e^{\mp\mu\lambda}, \tag{3.13}$$

where

$$\lambda = 2\sqrt{\frac{R}{t_f}}\tan^{-1}(\sqrt{\frac{R}{t_f}}\theta). \tag{3.14}$$

Applying equation (3.9) to the beginning of the rolling, we obtain

$$\frac{p_i}{2K} = 1 - \frac{\sigma_{x_i}}{2K},$$

where p_i is the roll pressure at the starting point. Using this equation in equation (3.13), we have

$$\frac{p_i}{2K} = (1 - \frac{\sigma_{x_i}}{2K}) = C^{-}\frac{t_i}{2R}e^{-\mu\lambda_i},$$

where

$$\lambda_i = 2\sqrt{\frac{R}{t_f}}\tan^{-1}\left(\sqrt{\frac{R}{t_f}}\theta_i\right)$$

and C^- is the value of the constant C before the neutral point is reached. Hence,

$$C^- = \frac{2R}{t_i}\left(1 - \frac{\sigma_{x_i}}{2K}\right)e^{+\mu\lambda_i}. \tag{3.15}$$

For the region beyond the neutral point, equation (3.13) can be written as

$$\frac{p}{2K} = C^+\frac{y}{R}e^{+\mu\lambda}.$$

Again, applying equation (3.9) to the end of the rolling, we get

$$\frac{p_f}{2K} = 1 - \frac{\sigma_{x_f}}{2K},$$

where p_f is the roll pressure at the exit point. So,

$$\left(1 - \frac{\sigma_{x_f}}{2K}\right) = C^+\frac{t_f}{2R}$$

since the value of θ at the end point is zero. Hence,

$$C^+ = \frac{2R}{t_f}\left(1 - \frac{\sigma_{x_f}}{2K}\right). \tag{3.16}$$

Using the values of C^- and C^+ from equations (3.15) and (3.16), respectively, in equation (3.13), the expressions for the nondimensional roll pressure $[p/(2K)]$ in the regions before and after the neutral point we obtain are

$$\left(\frac{p}{2K}\right)_{\text{before}} = \frac{2y}{t_i}\left(1 - \frac{\sigma_{x_i}}{2K}\right)e^{\mu(\lambda_i-\lambda)}, \tag{3.17}$$

$$\left(\frac{p}{2K}\right)_{\text{after}} = \frac{2y}{t_f}\left(1 - \frac{\sigma_{x_f}}{2K}\right)e^{\mu\lambda}. \tag{3.18}$$

The pressure at the neutral point can be determined from either equation (3.17) or equation (3.18). So, the value of λ corresponding to the neutral point (λ_n) is obtained by equating the right-hand sides of equations (3.17) and (3.18). Thus,

$$\lambda_n = \tfrac{1}{2}\left[\frac{1}{\mu}\ln\left\{\frac{t_f}{t_i}\left(\frac{1 - \dfrac{\sigma_{x_i}}{2K}}{1 - \dfrac{\sigma_{x_f}}{2K}}\right)\right\} + \lambda_i\right]. \tag{3.19}$$

The location of the neutral point (θ_n) can be obtained by using equations (3.14) and (3.19). Before going into a discussion on the roll separating force and the driving power, let us have a look into the nature of variation of the roll pressure. For typical values of the parameters in a rolling operation, we find that the roll pressure p increases continuously from the point of entry till the neutral point is reached. Thereafter, it decreases continuously, as is evident from equation (3.18). The typical distributions of pressure p are shown in Fig. 3.9. The peak pressure at the neutral point is normally called the friction hill. This peak pressure increases with increasing coefficient of friction.

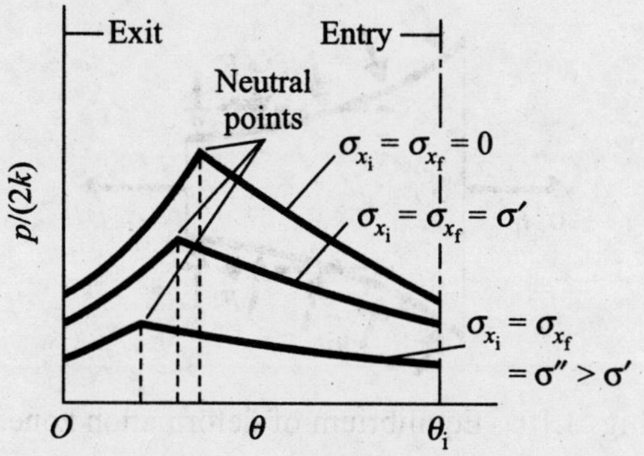

Fig. 3.9 Pressure distribution in rolling.

Determination of Roll Separating Force

Assuming that the width of the strip is unity, the total force F trying to separate the rolls can be obtained by integrating the vertical component of the force acting at the roll-strip interface. Since the angle θ_i is normally very small, the contribution of the roll-strip interface friction force is negligible in the vertical direction. Thus,

$$F = \int_0^{\theta_i} pR \cos\theta \, d\theta$$

$$\approx \int_0^{\theta_i} pR \, d\theta \qquad \text{(since } \theta \text{ is small)},$$

i.e.,

$$F = \int_0^{\theta_n} p_{\text{after}} R \, d\theta + \int_{\theta_n}^{\theta_i} p_{\text{before}} R \, d\theta. \tag{3.20}$$

The integrations in equation (3.20) are normally computed numerically.

Driving Torque and Power

The driving torque is required to overcome the torque exerted on the roll by the interfacial friction force. So, the driving torque corresponding to unit width can be expressed as

$$T = \int_0^{\theta_i} \mu p R^2 \, d\theta = -\int_0^{\theta_n} \mu p_{\text{after}} R^2 \, d\theta + \int_{\theta_n}^{\theta_i} \mu p_{\text{before}} R^2 \, d\theta. \tag{3.21}$$

It should be noted that friction resists the rotation of the roll before the neutral point, whereas it helps the rotation afterwards. By this method, the result comes out as the difference of two, nearly equal, large numbers, causing the numerical error to be significant. This limits the use of equation (3.21) in practice.

An alternative approach to determine T is to consider the horizontal equilibrium of the deformation zone of the strip. Figure 3.10 shows the deformation zone along with the forces acting on it, including an equivalent horizontal force F^e

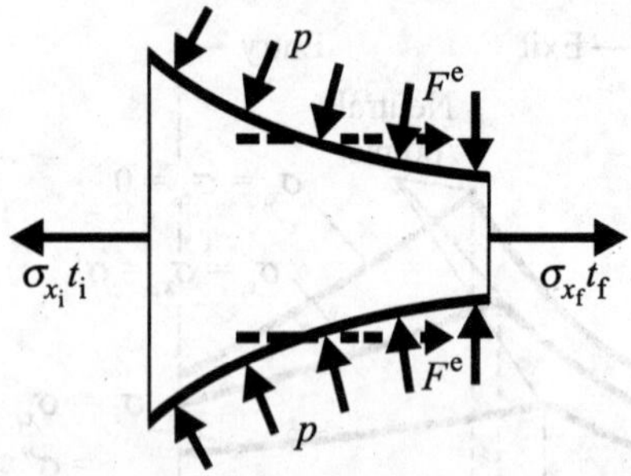

Fig. 3.10 Equilibrium of deformation zone.

which represents the net frictional interaction between each roll and the strip (the reaction $-F^e$ of F^e has to be overcome by the roll driving torque T). F^e can be determined by considering the horizontal equilibrium of the system. Thus,

$$F^e = \tfrac{1}{2}[(\sigma_{x_i} t_i - \sigma_{x_f} t_f) + 2\int_0^{\theta_i} pR \sin\theta \, d\theta]$$

$$\approx \tfrac{1}{2}(\sigma_{x_i} t_i - \sigma_{x_f} t_f) + \int_0^{\theta_i} pR^2 \theta \, d\theta. \tag{3.22}$$

Accordingly,

$$T \approx F^e R = \tfrac{1}{2}R(\sigma_{x_i} t_i - \sigma_{x_f} t_f) + \int_0^{\theta_i} pR^2 \theta \, d\theta.$$

Once the driving torque T is determined, the driving power per roll P_R is obtained as

$$P_R = T\omega, \tag{3.23}$$

where ω is the angular speed of the roll.

EXAMPLE 3.1 A strip with a cross-section of 150 mm × 6 mm is being rolled with 20% reduction of area, using 400-mm-diameter steel rolls. Before and after rolling, the shear yield stress of the material is 0.35 kN/mm^2 and 0.4 kN/mm^2, respectively. Calculate (i) the final strip thickness, (ii) the average shear yield stress during the process, (iii) the angle subtended by the deformation zone at the roll centre, and (iv) the location of the neutral point θ_n. Assume the coefficient of friction to be 0.1.

SOLUTION (i) As no widening is considered during rolling, 20% reduction in the area implies a longitudinal strain of 0.2 with consequent 20% reduction in the thickness. Therefore, the final strip thickness is given as

$$t_f = 0.8t_i = 0.8 \times 6 \text{ mm} = 4.8 \text{ mm}.$$

(ii) The average shear yield stress during the process is taken to be the arithmetic mean of the initial and the final values of the yield stress. So,

$$K = (K_i + K_f)/2 = \frac{0.75}{2} \text{ kN/mm}^2 = 0.375 \text{ kN/mm}^2.$$

(iii) From Fig. 3.8a, it is clear that

$$\theta_i = \sqrt{\frac{t_i - t_f}{R}}.$$

Substituting the values, we get

$$\theta_i = \sqrt{\frac{6 - 4.8}{200}} \text{ rad} = 0.0775 \text{ rad}.$$

(iv) To determine θ_n, first λ_n has to be calculated from equation (3.19). Since nothing has been mentioned regarding the forward and the back tension, σ_{x_i} and σ_{x_f} will be assumed zero. Hence,

$$\lambda_n = \tfrac{1}{2}[\frac{1}{\mu} \ln(\frac{t_f}{t_i}) + \lambda_i],$$

where

$$\lambda_i = 2\sqrt{\frac{R}{t_f}} \tan^{-1}(\sqrt{\frac{R}{t_f}}\theta_i).$$

Substituting the values of R, t_f, and θ_i in the expression for λ_i, we get $\lambda_i = 5.99$. Now, using this value of λ_i, we find that the value of λ_n is 1.88. θ_n can be expressed as

$$\theta_n = \sqrt{\frac{t_f}{R}} \tan [\frac{\lambda_n}{2}\sqrt{\frac{t_f}{R}}].$$

Hence,

$$\theta_n = \sqrt{\frac{4.8}{200}} \tan [\frac{1.88}{2}\sqrt{\frac{4.8}{200}}] = 0.023 \text{ rad}.$$

EXAMPLE 3.2 Assuming the speed of rolling to be 30 m/min, determine (i) the roll separating force, and (ii) the power required in the rolling process described in Example 3.1.

SOLUTION From equation (3.20), we see that to find out the roll separating force F, we need to know the pressure p (as a function of θ), θ_n, and θ_i. In Example 3.1, we have already calculated θ_i and θ_n. For the numerical integration of equation (3.20), we employ Simpson's rule, using four divisions after θ_n and eight divisions before θ_n. Thus,

$$\Delta\theta_{\text{after}} = \frac{\theta_n}{4} = 0.00575,$$

$$\Delta\theta_{\text{before}} = \frac{\theta_i - \theta_n}{8} = 0.00681.$$

With these intervals, the pressure at different station points can be computed, using equations (3.17) and (3.18). Next, we compile the following data. After the neutral point:

Station point	θ (rad)	$y = \frac{1}{2}(t_f + R\theta^2)$ (mm)	λ [see eqn (3.14)]	p_{after} [see eqn (3.18)] (kN/mm^2)
1	0.00000	2.40000	0.000	0.75
2	0.00575	2.40331	0.479	0.788
3	0.0115	2.4132	0.958	0.830
4	0.01725	2.4298	1.432	0.876
5	0.023	2.453	1.88	0.925

Before the neutral point:

Station point	θ (rad)	y (mm)	λ	p_{before} [see eqn (3.17)] (kN/mm^2)
1	0.023	2.453	1.88	0.925
2	0.02981	2.489	2.454	0.887
3	0.03662	2.534	2.997	0.855
4	0.04343	2.589	3.529	0.828
5	0.05024	2.65	4.049	0.804
6	0.05705	2.725	4.555	0.786
7	0.06386	2.81	5.048	0.772
8	0.07067	2.899	5.525	0.759
9	0.0775	3.000	5.99	0.75

Now, from equation (3.20), the roll separating force per unit width is

$$F = R\left[\int_0^{\theta_n} p_{\text{after}}\, d\theta + \int_{\theta_n}^{\theta_i} p_{\text{before}}\, d\theta\right],$$

$$F = 200\left[\frac{0.00575}{3}\{0.75 + 0.925 + 4(0.788 + 0.876) + 2(0.830)\}\right.$$

$$+ \frac{0.00681}{3}\{0.925 + 0.75 + 4(0.887 + 0.828 + 0.786 + 0.759)$$

$$\left. + 2(0.855 + 0.804 + 0.772)\}\right] \text{ kN/mm}$$

$$= 12.73 \text{ kN/mm}.$$

(i) Since the width of the strip is 150 mm, the roll separating force is 150 × 12.73 kN = 1909 kN.

(ii) The driving torque per unit width for each roll can be computed, using equation (3.21), as

$$T = R^2\mu[\int_{\theta_n}^{\theta_i} p_{\text{before}}\, d\theta - \int_0^{\theta_n} p_{\text{after}}\, d\theta].$$

Taking the values before and after the neutral point and using Simpson's rule, we find

$$T = 4000[0.0444 - 0.0192] \text{ kN} = 100.8 \text{ kN}.$$

So, the total driving torque for each roll is 100.8 × 150 N-m = 15,120 N-m. Now, the total power required to drive the (two) rolls is 2 × 15,120 × ω W, ω being the roll speed in rad/sec. The rolling speed (same as the peripheral speed of the rolls) is given to be 30 m/min. So,

$$\omega = \frac{30{,}000}{200 \times 60} \text{ rad/sec} = 2.5 \text{ rad/sec}.$$

Thus, the total power ($2P_R$) = 75.6 kW. The student is advised to find out the total power, using equation (3.22), and to verify the result.

Power Loss in Bearings

The friction in the bearings, supporting the rolls, obviously causes some power loss. An exact analysis of the power loss in bearings is too complicated. However, to estimate the approximate power requirement of the rolling mill, it is sufficient to assume that the power loss in each bearing is given by

$$P_b = \tfrac{1}{2}\mu_b F_b d_b \omega, \tag{3.24}$$

where

μ_b = coefficient of friction in the bearing (typical value is in the range 0.002–0.01),

F_b = radial load for each bearing,

d_b = bearing diameter, and

ω = angular speed.

Assuming each roll to be supported by two bearings, we see that $F_b = F/2$. Therefore, the total power loss in the mill is

$$P_L = \mu_b F d_b \omega. \tag{3.25}$$

Hence, the total power requirement of the mill is

$$P = 2P_R + P_L. \tag{3.26}$$

EXAMPLE 3.3 If the diameter of the bearings in Examples 3.1 and 3.2 is 150 mm and the coefficient of friction (μ_b) is 0.005, estimate the required mill power.

SOLUTION Substituting the values of μ_b, d_b, ω, and F in equation (3.25), we obtain

$$P_L = 0.005 \times 1909 \times 150 \times 2.5 \text{ W} = 3.58 \text{ kW}.$$

So, the mill power is given as

$$P = (2 \times P_R + 3.58) \text{ kW} = (75.6 + 3.58) \text{ kW} = 79.18 \text{ kW}.$$

3.4.2 FORGING

In this section, our analysis is mainly devoted to determining the maximum force required for forging a strip and a disc between two parallel dies. Obviously, it is a case of open die forging.

Forging of Strip

Figure 3.11a shows a typical open die forging of a flat strip. To simplify our analysis, we shall make the following assumptions:

(i) The forging force F attains its maximum value at the end of the operation.

(ii) The coefficient of friction μ between the workpiece and the dies (platens) is constant.

(iii) The thickness of the workpiece is small as compared with its other dimensions, and the variation of the stress field along the y-direction is negligible.

(iv) The length of the strip is much more than the width and the problem is one of plane strain type.

(v) The entire workpiece is in the plastic state during the process.

At the instant shown in Fig. 3.11a, the thickness of the workpiece is h and the width is $2l$. Let us consider an element of width dx at a distance x from the origin. [In our analysis, we take the length of the workpiece as unity (in the z-direction).] Figure 3.11b shows the same element with all the stresses acting on it. Considering the equilibrium of the element in the x-direction, we get

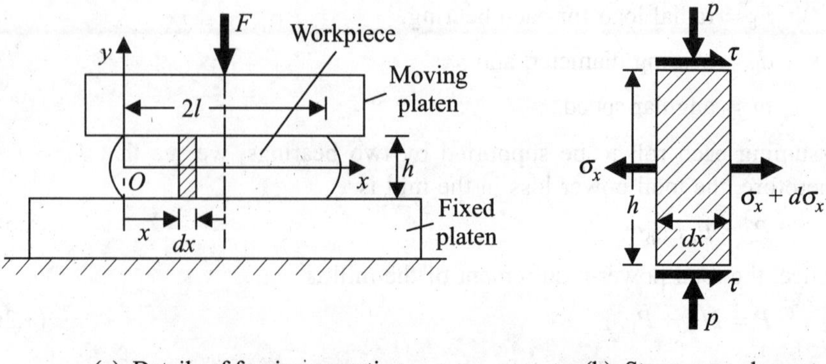

(a) Details of forging operation (b) Stresses on element

Fig. 3.11 Forces and stresses during forging.

$$h\,d\sigma_x + 2\tau\,dx = 0, \tag{3.27}$$

where τ is the frictional stress. To make the analysis simpler, $-p$ and σ_x are considered as the principal stresses[1]. The problem being of a plane strain type, equation (3.9) may be used as the yield criterion. Thus,

$$\sigma_x + p = 2K \quad \text{or} \quad d\sigma_x = -dp.$$

Substituting $d\sigma_x$ from the foregoing relation in equation (3.27), we get

$$dp = \frac{2\tau}{h}\,dx. \tag{3.28}$$

Near the free ends, i.e., when x is small (and also at $x \approx 2l$; the problem being symmetric about the midplane, we are considering only one-half in our analysis, i.e., $0 \leqslant x \leqslant l$), a sliding between the workpiece and the dies must take place to allow for the required expansion of the workpiece. However, beyond a certain value of x (in the region $0 \leqslant x \leqslant l$), say, x_s, there is no sliding between the workpiece and the dies. This is due to the increasing frictional stress which reaches the maximum value, equal to the shear yield stress, at $x = x_s$ and remains so in the rest of the zone, $x_s \leqslant x \leqslant l$. Hence, for $0 \leqslant x \leqslant x_s$,

$$\tau = \mu p \tag{3.29}$$

and, for $x_s \leqslant x \leqslant l$,

$$\tau = K. \tag{3.30}$$

For the sliding (nonsticking) zone, using equation (3.29) in equation (3.28) and integrating, we have

$$\int \frac{dp}{p} = \frac{2\mu}{h}\int dx + C_1 \qquad (0 \leqslant x \leqslant x_s)$$

or

$$\ln p = \frac{2\mu x}{h} + C_1.$$

Now, at $x = 0$, $\sigma_x = 0$, i.e., $p = 2K$ (from the yield criterion). So,

$$C_1 = \ln 2K$$

or

$$p = 2Ke^{2\mu x/h} \qquad (0 \leqslant x \leqslant x_s). \tag{3.31}$$

For the sticking zone, using equation (3.30) in equation (3.28) and integrating, we have

$$\int dp = \frac{2K}{h}\int dx + C_2 \qquad (x_s \leqslant x \leqslant l)$$

[1]However, it should be noted that this assumption is incorrect as the shear stresses τ act on the planes on which $-p$ is acting (Fig. 3.11b).

or

$$p = \frac{2Kx}{h} + C_2.$$

If $p = p_s$ at $x = x_s$, then $C_2 = p_s - 2Kx_s/h$. Thus,

$$p - p_s = \frac{2K}{h}(x - x_s). \tag{3.32}$$

Again, from equation (3.31),

$$p_s = 2K \exp(2\mu x_s/h)$$

or

$$p = 2K[\exp(2\mu x_s/h) + \frac{1}{h}(x - x_s)]. \tag{3.33}$$

At $x = x_s$, $\tau = \mu p_s = K$. Using this along with the expression for p_s, we get

$$\mu 2K \exp(2\mu x_s/h) = K$$

or

$$\frac{2\mu x_s}{h} = \ln(\frac{1}{2\mu})$$

or

$$x_s = \frac{h}{2\mu} \ln(\frac{1}{2\mu}). \tag{3.34}$$

Substituting this value of x_s in equation (3.33), we obtain

$$p = 2K[\frac{1}{2\mu}\{1 - \ln(\frac{1}{2\mu})\} + \frac{x}{h}], \qquad x_s \leqslant x \leqslant l. \tag{3.35}$$

The total forging force per unit length of the workpiece is given as

$$F = 2[\int_0^{x_s} p_1\, dx + \int_{x_s}^{l} p_2\, dx], \tag{3.36}$$

where p_1 and p_2 are the pressures given by equations (3.31) and (3.35), respectively.

EXAMPLE 3.4 A strip of lead with initial dimensions 24 mm × 24 mm × 150 mm is forged between two flat dies to a final size of 6 mm × 96 mm × 150 mm. If the coefficient of friction between the job and the dies is 0.25, determine the maximum forging force. The average yield stress of lead in tension is 7 N/mm^2.

SOLUTION First, let us determine the shear yield stress K for lead by using equation (3.6). Thus,

$$K = \frac{1}{\sqrt{3}}\sigma_Y = 4.04 \text{ N/mm}^2.$$

To use equation (3.36), the value of x_s is required. From equation (3.34),

$$x_s = \frac{6}{2 \times 0.25} \ln \left(\frac{1}{2 \times 0.25}\right) \text{ mm} = 8.3 \text{ mm}.$$

Now, from equations (3.31) and (3.35), the expressions for the pressures p_1 and p_2 (for the nonsticking and the sticking zones, respectively) can be found out. Thus,

$$p_1 = 8.08e^{0.083x} \text{ N/mm}^2 \qquad (0 \leqslant x \leqslant 8.3 \text{ mm}),$$

$$p_2 = 8.08(0.614 + 0.167x) \text{ N/mm}^2 \qquad (8.3 \text{ mm} \leqslant x \leqslant 48 \text{ mm}).$$

Using equation (3.36), the force per unit length we get is

$$F = 2\left[\int_0^{8.3} 8.08e^{0.083x}\, dx + \int_{8.3}^{48} 8.08(0.614 + 0.167x)\, dx\right] \text{ N/mm}$$

$$= 3602.5 \text{ N/mm}.$$

Since the length of the strip is 150 mm, the total forging force is 150 × 3602.5 N = 0.54×10^6 N.

EXAMPLE 3.5 Solve Example 3.4 when the coefficient of friction $\mu = 0.08$.

SOLUTION Using equation (3.34), we obtain

$$x_s = \frac{6}{0.16} \ln \frac{1}{0.16} = 68.72 \text{ mm}.$$

Since x_s is more than l, the entire zone is nonsticking, and, as a result, the expression for the pressure throughout the contact surface is given by equation (3.31). Thus,

$$p = 8.08e^{0.027x} \text{ N/mm}^2 \qquad (0 \leqslant x \leqslant 48 \text{ mm}).$$

So,

$$F = 2\int_0^{48} 8.08e^{0.027x}\, dx \text{ N/mm} = 1588.5 \text{ N/mm}.$$

The corresponding value of the total forging force is

$$150 \times 1588.5 \text{ N} = 0.238 \times 10^6 \text{ N}.$$

Forging of Disc

Figure 3.12 shows a typical open die forging of a circular disc at the end of the operation (i.e., when F is maximum) when the disc has a thickness h and a radius R. The origin of the cylindrical coordinate system r, θ, y is taken at the centre of the disc. An element of the disc, subtending an angle $d\theta$ at the centre, between the radii r and $r + dr$ is shown in Fig. 3.13 along with the stresses acting on it. In our analysis here, we make the same assumptions as in the forging of a strip, except (iv). Considering the cylindrical symmetry, it can be shown[1] that

[1]Avitzur, B., Metal Forming: Processes and Analysis, McGraw-Hill, New York, 1968.

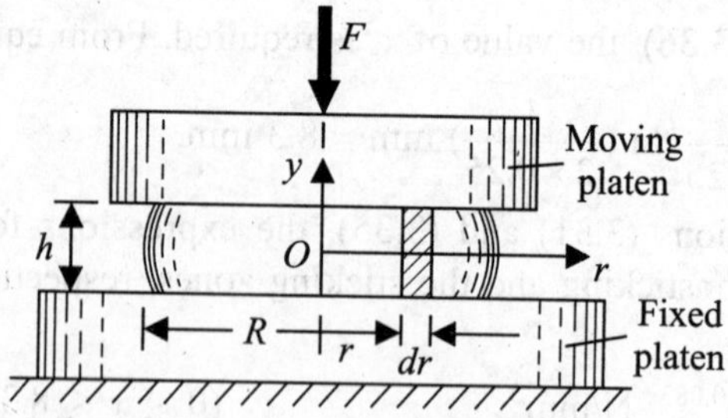

Fig. 3.12 Forging of disc.

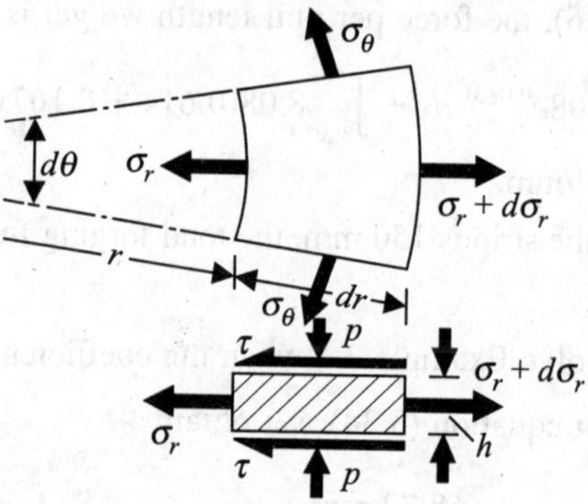

Fig. 3.13 Stresses on element during disc forging.

$\sigma_\theta = \sigma_r$ and both σ_θ and σ_r are independent of θ. Now, considering the radial equilibrium of the element, we have

$$(\sigma_r + d\sigma_r)h(r + dr)\, d\theta - \sigma_r hr\, d\theta - 2\sigma_\theta h\, dr \sin\left(\frac{d\theta}{2}\right) - 2\tau r\, d\theta\, dr = 0.$$

Neglecting the higher order terms and using $\sigma_\theta = \sigma_r$, we find the foregoing equation reduces to the form

$$h\, d\sigma_r - 2\tau\, dr = 0. \tag{3.37}$$

Again, to simplify the analysis, we take σ_r, σ_θ, and $-p$ as the principal stresses. Using equations (3.3) and (3.5) with $\sigma_1 = \sigma_r$, $\sigma_2 = \sigma_\theta\ (=\sigma_r)$, and $\sigma_3 = -p$, we obtain

$$\sigma_r + p = \sqrt{3}K \tag{3.38}$$

or

$$d\sigma_r = -dp \tag{3.39}$$

since the shear yield stress K is constant. Substituting $d\sigma_r$ in equation (3.37) from equation (3.39), we get

$$h\, dp + 2\tau\, dr = 0. \tag{3.40}$$

In this case also, beyond a certain radius, say, r_s, a sliding takes place at the interface to allow the radial expansion of the workpiece. Hence,

$$\tau = \mu p \qquad (r_s \leqslant r \leqslant R), \tag{3.41a}$$

$$\tau = K \qquad (0 \leqslant r \leqslant r_s). \tag{3.41b}$$

Thus, in these two zones, equation (3.40) takes the forms

$$\frac{dp}{p} + \frac{2\mu}{h}\,dr = 0 \qquad (r_s \leqslant r \leqslant R),$$

$$dp + \frac{2K}{h}\,dr = 0 \qquad (0 \leqslant r \leqslant r_s).$$

Integrating these two equations, we get

$$p = C_1 e^{-2\mu r/h} \qquad (r_s \leqslant r \leqslant R), \tag{3.42a}$$

$$p = C_2 - \frac{2K}{h}r \qquad (0 \leqslant r \leqslant r_s). \tag{3.42b}$$

As the periphery of the disc is free, at $r = R$, $\sigma_r = 0$. So, from equation (3.38),

$$p = \sqrt{3}K \qquad (\text{at } r = R). \tag{3.43}$$

Using equation (3.43) in equation (3.42a), we obtain

$$C_1 = \sqrt{3}Ke^{2\mu R/h}. \tag{3.44}$$

At $r = r_s$, equating the right-hand sides of equations (3.41a) and (3.41b) the value of p we obtain is K/μ. So, from equations (3.42a) and (3.44),

$$\frac{K}{\mu} = \sqrt{3}K \exp\left[\frac{2\mu}{h}(R - r_s)\right]$$

or

$$r_s = \left(R - \frac{h}{2\mu} \ln \frac{1}{\sqrt{3}\mu}\right). \tag{3.45}$$

Now, at $r = r_s$, using equation (3.42b) along with equation (3.45), we get

$$p = \frac{K}{\mu} = C_2 - \frac{2K}{h}r_s = C_2 - \frac{2K}{h}\left(R - \frac{h}{2\mu} \ln \frac{1}{\sqrt{3}\mu}\right)$$

or

$$C_2 = K\left[\frac{2R}{h} + \frac{1}{\mu}(1 + \ln \sqrt{3}\mu)\right]. \tag{3.46}$$

Finally, the expressions for the pressure in the nonsticking and the sticking zone can be written as

$$p = \sqrt{3}K \exp\left[\frac{2\mu}{h}(R - r)\right] \qquad (r_s \leqslant r \leqslant R), \tag{3.47a}$$

$$p = \frac{2K}{h}(R - r) + \frac{K}{\mu}[1 + \ln(\sqrt{3}\mu)] \qquad (0 \leqslant r \leqslant r_s). \tag{3.47b}$$

The total forging force is

$$F = 2\pi[\int_0^{r_s} p_2 r\,dr + \int_{r_s}^{R} p_1 r\,dr], \tag{3.48}$$

where p_1 and p_2 are the pressures given by equations (3.47a) and (3.47b), respectively. Or, finally,

$$F = 2\pi[[\frac{2K}{h}r^2(\frac{R}{2} - \frac{r}{3}) + \frac{Kr^2}{2\mu}\{1 + \ln(\sqrt{3}\mu)\}]_0^{r_s}$$

$$- \sqrt{3}K\frac{\exp\{\frac{2\mu}{h}(R-r)\}}{4\mu^2}h^2(1+\frac{2\mu r}{h})\Big|_{r_s}^{R}]. \tag{3.49}$$

EXAMPLE 3.6 A circular disc of lead of radius 150 mm and thickness 50 mm is reduced to a thickness of 25 mm by open die forging. If the coefficient of friction between the job and the die is 0.25, determine the maximum forging force. The average shear yield stress of lead can be taken as 4 N/mm^2.

SOLUTION Since the volume remains constant, the final radius of the disc is $R = \sqrt{2} \times 150$ mm = 212.1 mm. Now, using equation (3.45),

$$r_s = (212.1 - \frac{25}{0.5}\ln\frac{1}{\sqrt{3}\times 0.25})\text{ mm}$$

$$= (212.1 - 41.85)\text{ mm} = 170.25\text{ mm}.$$

The expressions for p_1 and p_2 are

$$p_1 = 6.93e^{0.02(212.1-r)}\text{ N/mm}^2 \qquad (170.25 \leqslant r \leqslant 212.1),$$

$$p_2 = 0.32(212.1 - r) + 22.93\text{ N/mm}^2 \qquad (0 \leqslant r \leqslant 170.25).$$

Substituting the values of K, h, R, μ, and r_s in equation (3.49), the total forging force we obtain is

$$F = 3.645 \times 10^6\text{ N}.$$

3.4.3 DRAWING

In a drawing operation, in addition to the work load and power required, the maximum possible reduction without any tearing failure of the workpiece is an important parameter. In the analysis that we give here, we shall determine these quantities. Since the drawing operation is mostly performed with rods and wires, we shall assume the workpiece to be cylindrical, as shown in Fig. 3.14. A typical drawing die consists of four regions, viz., (i) a bell-shaped entrance zone for proper guidance of the workpiece, (ii) a conical working zone, (iii) a straight and short cylindrical zone for adding stability to the operation, and (iv) a bell-shaped exit zone. The final size of the product is determined by the diameter of the stabilizing zone (d_f), the other important die dimension being

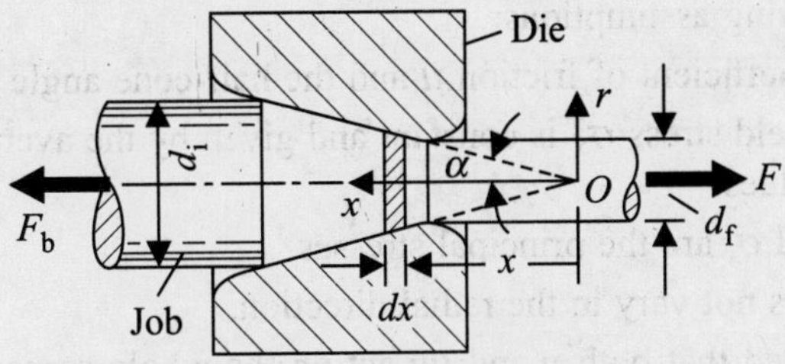

Fig. 3.14 Drawing of cylindrical rod.

the half-cone angle (α). Sometimes, a back tension F_b is provided to keep the input workpiece straight. The work load, i.e., the drawing force F, is applied on the exit side, as shown in Fig. 3.14. A die can handle jobs having a different initial diameter (d_i) which, in turn, determines the length of the job-die interface. The degree of a drawing operation (D) is normally expressed in terms of the reduction factor in the cross-sectional area. Thus,

$$D = \frac{A_i - A_f}{A_i} = \frac{d_i^2 - d_f^2}{d_i^2} = 1 - \left(\frac{d_f}{d_i}\right)^2$$

when the true strain is

$$\varepsilon = \ln\left(\frac{A_i}{A_f}\right) = \ln\left(\frac{1}{1-D}\right), \tag{3.50}$$

A_i and A_f being the initial and the final cross-sectional areas of the workpiece.

Determination of Drawing Force and Power

The case being an axisymmetric one, a cylindrical coordinate system r, x (θ being of no importance) is chosen with its origin O at the vertex of the die cone, as shown in Fig. 3.14. An element of length dx at a distance x, along with the stresses acting on it, is shown in Fig. 3.15a. To simplify our analysis, we

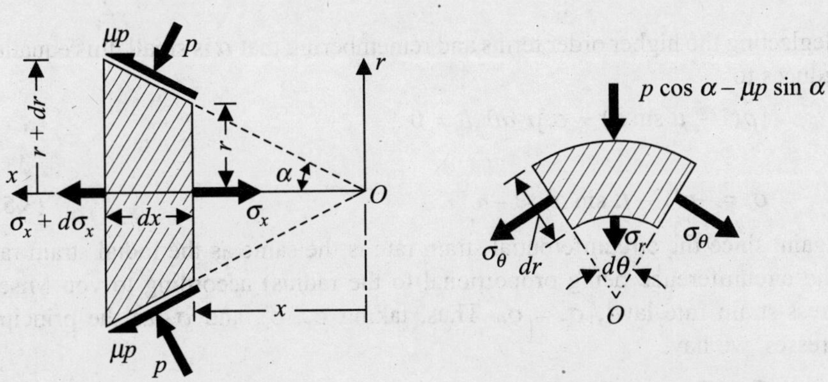

(a) Complete element (b) Segment of element

Fig. 3.15 Stresses on elements during drawing.

make the following assumptions:

(i) The coefficient of friction μ and the half-cone angle α are small.

(ii) The yield stress σ_Y is constant and given by the average of the initial and the final values.

(iii) $-p$ and σ_x are the principal stresses.

(iv) σ_x does not vary in the radial direction.

It should be noted that both p and μp act on the whole conical surface of the element. Now, from Fig. 3.15a,

$$\tan \alpha = \frac{(r+dr)-r}{dx} = \frac{dr}{dx}$$

or

$$dr = \tan \alpha \, dx. \tag{3.51}$$

Considering the equilibrium of the element in the x-direction, we have

$$(\sigma_x + d\sigma_x)\pi(r+dr)^2 - \sigma_x \pi r^2 + \mu p 2\pi r \frac{dx}{\cos \alpha} \cos \alpha$$

$$+ p 2\pi r \frac{dx}{\cos \alpha} \sin \alpha = 0.$$

Neglecting the higher order terms and using equation (3.51), we get

$$r \, d\sigma_x + 2[\sigma_x + p(1 + \frac{\mu}{\tan \alpha})] \, dr = 0, \tag{3.52}$$

where σ_x and p are related through the yield criterion. A segment of an annular element of thickness δr at the surface is shown in Fig. 3.15b. The resultant radial stress at the surface is composed of the radial components of both p and μp. Now, considering the radial equilibrium of the segment, we get

$$-(p \cos \alpha - \mu p \sin \alpha)(r + \delta r) \, d\theta \, dx - \sigma_r r \, d\theta \, dx - 2\sigma_\theta \sin \frac{d\theta}{2} \delta r \, dx = 0.$$

Neglecting the higher order terms and remembering that α is small, this equation reduces to

$$[p(1 - \mu \sin \alpha) + \sigma_r] r \, d\theta \, dx = 0$$

or

$$\sigma_r = -p(1 - \mu \sin \alpha) \approx -p. \tag{3.53}$$

Again, since the circumferential strain rate is the same as the radial strain rate (the circumference being proportional to the radius) according to von Mises' stress-strain rate law[1], $\sigma_r = \sigma_\theta$. Thus, taking σ_x, σ_r, and σ_θ as the principal stresses, we have

$$\sigma_1 = \sigma_x, \qquad \sigma_2 = \sigma_r = \sigma_\theta = \sigma_3 = -p. \tag{3.54}$$

[1] Avitzur, B., Metal Forming: Processes and Analysis, McGraw-Hill, New York, 1968.

Using the von Mises yield criterion given by equation (3.3), along with equations (3.4) and (3.54), we get

$$\sigma_x + p = \sigma_Y \quad \text{or} \quad p = \sigma_Y - \sigma_x.$$

Substituting this in equation (3.52), we obtain

$$r\,d\sigma_x + 2[\sigma_x + (\sigma_Y - \sigma_x)\phi]\,dr = 0, \tag{3.55}$$

where

$$\phi = (1 + \frac{\mu}{\tan\alpha}).$$

Therefore

$$\int \frac{dr}{r} = -\int \frac{d\sigma_x}{2[\phi\sigma_Y + (1-\phi)\sigma_x]}$$

or

$$\ln r = +\frac{1}{2(\phi-1)} \ln\,[\phi\sigma_Y - (\phi-1)\sigma_x] + C$$

when

$$r = \frac{d_i}{2}, \quad \sigma_x = \frac{F_b}{A_i}.$$

So,

$$C = \ln\frac{d_i}{2} - \frac{1}{2(\phi-1)} \ln\,[\phi\sigma_Y - (\phi-1)\frac{F_b}{A_i}].$$

Substituting C in the equation relating r and σ_x, we have

$$\ln\,(\frac{2r}{d_i}) = \frac{1}{2(\phi-1)} \ln\,[\frac{\phi\sigma_Y - (\phi-1)\sigma_x}{\phi\sigma_Y - (\phi-1)(F_b/A_i)}]$$

or

$$(\frac{2r}{d_i})^{2(\phi-1)} = \frac{\phi\sigma_Y - (\phi-1)\sigma_x}{\phi\sigma_Y - (\phi-1)(F_b/A_i)}.$$

Rearranging, we get

$$\frac{\sigma_x}{\sigma_Y} = \frac{F_b}{\sigma_Y A_i}(\frac{2r}{d_i})^{2(\phi-1)} - (\frac{\phi}{\phi-1})[(\frac{2r}{d_i})^{2(\phi-1)} - 1].$$

Hence, the drawing stress σ_{x_f} at $r = d_f/2$ is given by

$$\frac{\sigma_{x_f}}{\sigma_Y} = \frac{F_b}{\sigma_Y A_i}(\frac{d_f}{d_i})^{2(\phi-1)} + (\frac{\phi}{\phi-1})[1 - (\frac{d_f}{d_i})^{2(\phi-1)}]. \tag{3.56}$$

Finally, the drawing force is

$$F = \sigma_{x_f} A_f, \tag{3.57}$$

where σ_{x_f} is obtained from equation (3.56). If the drawing speed, i.e., the exit velocity, is V, the power required for the drawing operation is

$$P = FV. \tag{3.58}$$

Determination of Maximum Allowable Reduction

The maximum allowable reduction or the maximum degree of a drawing operation (D) is determined from the constraint that the pulling stress σ_{x_f} cannot be more than the tensile yield stress of the work material[1]. So, at this limiting condition,

$$\frac{\sigma_{x_f}}{\sigma_Y} = 1.$$

Using this in equation (3.56), the relation we obtain is

$$D_{max} = 1 - \frac{1}{\left[\phi - \dfrac{F_b(\phi - 1)}{\sigma_Y A_i}\right]^{1/(\phi-1)}}. \tag{3.59a}$$

EXAMPLE 3.7 A steel wire is drawn from an initial diameter of 12.7 mm to a final diameter of 10.2 mm at a speed of 90 m/min. The half-cone angle of the die is 6° and the coefficient of friction at the job-die interface is 0.1. A tensile test on the original steel specimen gives a tensile yield stress 207 N/mm². A similar specimen shows a tensile yield stress of 414 N/mm² at a strain of 0.5. Assuming a linear stress-strain relationship for the material, determine the drawing power and the maximum possible reduction with the same die. No back tension is applied.

SOLUTION To start with, we determine the average tensile yield stress of the material during the process. For the given operation, the strain is

$$\varepsilon = \ln\frac{A_i}{A_f} = 2\ln\frac{d_i}{d_f} = 2\ln\frac{12.7}{10.2} = 0.438.$$

Since the stress-strain relationship is linear, the value of the tensile yield stress at the end of the operation is given by

$$\sigma_{Y_f} = \left(207 + \frac{414 - 207}{0.5} \times 0.438\right)\ \text{N/mm}^2 = 388\ \text{N/mm}^2.$$

So, the average yield stress is

[1]For strain hardening a material, the value of σ_{x_f}, strictly speaking, can go up to the value of the tensile yield stress at the end of the operation (σ_{Y_f}) which may be considerably greater than the average value σ_Y. So, when $(\sigma_{x_f})_{max}$ is put equal to σ_{Y_f},

$$D_{max} = 1 - \left[\frac{1 - \sigma_Y(\phi - 1)/(\sigma_Y \phi)}{1 - F_b(\phi - 1)/(\sigma_Y A_i \phi)}\right]^{1/(\phi-1)}. \tag{3.59b}$$

When $\sigma_{Y_f}(\phi - 1)/(\sigma_Y \phi)$ approaches 1, there is no limit to the reduction.

$$\sigma_Y = \frac{207 + 388}{2} \text{ N/mm}^2 = 297.5 \text{ N/mm}^2.$$

From the given data,

$$\phi = 1 + \frac{\mu}{\tan \alpha} = 1 + \frac{0.1}{\tan 6°} \approx 2.$$

Now, from equation (3.56) with $F_b = 0$,

$$\sigma_{x_f} = 297.5 \times 2[1 - (\frac{10.2}{12.7})^2] \text{ N/mm}^2 = 211.2 \text{ N/mm}^2.$$

Using equations (3.57) and (3.58), the drawing power we obtain is

$$P = 211.2 \times \frac{\pi}{4} \times (10.2)^2 \times \frac{90}{60} \text{ W} = 25.887 \text{ kW}.$$

It is interesting to note that the drawing stress is more than the tensile yield stress of the original material. To find out the maximum allowable reduction, we use equation (3.59) with $F_b = 0$ and $\phi = 2$. Hence, $D_{max} = 1 - \frac{1}{2} = 0.5$. If we use equation (3.59a), then $D_{max} = 0.652$.

3.4.4 DEEP DRAWING

From the point of view of analysis, the process of deep drawing is very complex. In this process, various types of forces operate simultaneously. The annular portion of the sheet metal workpiece (see Fig. 3.4) between the blank holder and the die is subjected to a pure radial drawing, whereas the portions of the workpiece around the corners of the punch and the die are subjected to a bending operation. Further, the portion of the job between the punch and the die walls undergoes a longitudinal drawing. Though in this operation varying amount of thickening and thinning of the workpiece is unavoidable, we shall not take this into consideration in our analysis.

The major objectives of our analysis are (i) to correlate the initial and final dimensions of the job, and (ii) to estimate the drawing force F. Figure 3.16 shows the drawing operation with the important dimensions. The radii of the

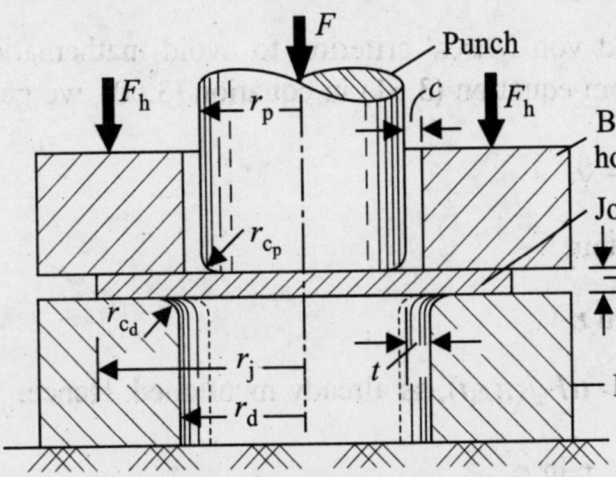

Fig. 3.16 Deep drawing.

punch, the job, and the die are r_p, r_j, and r_d, respectively. Obviously, without taking the thickening and thinning into account, the clearance between the die and the punch $(r_d - r_p)$ is equal to the job thickness t. The corners of the punch and the die are provided with radii r_{c_p} and r_{c_d}, respectively. A clearance (c) is maintained between the punch and the blank holder.

To start with, let us consider the portion of the job between the blank holder and the die. Figure 3.17a shows the stresses acting on an element in this region. It should be noted that the maximum thickening (due to the decreasing circumference of the job causing a compressive hoop stress) takes place at the outer periphery, generating a line contact between the holder and the job. As a result, the entire blank holder force F_h is assumed to act along the circumference (Fig. 3.17b). Thus, the radial stress due to friction can also be represented by an equivalent radial stress $2\mu F_h/(2\pi r_j t)$ at the outer periphery.

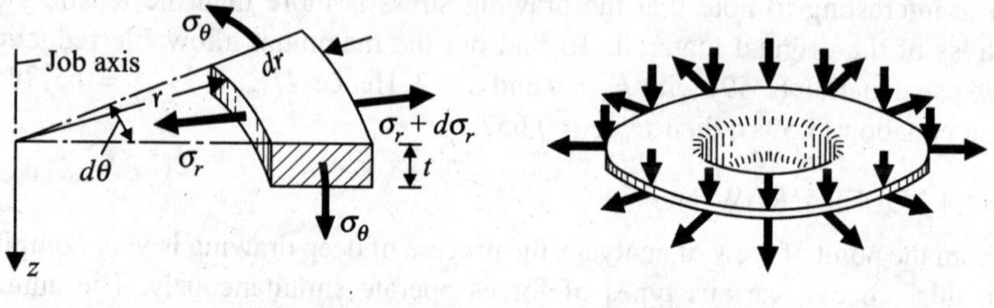

(a) Stresses acting on element during drawing

(b) Radial stress due to blank holding pressure

Fig. 3.17 Analysis of deep drawing operation.

Now, considering the radial equilibrium of the element (shown in Fig. 3.17a), we get

$$r\, d\sigma_r + \sigma_r\, dr - \sigma_\theta\, dr = 0. \tag{3.60}$$

As σ_r and σ_θ are the principal stresses, the equation we obtain by using Tresca's yield criterion is

$$(\sigma_r - \sigma_\theta) = 2K. \tag{3.61}$$

(We have not used von Mises' criterion to avoid mathematical complexity.) Substituting σ_θ from equation (3.61) in equation (3.60), we get

$$\frac{dr}{r} + \frac{d\sigma_r}{2K} = 0.$$

Integrating, we obtain

$$\frac{\sigma_r}{2K} = C - \ln r.$$

Now, at $r = r_j$, $\sigma_r = \mu F_h/(\pi r_j t)$, as already mentioned. Hence,

$$C = \frac{\mu F_h}{2\pi K r_j t} + \ln r_j.$$

Using this in the expression for σ_r, we have

$$\frac{\sigma_r}{2K} = \frac{\mu F_h}{2\pi K r_j t} + \ln \frac{r_j}{r}. \tag{3.62}$$

So, the radial stress at the beginning of the die corner (i.e., at $r = r_d = r_p + t$) is given by

$$\left.\frac{\sigma_r}{2K}\right|_{r=r_d} = \frac{\mu F_h}{2\pi K r_j t} + \ln \left(\frac{r_j}{r_d}\right). \tag{3.63}$$

As the job slides along the die corner, the radial stress, given by equation (3.63), increases to σ_z due to the frictional forces, as shown in Fig. 3.18. This increment

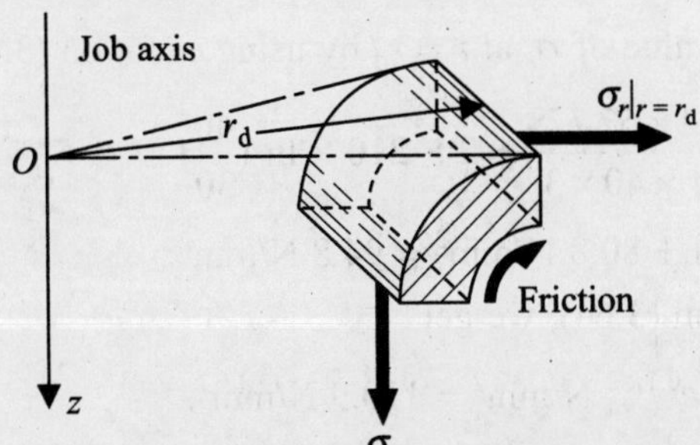

Fig. 3.18 Effect of friction at corners.

can be roughly estimated by using a belt-pulley analogy. Thus,

$$\frac{\sigma_z}{\sigma_r|_{r=r_d}} = e^{\mu\pi/2}, \tag{3.64}$$

where μ is the coefficient of friction between the workpiece and the die.

There is a further increase in the stress level around the punch corner due to bending. As a result, the drawn cup normally tears around this region. However, to avoid this, an estimate of the maximum permissible value of (r_j/r_d) can be obtained by using equations (3.64) and (3.63) with σ_z equal to the maximum allowable stress of the material. Since r_d is the final outside diameter of the product, it is easy to arrive at such an estimate. This estimate is based on the consideration of fracture of the material. However, to avoid buckling (due to the compressive hoop stress in the flange region), $(r_j - r_p)$ should not, for most materials, exceed $4t$.

Normally, the blank holder force is given as

$$F_h = \beta \pi r_j^2 K, \tag{3.65}$$

where β is between 0.02 and 0.08. An estimate of the drawing force F (neglecting the friction between the job and the die wall) can easily be obtained, using equation (3.64), as

$$F \approx \sigma_z 2\pi r_p t. \tag{3.66}$$

EXAMPLE 3.8 A cold rolled steel cup with an inside radius 30 mm and a thickness 3 mm is to be drawn from a blank of radius 40 mm. The shear yield stress and the maximum allowable stress of the material can be taken as 210 N/mm^2 and 600 N/mm^2, respectively.

(i) Determine the drawing force, assuming that the coefficient of friction $\mu = 0.1$ and $\beta = 0.05$.

(ii) Determine the minimum possible radius of the cup which can be drawn from the given blank without causing a fracture.

SOLUTION (i) We first calculate the blank holding force F_h from the given data as

$$F_h = 0.05\pi 40^2 \times 210 \text{ N} = 52{,}778 \text{ N}.$$

Next, we find the value of σ_r at $r = r_d$ by using equation (3.63). Thus,

$$\sigma_r|_{r=r_d} = \frac{0.1 \times 52{,}778}{\pi \times 40 \times 3} + 2 \times 210 \times \ln\left(\frac{40}{30}\right)$$

$$= 14 + 80.8 \text{ N/mm}^2 = 94.8 \text{ N/mm}^2.$$

Now, using equation (3.64), we get

$$\sigma_z = 94.8 \times e^{0.1\pi/2} \text{ N/mm}^2 = 110.9 \text{ N/mm}^2.$$

It should be noted that this is much less than the fracture strength though $r_j - r_p = 10 \text{ mm} = 3.33t$, i.e., very close to the limit set by the condition of plastic buckling. From equation (3.66), the drawing force is given as

$$F = 2\pi \times 30 \times 3 \times 110.9 \text{ N} = 62{,}680 \text{ N}.$$

(ii) In this case, $\sigma_z = 600$ N/mm^2. From equation (3.64), we have

$$\sigma_r|_{r=r_d} = \frac{600}{e^{0.05\pi}} \text{ N/mm}^2 = 512.8 \text{ N/mm}^2.$$

Using equation (3.63) and taking $F_h = 52{,}778$ N, we get

$$\ln\left(\frac{40}{r_p + 3}\right) = \frac{512.8}{420} - \frac{0.1 \times 52{,}778}{2\pi \times 210 \times 40 \times 3} = 1.221 - 0.033 = 1.188$$

or

$$r_p = \frac{40}{e^{1.188}} - 3 \text{ mm} = 9.2 \text{ mm}.$$

Again, it is interesting to note that $r_j - r_p = 30.8 \text{ mm} \gg 4t$, and this goes much beyond the limit set by the plastic buckling condition.

3.4.5 BENDING

In a bending operation, apart from the determination of work load, an estimate of the amount of elastic recovery (spring back) is essential. When the final shape is prescribed, a suitable amount of overbending is required to take care of this spring back. In this section, we shall work out these quantities and also illustrate how the stock size for a given job is computed.

Figure 3.19 shows a bending operation with characteristic dimensions. A radius r_p is provided at the nose of the punch and, accordingly, the die centre

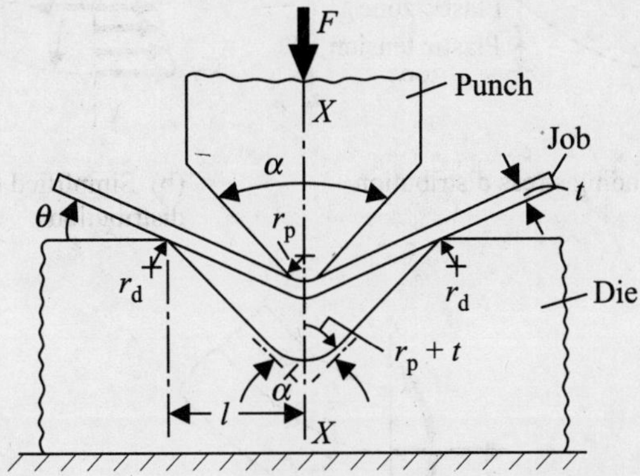

Fig. 3.19 Details in bending.

has a radius (r_p + t), where t is the job thickness. The portions of the die, in contact with the job during the operation, are also provided with some radius, say, r_d. The angle between the two faces of the punch and the die is α. At the instant shown, the angle between the two bent surfaces of the job is $(\pi - 2\theta)$. As we shall subsequently show, the bending force F is maximum at some intermediate stage, depending on the frictional characteristics. The degree of a bending operation is normally specified in terms of the strain in the outer fibre. The width of the job w (in the direction perpendicular to the plane of the paper) is much larger as compared with t, and hence a plane strain condition can be assumed. It is obvious that the stock length should be calculated on the basis of the length of the neutral plane of the job. Since the radius of curvature involved in a bending operation is normally small, the neutral plane shifts towards the centre of curvature. Usually, a shift of 5–10% of the thickness is assumed for the calculation of strain and stock length. Thus, the strain in the outer fibre of the bend is given as

$$\varepsilon_{max} = \ln \left[1 + \frac{(r_p + t) - (r_p + 0.45t)}{(r_p + 0.45t)}\right] = \ln \left[1 + \frac{1}{1.82(r_p/t) + 0.82}\right], \qquad (3.67)$$

assuming a 5% shift of the neutral plane. Depending on the ductility of the job material, ε_{max} has a limiting value beyond which a fracture takes place. So, from equation (3.67) and the limiting value of ε_{max} (=$\varepsilon_{fracture}$), we can determine the smallest punch radius for a given job thickness.

Determination of Work Load

Since the job undergoes plastic bending, the stress distribution at the cross-section along the centre line (XX) is as shown in Fig. 3.20a. This

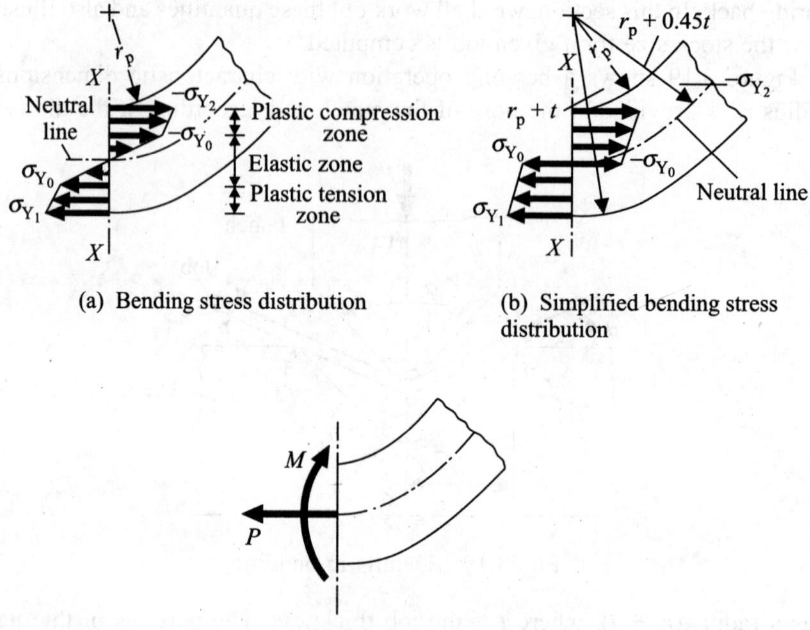

(a) Bending stress distribution

(b) Simplified bending stress distribution

(c) Forces and moments in bending

Fig. 3.20 Mechanics of bending.

distribution is obtained by neglecting all other effects of curvature except the shift of neutral line. It is obvious that in the zone on either side of the neutral plane the strain level is within the elastic range. When the strain (both in the tensile and the compressive zones) reaches the yield limit, plastic deformation starts. Assuming the yield stress to be σ_{Y_0} (same in both tension and compression) and linear strain hardening, the stress distribution will be as shown in the figure. The magnitude of σ_{Y_1} and σ_{Y_2} is different due to the shift of the neutral plane. For the sake of simplicity, the stress distribution for large plastic bending is idealized as shown in Fig. 3.20b. When the strain hardening rate is n, then

$$\sigma_{Y_1} = \sigma_{Y_0} + n\varepsilon_{max}, \tag{3.68a}$$

$$\sigma_{Y_2} = \sigma_{Y_0} + n \ln \left[1 + \frac{(r_p + 0.45t) - r_p}{r_p + 0.45t}\right]$$

$$= \sigma_{Y_0} + n \ln \left[1 + \frac{1}{2.22(r_p/t) + 1}\right]. \tag{3.68b}$$

The loading due to this stress distribution can be represented by a bending moment M and a force P (per unit width of the job), shown in Fig. 3.20c, expressed as

$$M = (0.55t)^2\left(\frac{\sigma_{Y_0}}{6} + \frac{\sigma_{Y_1}}{3}\right) + (0.45t)^2\left(\frac{\sigma_{Y_0}}{6} + \frac{\sigma_{Y_2}}{3}\right), \tag{3.69}$$

$$P = \frac{t}{2}[0.1\sigma_{Y_0} + 0.55\sigma_{Y_1} - 0.45\sigma_{Y_2}]. \tag{3.70}$$

Now, let us consider the right half of the job (of unit width) and the forces and moments acting on it (see Fig. 3.21). Since P arises from the shift of the

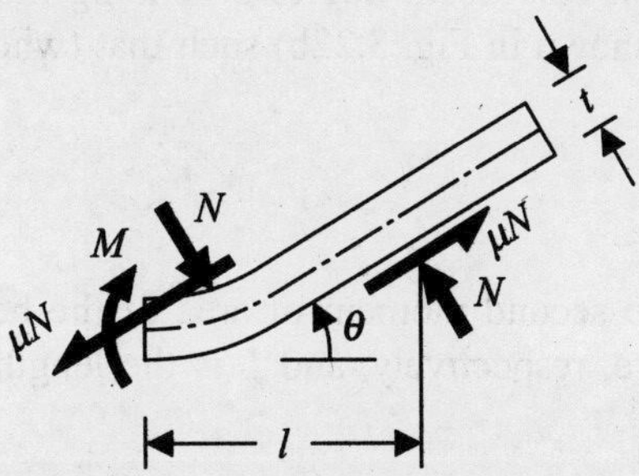

Fig. 3.21 Free body diagram of half the job.

neutral plane which is very small, it can be neglected in comparison with the other forces. The normal and frictional forces exerted by the die and the punch at their contact lines (since r_p is small as compared with the other dimensions, the finite contact between the job and the punch can be idealized as a line) are N and μN, respectively. As t is small, the moment due to μN is negligible. Hence, $M = Nl/\cos\theta$. One-half of the bending force per unit width is given as

$$\frac{F}{2} = N\cos\theta + \mu N\sin\theta \quad \text{or} \quad F = 2N(\cos\theta + \mu\sin\theta).$$

Substituting N in terms of M, we obtain

$$F = \frac{2M}{l}(\cos^2\theta + \mu\sin\theta\cos\theta). \tag{3.71}$$

Now, differentiating F with respect to θ, we get

$$\frac{dF}{d\theta} = \frac{2M}{l}(-\sin 2\theta + \mu\cos 2\theta)$$

since M is independent of θ. It is obvious that F reaches a maximum when

$$\theta = \theta_{cr} = \tfrac{1}{2}\tan^{-1}\mu. \tag{3.72}$$

Thus, the maximum work load per unit width is given as

$$F_{max} = \frac{M}{l}[1 + \cos(\tan^{-1}\mu) + \mu\sin(\tan^{-1}\mu)]. \tag{3.73}$$

Estimation of Spring Back

Figure 3.22a shows the stress-strain characteristics of a linearly strain hardened material. When the material is unloaded from the point A, the path of unloading is given by the line AB, as shown. The amount of recovered strain obviously is

$$\varepsilon_{rec} = BC = \frac{\sigma_A}{E}.$$

So, the amount of bending strain recovered is given by the elastic bending strain resulting from a bending moment M which is removed when the operation is over. The elastic strain of a beam due to a bending moment M results in an included angle ϕ (as shown in Fig. 3.22b) such that (when the curvature effect is neglected)

$$\frac{M}{I} = \frac{E}{R} = \frac{E\phi}{L},$$

where I and R are the second moment of area of the beam cross-section and the radius of curvature, respectively, and L is the length of the neutral plane.

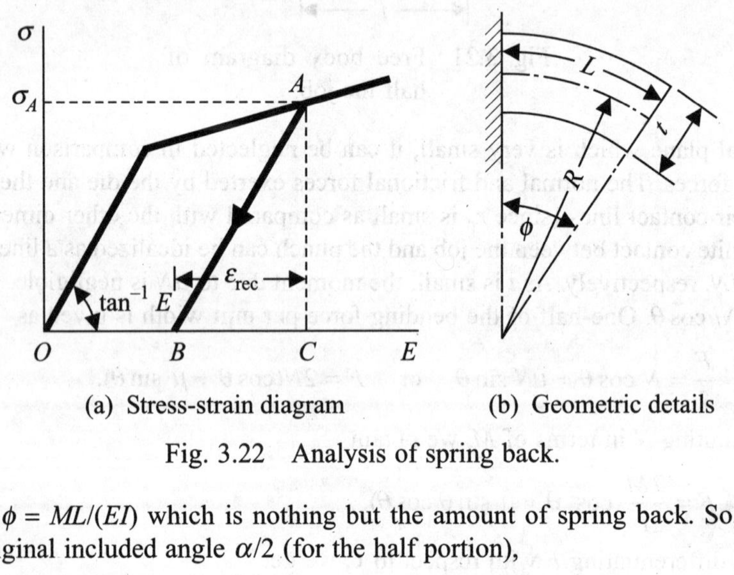

(a) Stress-strain diagram (b) Geometric details

Fig. 3.22 Analysis of spring back.

Thus, $\phi = ML/(EI)$ which is nothing but the amount of spring back. So, with the original included angle $\alpha/2$ (for the half portion),

$$\frac{2\phi}{\alpha} = \frac{2ML}{EI\alpha}.$$

Since (with a 5% shift in the neutral axis)

$$\frac{\alpha}{2} = \frac{L}{(r_p + 0.45t)},$$

we have

$$\frac{2\phi}{\alpha} = \frac{M(r_p + 0.45t)}{EI}. \tag{3.74}$$

Now, for a rectangular cross-section with unit width,

$$I = \tfrac{1}{12}t^3.$$

Using this in equation (3.74), the total spring back we finally get is

$$\frac{2\phi}{\alpha} = \frac{12(r_p + 0.45t)M}{Et^3}. \tag{3.75}$$

Hence, the required punch angle should be $(\alpha - 2\phi)$ to produce a bend with the included angle α.

EXAMPLE 3.9 For the mild steel product shown in Fig. 3.23, what is the minimum possible corner radius r if the fracture strain $\varepsilon_{frac} = 0.2$ for the given material? Using the data $E = 207$ kN/mm², $\mu = 0.1$, $l = 25$ mm, $\sigma_{Y_0} = 345$ N/mm², $n = 517$ N/mm², and considering r to be minimum, determine (i) the maximum bending force, (ii) the required punch angle, and (iii) the stock length.

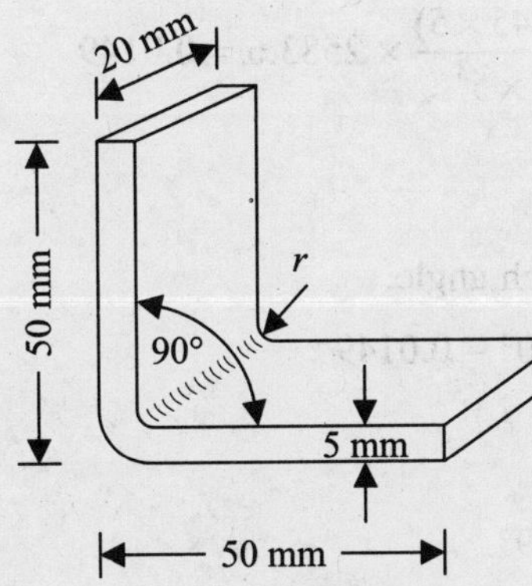

Fig. 3.23 90° mild steel angle.

SOLUTION To determine the minimum possible radius of the corner, we use equation (3.67). Thus,

$$0.2 = \ln\left[1 + \frac{1}{1.82(r_{P_{min}}/5) + 0.82}\right]$$

or

$$r_{P_{min}} = 10.16 \text{ mm} \approx 10.2 \text{ mm}.$$

Now onwards, we will consider $r_p = r_{P_{min}} = 10.2$ mm.

(i) To determine σ_{Y_1} and σ_{Y_2}, we use equations (3.68a) and (3.68b). Substituting the given values, we get

$$\sigma_{Y_1} = (345 + 517 \times 0.2) \text{ N/mm}^2 = 448.4 \text{ N/mm}^2,$$

$$\sigma_{Y_2} = \left[345 + 517 \ln\left(1 + \frac{1}{\frac{2.22 \times 10.2}{5} + 1}\right)\right] \text{ N/mm}^2 = 431.2 \text{ N/mm}^2.$$

With these values, equation (3.69) gives the bending moment M per mm width as

$$M = (0.55 \times 5)^2\left(\frac{345}{6} + \frac{448.4}{3}\right) + (0.45 \times 5)^2\left(\frac{345}{6} + \frac{431.2}{3}\right) \text{ N}$$

$$= 2583.6 \text{ N.}$$

So, the maximum bending force per mm width of the job is obtained from equation (3.73) with this value of *M*. Hence,

$$F_{max} = \frac{2583.6}{25}(1 + 0.995 + 0.1 \times 0.0995) \text{ N/mm} = 207.2 \text{ N/mm.}$$

Therefore, the actual maximum bending force will be

$$F_{max} \times 20 \text{ N} = 207.2 \times 20 \text{ N} = 4144 \text{ N.}$$

(ii) To find out the punch angle in order to obtain the 90° bend, we must determine the spring back angle 2ϕ. From equation (3.75),

$$\frac{2\phi}{\alpha} = \frac{12(10.2 + 0.45 \times 5)}{207 \times 10^3 \times 5^3} \times 2583.6 = 0.0149$$

or

$$2\phi = 0.0149\alpha.$$

Now, since α is the punch angle,

$$\alpha = 90° - 2\phi = 90° - 0.0149\alpha$$

or

$$1.0149\alpha = 90°.$$

Finally, we get $\alpha = 88.69°$.

(iii) It is easy to show that the total length of the neutral plane, which is equal to the required stock length, is

$$L_s = 2[(r_p + 0.45t)\frac{\pi}{4} + 50 - (r_p + t)] \text{ mm} = 89.2 \text{ mm.}$$

EXAMPLE 3.10 If the job described in Example 3.9 has to be produced in a machine where the maximum available force is 3000 N, what should be the minimum value of the die length, $2l$? Also, what should be the minimum capacity of the machine for a successful production of the same job?

SOLUTION It is clear from equation (3.73) that, for a given job, the product $F_{max}l$ is constant. From Example 3.9,

$$F_{max}l = 4144 \times 25 \text{ N-mm}$$

$$= 103{,}600 \text{ N-mm.}$$

Since here F_{max} is limited to 3000 N, the minimum possible value of l is given by

$$l_{min} = \frac{103{,}600}{3000} \text{ mm} = 34.53 \text{ mm}$$

or

$$2l_{min} = 69.1 \text{ mm.}$$

To find out the minimum capacity of the machine, we must determine the maximum possible value of l. It is obvious that when $2l$ exceeds the stock length L, the operation will not be satisfactory. Thus,

$$2l_{max} = L = 89.2 \text{ mm} \quad \text{or} \quad l_{max} = 44.6 \text{ mm}.$$

Hence, the minimum required capacity of the machine is

$$(F_{max})_{min} = \frac{103{,}600}{44.6} \text{ N} = 2323 \text{ N}.$$

3.4.6 EXTRUSION

The basic nature of the deformation in extrusion is, to some extent, similar to that in drawing. Here instead of applying a tensile load at the exit end, a compressive load is applied at the other end. However, a number of complexities arise as the die is commonly flat-face (i.e., the equivalent half-cone angle is very large unlike in the drawing die). Consequently, with the same assumptions as in drawing, the results become highly inaccurate. In our analysis here, we shall determine the work load and the frictional power loss for a simple forward extrusion with a flat-face die. For doing this, we shall use two approaches; of these, one is in line with that used for drawing, whereas the other is based on the energy consideration. Since both involve rather drastic assumptions, we shall compare the results obtained from the two approaches.

Determination of Work Load from Stress Analysis

With a flat-face die and high friction between the material and the container wall, a dead zone, shown in Fig. 3.24a, develops where no flow of material takes place. We assume that the dead zone can be approximated by a half-cone angle of 45°. The material undergoing deformation can be divided into two regions, namely, (i) section AA to BB, where the flow of material is considered as a rigid body motion, and (ii) section BB to CC, where the flow is analogous to that in a drawing operation (of course, with a compressive load). Figure 3.24b shows an element in the region BB-CC along with the stresses acting on it. Comparing

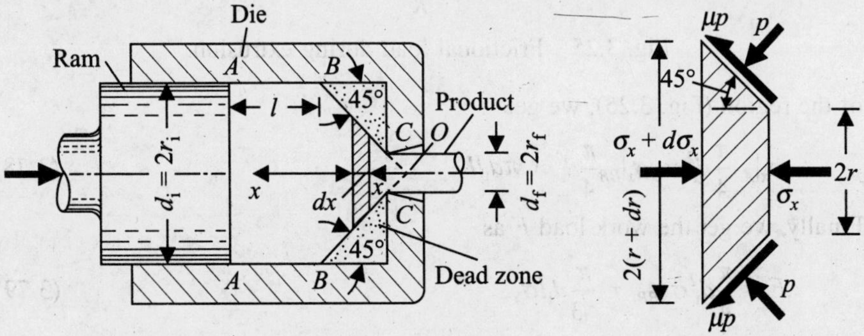

(a) Details of extrusion process (b) Stresses on element

Fig. 3.24 Analysis of extrusion process.

Fig. 3.24b with Fig. 3.15a, the similarity between extrusion and drawing is easily discernible. The only change here is that σ_x is compressive. Therefore, following the same analysis as in drawing, equation (3.56) can be rewritten as[1]

$$\frac{-\sigma_{x_f}}{\sigma_Y} = \frac{-F_b}{\sigma_Y A_i}\left(\frac{d_f}{d_i}\right)^{2(\phi-1)} + \frac{\phi}{\phi-1}\left[1-\left(\frac{d_f}{d_i}\right)^{2(\phi-1)}\right],$$

where $\sigma_{x_f} = 0$ and $-F_b/A_i$ is nothing but the compressive stress at the section BB. Thus, considering the compressive nature here, we have

$$\frac{\sigma_x|_{BB}}{\sigma_Y} = \frac{\phi}{\phi-1}\left[\left(\frac{d_i}{d_f}\right)^{2(\phi-1)} - 1\right], \tag{3.76}$$

where $\phi = 1 + \mu$ (because $\tan\alpha = \tan 45° = 1$).

Let us now consider the stresses acting on the boundaries of the region between the sections AA and BB. The frictional stress at the container wall is assumed as the shear yield stress K, i.e., $\mu p = K$. At the section BB, the value of p is given by $p = \sigma_x|_{BB} + \sigma_Y$ (considering von Mises' yield criterion). Hence [using equation (3.76) and taking $K = \sigma_Y/\sqrt{3}$],

$$\mu = \left(\frac{K}{\sigma_x|_{BB} + \sigma_Y}\right) = \frac{1/\sqrt{3}}{(1+1/\mu)[(d_i/d_f)^{2\mu} - 1] + 1} \tag{3.77}$$

which can be solved to obtain the value of μ. Considering the axial equilibrium

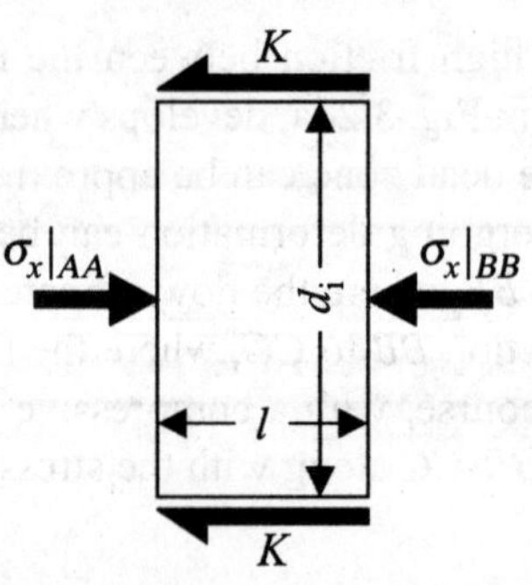

Fig. 3.25 Frictional load during extrusion.

of the region (Fig. 3.25), we get

$$\sigma_x|_{AA}\frac{\pi}{4}d_i^2 = \sigma_x|_{BB}\frac{\pi}{4}d_i^2 + \pi d_i l K. \tag{3.78}$$

Finally, we get the work load F as

$$F = \frac{\pi}{4}d_i^2\sigma_x|_{BB} + \frac{\pi}{\sqrt{3}}d_i l \sigma_Y. \tag{3.79}$$

[1]It should be noted that equation (3.56) was derived assuming that the half-cone angle α is small, whereas here we have taken $\alpha = 45°$. This introduces some inaccuracy in the analysis but the solution becomes unwieldy without this assumption.

Determination of Work Load from Energy Consideration

The mode of material deformation here is assumed to be the same as that in the foregoing analysis, the only exception being that the frictional stress between the dead zone and the conical work surface is taken as the shear yield stress K. Assume the axial velocity at the section BB as V_B, which is the same as the ram velocity. Hence, the rate of energy input to the conical working zone across the section BB is

$$\dot{W}_{BB} = \sigma_x|_{BB}\frac{\pi}{4}d_\mathrm{i}^2V_B. \tag{3.80}$$

The frictional power loss at the surface of the element shown in Fig. 3.24b (with $\mu p = K$) is given as

$$d\dot{W}_\mathrm{f} = K\pi d\sqrt{2}\ dr\ \sqrt{2}V,$$

where V is the axial velocity across the element and is given by the continuity equation

$$V\frac{\pi}{4}d^2 = V_B\frac{\pi}{4}d_\mathrm{i}^2.$$

Using this, we get

$$d\dot{W}_\mathrm{f} = \pi K d_\mathrm{i}^2 V_B\frac{dr}{r} \qquad \text{(because } d = 2r\text{)}.$$

Integrating over the conical working zone, we obtain

$$\dot{W}_\mathrm{f} = \pi K d_\mathrm{i}^2 V_B \ln\left(\frac{d_\mathrm{i}}{d_\mathrm{f}}\right) = \frac{\pi}{\sqrt{3}}\sigma_\mathrm{Y} d_\mathrm{i}^2 V_B \ln\left(\frac{d_\mathrm{i}}{d_\mathrm{f}}\right). \tag{3.81}$$

Now, we shall determine the rate of energy expenditure in the plastic deformation process in the conical zone. It is easy to see that during a deformation process the work done per unit volume is

$$W_\mathrm{d} = \int_0^\varepsilon \sigma\, d\varepsilon.$$

The change in the cross-sectional area from BB to CC is related to the total strain as

$$\varepsilon = \ln\left(\frac{A_\mathrm{i}}{A_\mathrm{f}}\right) = 2\ln\left(\frac{d_\mathrm{i}}{d_\mathrm{f}}\right),$$

and the average value of σ during this plastic deformation process can be assumed to be σ_Y. Hence,

$$W_\mathrm{d} = \int_0^{2\ln(d_\mathrm{i}/d_\mathrm{f})} \sigma_\mathrm{Y}\, d\varepsilon = 2\sigma_\mathrm{Y}\ln\left(\frac{d_\mathrm{i}}{d_\mathrm{f}}\right).$$

Now, the volume of the material undergoing deformation per unit time is $V_B(\pi/4)d_\mathrm{i}^2$. So, the total rate of energy expenditure in the deformation process is

$$\dot{W}_p = V_B \frac{\pi}{4} d_i^2 W_d = \frac{\pi}{2} \sigma_Y V_B d_i^2 \ln\left(\frac{d_i}{d_f}\right). \tag{3.82}$$

Equating the input power with the power used in plastic deformation and overcoming the friction, we get

$$\sigma_x|_{BB} \frac{\pi}{4} d_i^2 V_B = \frac{\pi}{\sqrt{3}} \sigma_Y d_i^2 V_B \ln\left(\frac{d_i}{d_f}\right) + \frac{\pi}{2} \sigma_Y d_i^2 V_B \ln\left(\frac{d_i}{d_f}\right)$$

or

$$\frac{\sigma_x|_{BB}}{\sigma_Y} = 4.31 \ln\left(\frac{d_i}{d_f}\right). \tag{3.83}$$

Using equation (3.83) instead of equation (3.76) in equation (3.79), we can find out the work load required for the extrusion process.

Determination of Frictional Power Loss

The total frictional power loss can be found out by summing up the frictional power losses in the conical and the cylindrical regions of the work material. The contribution from the second region can be expressed as

$$P_{f_2} = \pi d_i l K V_B = \frac{\pi}{\sqrt{3}} \sigma_Y d_i V_B l. \tag{3.84}$$

To determine the loss in the first region, we can use either of the two approaches followed in computing the work load. According to the first of these approaches, the frictional power loss in the conical region is given as

$$P_{f_1} = \int_{r_f}^{r_i} \sqrt{2}\, \pi d \mu p \sqrt{2} V \, dr,$$

where

$$V = V_B \left(\frac{d_i}{d}\right)^2.$$

The yield criterion can be written as $p = \sigma_x + \sigma_Y$, where σ_x is expressed, using equation (3.76) with $\phi = 1 + \mu$, as

$$\sigma_x = \sigma_Y \left(\frac{1+\mu}{\mu}\right)\left[\left(\frac{d}{d_f}\right)^{2\mu} - 1\right].$$

The value of μ is obtained by solving equation (3.77). Making use of the foregoing relations in the expression for P_{f_1}, we get

$$P_{f_1} = \int_{r_f}^{r_i} 2\pi V_B d_i^2 \mu \frac{\sigma_Y}{d} \left[1 + \frac{1+\mu}{\mu}\left\{\left(\frac{d}{d_f}\right)^{2\mu} - 1\right\}\right] dr$$

$$= 2\pi\mu\sigma_Y V_B d_i^2 \int_{r_f}^{r_i} \frac{1}{2r}\left[\frac{1+\mu}{\mu}\left(\frac{r}{r_f}\right)^{2\mu} - \frac{1}{\mu}\right] dr.$$

Finally,

$$P_{f_1} = \pi\sigma_Y V_B d_i^2\left[\frac{1+\mu}{2\mu}\left\{\left(\frac{d_i}{d_f}\right)^{2\mu} - 1\right\} - \ln\left(\frac{d_i}{d_f}\right)\right]. \tag{3.85}$$

So, the total power loss in friction is

$$P_f = P_{f_1} + P_{f_2}. \tag{3.86}$$

The computation of P_{f_1} from the second approach is obviously much simpler and is given by equation (3.81). So, the total power loss is

$$P_f = P_{f_2} + \frac{\pi}{\sqrt{3}}\sigma_Y d_i^2 V_B \ln\left(\frac{d_i}{d_f}\right)$$

$$= \frac{\pi}{\sqrt{3}}\sigma_Y d_i V_B\left[l + d_i \ln\left(\frac{d_i}{d_f}\right)\right]. \tag{3.87}$$

EXAMPLE 3.11 Estimate the maximum force required for extruding a cylindrical aluminium billet of 50 mm diameter and 75 mm length to a final diameter of 10 mm. The average tensile yield stress for aluminium is 170 N/mm^2. What per cent of the total power input will be lost in friction at the start of the operation?

SOLUTION First approach. The ram force is obviously maximum just after the start of the extrusion process as l is maximum. For the given dimensions and assuming a 45° dead zone, the value of l at the beginning is given as

$$l = \left[75 - \left(\frac{d_i - d_f}{2}\right)\cot 45°\right] \text{ mm} = 75 - 20 \text{ mm} = 55 \text{ mm}.$$

Next, the value of μ is determined by solving equation (3.77) through a trial-and-error method. Thus, we get $\mu = 0.15$. Now, $\sigma_x|_{BB}$ is found out from equation (3.76) with this value of μ. Hence,

$$\sigma_x|_{BB} = 170 \times \frac{1.15}{0.15}[(5)^{0.3} - 1] \text{ N/mm}^2 = 809 \text{ N/mm}^2.$$

Finally, using equation (3.79), we have

$$F = \left[\frac{\pi}{4} \times (50)^2 \times 809 + \frac{\pi}{\sqrt{3}} \times 50 \times 55 \times 170\right] \text{ N}$$

$$= (1{,}588{,}468 + 847{,}976) \text{ N} = 2{,}436{,}444 \text{ N}.$$

If the velocity of the ram is V_B mm/sec, then the frictional power loss [using equations (3.84) and (3.86)] is given as

$$P_f = \pi \times 170 \times (50)^2\left[\frac{1.15}{0.3}\{(5)^{0.3} - 1\} - \ln 5\right]V_B$$

$$+ \frac{\pi}{\sqrt{3}}170 \times 50 \times 55 \times V_B \text{ N-mm/sec}$$

$$= (1027 + 848)V_B \text{ W} = 1875 V_B \text{ W}.$$

Now, the total input power is

$$P = FV_B = 2436.4V_B \text{ W}.$$

Therefore, the power loss in friction is

$$\frac{1875}{2436.4} \times 100 = 76.95\%.$$

Second approach. The determination of $\sigma_x|_{BB}$ is much simpler with this approach. Using equation (3.83), we directly find out the value of $\sigma_x|_{BB}$ as

$$\sigma_x|_{BB} = 4.31 \times 170 \ln 5 \text{ N/mm}^2 = 1179 \text{ N/mm}^2.$$

Therefore, the work load (when l is maximum and equal to 55 mm) is

$$F = [\frac{\pi}{4} \times (50)^2 \times 1179 + \frac{\pi}{\sqrt{3}} \times 50 \times 55 \times 170] \text{ N}$$
$$= (2,314,961 + 847,976) \text{ N} = 3,162,937 \text{ N}.$$

To determine the percentage of the total input power lost in friction, we again assume a ram velocity V_B mm/sec. The frictional power loss in the conical working zone is given by equation (3.81) as

$$P_{f_1} = \dot{W}_f = \frac{\pi}{\sqrt{3}} \times 170 \times (50)^2 \times \ln 5 \times V_B \text{ N-mm/sec}$$
$$= 1,240,695V_B \text{ N-mm/sec} = 1240.7V_B \text{ W}.$$

The power lost in friction in the cylindrical zone of the work material is the same as that found out in the previous calculations. So, $P_{f_2} = 848V_B$ W. Hence, the total power loss is

$$P_f = P_{f_1} + P_{f_2} = 2088.7V_B \text{ W}.$$

The input power at the ram is

$$P = FV_B = 3162.94V_B \text{ W}.$$

Thus, the power loss in friction is

$$\frac{2088.7}{3162.94} \times 100 = 66.04\%.$$

3.4.7 PUNCHING AND BLANKING

As we have already noted, the punching and blanking processes cannot, strictly speaking, be grouped under the forming operations. In these processes, a finite volume from a sheet metal is removed *en bloc* by using a die and a punch. The shape and size of the portion removed are determined by the geometry of the die and the punch. If the final product happens to be the removed portion, then the operation is termed as blanking. On the other hand, if the pierced sheet metal is the final product, then the operation is called punching. Since the basic mechanics of material removal is the same in both the operations, we shall discuss these under a single heading, namely, punching.

Figure 3.26 shows a simple punching operation. As in deep drawing, so too here the job is held by job holders to prevent any distortion and to provide a support. It should be noted that the punch and die corners are not provided with any radius (unlike in the deep drawing operation) as the objective in this process is to cause a rupture of the material. A clearance c is provided between the punch and the die. Hence, the die diameter $d_d = d_p + 2c$, where d_p is the diameter of the punch.

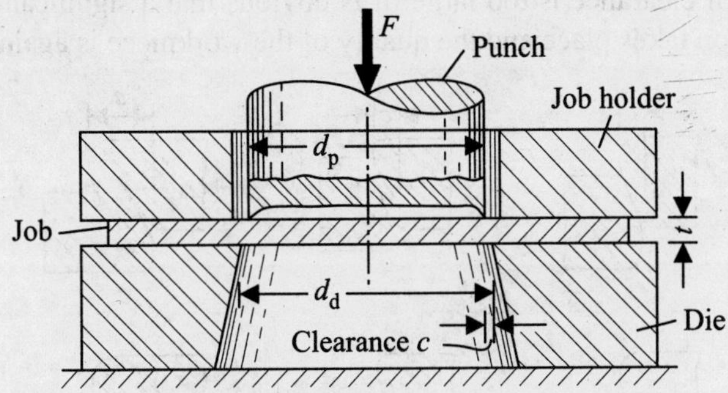

Fig. 3.26 Details of punching process.

Mode of Metal Deformation and Failure

To develop a mathematical model for the punching and blanking operations, it is necessary to study the nature of metal deformation with the progress of the punch, and the mechanism of ultimate failure of the material. Figure 3.27 shows the nature of metal deformation as the punch penetrates the workpiece. It is clear

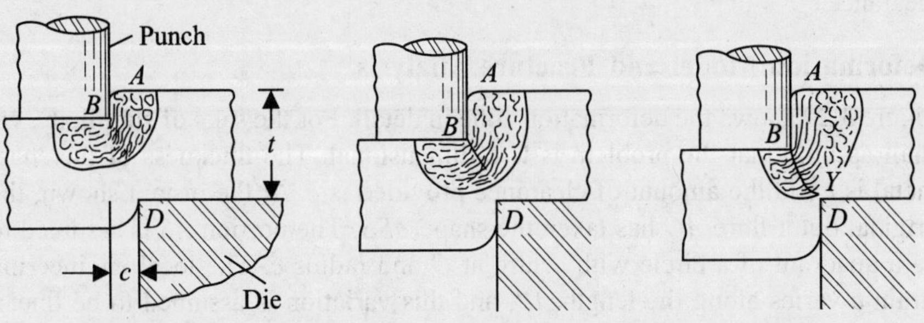

Fig. 3.27 Deformation of workpiece with punch travel.

that the workpiece bends and is pulled down by the punch movement and the grains elongate near the punch corner B. A similar type of deformation takes place near the die corner D. When the grain elongation or the local natural strain in the surface fibre AB (of the workpiece) reaches a limiting value, the fibre ruptures. Since the local strain is maximum at the corner, a crack opens up just ahead of the punch corner. After this, with a slight movement of the punch, the inner fibres also get ruptured. Thus, the fracture line BY propagates, following a path along which the successive inner fibres attain the fracture strain. A similar crack also propagates from the die corner D (because of the symmetry of the deformation geometry). It should be remembered that the velocity of propagation of the fracture line is very high. Now, if the amount of clearance c is optimum, then the two fracture lines meet and a clean edge is obtained after the operation (Fig. 3.28a). If the clearance is too small, then the fracture lines miss each other and a secondary deformation takes place, resulting in an unclean edge as shown in Fig. 3.28b. Figure 3.28c shows what happens when the amount of clearance is too large. It is obvious that a significant amount of drawing action takes place and the quality of the workpiece is again quite poor.

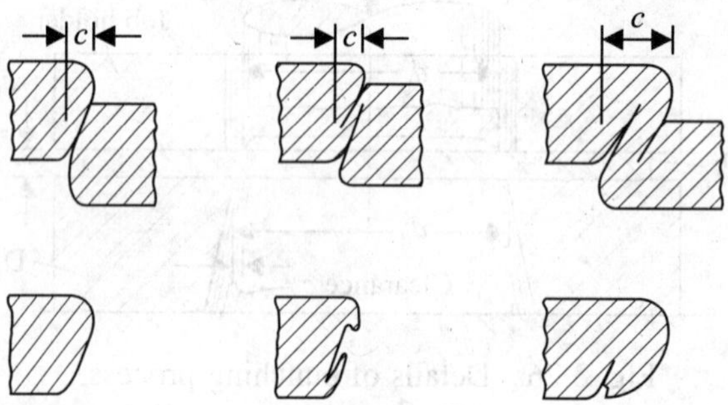

(a) Correct clearance (b) Too little clearance (c) Too large clearance

Fig. 3.28 Fracture lines with different clearances.

In what follows, we shall give an analysis to approximately determine the fracture line. Subsequently, we shall use this to estimate the correct amount of clearance.

Deformation Model and Fracture Analysis[1]

Figure 3.29 shows the deformation zone in detail. For the sake of simplicity, we shall assume that the problem is two-dimensional. The thickness of the sheet metal is t and the amount of clearance provided is c. At the instant shown, the original outer fibre AQ has taken the shape ASB. The portion AS is assumed to be a quadrant of a circle with centre at O and radius c. The local engineering strain e varies along the length AB, and this variation is assumed to be linear. As already mentioned, the criterion for fracture is taken as the maximum tensile strain rather than the stress. Thus, if the strain at B reaches the fracture limit ε_f, the outer fibre ruptures and tears open. Similarly, the inner fibre ER, originally at an infinitesimal depth δ, takes the shape ETV. Again, the portion ET is assumed to be a quadrant of a circle with centre at O' and radius r. When the strain at V reaches the fracture limit, this inner fibre tears at V. Thus, the direction of the fracture can be considered as the line joining B and V. The fracture operation

[1]Ghosh, A. and Mallik, A.K., Mechanics of Stamping and Blanking, Proceedings of the 9th All India Machine Tool Design and Research Conference, Kanpur, 1980.

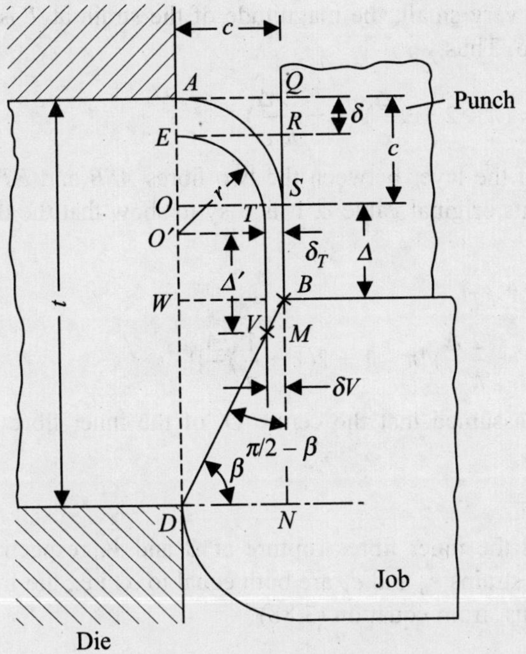

Fig. 3.29 Geometry of deformation of two fibres at a distance δ apart.

is very quick and is completed with almost an imperceptible movement of the punch. With the optimum amount of clearance c, the fracture line BV, when extended, should pass through the die corner D.

Since the engineering strain has been assumed to be varying linearly, the elongation of the outer fibre AQ (of length c) to ASB is

$$ASB - AQ = AS + SB - AQ = \frac{\pi}{2}c + \Delta - c$$

$$= \tfrac{1}{2}e_B c$$

or

$$\tfrac{1}{2}e_B = \left(\frac{\pi}{2} - 1\right) + \frac{\Delta}{c}. \tag{3.88}$$

Again, using the linear variation, we find the strain at the point S is

$$e_S = e_B\, AS/ASB$$

$$= e_B[(\pi/2)c]/[(\pi/2)c + \Delta]$$

$$= e_B/[1 + (2/\pi)(\Delta/c)].$$

Substituting for e_B from equation (3.88), we get

$$e_S = \left(\pi - 2 + 2\frac{\Delta}{c}\right)/\left(1 + \frac{2}{\pi}\frac{\Delta}{c}\right). \tag{3.89}$$

Now, since δ is very small, the magnitude of the strain at T is approximately equal to that at S. Thus,

$$e_T \approx e_S = (\pi - 2 + 2\frac{\Delta}{c})/(1 + \frac{2}{\pi}\frac{\Delta}{c}). \tag{3.90}$$

The thickness of the layer between the two fibres ASB and ETV continuously decreases from its original value δ. It is easy to show that the thickness at T is given by

$$\delta_T = \delta/(1 + e_T)$$

$$= \delta[(1 + \frac{2}{\pi}\frac{\Delta}{c})/\{\pi - 1 + 2(1 + \frac{1}{\pi})\frac{\Delta}{c}\}]. \tag{3.91}$$

Since we have assumed that the centre O' of the inner fibre ET lies on the line AD,

$$r = c - \delta_T. \tag{3.92}$$

If the outer and the inner fibres rupture at B and V, respectively, it is then obvious that the strains e_B and e_V are both equal to e_f, i.e., the fracture strain of the material. Thus, from equation (3.88),

$$\frac{\Delta}{c} = \tfrac{1}{2}e_f - [\frac{\pi}{2} - 1]. \tag{3.93}$$

Again, considering the extension of the fibre ER, we have

$$\tfrac{1}{2}e_f = [\frac{\pi}{2}r + \Delta' - c]/c, \tag{3.94}$$

where Δ' is as indicated in Fig. 3.29. Using equations (3.91), (3.92), and (3.94), we obtain

$$\frac{\Delta'}{c} = 1 + \tfrac{1}{2}e_f - \frac{\delta}{2}[1 - \frac{\Delta}{c}(1 + \frac{2}{\pi}\frac{\Delta}{c})/\{\pi - 1 + 2(1 + \frac{1}{\pi})\frac{\Delta}{c}\}]. \tag{3.95}$$

Now, the thickness of the layer at V (Fig. 3.29) is given as

$$VM = \delta V = \delta/(1 + e_V) = \delta/(1 + e_f),$$

and

$$BM = OO' + \Delta' - \Delta = r + \delta - c + \Delta' - \Delta.$$

Using equations (3.91), (3.92), and (3.95), we get

$$BM = \delta[(2.71 + \frac{3\Delta}{c})/(2.14 + 2.64 \times \frac{\Delta}{c})].$$

Finally, the direction of the fracture line is obtained in terms of the angle β as

$$\tan\beta = BM/VM = [(2.71 + 3 \times \frac{\Delta}{c})/(2.14 + 2.64 \times \frac{\Delta}{c})](1 + e_f).$$

Using equation (3.93), we have

$$\tan \beta = [(1 + 1.15e_f)/(0.63 + 1.32e_f)](1 + e_f). \tag{3.96}$$

Now, from Fig. 3.29, $\tan \beta$ can also be expressed as

$$\tan \beta = \frac{BN}{DN}(t - c - \Delta)/c. \tag{3.97}$$

The optimum clearance is obtained, using equations (3.93), (3.96), and (3.97), as

$$\frac{t}{c_o} = \frac{(1+1.5e_f)}{(0.63+1.32e_f)}(1 + e_f) + 0.5e_f + 0.43. \tag{3.98}$$

Here and now onwards, c_o represents the optimum clearance. The penetration of the punch $(\Delta + c_o)/t$ can then be obtained, using equations (3.93) and (3.98), as

$$\frac{\Delta + c_o}{t} = \frac{c_o}{t}\left(\frac{\Delta}{c_o} + 1\right) = \frac{(0.43 + 0.5e_f)}{\dfrac{(1+1.5e_f)(1+e_f)}{(0.63+1.32e_f)} + 0.5e_f + 0.43}. \tag{3.99}$$

The engineering fracture strain e_f, used in equations (3.98) and (3.99), can be substituted in terms of the natural fracture strain ε_f, using the relation

$$\varepsilon_f = \ln (e_f + 1). \tag{3.100}$$

Using equation (3.100) in equations (3.98) and (3.99) and simplifying, we finally get

$$\frac{t}{c_o} = 1.36 \exp(\varepsilon_f) \frac{2.3 \exp(\varepsilon_f) - 1}{2 \exp(\varepsilon_f) - 1}, \tag{3.101}$$

$$\frac{\Delta + c_o}{t} = \frac{1}{2.45}\left[\frac{1.9 \exp(\varepsilon_f) - 1}{2.56 \exp(\varepsilon_f) - 1}\right]. \tag{3.102}$$

The magnitude of ε_f at which the fibres rupture depends on the material and the magnitude of the hydrostatic stress, as shown in Fig. 3.30[1]. The values of optimum clearance (c_o) and penetration depth ($\Delta + c_o$), in terms of the thickness of the workpiece (t), for various values of ε_f are calculated from equations (3.101) and (3.102), respectively. These values are as presented here:

ε_f	1	1.5	2.0	2.5
c_o/t	0.215	0.133	0.082	0.05
$(\Delta + c_o)/t$	0.278	0.289	0.296	0.3

We see from these values that depending on the ductility of the material the clearance varies from 5% to 20% of the sheet thickness. A smaller clearance is required for a more ductile material. The percentage penetration is about 30% of the sheet thickness and increases very slowly with the ductility of the material being worked. It may be noted that the compressive transverse stresses σ_2 and

[1]American Society of Metals, Ductility, Proceedings of the Seminar, October, 1967.

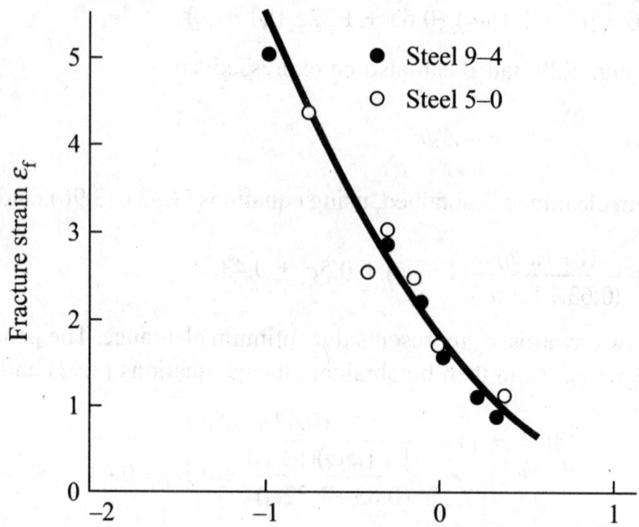

Fig. 3.30 Variation of fracture stress with hydrostatic component of stress.

σ_3 are present at the die and punch corners. Thus, for a material which shows the value of $\varepsilon_f = 1$ in the tensile test (with $\sigma_2 = \sigma_3 = 0$), the effective value of ε_f, while the material is being punched, will be somewhat higher. This, in turn, means that the optimum clearance for such a material is likely to be closer to 10% of the sheet thickness.

Determination of Working Force

Once the ruptures start at B and D, a slight progress of the punch causes a complete rupture of all the fibres, thus separating the blank (Fig. 3.31). Of course, in a very ductile material, some more punch travel, after the ruptures start at B and D, is needed to complete the process. So, the percentage penetration here is somewhat higher than the calculated values. The maximum force is obtained by determining the force required to cause the rupture of the area ($c_o \times L$), where L is the length of the cut (equal to πD, D being the punch diameter for a cylindrical punch). Thus,

$$F_{max} = \sigma_f c_o L, \tag{3.103}$$

where σ_f is the true rupture stress and c_o is the optimum clearance given by equation (3.101).

Figure 3.32 shows the nature of variation of the force F with the travel of the punch. So, the work required (W) for the punching operation can be expressed as

$$W = \tfrac{1}{2} F_{max} p, \tag{3.104}$$

where p is the depth of penetration $= \Delta + c_o$.

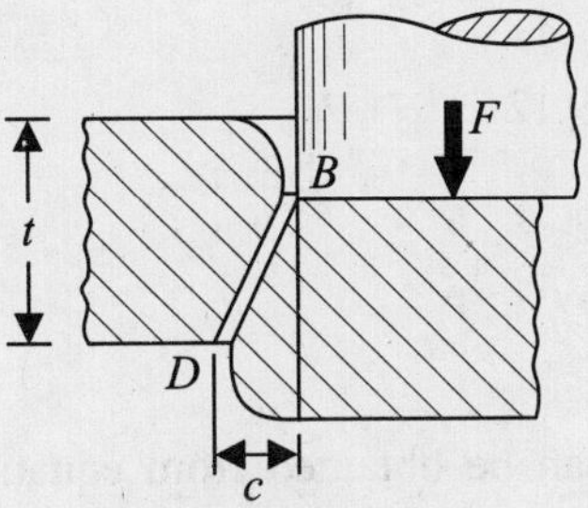

Fig. 3.31 Separation of blank from workpiece.

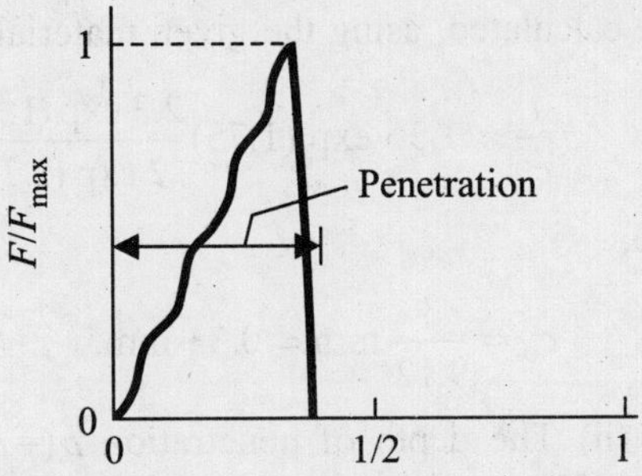

Fig. 3.32 Variation of load with punch travel (for optimum clearance).

The maximum punching force can be reduced by avoiding the simultaneous failure of the total area. This can be achieved by providing an angle (commonly known as shear) to the punch edge. We now explain the effectiveness of shear assuming the availability of a punch having a straight edge. In this case, the operation is called shearing instead of punching. Figure 3.33 shows that when a shear is provided, at any instant the width of the job undergoing deformation is

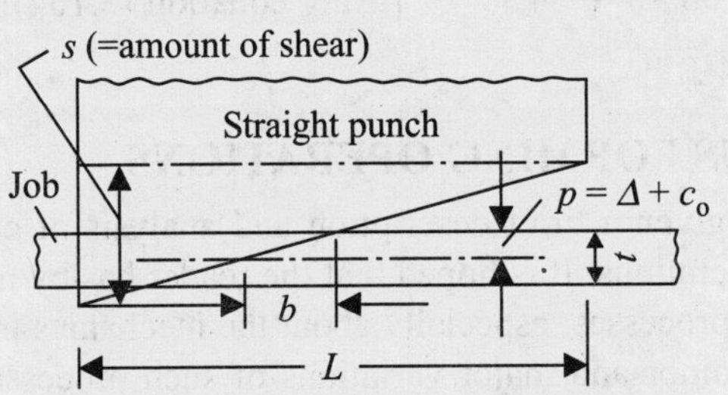

Fig. 3.33 Geometry of cutting, using straight punch with shear.

b which is much smaller than the total punch width L. It should be remembered that the maximum force occurs when the punch travels up to the penetration depth p. Therefore, the average force for a width of cut b is given by the half of the maximum force for a width b, which itself is the maximum punch force with shear. The amount of shear is commonly expressed as the difference in the height of the two extreme ends of the punch. For a straight punch, we finally get

$$F_{\max} = \sigma_f c_o b. \tag{3.105}$$

EXAMPLE 3.12 Determine (i) the proper clearance between the die and the punch, and, assuming that no shear is provided to the punch, also determine the (ii) maximum punching force, and (iii) energy required to punch a hole of 50 mm diameter in a 3-mm-thick steel sheet. The true fracture strain and stress for the material can be taken as 1.75 and 2.1 kN/mm^2, respectively.

SOLUTION (i) From equation (3.101), the proper amount of clearance can be calculated, using the given material properties. Thus,

$$\frac{t}{c_0} = 1.36 \exp(1.75) \frac{2.3 \exp(1.75) - 1}{2 \exp(1.75) - 1} = 9.12.$$

So,

$$c_0 = \frac{3}{9.12} \text{ mm} = 0.33 \text{ mm}.$$

(ii) The depth of penetration $p(=\Delta + c_0)$ can be obtained from equation (3.102) as

$$p = t \frac{1}{2.45} \frac{1.9 \exp(1.75) - 1}{2.56 \exp(1.75) - 1} = 0.28t = 0.84 \text{ mm}.$$

Since the punch is not provided with any shear, the maximum punching force is found from equation (3.103). Thus,

$$F_{max} = 2100 \times 0.33 \times \pi \times 50 \text{ N} = 108.9 \text{ kN}.$$

(iii) Now, the energy required to punch the hole can be given as

$$W = \tfrac{1}{2} \times 108.9 \times 0.84 \text{ J} \qquad \text{[using equation (3.104)]}$$

$$= 45.74 \text{ J}.$$

3.5 VARIOUS FORMING OPERATIONS

So far, we have given a brief description and analysis of each of the various basic forming operations. It is hoped that the reader has by now acquired some idea about these processes, especially about the mechanics involved. However, there are many minor and major variations of such processes. In this section, we shall discuss some of these along with the associated technological aspects.

3.5.1 ROLLING

It is seldom possible to achieve the final cross-section in one step. Generally, rolling is performed with a number of passes, using different rolling equipment, in a continuous manner. The whole shop is usually called a rolling mill. When rolling flat strips, it is possible to perform the successive stages, using the same pair of rolls. The upper roll is normally adjusted to control the gap after each pass. To avoid the problem of extensive material handling, it is desirable to have the provision of reversing the direction of roll rotations. As a result, the workpiece moves back and forth in successive passes. Sometimes, the space can be optimized by using a three-high rolling mill, as shown in Fig. 3.34a. During hot rolling, the lapse of time should be minimized as the job continuously cools down. This should be one of the major considerations in the layout of a rolling mill. Normally, the job movement is facilitated by providing support rolls (see Fig. 3.1). If a job is sufficiently long and flexible, a three-high rolling mill can be provided with some arrangement for feeding the second pass even before the first pass is completed. This is achieved by what is commonly known as a looping mill (Fig. 3.34b). The looping can be done mechanically by using a bent tube or trough, known as repeater. A continuous multipass rolling can also be performed for a flexible, long job by suitably arranging the rolling equipment with one roll pass near the other.

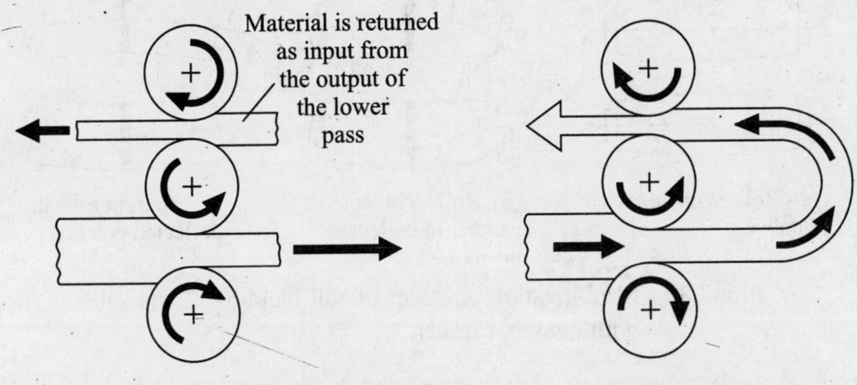

(a) Separate workpieces (b) Same workpiece

Fig. 3.34 Three-high rolling mill.

For a given reduction in area, commonly known as the draft, the roll separating force, which tends to bend the rolls, increases linearly with roll radius R given by equation (3.20). Hence, the bending deflection of the rolls cannot be very effectively and economically controlled by using large drive rolls. A better and more economical way to reduce the roll deflection is to use backing rolls, as indicated in Fig. 3.35. In this figure, two different methods of using the backing rolls are shown. Since the roll separating force depends on the radius of the drive

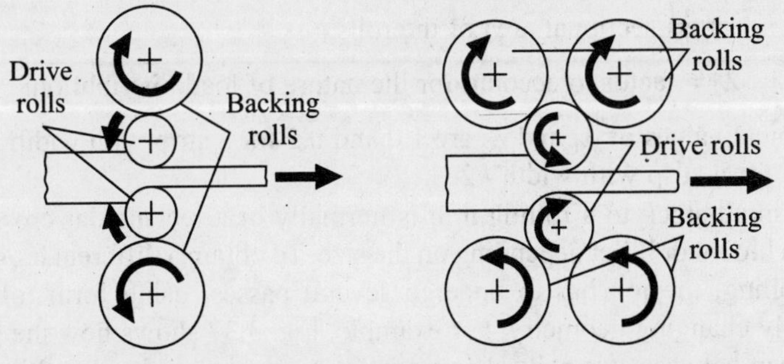

(a) Single backing rolls (b) Double backing rolls

Fig. 3.35 Use of backing rolls to reduce roll deflection.

rolls, these are always kept small in size, whereas the backing rolls are provided with a larger radius to increase the rigidity. However, a certain amount of roll bending is unavoidable, but this can be taken care of by having noncylindrical rolls (Fig. 3.36a) which, under the roll separating force, bend, thus providing a uniform gap between the rolls (Fig. 3.36b). The rolls shown in Fig. 3.36a are called rolls with convex camber. With uncambered rolls, the thickness of the rolled strip is more at the centre, as explained in Fig. 3.36c. Considering the

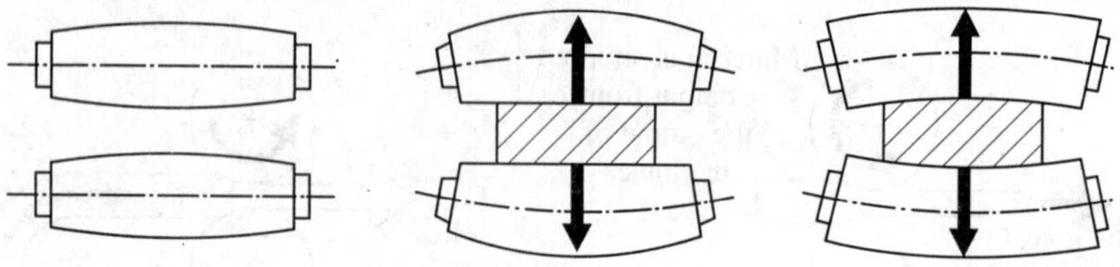

(a) Rolls with convex camber

(b) Rolls with convex camber in deflected position

(c) Straight rolls in deflected position

Fig. 3.36 Minimization of effect of roll bending, using rolls with convex camber.

rolls as thick, short beams simply supported at the ends, the deflection at the centre can be expressed as

$$\delta = \lambda_1 \frac{Fl^3}{EI} + \lambda_2 \frac{Fl}{GA}, \tag{3.106}$$

where

F = roll separating force,

l = length of each roll,

E, G = modulus of elasticity and shear, respectively, of the roll material,

I = second moment of area of the roll cross-section about a diameter,

A = cross-sectional area of the rolls,

λ_1, λ_2 = factor to account for the nature of load distribution.

The typical values of λ_1 and λ_2 are 1.0 and 0.2 for a strip with width l, and 0.5 and 0.1 for a strip with width $l/2$.

The input stock to a rolling mill is normally of a rectangular cross-section, called a bloom or billet depending on the size. To obtain a different cross-section after rolling, the job has to undergo several passes, using form rolls with a gradually changing geometry. For example, Fig. 3.37 shows how the geometry of the gap between two rolls changes while producing a circular, thin rod from a square billet.

Rolls are normally made of cast or forged steel. Alloyed cast irons are sometimes used to lower down the cost. Superior strength and rigidity characteristics can be obtained by using special alloy steel which, obviously, is costlier. Hot rolls are roughened (even notched sometimes) to provide a good bite on the job, whereas cold rolls are ground to provide a fine surface to impart a good finish to the final product.

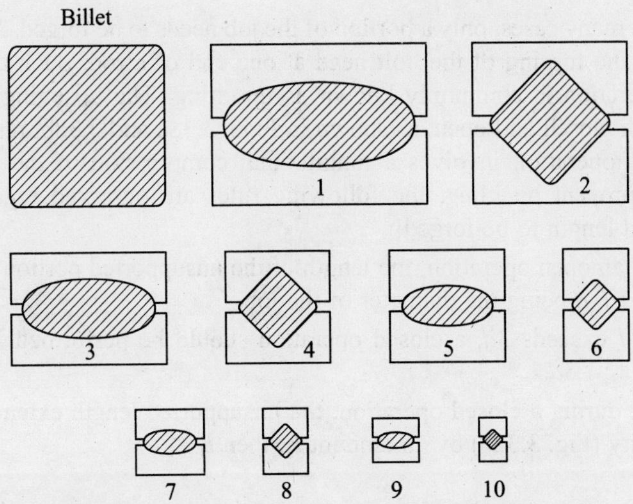

Fig. 3.37 Passes for reducing billet to round bar.

The main parameters of rolling include (i) the temperature range (in hot rolling), (ii) the rolling speed schedule, and (iii) the allotment of reductions to various passes. All these, in turn, influence the dimensional accuracy of the product and also its physical and mechanical properties.

3.5.2 FORGING

There are many variations of the basic forging operation, and the most commonly practised are (i) smith forging, (ii) drop forging, (iii) press forging, (iv) upset forging, (v) swaging, and (vi) roll forging.

(i) Smith forging is probably the most ancient metal working process. Here, a hot workpiece is given the desired shape by using hand-held tools and hammers. Nowadays, power-driven hammers are used to impart the repeated blows. The anvil and the hammer are mostly flat and the desired shape (of course with limited varieties) is obtained by a manipulation of the job between the blows.

(ii) In drop forging, the impact loads (blows) are applied to the workpiece to cause metal flow for filling up the cavity formed by the two halves of the closed die. To ensure complete filling, normally an excess amount of material is provided. This excess material flows out circumferentially to form a flash (Fig. 3.2b) which is subsequently trimmed. When the product geometry is complicated, a set of dies may be necessary to obtain the final form.

(iii) Instead of the repeated blows, a gradual force is applied in press forging. However, depending on the complexity of the job, a set of dies may be required to obtain the final product. It is obvious that here the alignment of the two halves of the die poses a lesser problem than in drop forging. Since the operation is completed in one stroke, a provision must be made for the air and excess die lubricant to escape.

(iv) In many cases, only a portion of the job needs to be forged. A common example is the forging of the bolt head at one end of a rod. Such a localized forging operation is commonly known as upsetting. The upsetting operation may be both closed and open, as shown in Figs. 3.38a and 3.38b, respectively. Clearly, the operation involves a longitudinal compression of the bar stock. Hence, to prevent buckling, the following rules are observed regarding the unsupported length to be forged:

(a) In an open operation, the length of the unsupported portion (l) should not exceed $3d$, d being the diameter of the job.

(b) If l exceeds $3d$, a closed operation should be performed with a die diameter $D \leqslant 1.5d$.

(c) If, during a closed operation, the unsupported length extends beyond the die cavity (Fig. 3.38c) by an amount l_1, then $l_1 \leqslant d$.

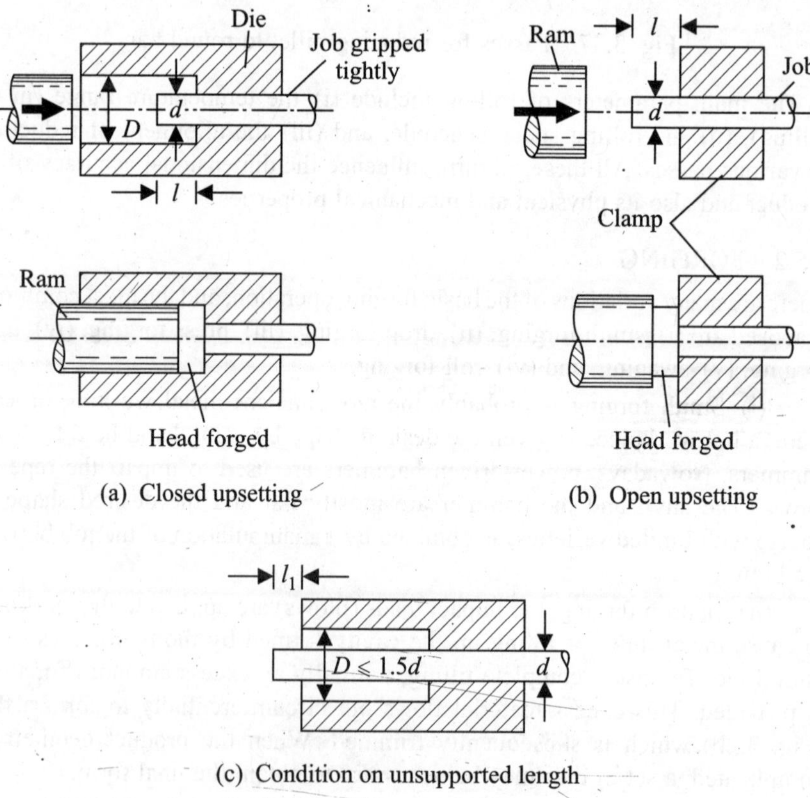

(a) Closed upsetting (b) Open upsetting

(c) Condition on unsupported length

Fig. 3.38 Upsetting operation.

(v) Swaging is a special variation of impact forging where the repeated blows are obtained by a radial movement of shaped dies, as explained in Fig. 3.39. This operation is generally used for reducing the diameters and tapering of bars and tubes.

(vi) Roll forging is performed with two semicircular, grooved rolls held by two parallel shafts, as shown in Fig. 3.40. The process is used for reducing the diameter of rods. The heated workpiece is placed between the dies in an open position. After a half revolution of the rolls, the workpiece is rolled out. It is then put in the smaller groove and the operation continued until the desired dimension is achieved.

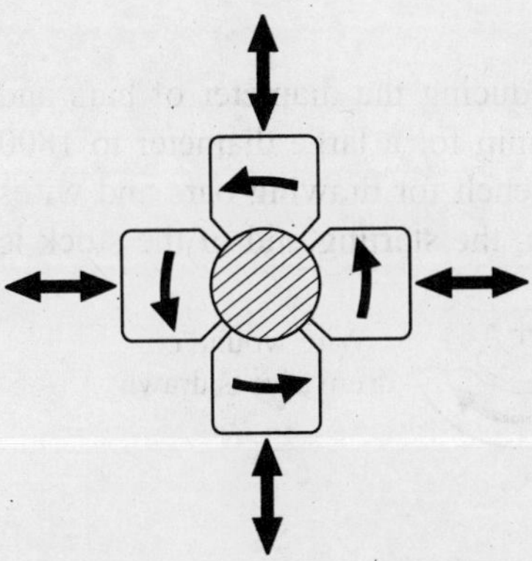

Fig. 3.39 Principle of rotary swaging.

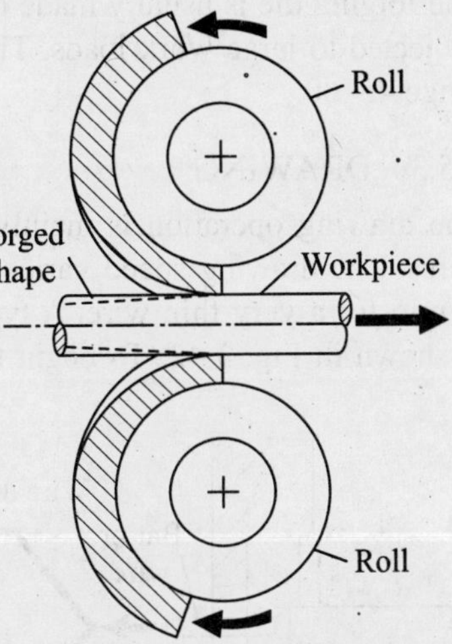

Fig. 3.40 Principle of roll forging.

It is obvious that the die is one of the most critical components of the forging operation, and therefore the success of the process depends considerably on the design of the die. The basic features a forging die should have are as follows (see also Fig. 3.41):

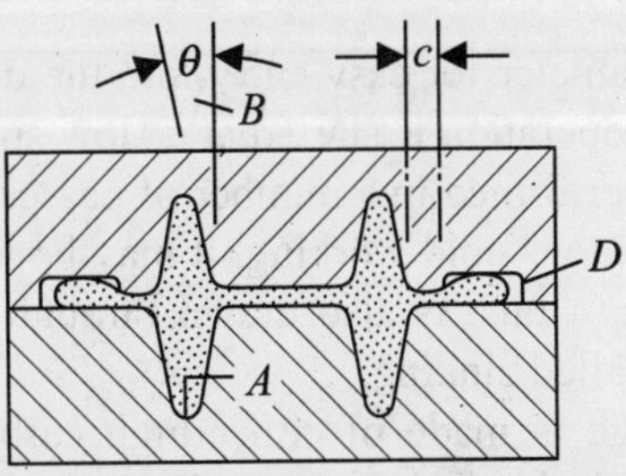

Fig. 3.41 Basic features of forging die.

(i) To ease the flow of metal around the corners, a proper fillet radius should always be provided. This also helps in preventing excessive die wear and fracture of metals near the corners.

(ii) As in a moulding pattern, so too here all vertical surfaces should be given a suitable draft for easy removal of the job from the die.

(iii) As already mentioned, a space around the die edges should be provided to accommodate the excess material, known as flash. To receive this flash, it is recommended that a flash gutter be provided.

In hot forging, the die dimensions should include the shrinkage allowance (to compensate for the contraction of the product after cooling) as the forged product is normally not subjected to any subsequent overall finishing operation. The forging die is usually made of a high or medium carbon alloy steel as it is subjected to large work loads. The hardness (R_C) of the die is normally in the range 45–60.

3.5.3 DRAWING

The drawing operation is mainly used for reducing the diameter of bars and wires. The drawing speed varies from 10 m/min for a large diameter to 1800 m/min for a very thin wire. A typical draw bench for drawing bars and wires is shown in Fig. 3.42. To begin the operation, the starting end of the stock is

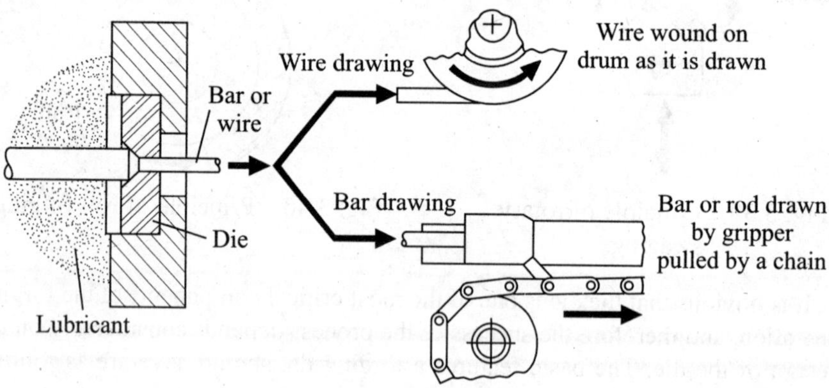

Fig. 3.42 Cold drawing of wire or bar.

swaged to a smaller diameter for easy entry into the die. Moreover, to prevent any impact action, the operation is started at a slow speed. In large reductions, the operation may be performed in a number of passes. Since sufficient heat is generated due to continuous cold working, it may be necessary to cool the die with water. Sometimes, a tube is also drawn through a draw die, and in this case, the operation is called sinking.

Normally, a large die is made of either high carbon or high speed steel, whereas tungsten carbide is used for a medium-size die. For drawing a fine wire, the die is made of diamond.

3.5.4 DEEP DRAWING

It is evident from our description of the mechanics of the deep drawing process that an attempt should be made to draw the sheet metal into the die as much as possible. This helps in minimizing the thinning of the cup wall. Consequently, the outer circumference of the blank reduces, causing a compressive hoop stress which, when exceeds a limit, may result in a plastic wrinkling of the cup flange as shown in Fig. 3.71. These wrinkles cannot be ironed out afterwards but can be avoided by using a blank holder. However, an excessive pressure from the blank holder resists an easy drawing of the material into the die. If the drawing ratio (defined as r_j/r_d) is not more than 1.2, the operation can be conducted even without a blank holder. Higher values of the drawing ratio can be achieved depending on the thickness of the blank and die profile, as shown in Fig. 3.43.

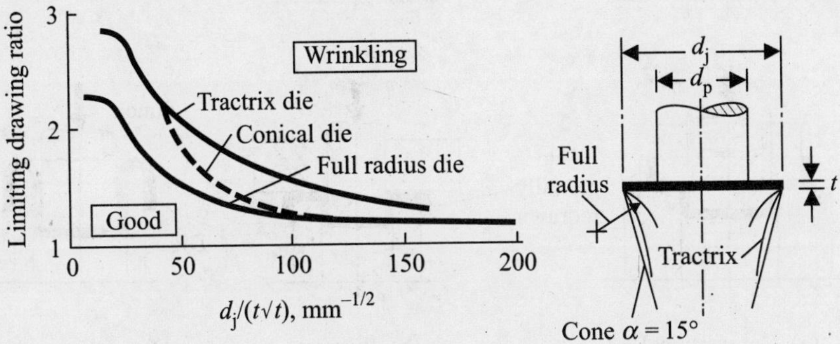

Fig. 3.43 Limiting drawing ratio in drawing low carbon steel without blank holder (with different die shapes). (After Shawki, G.S.A., *Werkstattstechnik*, **53**, 1, 12, 1963.)

When the ratio of the blank diameter and the final cup diameter is too large, the operation is performed in more than one stage. The successive drawing operations after the first one are known as redrawing. Figures 3.44a and 3.44b show two typical redrawing operations. The operation shown in Fig. 3.44b is termed as reverse redrawing, because, in this, the initially drawn cup is turned inside out. This operation appears to involve a more severe working of the material than the conventional redrawing operation. However, the real situation is just the opposite, as now explained. In conventional redrawing (Fig. 3.44a), the material bends in the opposite directions around the blank holder and the die corners. On the other hand, in reverse drawing (Fig. 3.44b), the material bends in only one direction, namely, along the outer and the inner die corners. In an extreme case, the die can be provided with a round edge, as shown in Fig. 3.44c, resulting in a less severe working of the material. Since some amount of strain hardening takes place during the initial operation, annealing is normally advised (to restore the ductility) before commencing the redrawing operation.

In general, the flow of metal is not uniform throughout the workpiece and in most cases the drawn parts have to be trimmed to remove the undesired metal. Such a trimming can be done either by a hand-guided operation or by using a separate trimming die.

The stripping of the job from the punch can be achieved by machining a slight recess into the underside of the draw die, as shown in Fig. 3.45a. During the return stroke, the punch pressure is removed from the cup; as a result, the drawn cup tends to spring back. Due to this action, the recess prevents the drawn

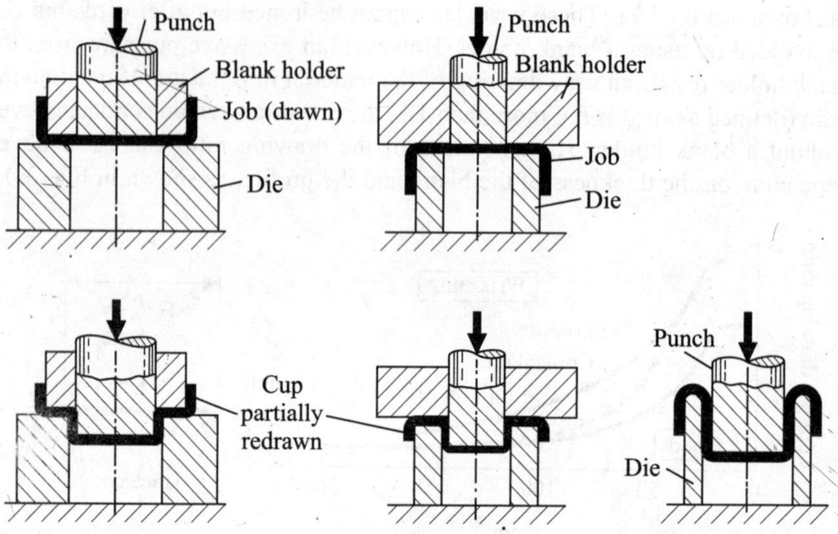

(a) Conventional (b) Reverse (c) Die with rounded edge in reverse redrawing

Fig. 3.44 Redrawing operation.

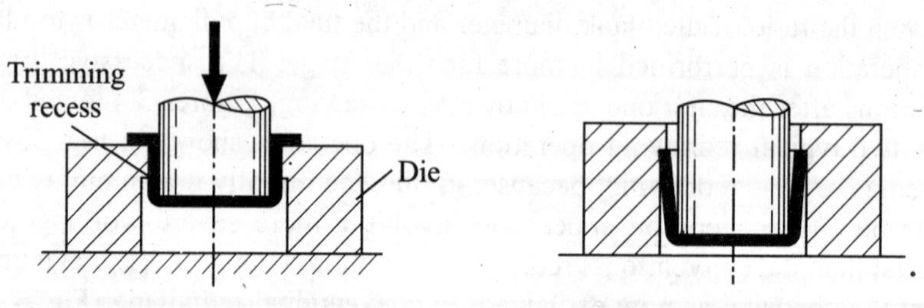

(a) Deep drawing with die having trimming recess (b) Action of trimming recess at end of deep drawing operation

Fig. 3.45 Stripping of job by using trimming recess.

cup from moving along with the punch during its upward stroke, as illustrated in Fig. 3.45b.

3.5.5 BENDING

The analysis of the bending operation we have given is applicable only when corners are required to be produced in a sheet metal. However, more complicated shapes can also be obtained by this operation. In general, such an operation may need more than one stage (see Fig. 3.46). For producing a complex shape, the bending operation is performed continuously, using a series of contoured rolls (see Fig. 3.47). Idle rollers are used when necessary for pressing the job from the side during the production of such a shape.

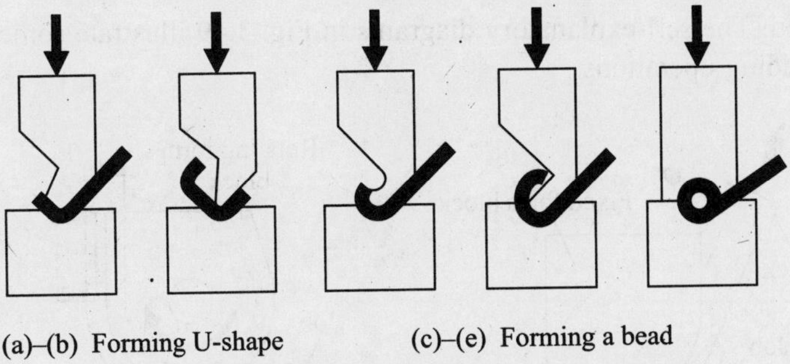

Fig. 3.46 Bending requiring more than one stage.

Fig. 3.47 Continuous bending operation, using series of contoured rolls.

Tubes and other hollow sections can also be bent by wrapping the job around a form block through the use of a wiper roll. If the wiper roll has a constant curvature, it may be hinged at the centre of the curvature to be produced. Figure 3.48 explains such an operation for bending a tube. The tube can be prevented from collapsing by filling the inside space with some filling material,

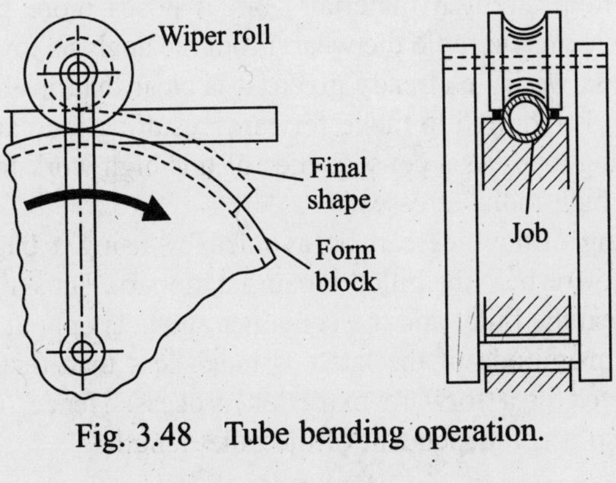

Fig. 3.48 Tube bending operation.

e.g., sand. The self-explanatory diagrams in Fig. 3.49 illustrate some common tube bending operations.

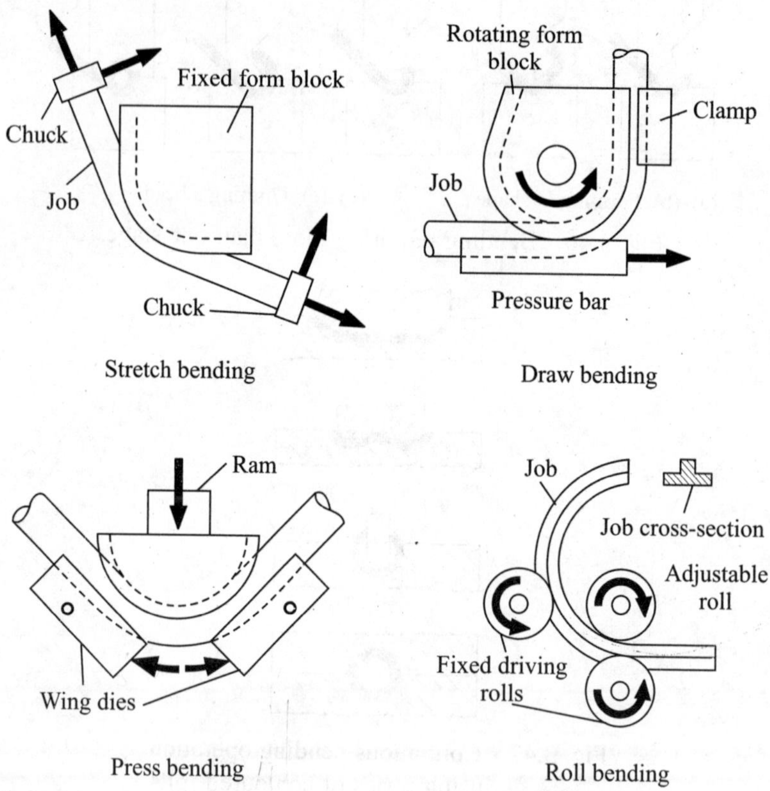

Fig. 3.49 Various bending operations.

3.5.6 EXTRUSION

Extrusion is one of the most potential and useful metal working processes and has a large number of variations in the mode of application. It can be performed under both hot and cold conditions. Hot extrusion helps reduce the work load (specially for high strength materials) but it poses more problems such as cooling arrangement and rapid die wear. From the analysis for a simple forward extrusion process we have already given, it is clear that, in this direct process, the whole billet is required to move forward, resulting in a large frictional loss and high working load. As a consequence of this high work load, the container is subjected to high radial stresses.

The foregoing difficulties can be avoided by using a backward extrusion process (Fig. 3.50) where the billet remains stationary. Thus, the frictional force is absent between the billet and the container and acts only at the die-container interface. The magnitude of the latter is much less than that of the frictional force encountered in a forward extrusion process. Hence, the work load is reduced and also it is independent of the billet length.

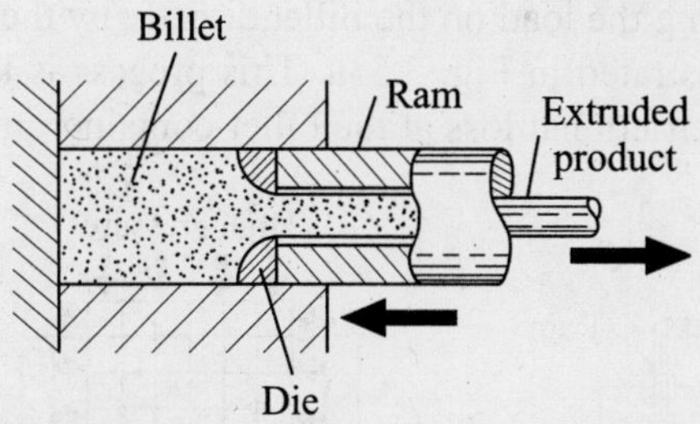

Fig. 3.50 Principle of backward extrusion.

Tubular sections can also be extruded by using a mandrel along with the ram, as illustrated in Fig. 3.51. Both open (Fig. 3.51a) and closed (Fig. 3.51b) end products can be obtained depending on the initial blank shape. The mandrel may either be fixed to the ram or to a separate body, as indicated in Fig. 3.51c.

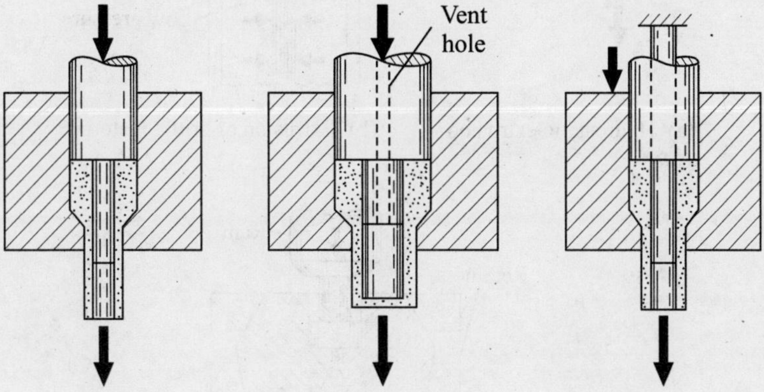

(a) Open-ended product

(b) Extruding closed products using mandrel fixed to ram

(c) Extruding closed products using mandrel fixed to frame

Fig. 3.51 Methods of extruding tubular section.

Thin-walled cans may be obtained by using impact extrusion. This process is limited to soft and ductile materials and is normally performed under cold conditions. The process is schematically shown in Fig. 3.52.

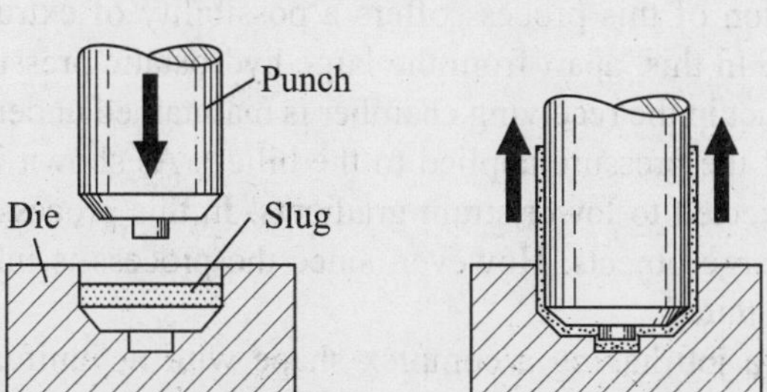

Fig. 3.52 Impact extrusion process.

Instead of applying the load on the billet directly by the ram, a fluid medium can be used, as illustrated in Fig. 3.53a. This process is known as hydrostatic extrusion; here, the frictional loss at the billet-container interface is eliminated.

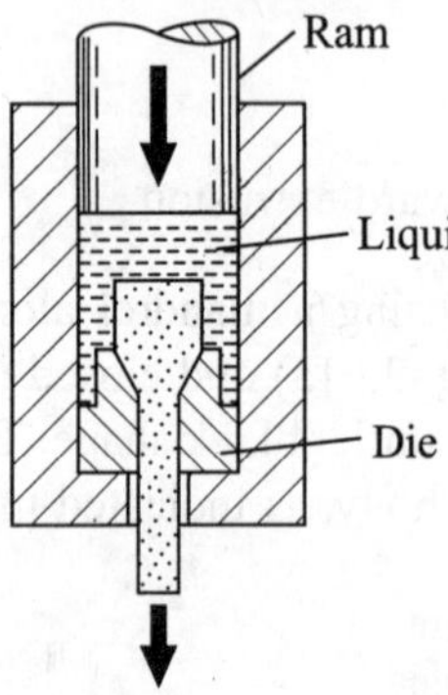

(a) Hydrostatic extrusion

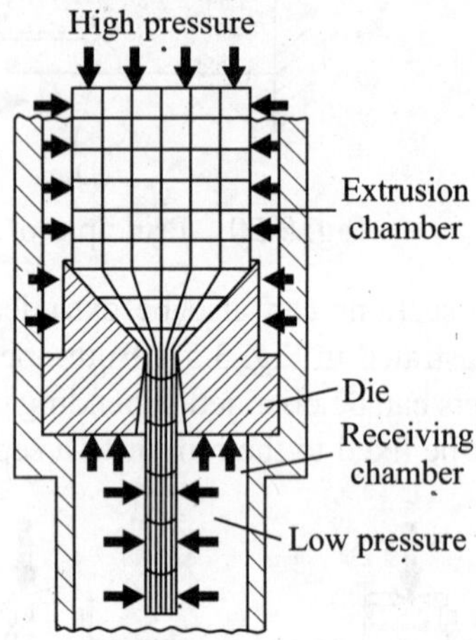

(b) Extrusion of brittle material

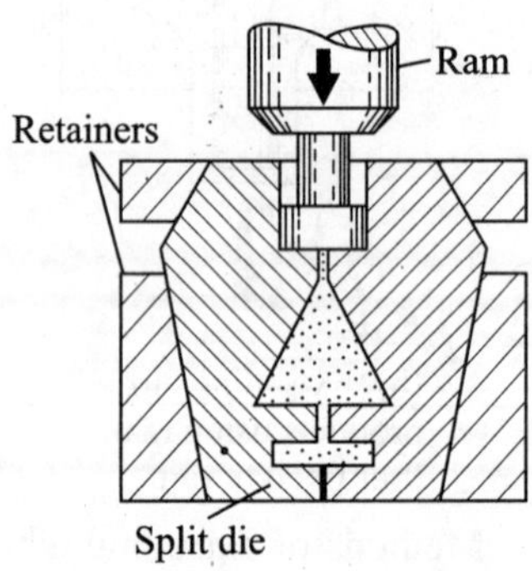

(c) Closed cavity extrusion

Fig. 3.53 Various extrusion processes.

A slight variation of this process offers a possibility of extruding a relatively brittle material. In this, apart from the large hydrostatic pressure applied to the billet, the product in the receiving chamber is maintained under a lower pressure (about one-half the pressure applied to the billet). As shown in Fig. 3.53b, the material is subjected to lower strain gradients. In this process, it is possible to produce very large objects. However, since the process is inherently slow, its application is limited.

To produce a job having a complex shape with nonuniform cross-section, closed cavity extrusion with a split die can be used. The process is similar to closed die forging and is illustrated in Fig. 3.53c.

All billets are usually covered with an oxide layer. During a normal extrusion process, this oxide layer may be drawn into the core of the product (reducing its strength characteristics) unless a laminar flow during the plastic deformation is ensured. Lubricants should be used between the billet, die, and container not only to reduce the work load but also to keep the flow laminar. As a result, the outer surface of the billet forms the skin of the product. This principle of maintaining the surface layer is also used in a hot extrusion of high strength materials and clad products as now discussed.

The temperature range of the billet during the hot extrusion of steels is 1200–1500°C. The die must be kept at a lower temperature (approximately 200°C) to avoid excessive wear rate. Glass fibres (or powders) are normally used as lubricants since the viscosity of glass is sensitive to temperature. Thus, the viscosity is high at the die surface, providing a good protection to die wear and facilitating the formation of a glass skin (about 0.025 mm thick) on the product. At the same time, the work load is reduced since the viscosity of glass is much lower at the billet-container interface.

Another useful application of this cladding process is in the production of a radioactive nuclear fuel rod of, for example, uranium and thorium. The rod is canned in copper or brass, both of which are less reactive to the atmospheric gases and protect the fuel rod from oxidation and other types of contamination. The billet is prepared with the cover made of a cladding material.

3.5.7 PUNCHING AND BLANKING

Though punching and blanking are the most common sheet metal operations involving shearing of the metal strips, there are other similar operations (see Fig. 3.54) such as (i) notching, (ii) lancing, (iii) slitting, (iv) nibbling, and (v) trimming.

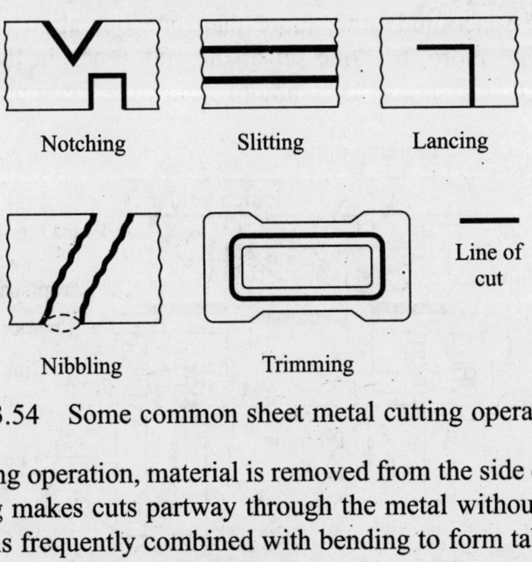

Fig. 3.54 Some common sheet metal cutting operations.

In the notching operation, material is removed from the side of a sheet metal, whereas lancing makes cuts partway through the metal without producing any scrap. Lancing is frequently combined with bending to form tabs. Slitting is an operation to cut a coiled sheet metal lengthwise to produce narrower strips. In the nibbling operation, complicated shapes are cut out from a sheet metal by producing overlapping notches, starting either from the outer boundary or from a punched hole. Without using any special tool, a simple, round or triangular punch of small dimensions is reciprocated at a fixed location. The sheet metal is guided to obtain the desired shape of the cut. Trimming refers to the removal of the excess material in a flange or flash.

In reducing the operation time and cost, the design of the die and punch for blanking plays an extremely important role. A typical simple die-punch combination is shown in Fig. 3.55. An accurate relative location of the punch and the die is maintained with the help of a set of guide posts. The stripper helps

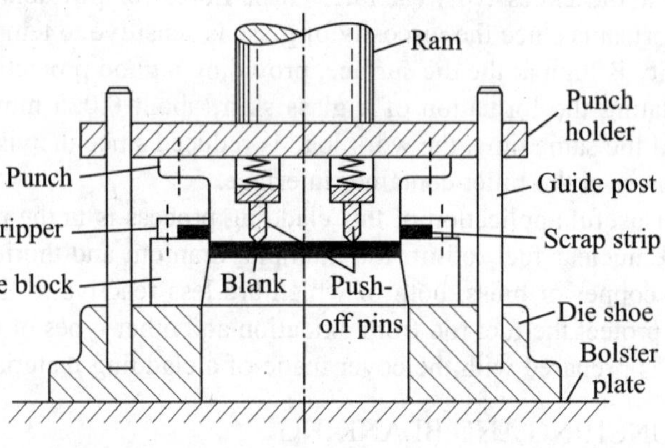

Fig. 3.55 Typical die-punch combination.

in removing the sheet metal workpiece from the punch during the return stroke, whereas the spring loaded push-off pins help in removing the blank from the punch face. The stripper also acts as a blank holder to prevent drawing.

To optimize space and time, more than one operation can be performed in a stroke, using more than one set of die and punch in the same assembly (Fig. 3.56). Such an assembly is commonly known as a compound die. It should

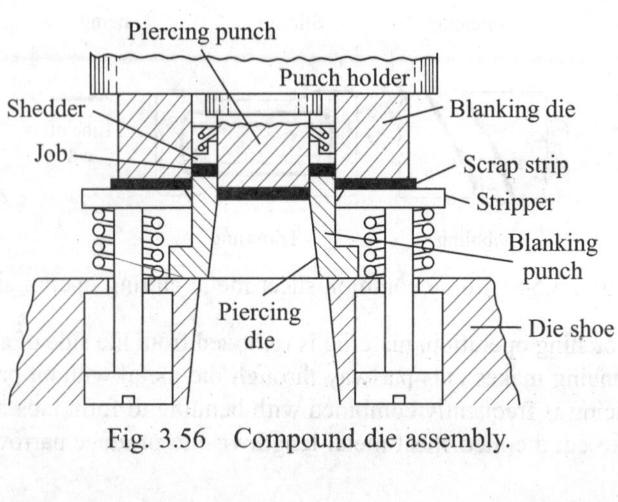

Fig. 3.56 Compound die assembly.

be noted that the blanking punch and die are in the inverted position in Fig. 3.56. It is obvious that piercing of the inner hole has to be performed before blanking. Sometimes, a combination of drawing (or bending) and blanking is also used for economy.

In the foregoing situation, more than one operation is performed in only one location. However, it is also possible to use a series of die-punch elements at different locations. Here, one operation is performed at each station and the metal stock is advanced to the next station. Thus, a continuous operation is possible. Such an assembly of dies is called a progressive die (Fig. 3.57).

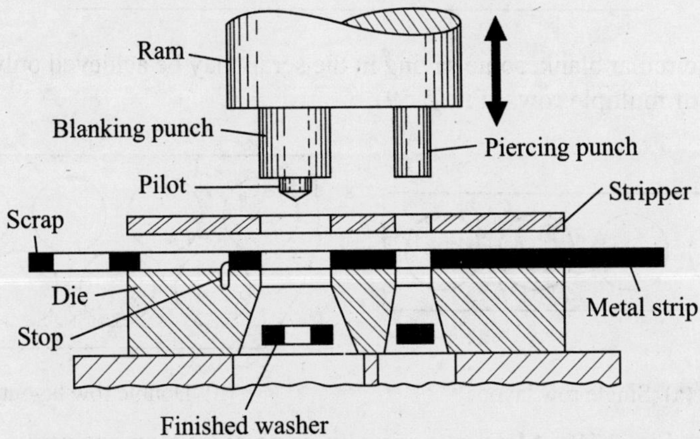

Fig. 3.57 Progressive die for producing washers.

Another important aspect of the blanking operation is to minimize the scrap by an optimum layout design (also known as nesting). This is schematically represented in Fig. 3.58. The restrictions on the layout are shown in Fig. 3.58b. The minimum gap between the edge of the blank and the side of the strip is

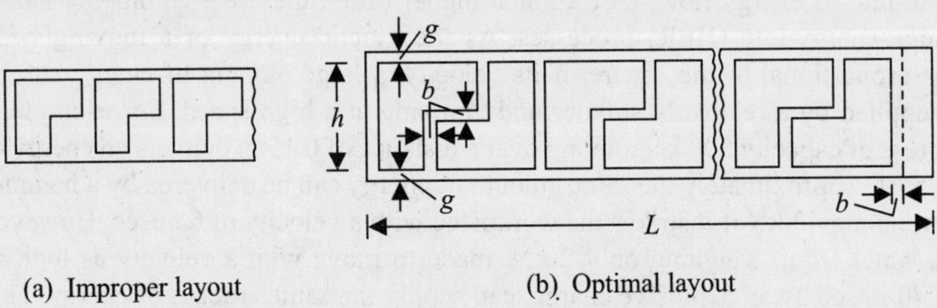

Fig. 3.58 Scrap minimization by optimal layout.

given as $g = t + 0.015h$, where t is the thickness of the strip and h is the width of the blank. The gap between the edges of two successive blanks (b) depends on the strip thickness t. Table 3.1 shows the various values of b. Sometimes, the relative direction of grain flow (when a rolled strip is used as stock) with respect to the blank is specified. In such a case, the freedom of nesting is nearly

Table 3.1 Gap between blank edge and strip boundary for various strip thicknesses

Strip thickness t (mm)	Gap b (mm)
$t \leqslant 0.8$	0.8
$0.8 \leqslant t \leqslant 3.2$	t
$t > 3.2$	3.2

lost. In a circular blank, some saving in the scrap may be achieved only through a choice of multiple rows (Fig. 3.59).

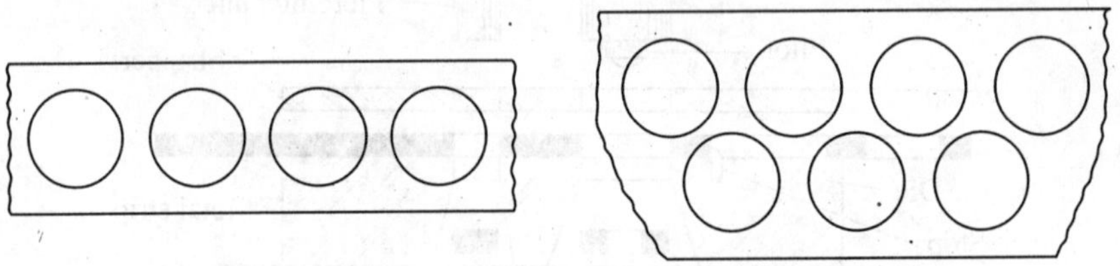

(a) Single row layout (b) Double row layout

Fig. 3.59 Material saving by using more than one row.

3.5.8 HIGH-ENERGY-RATE FORMING PROCESSES

In all the metal forming processes we have discussed, the conventional energy sources are used. In addition to these, energy sources such as chemical, magnetic, and electrical discharge can be used. Since, in all such processes, the rate of energy flow is of a much higher order, these are commonly called high-energy-rate (HER) processes. As the kinetic energy of a moving body is proportional to the square of its velocity, a large amount of energy can be supplied by a relatively smaller body moving at a high speed. For example, a press of capacity 500 kN moving over a distance of 0.15 m delivers an energy of 75 kJ. Approximately the same amount of energy can be delivered by a hammer weighing 42 kN if it strikes the workpiece with a velocity of 6 m/sec. However, a water front, weighing only 26 N, made to move with a velocity as high as 240 m/sec by an explosive charge, can supply the same amount of energy. This principle can be used in making small machines and equipment. Now, let us consider the rate of energy release in the three cases we have mentioned. In the first case, the typical time consumed is about 0.5 sec, indicating a power of 150 kW. The drop hammer takes about 0.06 sec to come to rest and the power involved turns out to be 1.25 MW. The explosive operation is completed in about 0.0007 sec, implying a power of 107 MW. This indicates that the last case results in not only the most compact but also the most powerful machine. High velocity forming operations, viz., explosive and electric discharge forming, are based on the foregoing principle.

We now discuss the three common HER processes.

Explosive Forming

Figure 3.60 shows two schemes of explosive forming. In both, a shock wave in the fluid medium (normally water) is generated by detonating an explosive

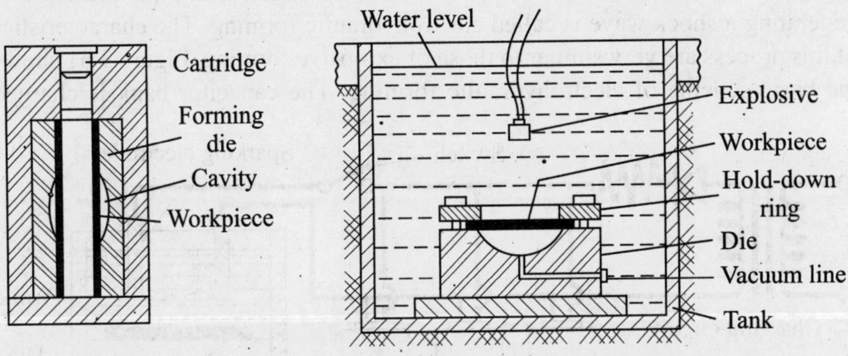

(a) Confined (b) Unconfined

Fig. 3.60 Explosive forming.

charge. For a small part, the entire shock wave front is utilized in a confined space, whereas for a large object, only a part of the wave front is used. Obviously, the unconfined operation is less efficient. However, there is a greater hazard of die failure in the confined operation due to the inevitable lack of control in explosive forming.

The typical explosives include TNT and dynamite for higher energy, and gun powder for lower energy. With high explosives placed directly over the workpiece, pressures up to 35 kN/mm^2 can be generated. With low explosives, pressures are limited to 350 N/mm^2.

With water as the transmitting medium, the peak pressure p obtained is given by

$$p = CW^{n/3}D^{-n} \text{ N/mm}^2, \tag{3.107}$$

where W is the weight of the explosive in newtons and D is the distance of the workpiece from the explosive (stand-off distance) in cm. The typical value of n is around 1.15. The constant of proportionality C has varying values for different types of explosives, as given here:

	Pentolite	TNT	Tetryl
C	4500	4320	4280

The distance between the explosive charge and the free surface of water (in unconfined forming) should be at least twice the stand-off distance. Otherwise much energy is lost, lowering down the efficiency of the operation. Using various types of tooling, we can form a variety of shapes. Generally, the effects of the process on material properties are similar to those in conventional forming.

Electrohydraulic Forming

Electric discharge in the form of sparks, instead of explosives, can also be used to generate a shock wave in a fluid. An operation using this principle of generating a shock wave is called electrohydraulic forming. The characteristics of this process are very similar to those of explosive forming. Figure 3.61 shows the basic scheme of electrohydraulic forming. The capacitor bank is charged

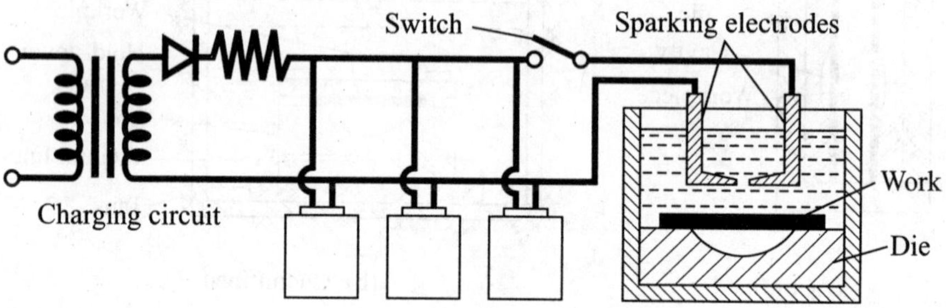

Fig. 3.61 Electrohydraulic forming.

through the charging circuit; subsequently, the switch is closed, resulting in a spark within the electrode gap to discharge the capacitors. The energy level in this process is lower than that in explosive forming. The peak pressure developed over the workpiece is a function of the amount of energy discharged (through the spark) and the stand-off distance.

Electromagnetic Forming

Just as in electrohydraulic forming, so too in electromagnetic forming, the electrical energy is first stored in a capacitor bank. This energy is then discharged through a coil by closing the switch (Fig. 3.62). The coil produces a magnetic field; the intensity of this field depends on the value of the current. Since the metallic workpiece is in this magnetic field (varying with time), a current is induced in the job which sets up its own magnetic field. The directions of these fields are such that the rigidly held coil repels the workpiece into the die. The workpiece obviously has to be electrically conductive but need not be magnetic. Short life of the coil is the major problem in such an operation.

3.5.9 MISCELLANEOUS FORMING PROCESSES

In this section, we shall briefly discuss some of the common metal forming processes other than those we have already considered.

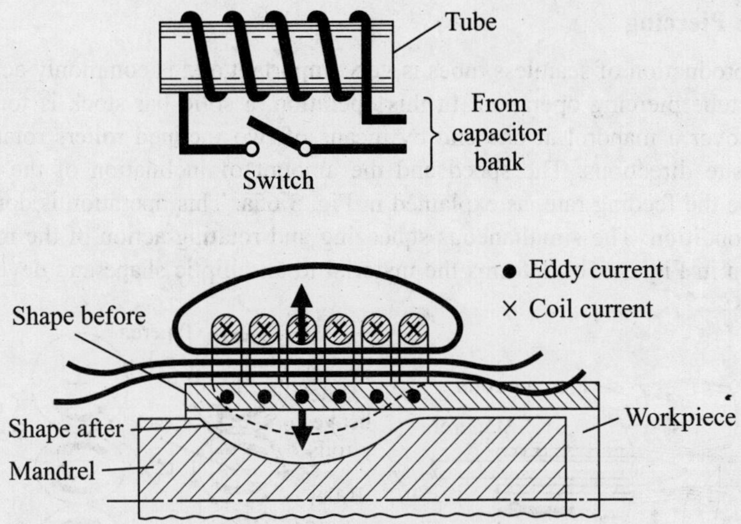

Fig. 3.62 Electromagnetic forming.

Coining

Coining is a closed die forging operation which imparts the desired variation in the thickness (because of lateral constraints) to thin and flat workpieces. As the name implies, this process is widely used in producing coins and also other similar objects requiring a well-defined impression of the die face. The process is illustrated in Fig. 3.63.

Thread Rolling

For a mass production of threaded objects, e.g., bolts and screws, two flat, reciprocating dies (or threaded rolls rotating in opposite directions) can be used to obtain the thread in the workpiece through plastic deformation. This is basically a rolling operation, and hence the name thread rolling. Figure 3.64 explains this process schematically.

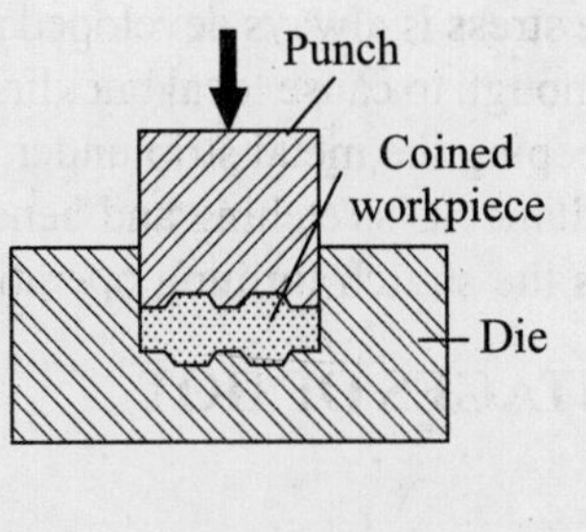

Fig. 3.63 Coining operation.

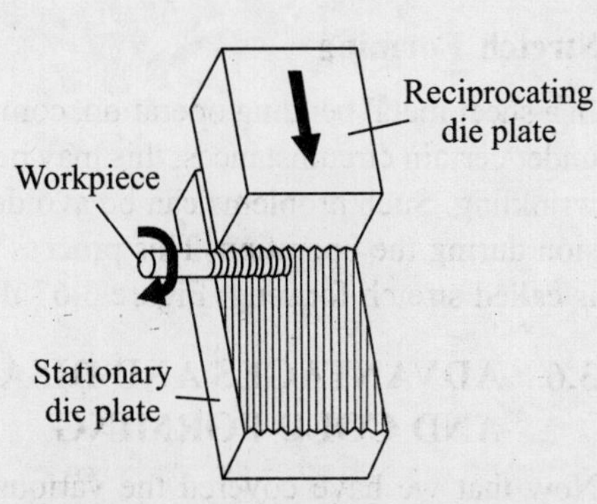

Fig. 3.64 Thread rolling with flat dies.

Tube Piercing

The production of seamless tubes is very important and is commonly achieved by a tube piercing operation. In this operation, a solid bar stock is forced to flow over a mandrel at one end by means of two inclined rollers rotating in opposite directions. The speed and the amount of inclination of the rollers decide the feeding rate, as explained in Fig. 3.65a. This operation is done in a hot condition. The simultaneous squeezing and rotating action of the rolls, as shown in Fig. 3.65b, deforms the material to an elliptic shape and develops a

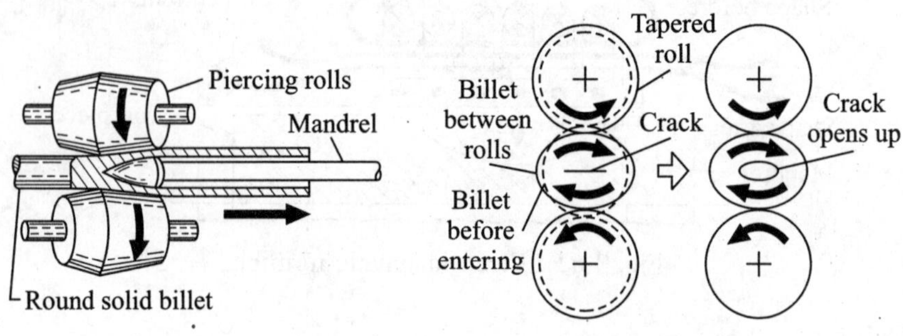

(a) Operational details (b) Deformation characteristics

Fig. 3.65 Principle of tube piercing.

crack along the major axis. A further rotation of the deformed material causes the crack to expand and transform into a hole which is finally shaped and sized by the mandrel.

Spinning

In the spinning process, an object with surface of revolution is produced from a sheet metal. The blank is held against a form die which is rotated and the sheet metal blank is laid over this die, using a specially-shaped tool or roller. If a simultaneous thinning of the sheet metal takes place during the operation, the process then is called shear spinning. Both these operations are shown in Fig. 3.66.

Stretch Forming

In a sheet metal bending operation, compressive stress is always developed and, under certain circumstances, this may be large enough to cause local buckling or wrinkling. Such problems can be avoided by keeping the metal strip under tension during the operation. This process of simultaneous stretching and bending is called stretch forming. Figure 3.67 illustrates the stretch forming operation.

3.6 ADVANTAGES AND DISADVANTAGES OF HOT AND COLD FORMING

Now that we have covered the various types of metal working operations, it would only be appropriate that we provide an overall evaluation of the hot and

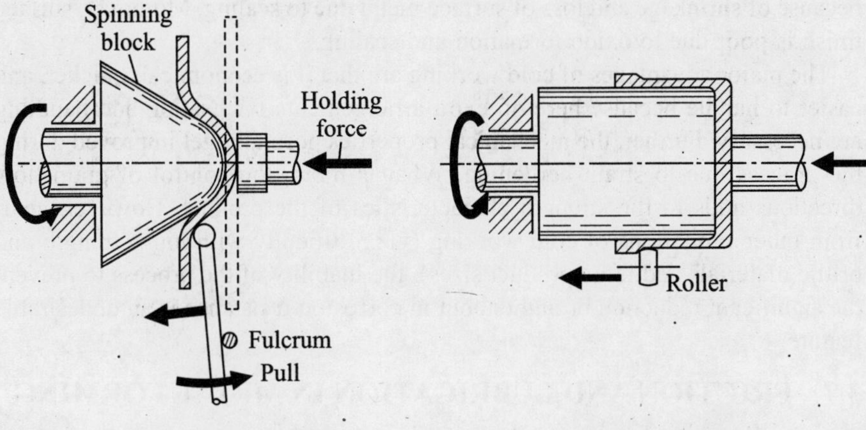

(a) Spinning (b) Shear spinning

Fig. 3.66 Spinning and shear spinning operations.

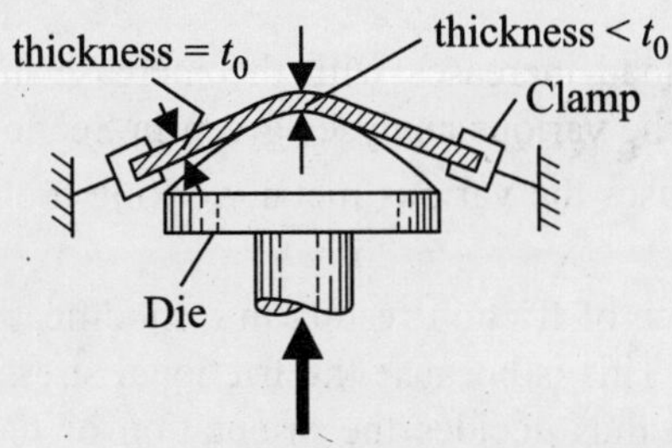

Fig. 3.67 Principle of stretch forming.

cold working processes. Such a discussion will help in choosing the proper working conditions for a given situation.

During hot working, a proper control of the grain size is possible since active grain growth takes place in the range of the working temperature. As a result, there is no strain hardening, and therefore there is no need of expensive and time-consuming intermediate annealing. Of course, strain hardening is advisable during some operations (e.g., drawing) to achieve an improved strength; in such cases, hot working is less advantageous. Apart from this, strain hardening may be essential for a successful completion of some processes (for example, in deep drawing, strain hardening prevents the rupture of the material around the bottom circumference where the stress is maximum). Large products and high strength materials can be worked upon under hot conditions since the elevated temperature lowers down the strength and, consequently, the work load. Moreover, for most materials, the ductility increases with temperature and, as a result, brittle materials can also be worked upon by the hot working operation. It should, however, be remembered that there are certain materials (viz., steels containing sulphur) which become more brittle at elevated temperatures. When a very accurate dimensional control is required, hot working is not advised because of shrinkage and loss of surface metal due to scaling. Moreover, surface finish is poor due to oxide formation and scaling.

The major advantages of cold working are that it is economical, quicker, and easier to handle because here no extra arrangements for heating and handling are necessary. Further, the mechanical properties normally get improved during the process due to strain hardening. What is more, the control of grain flow directions adds to the strength characteristics of the product. However, apart from other limitations of cold working (viz., difficulty with high strength and brittle materials and large product sizes), the inability of the process to prevent the significant reduction brought about in corrosion resistance is an undesirable feature.

3.7 FRICTION AND LUBRICATION IN METAL FORMING

Friction plays an important role in all metal working processes. Though, in general, the presence of friction increases the work load and the loss of energy, it has certain beneficial aspects. We shall now describe the major effects of friction.

(i) The work load increases with the increase in friction. This can be easily verified from the various analyses given in Section 3.4.

(ii) Friction causes the various metal working tools, e.g., die and roll, to wear out.

(iii) The presence of friction results in a modification of the deformation pattern of the metal. This is because the frictional stress changes the principal stress directions and thus decides the orientation of the atomic planes along which the slip occurs.

(iv) Friction introduces macroscopic inhomogeneity since it occurs only at the surface. High friction also causes the surface layer to mix with the inner core. As the surfaces are normally contaminated, this phenomenon results in inhomogeneous and weaker products. The frictional drag on the surface layer may also result in the development of micro cracks on the surface, considerably affecting the fatigue strength of the product.

(v) High friction also increases the tendency of the formation of dead zones (especially in extrusion) and built-up edges which, in turn, spoil the surface finish of the product. However, it has been observed that a matted surface is produced when the lubricant film is too thick. Thus, to produce a bright surface, it is not desirable to maximize the lubrication efficiency (though it seems reasonable from the point of view of friction considerations).

(vi) To investigate the role of friction in strip rolling, let us consider Fig. 3.68 which shows the beginning of the operation. It is clear that if the strip has to enter the rolls unaided, the horizontal component of the friction (per unit width) force must at least be equal to the horizontal component of the roll separating force $p \cdot \Delta l$. Thus,

$$\mu p \cdot \Delta l \cos \theta_i \geqslant p \cdot \Delta l \sin \theta_i$$

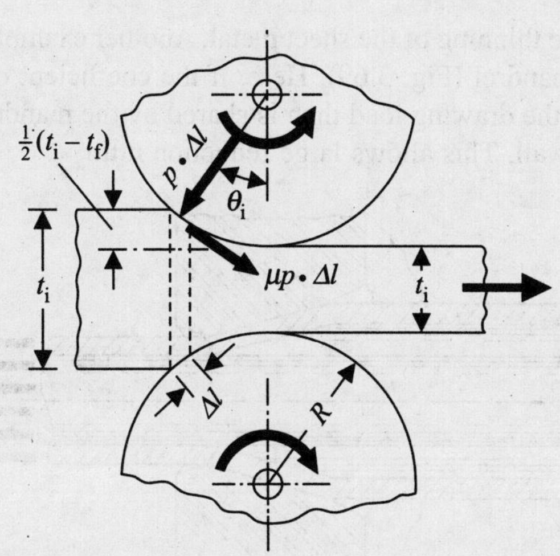

Fig. 3.68 Role of friction in strip rolling.

or

$$\mu \geqslant \tan \theta_i$$

or

$$\mu \geqslant \theta_i \quad \text{(for small } \theta_i\text{)}.$$

Now, it can be easily shown that (for small θ_i)

$$\theta_i = \sqrt{\left(\frac{t_i - t_f}{R}\right)}.$$

Thus,

$$\mu \geqslant \sqrt{\left(\frac{t_i - t_f}{R}\right)}.$$

So,

$$(t_i - t_f) \leqslant \mu^2 R.$$

Therefore,

$$(t_i - t_f)_{max} = \mu^2 R.$$

Hence, it is clear that high friction increases the maximum draft possible. Moreover, for a specified draft $(t_i - t_f)$, the minimum coefficient of friction, required for the unaided bite to be possible, is given by the relation

$$\mu_{min} = \sqrt{\left(\frac{t_i - t_f}{R}\right)}.$$

(vii) Sometimes, a proper distribution of friction improves the deformation process. For example, in deep drawing, the coefficient of friction around the punch corner should be high and that around the die corner should be low in order to prevent the thinning of the sheet metal. Another example is the drawing of a tube over a mandrel (Fig. 3.69). Here, if the coefficient of friction at the interfaces is high, the drawing load then is shared by the mandrel, reducing the stress in the tube wall. This allows large reduction ratio.

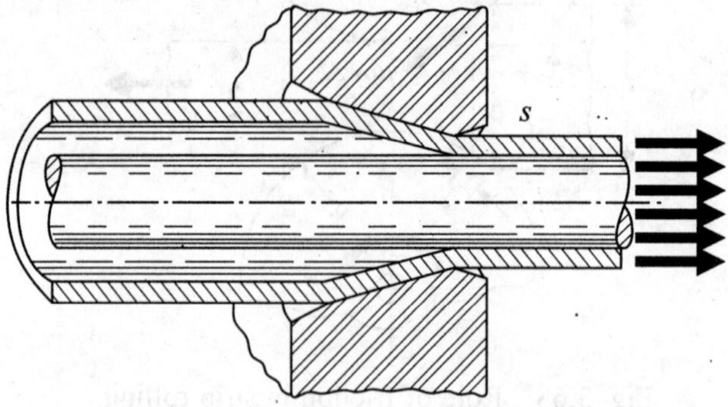

Fig. 3.69 Tube drawing over mandrel.

One of the most important requirements for a metal working operation to be successful is the proper lubrication of the contacting surfaces in relative motion. The three important purposes served by lubrication are (i) prevention of pick up (this means transfer of metal from one body to another due to direct contact), (ii) reduction of die wear, and (iii) reduction of friction (this is the least important among the three). To prevent pick up, a separating film of the lubricant must be maintained between the two metal surfaces. This can be achieved through hydrodynamic lubrication where the viscous lubricating fluid separates the surfaces. In principle, any material, including a solid one, having a very low shear strength can be used for lubrication. Some solid materials, e.g., graphite and molybdenum disulphide, satisfy this requirement; in addition, such materials provide a low coefficient of friction even at a high temperature. For a list of the typical lubricants used in various metal working operations for different job materials, see the well-known texts[1].

3.8 DEFECTS IN METAL FORMING

It is not difficult to realize that without proper precautions the products of the various metal working processes may not be free from defects. Often, the defects in a given stock may sustain or even aggravate during a metal working operation. But sometimes new defects may also develop depending on the characteristics of the operation. Therefore, some understanding of the basic reasons for the origin of the defects is necessary to find the remedial measures.

The sources of defects in a typical rolling operation are mainly surface irregularities, nonmetallic inclusions, and internal blow holes in the stock. The first source may lead to a trapping of scales which remain inside as laps.

[1]Schey, J.A., Introduction to Manufacturing Processes, McGraw-Hill, New York, 1977.

The second source produces, especially during the hot rolling of a thick slab, crocodile cracks, separating the product into two halves. The third source results in an elongation of the blow holes and the product becomes weaker. During the rolling of a complex shape, nonuniform deformation (as some portions elongate due to direct compression, whereas the adjoining sections, free from direct compression, elongate due to the dragging action of the portions under direct compression) gives rise to the development of secondary tensile stresses, resulting in internal cracks. If the deformation takes place mostly at the surface, the product may be left with residual stresses. The foregoing defects can be minimized by a careful inspection of the billet and by keeping the roll smooth and clean. To avoid the internal cracks, it is necessary to properly design the roll pass.

The faults in forged products originate from various sources, e.g., defects in the stock, improper heating, and poor die design. The cracks around the corners and at right angles to the forged surface, known as cold shuts, arise due to the misplacement of the metal in the die. In closed die forging, sometimes unfilled sections result mainly due to less amount of metal, improper heating, and misplacement of the metal. If the scales are not removed from the die surface, these are transferred to the subsequent forgings, which, when removed, leave scale pits. If the two halves of the die are misaligned, mismatched forging may result. When the lubrication is insufficient, too much barrelling (bulging of the free surface due to the pressure from the internal flowing material during open die forging) along with cracks on the free barrelled surface may develop.

The typical surface defects in rod and wire drawing are due to a ploughing by hard particles and local breakdown of the lubricating film. The other kinds of defects include the formation of a bulge, ahead of the die, with low reduction and high die angle, and the development of a centre burst with too large a deformation gradient along the cross-section. Figure 3.70a shows a case of drawing where the half die angle α is large and the reduction ratio is low, i.e., h/L is large. In such a situation, the deformation is localized near the surface and the relatively undeformed central portion is dragged forward by the upper layer. This causes the development of secondary tensile stresses, resulting in the centre-burst defect. Figure 3.70b shows the centre-burst defect during extrusion. When h/L is less than 2, the danger is less.

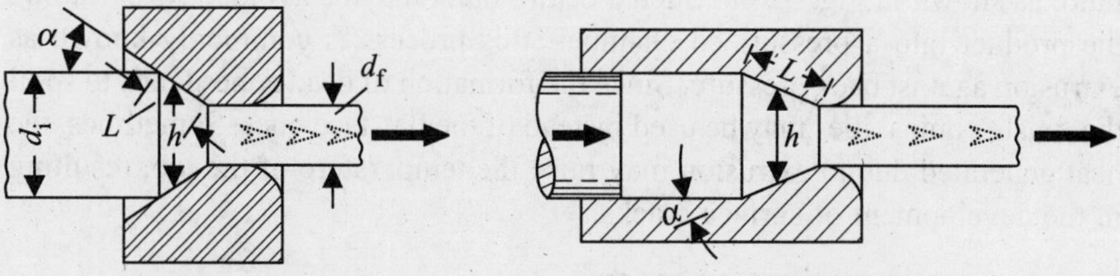

(a) In drawing (b) In extrusion

Fig. 3.70 Development of centre-burst defect.

As already noted, many types of deformations are involved in the deep drawing operations on sheet metals. Figure 3.71 illustrates the various types of common deep drawing defects. An insufficient blank holder pressure causes

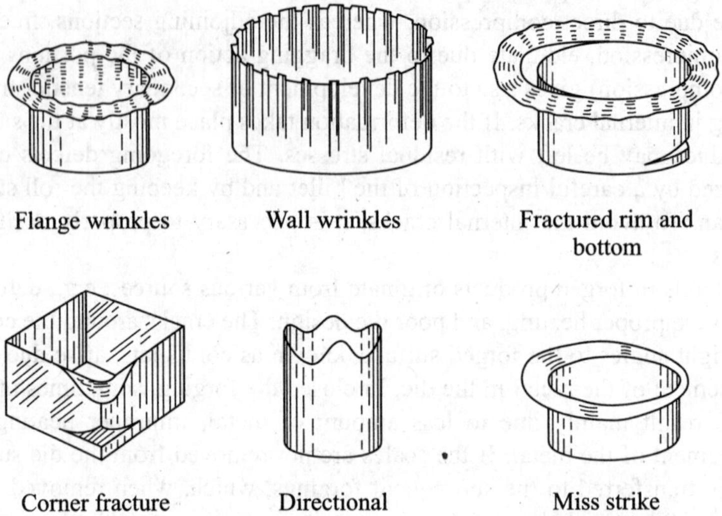

Fig. 3:71 Some common deep drawing defects.

wrinkles to develop on the flange, which may also extend to the wall of the cup. Further, too much of a blank holder pressure and friction may cause a thinning of the walls and a fracture at the flange, bottom, and the corners (if any). While drawing a rolled stock, ears or lobes tend to occur because of the anisotropy induced by the rolling operation. Moreover, due to the misplacement of the stock, unsymmetrical flanges may result. This type of defect is commonly referred to as a miss strike. The effect of a large grain size is to produce a dull surface (orange peel effect). This effect is also common in the bending operations.

The most common defect in extrusion (known as the extrusion defect) arises from the back flow of the material, pushing the end face of the billet into the core of the product. Such a defect weakens the product since the surface layer is normally contaminated by oxides. A centre burst is also possible with large h/L ratio, as shown in Fig. 3.70b. Such a centre burst may be avoided by extruding the product into a pressurized chamber; this process is commonly known as extrusion against back pressure. Since the formation of dead zones tends to spoil the finish, conical dies may be used instead of the flat-face ones. Sometimes, the heat generated due to extrusion may raise the temperature of the job, resulting in the development of surface cracks.

3.9 EXERCISE PROBLEMS

3.1 A 5-mm-thick aluminium alloy strip is rolled to a thickness of 4 mm, using steel rollers of radius 100 mm. The tensile yield stress of aluminium is

0.28 kN/mm^2. Determine (i) the minimum coefficient of friction μ_{min} between the workpiece and the rolls for an unaided bite to be possible, (ii) the angle subtended by the contact zone at the roll centre, and (iii) the location of the neutral point with $\mu = \mu_{min}$.

3.2 (i) Assuming a rolling speed of 30 m/min, determine (a) the roll separating force, and (b) the power required for the rolling operation described in Exercise 3.1. The width of the strip is 200 mm. The coefficient of friction μ_{min} may be taken as calculated in Exercise 3.1. Neglect strain hardening.

(ii) Determine the deflections of the rolls at their midpoints if the length of the rolls is 400 mm. Use the standard values for the properties of steel. How can this information be used in designing the roll profiles?

3.3 In Exercise 3.2, recalculate the power requirement after accounting for the frictional losses in the bearings. The coefficient of bearing friction = 0.005 and the bearing diameter = 75 mm.

3.4 A 20 mm × 20 mm × 160 mm copper plate is forged between two flat dies to a final size of 10 mm × 40 mm × 160 mm. Determine the peak forging force, assuming the coefficient of friction to be 0.2. The tensile yield stress of copper can be taken as 70 N/mm^2. Assume no strain hardening.

3.5 Solve Exercise 3.4 with $\mu = 0.2$ and the strain hardening characteristics for copper given as

$$\sigma_Y = 70 + 30\varepsilon^{0.33} \text{ N/mm}^2.$$

3.6 An aluminium disc with 200 mm diameter and 25 mm thickness is forged to a final thickness of 15 mm. Estimate the maximum forging force when the coefficient of friction = 0.3 and the tensile yield stress = 25 N/mm^2. Neglect strain hardening.

3.7 Let the drawing operation described in Example 3.7 be carried out by using two dies in series. Assume that the reduction achieved is the same in both the dies (keeping the total reduction the same). Using the same data, determine the drawing power and compare it with the value obtained in Example 3.7.

3.8 For Exercise 3.7, determine the worst distribution of reduction between the dies.

3.9 A channel section is formed from a sheet metal, as shown in Fig. 3.72. Each of the corner radii of the rectangular punch is r_c. Derive an expression for the forming force F. The coefficient of friction and the shear yield stress of the material can be taken as μ and K, respectively. Further, the clearance between the die and the punch is equal to the sheet thickness.

3.10 Figure 3.73 shows a tubular extrusion process. Making suitable assumptions, derive an expression for the force required to perform the operation.

3.11 A cylindrical lead alloy billet of 50 mm diameter and 100 mm length is extruded to a final diameter of 25 mm by using a direct extrusion process. The average tensile yield stress for the alloy is 12 N/mm^2. Estimate the maximum force required and the fraction of the total power lost in friction for this operation.

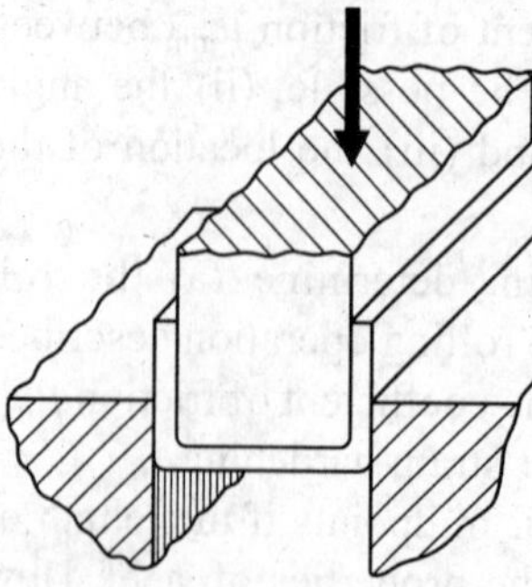

Fig. 3.72 Forming channel section.

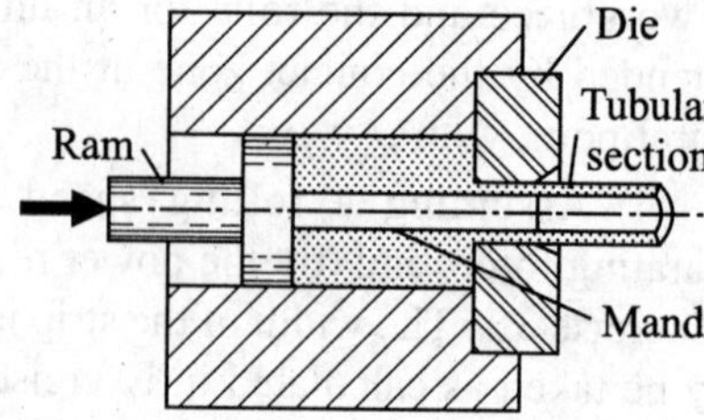

Fig. 3.73 Tubular extrusion.

3.12 If the operation in Exercise 3.11 is performed by a backward extrusion process, determine the maximum force required. Make suitable assumptions to first derive the expression for the force.

3.13 A cup of inside radius 40 mm and thickness 5 mm is to be drawn from a blank with a radius of 50 mm. The shear yield stress and the maximum allowable stress in the material can be taken as 14 N/mm^2 and 50 N/mm^2, respectively. Determine (i) the drawing force, and (ii) the minimum possible radius of the cup which can be drawn from the given blank without causing a fracture. Given $\mu = 0.1$ and $\beta = 0.05$.

3.14 An aluminium strip of 50 mm width and 3 mm thickness is bent into a 90° angle. The data given is

fracture strain = 0.25, $\mu = 0.1$, $\sigma_{Y_0} = 14$ N/mm^2,

$E = 70$ kN/mm^2, die opening = 30 mm,

coefficient of linear strain hardening (n) = 30 N/mm^2.

Determine (i) the minimum possible bending radius, (ii) the angle of the bending punch, taking care of the spring back, and (iii) the peak bending force.

3.15 For the bending operation described in Exercise 3.14, plot the variation of peak bending force with die opening.

3.16 Circular aluminium blanks with 25 mm diameter and 3 mm thickness are to be produced as the starting material for the tooth paste tube extrusion. The press has a capacity of 500 kN. The fracture strain and stress of the material are 2 and 80 N/mm^2. No shear is provided to the punch. Determine (i) the optimum clearance between the die and the punch, (ii) the maximum number of blanks which can be punched simultaneously per stroke, and (iii) the power required if the punch speed is 60 cycles/min.

3.17 A charge of TNT weighing 10 N is used in an unconfined explosive forming operation. Plot the peak pressure over the work surface with stand-off distance.

3.18 When extruding 40-mm-diameter billets through a die with a 28-mm opening 45°-half-cone angle, the centre-burst defect is found to be very frequent. Suggest a remedy along with the justification.

4 Machining Processes

4.1 INTRODUCTION

In Chapters 2 and 3, we discussed two methods, viz., casting and forming, for obtaining the desired shape, size, and finish. Both these methods require a suitable movement of the material particles either in the liquid or in the solid state. However, both the methods have limitations and, in a very large number of situations, neither is suitable. Casting imposes severe problems from the point of view of material properties and accuracy, whereas forming becomes impractical when the job is either very large (requiring very large forming forces and huge machines) or the material is not suitable for the forming operation. Apart from these, the geometric complexity of the final job may be such that these methods, again, are of no use.

The foregoing problems can be overcome if the method used is such that the desired shape, size, and finish are obtained through the removal of excess material (from the original workpiece of a suitable size and shape) in the form of small chips. In most cases, a large scale removal is not possible either for geometric reasons or for the size involved. This process is termed as machining, and is perhaps the most versatile manufacturing process. The body which removes the excess material through a direct mechanical contact is called the cutting tool and the machine which provides the necessary relative motions between the work and the tool is commonly known as the machine tool.

Since the removal of material takes place only in the form of small chips, the machining of a finite area requires a continuous feeding of the uncut portion at a suitable rate. The relative motion (between the tool and the work) responsible for the cutting action is known as the primary or cutting motion, and that responsible for gradually feeding the uncut portion is termed as the secondary or feed motion.

Depending on the nature of the two relative motions, various types of surfaces can be produced. To explain this in more detail, let us consider a point P (Fig. 4.1) where the material is being cut at a particular instant. Now, if the cutting motion is rectilinear and the feed (provided after the completion of each cutting stroke) is also rectilinear (Fig. 4.1a), the machined surface will be plain (Fig. 4.1b). The line generated by the cutting motion is called the generatrix and the line from the feed motion is termed as the directrix (Fig. 4.1c). So, various geometries can be obtained depending on the shapes of the generatrix and the directrix and their relative directions. To understand why the relative directions are important, let us consider the two situations in Fig. 4.2. Figure 4.2a

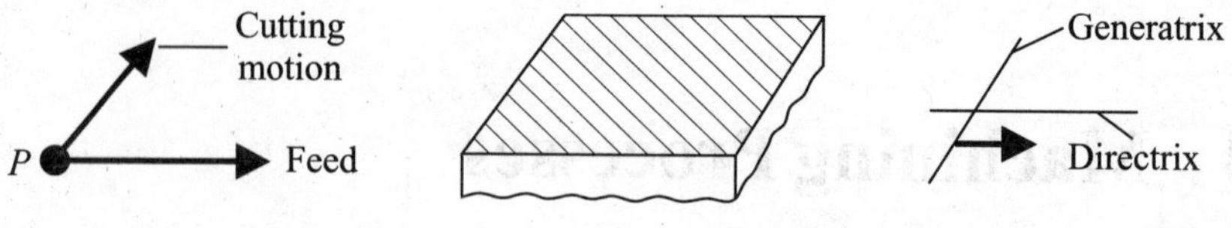

(a) Motions of cutting point (b) Production of surface (c) Generatrix and directrix

Fig. 4.1 Concept of generatrix and directrix.

shows a case with the generatrix as a circle and the directrix as a straight line perpendicular to the plane of the generatrix. It is seen that the surface obtained is cylindrical. If the same circular generatrix and the rectilinear directrix lie in the same plane, two straight lines are generated, as explained in Fig. 4.2b. So, a series of such generatrix-directrix combinations, placed side by side in a direction perpendicular to the plane of the generatrix, result in a plain surface. In Fig. 4.2a, the surface is generated by a direct tracing of the generatrix, and in Fig. 4.2b, the final surface geometry is given by the envelope of the generatrices,

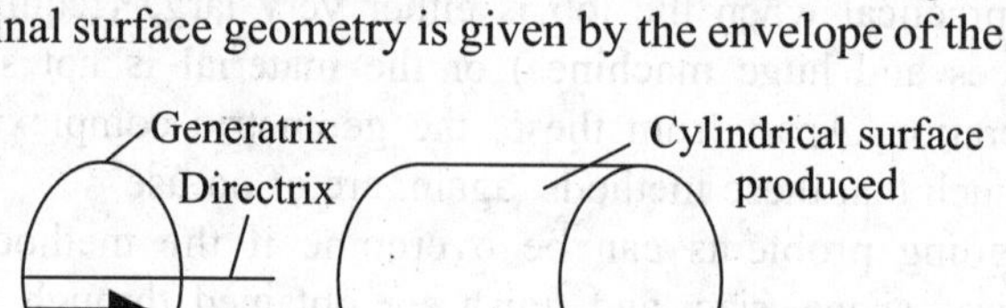

(a) Directrix perpendicular to plane of generatrix

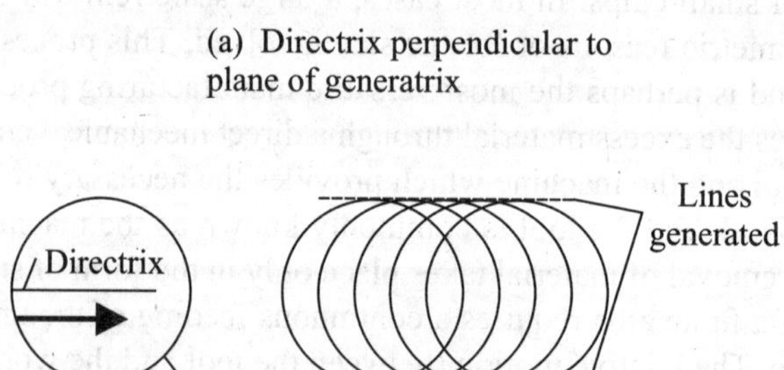

(b) Directrix in the plane of generatrix

Fig. 4.2 Cases with circular generatrix and straight directrix.

and the process is known as generation. Table 4.1 gives common examples of surface generation. In practice, cutting is really never done at a point but along a cutting edge. Thus, several generatrices and directrices are involved and generation yields not a line but a surface.

We shall now give brief descriptions of the common machining processes.

(i) *Shaping and planing* In shaping and planing, the surface obtained is plain (Number 1 in Table 4.1). In shaping, the cutting tool is given a reciprocating

Table 4.1 Generation of various surfaces

Number	Generatrix	Directrix	Process	Surface obtained
1	Straight line	Straight line	Tracing	Plain
2	Circular	Straight line	Tracing	Cylindrical
3	Plain curve	Circular	Tracing	Surface of revolution
4	Circular	Straight line	Generation	Straight line (plain surface in practice)

motion, and after every cutting stroke, the work is fed (Fig. 4.3) perpendicularly (during the return stroke, the work is advanced by a small distance) in order to provide a layer of the uncut material to the tool. Since here the cutting is not continuous, the machining is known as an intermittent cutting operation.

For a long job, it becomes inconvenient to provide long cutting strokes with the mechanism used in a shaping machine. In such a case, the work is provided with the cutting motion, whereas the feed is given to the tool; this operation is known as planing. The basic geometry of the machining operation is the same as that of shaping.

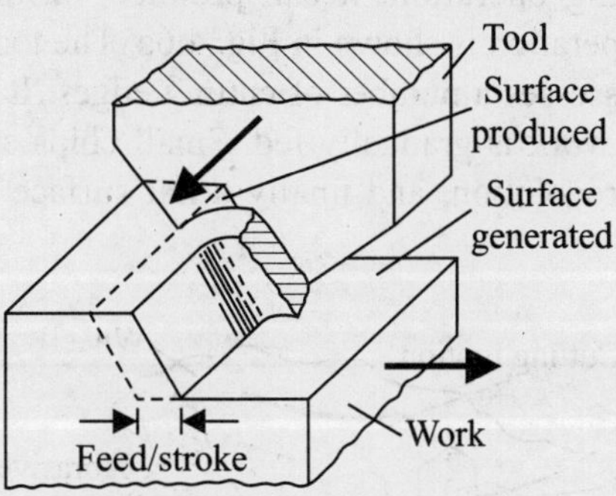

Fig. 4.3 Shaping operation.

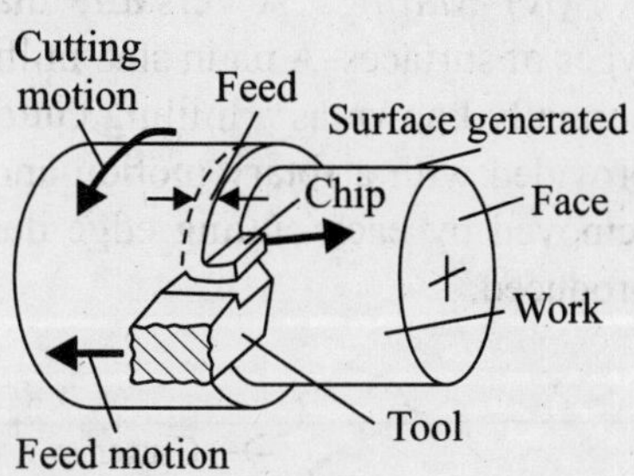

Fig. 4.4 Turning operation.

(ii) *Turning* This is a very basic operation and produces a cylindrical surface. Of course, by face turning, a flat surface can also be obtained. The machine tool used for this type of an operation is known as a lathe. Figure 4.4 shows a typical turning operation where a workpiece in the form of a cylindrical bar is rotated about the axis of symmetry. The tool is provided with a feed motion parallel to the work axis. Thus, it is easy to see that with respect to the work the tool has a helical motion and always encounters an uncut layer of the workpiece. Here, the machining operation is continuous. This operation results in a reduced work diameter and a new cylindrical surface. □hen the tool is fed in the radial direction along the face (Fig. 4.4), a flat surface is produced and the length of the workpiece gets reduced.

(iii) *Drilling* This is used for making a hole in a solid body. Figure 4.5 shows the operation schematically. The cutting motion is provided to the two cutting edges (lips) by rotating the drill, and the feeding is done by giving a rectilinear motion to the drill in the axial direction. The final surface obtained is an internal cylindrical surface.

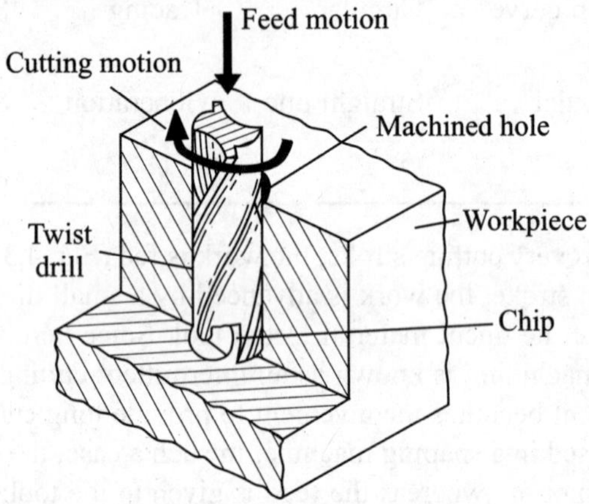

Fig. 4.5 Drilling operation.

(iv) *Milling* A versatile machining operation, it can produce various types of surfaces. A plain slab milling operation is shown in Fig. 4.6a. The tool, normally known as a milling cutter, possesses a number of cutting edges. It is provided with a rotary motion and the work is gradually fed. Small chips are removed by each cutting edge during revolution, and finally a flat surface is produced.

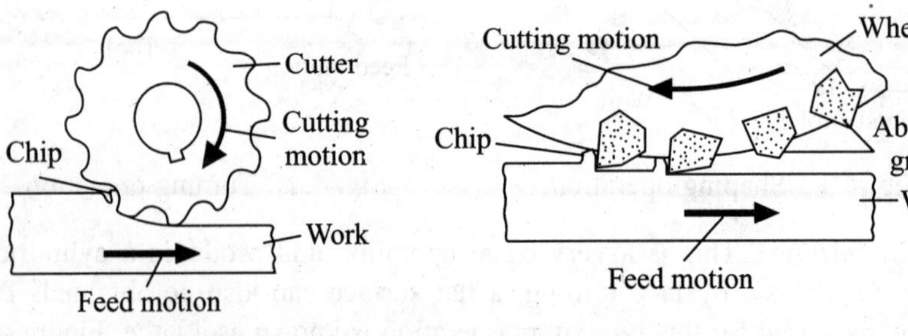

(a) Scheme of milling operation (b) Scheme of grinding operation

Fig. 4.6 Machining with multipoint tools.

(v) *Grinding* In grinding, the cutting tools are the sharp edges of the abrasive grains of the grinding wheel. These grains are very large in number and have a random orientation and distribution. However, if a particular grain is observed, its action would be as shown in Fig. 4.6b. Of course, the size of the chips removed by a grain is exceedingly small.

4.2 MECHANICS OF BASIC MACHINING OPERATION

Before we start our discussion on the mechanics of the machining operation, it is advisable that we study the basic similarity in the nature of material removal during the different types of machining operations (Fig. 4.7). The sectional view *XX* of the actual cutting zone in Figs. 4.7a, 4.7b, and 4.7c and an enlarged view of the cutting zone in Fig. 4.7d show that the basic nature of material removal

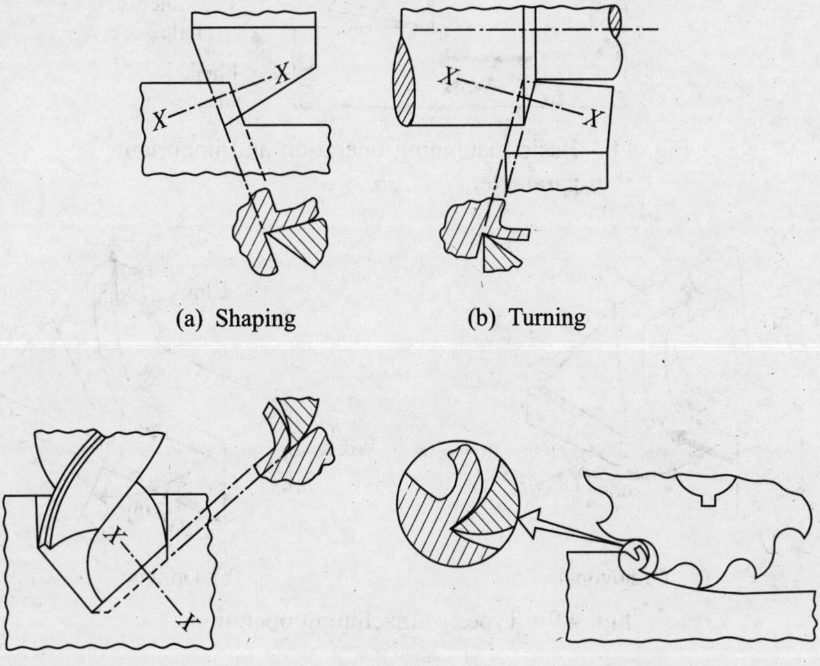

(a) Shaping (b) Turning

(c) Drilling (d) Milling

Fig. 4.7 Simulation of actual machining processes.

in each of these operations is similar and can be represented as in Fig. 4.8. The important parameters involved are (i) the thickness of the uncut layer (t_1), (ii) the thickness of the chips produced (t_2), (iii) the inclination of the chip-tool interface with respect to the cutting velocity (the face of the tool in contact with the chip is commonly known as the rake face), i.e., the rake angle (α), and (iv) the relative velocity of the workpiece and the tool (v). To make cutting possible, a clearance angle between the job and the flank surfaces is also provided. It should be noted that the machining operation can be represented by a two-dimensional figure (see Fig. 4.8) only when all the work and chip material particles move in the planes parallel to the plane of the paper. There is no component of velocity or motion in the direction perpendicular to the plane of the paper. Such a situation is realized when the cutting edge is straight and the relative velocity of the work and the tool is perpendicular to the cutting edge. Figure 4.9a shows such an idealized case. This type of machining is known

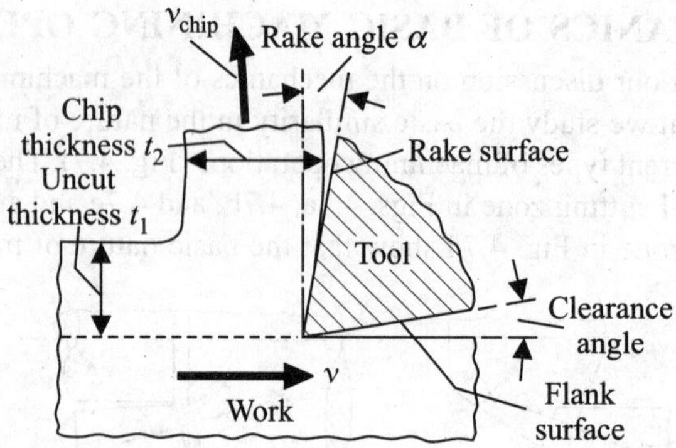

Fig. 4.8 Basic machining operation and important parameters.

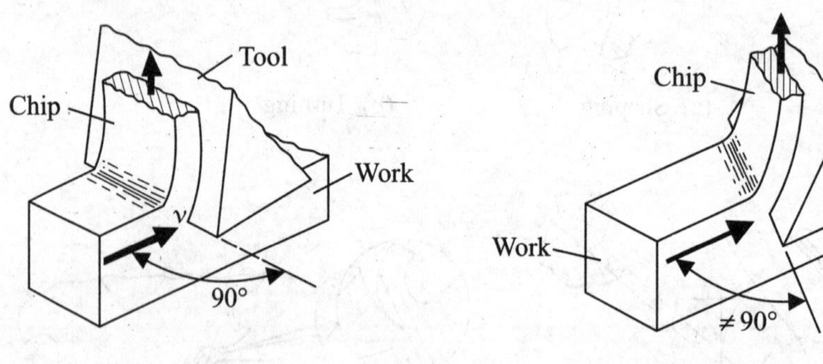

(a) Orthogonal (b) Oblique

Fig. 4.9 Types of machining operations.

as orthogonal machining. On the other hand, when the relative velocity of the work and the tool is not perpendicular to the cutting edge (Fig. 4.9b), all the work and chip material particles do not move in parallel planes, and thus a two-dimensional representation of the operation is not possible. Such a machining is termed as an oblique machining. It is obvious that the study and analysis of the orthogonal machining operation is much simpler. We shall therefore discuss examples from this category when dealing with the basic mechanics of the machining operation. However, in a large number of situations, the conditions for orthogonal machining are not satisfied, but to get an approximate result, the orthogonal analysis may be sufficient. For an analysis of the oblique machining, the interested reader may consult any standard book on metal cutting.

4.2.1 MECHANISM OF CHIP FORMATION

When the zone under the cutting action is carefully examined, the following observations can be made. The uncut layer deforms into a chip after it goes through a severe plastic deformation in the primary shear zone (Fig. 4.10a). Just after its formation, the chip flows over the rake surface of the tool and the

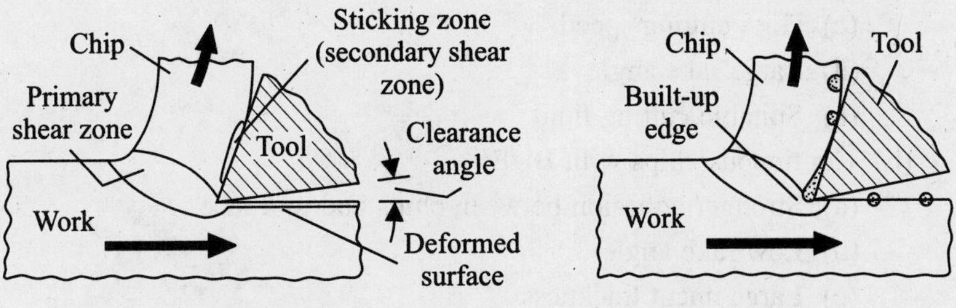

(a) Shear zones (b) Built-up edge formation

Fig. 4.10 Material deformation and associated phenomena in machining.

strong adhesion between the tool and the newly formed chip surface results in some sticking. Thus, the chip material at this surface (and the adjacent layers) undergoes a further plastic deformation since, despite the sticking, it flows. This zone is referred to as the secondary shear zone. Under suitable conditions, the machining operation is smooth and stable and produces continuous ribbon-like chips. As a result, the surface produced is smooth and the power consumption is not unnecessarily high.

At a somewhat high speed, the temperature increases and the tendency of the plastically deformed material to adhere to the rake face increases and a lump is formed at the cutting edge (Fig. 4.10b). This is called a built-up edge (BUE); it grows up to a certain size but ultimately breaks due to the increased force exerted on it by the adjacent flowing material. After it breaks, the broken fragments adhere to the finished surface and the chip surface results in a rough finish. With a further increase in the cutting speed or when a cutting fluid is used, the built-up edge disappears.

When the machining is performed at a very low speed or the work material is brittle, the shearing operation on the work material does not continue without causing a fracture. The ruptures occur intermittently, producing discontinuous chips. Figure 4.11 shows the progress of the formation of discontinuous chips. The resulting surface is rough.

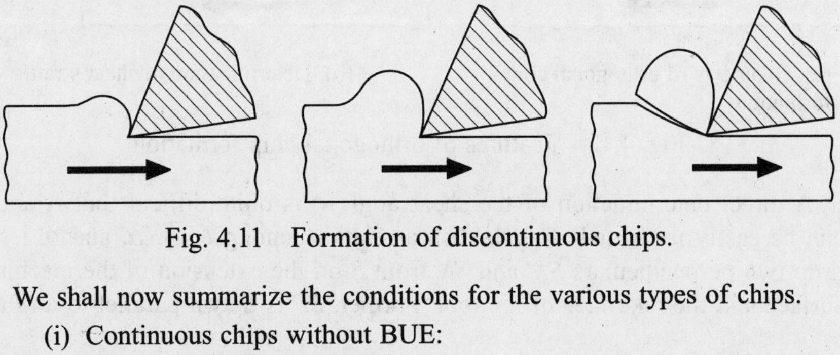

Fig. 4.11 Formation of discontinuous chips.

We shall now summarize the conditions for the various types of chips.

(i) Continuous chips without BUE:

(a) Ductile material

(b) Small uncut thickness

(c) High cutting speed
(d) Large rake angle
(e) Suitable cutting fluid

(ii) Continuous chips with BUE:
(a) Stronger adhesion between chips and tool face
(b) Low rake angle
(c) Large uncut thickness

(iii) Discontinuous chips:
(a) Low cutting speed
(b) Brittle work material
(c) Small rake angle
(d) Large uncut thickness

It is obvious that continuous chips without BUE are most desirable, and the machining is steady. We shall confine our discussion to such cases.

An analysis of the actual machining operation, schematically shown in Fig. 4.10a, is very difficult. But it is observed that under normal machining conditions at moderate and high speeds the thickness of the shear zone is very small and it can be idealized as a plane (Fig. 4.12a). The plane *OS* where the shear[1] occurs is known as the shear plane and its inclination with the machined surface (ϕ) is called the shear angle.

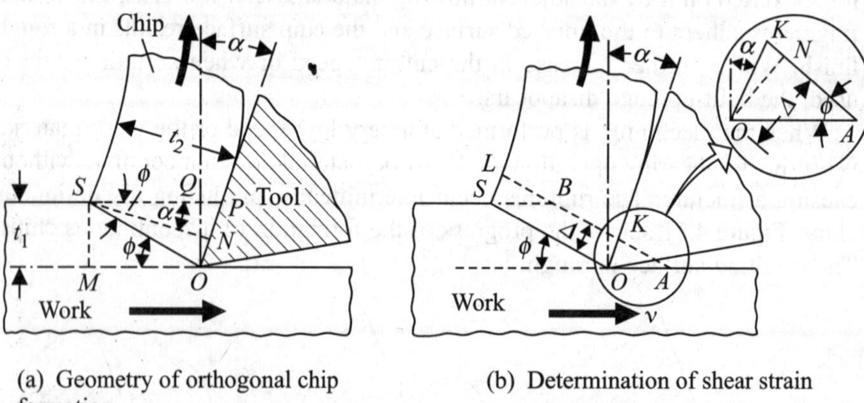

(a) Geometry of orthogonal chip formation

(b) Determination of shear strain

Fig. 4.12 Features of orthogonal chip formation.

A direct determination of the shear angle ϕ is quite difficult but t_1 and t_2 can be easily measured. To establish a relation among t_1, t_2, α, and ϕ, let us drop two perpendiculars *SM* and *SN* from *S* on the extension of the machined surface and the rake face of the tool. Further, *SP* is drawn parallel to *OM* and

[1]The nature of plastic deformation has been experimentally investigated and it has been established that it is of shear type.

Q is the intersection of SP with the normal drawn at O to OM. Considering the two right-angled triangles $\Delta\, SNO$ and $\Delta\, QPO$, we get

$$\angle PSN = \angle POQ = \alpha$$

or

$$\angle NSO = \angle PSO - \angle PSN = \phi - \alpha.$$

Again,

$$OS = SN/\cos(\phi - \alpha) = \frac{t_2}{\cos(\phi - \alpha)} = SM/\sin\phi = \frac{t_1}{\sin\phi}.$$

Hence,

$$\frac{t_1}{t_2} = \frac{\sin\phi}{\cos(\phi - \alpha)} = r, \tag{4.1}$$

where r is known as the cutting ratio. Equation (4.1) can be expressed also in the form

$$\tan\phi = \frac{r\cos\alpha}{1 - r\sin\alpha}. \tag{4.2}$$

The magnitude of the shear strain involved in the process can be easily determined from Fig. 4.12b. Let us consider an element of the undeformed work material $ABSO$ of thickness Δ. Due to the presence of the tool, this is sheared to the shape $KLSO$. Actually, the chip formation process may be considered as a succession of such a shear of the elements of the work material. From the definition, the magnitude of the shear strain is given by $\gamma = AK/\Delta$. An enlarged view of a part of the cutting zone is also shown in Fig. 4.12b where a perpendicular ON has been dropped on AK from O whose length is equal to Δ. Obviously, $\angle KAO = \phi$. Thus,

$$\frac{\pi}{2} + \alpha = \angle OKA + \phi \quad \text{or} \quad \angle OKA = \frac{\pi}{2} + \alpha - \phi.$$

Now,

$$\gamma = \frac{AK}{\Delta} = \frac{AN + NK}{ON} = \cot\phi + \tan\angle KON.$$

Hence,

$$\gamma = \cot\phi + \tan(\phi - \alpha). \tag{4.3}$$

EXAMPLE 4.1 During orthogonal machining with a cutting tool having a 10° rake angle, the chip thickness is measured to be 0.4 mm, the uncut thickness being 0.15 mm. Determine the shear plane angle and also the magnitude of the shear strain.

SOLUTION The cutting ratio is first calculated to be

$$r = \frac{0.15}{0.4} = 0.38.$$

Using equation (4.2), we obtain

$$\phi = \tan^{-1}\left(\frac{0.38 \cos 10°}{1 - 0.38 \sin 10°}\right) = 21.8°.$$

To determine the shear strain, we use equation (4.3). Thus,

$$\gamma = \cot 21.8° + \tan (21.8° - 10°) = 2.5 + 0.21 = 2.71.$$

4.2.2 MECHANICS OF CHIP FORMATION

The first scientific treatment of the problem was proposed by Ernst and Merchant[1]. They considered the idealized case of a single shear plane. Later, more accurate and exhaustive analyses were carried out by various researchers. However, the simple theory, based on the idealized single shear plane model, is good enough to predict the approximate values of power consumption, and this will be enough for our purposes.

If the chip above the shear plane is considered as a rigid body moving with a constant velocity, the resultant of the forces acting on it from the rake surface of the tool R and the work surface along the shear plane (R') must be zero (Fig. 4.13a). The total force R can be resolved into two components N and F

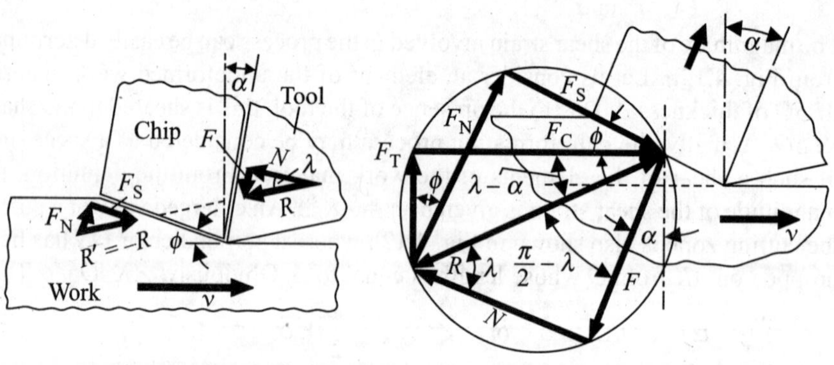

(a) Equilibrium of chip

(b) Merchant's circle diagram

Fig. 4.13 Forces in chip formation.

normal to and along the rake surface, respectively. Since F must be the friction force due to the existence of the normal load N, we have, according to the usual convention,

$$\frac{F}{N} = \mu, \tag{4.4}$$

where μ is the average coefficient of friction between the chip and the tool. The coefficient of friction can be expressed also as

$$\mu = \tan \lambda, \tag{4.5}$$

[1]Ernst, H. and Merchant, M.E., Chip formation, friction and high quality machine surfaces, in surface treatment of metals, *Trans. ASME*, **29**, 299, 1941.

where λ is the friction angle (Fig. 4.13a). Similarly, R' can also be resolved into the components along the directions normal and parallel to the shear plane, and let these components be F_N and F_S (Fig. 4.13a).

Now, since the inclinations of the shear plane and the rake surface vary, they do not suffice to provide some standard invariant directions. For this purpose, the directions along and perpendicular to the cutting motion are quite suitable. So, the force acting on the tool (which is equal to R in magnitude and opposite to the direction of R) can be resolved into two components F_C and F_T along and normal to the direction of the cutting velocity v. F_C and F_T are normally called the cutting and thrust components, respectively. Obviously, F_C is the component responsible for the energy consumption since it is along the direction of motion.

The relationship among the different components and the resultant cutting force can be best understood with the help of a diagram, first proposed by Merchant and commonly known as Merchant's circle diagram (Fig. 4.13b). Since the resultants of F_C, F_T and F_N, F_S are the same and those of F and N are the same in magnitude, the tips of all these force vectors must lie on an imaginary circle of diameter R, as shown in Fig. 4.13b.

Using Fig. 4.13b, the relations that we can write are

$$F_C = F_S \cos \phi + F_N \sin \phi, \tag{4.6a}$$

$$F_T = F_N \cos \phi - F_S \sin \phi, \tag{4.6b}$$

$$F = F_C \sin \alpha + F_T \cos \alpha, \tag{4.7a}$$

$$N = F_C \cos \alpha - F_T \sin \alpha, \tag{4.7b}$$

$$F_S = F_C \cos \phi - F_T \sin \phi, \tag{4.8a}$$

$$F_N = F_C \sin \phi + F_T \cos \phi, \tag{4.8b}$$

$$R = \frac{F_S}{\cos (\phi + \lambda - \alpha)}, \tag{4.9}$$

$$F_C = R \cos (\lambda - \alpha), \tag{4.10a}$$

$$F_T = R \sin (\lambda - \alpha). \tag{4.10b}$$

The measurement of F_C and F_T is easily done with the help of a tool dynamometer. Once F_C and F_T have been experimentally found, F_S, F_N, F, and N can be easily determined with the help of equations (4.7) and (4.8). Consequently, μ can also be calculated from the relation

$$\mu = \frac{F}{N} = \frac{F_C \sin \alpha + F_T \cos \alpha}{F_C \cos \alpha - F_T \sin \alpha}. \tag{4.11}$$

However, the foregoing analysis is of academic interest and we should try to find out a method for determining the components F_C and F_T theoretically when the material properties and the other necessary data are given. Merchant attempted to do this in the following way.

If τ_s is the ultimate shear stress of the work material, then the shear force F_S along the shear plane can be written as

$$F_S = \frac{wt_1\tau_s}{\sin\phi}, \tag{4.12}$$

where w is the width of the workpiece under cutting and t_1 is the uncut thickness. From equations (4.9) and (4.10a), we have

$$F_C = \frac{F_S\cos(\lambda-\alpha)}{\cos(\phi+\lambda-\alpha)}.$$

Using equation (4.12) in the foregoing equation, we obtain

$$F_C = wt_1\tau_s\cos(\lambda-\alpha)\left[\frac{1}{\sin\phi\cos(\phi+\lambda-\alpha)}\right]. \tag{4.13}$$

The power consumption during machining is given by

$$W = F_C v = vwt_1\tau_s\cos(\lambda-\alpha)\left[\frac{1}{\sin\phi\cos(\phi+\lambda-\alpha)}\right]. \tag{4.14}$$

Now, it is known that nature always tries to take the path of least resistance, and so, during a cutting operation, ϕ takes a value such that the least amount of energy is consumed or W is minimum. As v, w, t_1, and α are given, and if we assume that τ_s and λ do not change when ϕ varies and that W is a function only of ϕ and is of the form

$$W(\phi) = \frac{\text{constant}}{\sin\phi\cos(\phi+\lambda-\alpha)},$$

then $W(\phi)$ will be minimum when the denominator is maximum. Differentiating the denominator with respect to ϕ and equating it to zero, we get

$$\cos\phi\cos(\phi+\lambda-\alpha) - \sin\phi\sin(\phi+\lambda-\alpha) = 0$$

or

$$\cos(2\phi+\lambda-\alpha) = 0$$

or

$$2\phi+\lambda-\alpha = \frac{\pi}{2}, \tag{4.15}$$

where $\lambda = \tan^{-1}\mu$. Therefore, equation (4.15) provides a way to determine ϕ for a given α and a prescribed coefficient of friction between the chip and the tool. Using equation (4.15) in equation (4.13), we get

$$F_C = \frac{2wt_1\tau_s\cos(\lambda-\alpha)}{1-\sin(\lambda-\alpha)}, \tag{4.16}$$

and the power consumption can be calculated when the speed v is given.

Merchant found that this theory yields quite agreeable results when cutting synthetic plastics but agrees poorly with the results of machining metals. It was realized that τ_s is not completely independent of the normal stress. P.W. Bridgman showed that τ_s can be expressed as

$$\tau_s = \tau_{s_0} + k_1\sigma, \tag{4.17}$$

where k_1 is a constant and σ is the normal stress acting on the shear plane. During machining, σ is given by

$$\sigma = \frac{F_N}{wt_1/\sin\phi}.$$

So, the shear stress τ_s can be expressed as

$$\tau_s = \tau_{s_0} + k_1\frac{F_N}{wt_1/\sin\phi}.$$

From the circle diagram (Fig. 4.13b), we can write

$$\frac{F_N}{F_S} = \tan(\phi + \lambda - \alpha)$$

or

$$F_N = F_S \tan(\phi + \lambda - \alpha).$$

Using this in the expression for τ_s and writing τ_s in terms of F_S, we get

$$\frac{F_S}{wt_1/\sin\phi} = \tau_{s_0} + k_1\frac{F_S\tan(\phi + \lambda - \alpha)}{wt_1/\sin\phi}$$

or

$$\frac{F_S}{wt_1/\sin\phi}[1 - k_1\tan(\phi + \lambda - \alpha)] = \tau_{s_0}$$

or

$$F_S = \frac{wt_1\tau_{s_0}}{\sin\phi\,[1 - k_1\tan(\phi + \lambda - \alpha)]}.$$

Using equations (4.9) and (4.10a) along with the foregoing equation, we obtain

$$F_C = \frac{wt_1\tau_{s_0}\cos(\lambda - \alpha)}{\sin\phi\,[\cos(\phi + \lambda - \alpha) - k_1\sin(\phi + \lambda - \alpha)]}. \tag{4.18}$$

Now, applying the principle of minimum energy consumption, we finally get

$$2\phi + \lambda - \alpha = C_m, \tag{4.19}$$

where $C_m = \cot^{-1} k_1$ and is a constant for the work material. C_m is sometimes also called the machining constant.

This modified theory of Merchant agrees better with the experiments. It is seen (Fig. 4.14) that the plots of ϕ versus $(\lambda - \alpha)$ yield different straight lines of the same slope (obtained theoretically from Merchant's original theory) and can match the experimental results in the case of mild steel, copper, and lead.

From equation (4.19), it is clear that when α increases, ϕ also increases, and when μ (and so λ) increases, ϕ decreases. Further, when the machining speed increases, the coefficient of friction μ decreases. Figure 4.15 shows the

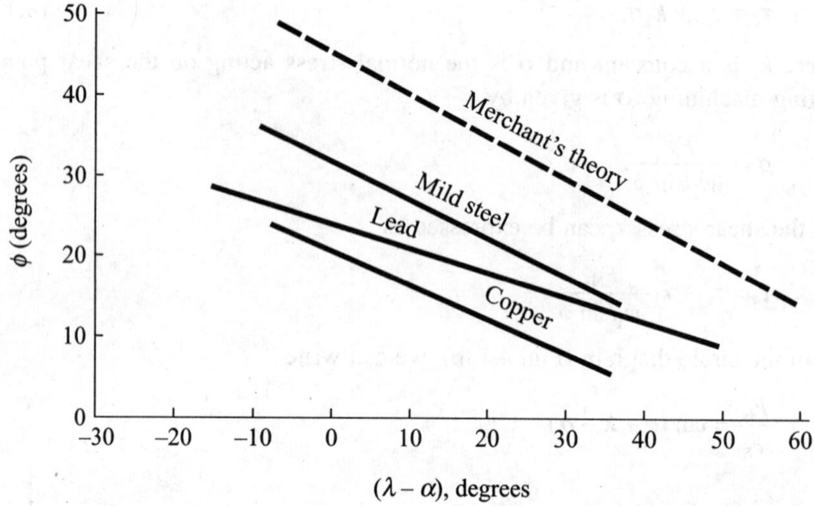

Fig. 4.14 Theoretical and experimental shear angle relationships. (After Pugh, H.D., Mechanics of the cutting process, Proceedings Institution of Mechanical Engineers Conference on Technology of Engineering Manufacture, London, 1958.)

variations of F_C, r with v, α. It has been experimentally observed that the nature of variation of F_C with the cutting speed v and the rake angle α tallies with that indicated in Fig. 4.15.

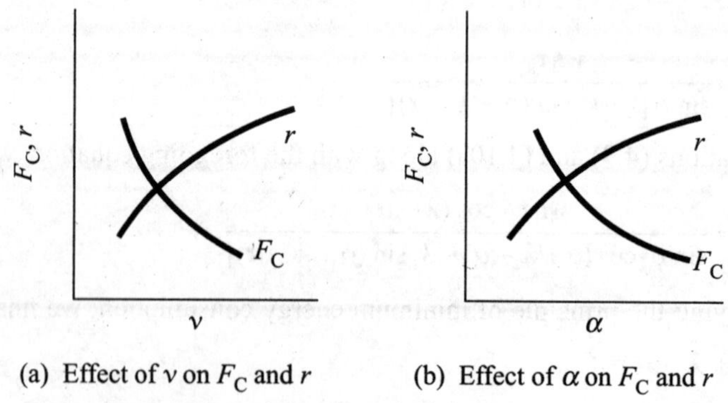

(a) Effect of v on F_C and r (b) Effect of α on F_C and r

Fig. 4.15 Effect of cutting parameters on chip formation.

EXAMPLE 4.2 During an orthogonal machining operation on mild steel, the results obtained are

$$t_1 = 0.25 \text{ mm}, \quad t_2 = 0.75 \text{ mm}, \quad w = 2.5 \text{ mm},$$
$$\alpha = 0°, \quad F_C = 950 \text{ N}, \quad F_T = 475 \text{ N}.$$

(i) Determine the coefficient of friction between the tool and the chip.

(ii) Determine the ultimate shear stress τ_s of the work material.

SOLUTION (i) To determine the coefficient of friction μ, we directly use equation (4.11). Thus,

$$\mu = \frac{F_C \sin\alpha + F_T \cos\alpha}{F_C \cos\alpha - F_T \sin\alpha} = \frac{475}{950} = 0.5.$$

(ii) If the shear plane area is A_S and the shear plane angle is ϕ, then $A_S = t_1 w/\sin\phi$.

Now, from equation (4.2),

$$\tan\phi = \frac{r\cos\alpha}{1 - r\sin\alpha} = r = \frac{t_1}{t_2} = \frac{1}{3}.$$

So, $\phi = \tan^{-1}(\frac{1}{3}) = 18.4°$. Hence,

$$A_S = \frac{0.25 \times 2.5}{\sin 18.4°}\ \text{mm}^2 = 1.98\ \text{mm}^2.$$

The component of force F_S from equation (4.8a) becomes

$$F_S = F_C \cos\phi - F_T \sin\phi = (950 \times 0.95 - 475 \times 0.316)\ \text{N} = 751.3\ \text{N}.$$

Finally, the ultimate shear stress τ_s is given by

$$\tau_s = \frac{F_S}{A_S} = \frac{751.3}{1.98}\ \text{N/mm}^2 = 379.4\ \text{N/mm}^2.$$

EXAMPLE 4.3 Mild steel is being machined at a cutting speed of 200 m/min with a tool of rake angle 10°. The width of cut and the uncut thickness are 2 mm and 0.2 mm, respectively. If the average value of the coefficient of friction between the tool and the chip is 0.5 and the shear stress τ_s of the work material is 400 N/mm^2, determine (i) the shear angle, and (ii) the cutting and the thrust components of the machining force.

SOLUTION In the absence of any data on the machining constant, we shall use Merchant's first solution to find out the shear plane angle. So, from equation (4.15),

$$2\phi + \lambda - \alpha = 90°,$$

where $\lambda = \tan^{-1}\mu = \tan^{-1} 0.5 = 26.57°$, and $\alpha = 10°$.

(i) Thus,

$$\phi = (90 + 10 - 26.57)/2\ \text{degrees} = 36.7°.$$

(ii) Using equation (4.12), we have

$$F_S = \frac{2 \times 0.2 \times 400}{\sin 36.7°} = 262.3\ \text{N}.$$

From equation (4.9), the resultant machining force works out to be

$$R = 262.3/\cos(36.7° + 26.57° - 10°)\ \text{N} = 438.6\ \text{N}.$$

Now that R has been determined, F_C and F_T can be calculated, using equation (4.10), as

$$F_C = R \cos (26.57° - 10°) = 420 \text{ N},$$

$$F_T = R \sin (26.57° - 10°) = 125 \text{ N}.$$

It should be remembered that the value of ϕ obtained from Merchant's first solution is always higher (in the case of metals) than the exact value, and consequently the forces calculated are lower.

The machining constants for various types of ferrous work materials are given in Table 4.2. A machining constant tends to increase with cold work.

Table 4.2 Machining constant C_m (degrees)

Work material (hot rolled steel)		C_m (degrees)
AISI	1010	69.8
AISI	1020	69.6
AISI	1045	78.0
AISI	2340	76.2
AISI	3140	70.6
AISI	4340	74.5
Stainless	303	92
Stainless	304	82

Various other shear angle relationships have been developed by later researchers. The important ones are given in Table 4.3.

Table 4.3 Shear angle relations

Source	Result
Ernst and Merchant	$2\phi + \lambda - \alpha = \pi/2$
Merchant's second solution	$2\phi + \lambda - \alpha = C_m$
Lee and Shaffer	$\phi + \lambda - \alpha = \pi/4$
Stabler	$\phi + \lambda - \alpha/2 = \pi/4$

EXAMPLE 4.4 Solve Example 4.3 assuming that the machining constant is 70° for the work material. Also, find out the results using the Lee and Shaffer relation and compare these.

SOLUTION From equation (4.19),

$$\phi = \frac{C_m - \lambda + \alpha}{2}$$

$$= \frac{70° - 26.57° + 10°}{2}$$

$$= 26.7°.$$

Using equation (4.12), we have

$$F_S = \frac{2 \times 0.2 \times 400}{\sin (26.7°)} \text{ N}$$

$$= 355.6 \text{ N}.$$

Then, we get

$$R = \frac{355.6}{\cos (26.7° + 26.57° - 10°)} \text{ N}$$

$$= 508 \text{ N},$$

$$F_C = 508 \times \cos (26.57° - 10°) \text{ N}$$

$$= 486.9 \text{ N},$$

$$F_T = 508 \times \sin (26.57° - 10°) \text{ N}$$

$$= 144.9 \text{ N}.$$

Now, using the Lee and Shaffer relation, we obtain

$$\phi = 45° - 26.57° + 10° = 28.43°.$$

Repeating the calculations for F_C and F_T, we find the final results become

$$F_C = 508 \times \cos (28.43° - 10°) \text{ N} = 481.9 \text{ N},$$

$$F_T = 508 \times \sin (28.43° - 10°) \text{ N} = 160.6 \text{ N}.$$

It is found that the results obtained by Merchant's second solution and the Lee and Shaffer method do not differ much.

If F_C is the cutting force when a machining is done at a velocity v, the power consumption is given by

$$W = F_C v. \tag{4.20}$$

At the same time, if t_1 and w are the uncut thickness and the width of cut, respectively, the volume of material removed per unit time is then, obviously, given by wt_1v. The energy consumption per unit volume of material removal, commonly known as specific energy, is given by

$$U_c = \frac{F_C}{wt_1}, \tag{4.21a}$$

$$W = U_c Q, \tag{4.21b}$$

where Q is the volume rate of material removal. This specific energy is a very convenient basis for judging the power required for a given process. An examination of the various published data shows that U_c fits the power law

$$U_c = U_0 \tilde{t}_1^{-0.4}, \tag{4.22}$$

where $\tilde{t}_1$ is the magnitude of the uncut thickness when measured in mm. The value of U_0 depends on the material; some typical values are given in Table 4.4. Substituting U_c from equation (4.22) in equation (4.21), we get

$$F_C = 1000 \times t_1 w U_0 \tilde{t}_1^{-0.4} \text{ N}, \tag{4.23}$$

where t_1 and w are in mm, U_0 is in J/mm^3, and $\tilde{t}_1$ is the magnitude of uncut thickness in mm. Using equation (4.23), we can directly obtain the cutting component of force, and with the speed given, the required power can also be estimated. Of course, the results are approximate and their dependence on parameters such as α and μ cannot be observed.

Table 4.4 Values of U_0 for various materials

Material	Hardness		U_0 (J/mm^3)
	BHN	R_C	
Steel	85–200		1.4
		35–40	1.6
		40–50	1.9
		50–55	2.4
		55–58	4.0
Stainless steel	135–275		1.4
		30–45	1.6
Cast iron	110–190		0.8
	190–320		1.6
Al alloys	30–150		0.35
Copper		80 R_B	1.2
Copper alloys		10–80 R_B	0.8
		80–100 R_B	1.2

EXAMPLE 4.5 Find out the order of magnitude of the cutting component of the machining force during orthogonal machining of mild steel with an uncut thickness of 0.25 mm, the width of cut being 2.5 mm.

SOLUTION To start with, we assume a suitable value of U_0. As nothing special has been mentioned, U_0 for the mild steel workpiece is assumed to be

1.4 J/mm^3 (from Table 4.4). Then, from equation (4.23),

$$F_C = 1000 \times 0.25 \times 2.5 \times 1.4 \times (0.25)^{-0.4} \text{ N}$$
$$= 3500 \times (0.25)^{0.6} \text{ N} \approx 1517 \text{ N}.$$

To compare the order of this value with the result obtained by the previous method, let us assume $\alpha = 0°$, $\mu = 0.5$, and the Lee and Shaffer shear angle relationship. Thus,

$$\phi = \frac{\pi}{4} + \alpha - \tan^{-1}\mu = 45° - 26.57° = 18.43°.$$

Then, from equation (4.13) with $\tau_s = 400$ N/mm^2,

$$F_C = \frac{2.5 \times 0.25 \times 300 \cos 26.57°}{\sin 18.43° \cos 45°} \text{ N}$$

or

$$F_C \approx 1000 \text{ N}.$$

4.2.3 HEAT GENERATION AND CUTTING TOOL TEMPERATURE

When a material is deformed elastically, some energy is spent to increase its strain energy which is returned during unloading. But in plastic deformation, most of the energy thus spent is converted into heat. During machining, the plastic deformation is large and takes place at a very high rate, and under such conditions, almost 99% of the energy is converted into heat. For cutting a low strength material, heat generation and the consequent increase in temperature of the tool at the cutting zone is not a big problem. However, when ferrous and other high strength materials are machined, the temperature rises with the speed and the tool strength decreases, leading to a faster wear and failure. So, though machining at a high speed is desirable, for higher productivity, the faster tool wear due to the high temperature puts a limit to the cutting speed. Thus, an understanding of the phenomena of heat generation and temperature rise during machining is very important.

The sources of heat generation during machining are as shown in Fig. 4.16.

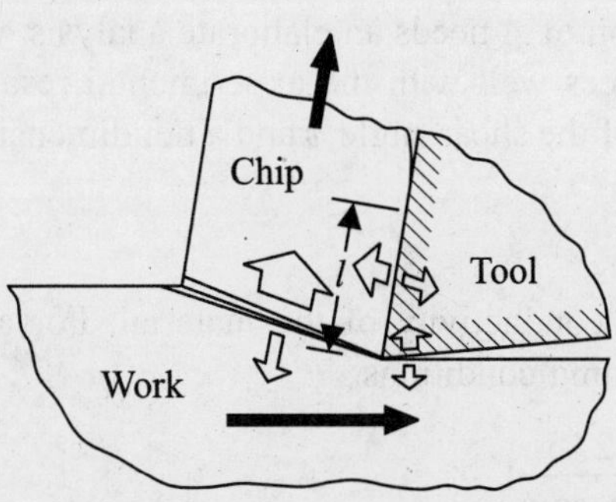

Fig. 4.16 Generation and distribution of heat during machining.

The major plastic deformation takes place in the shear zone and this heat source is the primary heat source. A large fraction of this heat goes to the chip. The sliding motion of the chip on the rake surface of the tool also generates heat, and this is the secondary heat source. From this source also, the chip takes away the major portion of the heat. There is another source of heat where the job rubs against the flank surface of the tool. But with sharp tools, the contribution of this source to the heating phenomenon is insignificant.

Now, the total power consumption (≡ total rate of heat generation) during machining is $W = F_C v$. If the rates of heat generation in the primary and the secondary deformation zones are W_P and W_S, respectively, then

$$W = W_P + W_S. \tag{4.24}$$

Again,

$$W_S = F v_c = F r v, \tag{4.25}$$

where v_c is the chip velocity. From relations (4.24) and (4.25), we can write

$$W_P = F_C v - F r v. \tag{4.26}$$

Thus, when enough information is available, the rate of heat generation in the primary deformation zone (i.e., the shear plane) and the secondary deformation zone (i.e., the rake face) can be found out.

When a material particle moves across the primary deformation zone, the temperature rise is given by[1]

$$\theta_P = \frac{(1 - \Lambda) W_P}{\rho c v t_1 w}, \tag{4.27}$$

where

Λ = fraction of primary heat which goes to the workpiece,

ρ = density of the material,

c = specific heat of the material,

t_1, w = uncut thickness, width of cut, respectively.

Since the computation of Λ needs an elaborate analysis, we will give here only the result which agrees well with the experimental results. It has been found that Λ is a function of the shear angle ϕ and a nondimensional quantity, namely,

$$\Theta = \frac{\rho c v t_1}{k}, \tag{4.28}$$

k being the thermal conductivity of the material. For a wide range of work materials and machining conditions,

$$\Lambda = 0.15 \ln \left(\frac{27.5}{\Theta \tan \phi}\right). \tag{4.29}$$

[1]Boothroyd, G., Temperatures in orthogonal metal cutting, *Proc. IME*, **177**, 789, 1963.

Hence, if ϕ is known or determined, Λ can be calculated using equation (4.29), and so θ_P can be estimated with the help of equation (4.27).

The maximum temperature rise θ_S when the material particle passes through the secondary deformation zone along the rake face of the tool can be approximately expressed as[1]

$$\theta_S \approx 1.13\sqrt{\frac{\Theta t_2}{l}}\left(\frac{W_S}{\rho c v w t_1}\right), \tag{4.30a}$$

where l is the length of contact between the tool and chip. The corresponding average temperature rise is obtained from the equation

$$\theta_{S_{av}} = \frac{W_S}{\rho c v w t_1}. \tag{4.30b}$$

It has been found[2] that

$$\frac{l}{t_2} = [1 + \tan(\phi - \alpha)].$$

Using this relation in equation (4.30a), we obtain

$$\theta_S = 1.13\sqrt{\frac{1}{\rho c v t_1 k[1 + \tan(\phi - \alpha)]}}\,\frac{W_S}{w}. \tag{4.30c}$$

The final temperature is given as

$$\theta = \theta_0 + \theta_P + \theta_S, \tag{4.30d}$$

where θ_0 is the initial temperature of the workpiece. This maximum temperature is along the rake face of the tool.

EXAMPLE 4.6 Determine the maximum temperature along the rake face of the tool when machining mild steel (ms), given

work material (ms) shear stress = 400×10^6 N/m^2,

$\alpha = 0°$, $v = 2$ m/sec, $t_1 = 0.25$ mm,

$w = 2$ mm, $\mu = 0.5$, $\rho = 7200$ kg/m^3,

$k = 43.6$ W/m-°C, $c = 502$ J/kg-°C, $\theta_0 = 40$°C.

Use the Lee and Shaffer shear angle relationship.

SOLUTION First, we have to determine the shear angle ϕ. According to the Lee and Shaffer condition,

$$\phi = 45° - \tan^{-1}(0.5) + 0° = 18.43°.$$

[1]Rapier, A.C., A theoretical investigation of the temperature distribution in the metal cutting process, *Brit. J. Applied Physics*, **5**, 11, 400, 1954.

[2]Sen, G.C. and Bhattacharyya, A., Principles of Metal Cutting, New Central Book Agency, Calcutta, 1969.

Then,

$$F_S = \frac{2 \times 0.25 \times 400 \times 10^6}{\sin 18.43° \times 10^6} \text{ N} = 632.6 \text{ N} \qquad \text{[from equation (4.12)].}$$

Using equations (4.9) and (4.10a), we get

$$F_C = \frac{632.6 \cos 26.57°}{\cos 45°} = 800 \text{ N.}$$

From equation (4.1),

$$r = \frac{\sin \phi}{\cos \phi} \qquad (\text{since } \alpha = 0)$$

$$= \tan \phi = \tan 18.43° = 0.333.$$

Again,

$$F_T = F_C \tan (\lambda - \alpha) = F_C \tan \lambda = \mu F_C = 0.5 \times 800 \text{ N} = 400 \text{ N.}$$

Using equation (4.7a), we obtain

$$F = F_C \sin \alpha + F_T \cos \alpha = F_T = 400 \text{ N.}$$

To calculate W_P and W_S, we use equations (4.25) and (4.26). Thus,

$$W_P = (800 - 400 \times 0.333) \times 2 \text{ W} = 1333 \text{ W,}$$

$$W_S = 400 \times 0.333 \times 2 \text{ W} = 266.7 \text{ W.}$$

The thermal number Θ is given by equation (4.28). So,

$$\Theta = \frac{7200 \times 502 \times 2 \times 0.25 \times 10^{-3}}{43.6} = 41.5,$$

$$\Theta \tan \phi = 41.5 \times 0.333 = 13.8.$$

From equation (4.29),

$$\Lambda = 0.15 \ln \left(\frac{27.5}{13.8}\right) = 0.1.$$

Now, the temperature rise in the primary zone is [from equation (4.27)]

$$\theta_P = \frac{(1 - 0.1) \times 1333 \times 10^3 \times 10^3}{7200 \times 502 \times 2 \times 0.25 \times 2} \text{°C} = 332\text{°C}.$$

The maximum temperature rise along the rake face due to the secondary source is given by [using equation (4.30c)]

$$\theta_S = 1.13 \sqrt{\frac{10^3 \times 1}{7200 \times 502 \times 2 \times 0.25 \times 43.6 \times 1.333}} \times \frac{262 \times 10^3}{2} \text{°C}$$

$$\approx 458\text{°C}.$$

Finally, the maximum temperature along the rake face of the tool becomes

$$\theta \approx (40 + 325 + 458)\ °\text{C} \approx 823°\text{C}.$$

The nature of variation in the overall chip-tool interface temperature with the major parameters can be approximately determined through dimensional analysis[1]. The result is

$$\theta_{\text{ov}} \propto U_{\text{c}}\sqrt{\frac{vt_1}{k\rho c}}, \tag{4.31}$$

where θ_{ov} is the overall temperature rise and U_{c} is the specific energy. The most important information we get from this result is that the overall interface temperature rise is proportional to the square root of the cutting speed. This is found to agree quite well with the experimental results. Figure 4.17 shows the variation of the experimentally measured cutting temperature when machining

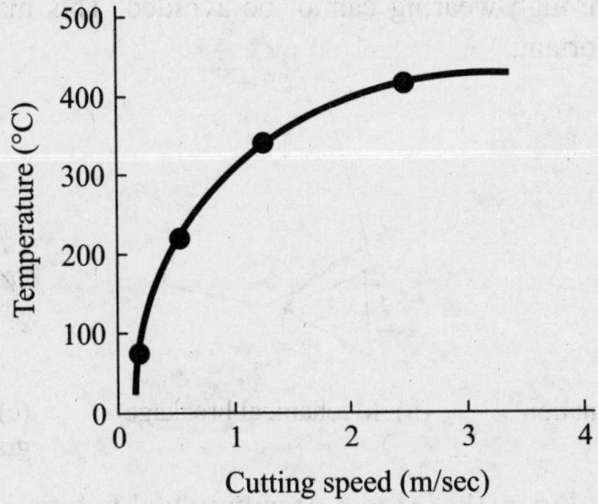

Fig. 4.17 Variation of temperature with cutting speed. (After Shaw, M.C., Metal Cutting Principles, MIT Press, Cambridge, Massachusetts, 1957.)

the SAE B1113 steel with the K2SWC tool having a 20° rake angle and a 0.06-mm uncut thickness.

As we shall see in Sections 4.2.5 and 4.2.6, the life of the tool, machinability, and other important factors depend heavily on the cutting temperature. An understanding of the thermal aspects during machining is therefore extremely important.

4.2.4 FAILURE OF CUTTING TOOL AND TOOL WEAR

It is not very difficult to realize that the success of a machining process depends on the sharpness of the tool. Even common sense tells us that the use of a blunt

[1]Sen, G.C. and Bhattacharyya, A., Principles of Metal Cutting, New Central Book Agency, Calcutta, 1969.

tool results in a large power consumption and deteriorated surface finish. When a cutting tool is unable to cut, consuming reasonable energy, and cannot produce an acceptable finish, it is considered to have failed. The failure of a cutting tool may be due to one or a combination of the following modes:

(i) Plastic deformation of the tool due to high temperature and large stress (Fig. 4.18a).

(ii) Mechanical breakage of the tool due to large force and insufficient strength and toughness (Fig. 4.18b).

(iii) Blunting of the cutting edge of the tool through a process of gradual wear (Fig. 4.18c).

By a proper selection of the cutting tool material, tool geometry, and cutting conditions, plastic deformation and mechanical failure can be prevented. However, the gradual wearing process cannot be totally stopped and ultimately the tool failure through wearing cannot be avoided. This makes the study of wear so very important.

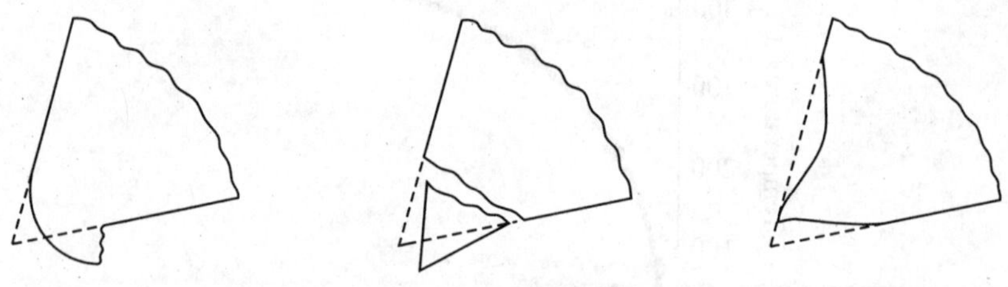

(a) Plastic deformation (b) Mechanical breakage (c) Failure through gradual wear

Fig. 4.18 Modes of cutting tool failure.

It is clear that the wearing action takes place on those surfaces along which there is a relative sliding with other surfaces. Thus, the wear takes place on the rake surface where the chip flows over the tool, and on the flank surface where rubbing between the work and the tool occurs (Fig. 4.19a). These wears are called the crater and the flank wears, respectively.

In Fig. 4.19b, we see that the width h_f can be taken as the measure of the flank wear. But the measure of a crater wear is not so simple. Quite often a nondimensional quantity h_k given by

$$h_k = \frac{e}{l/2 + f} \tag{4.32}$$

is taken as a measure of the crater wear.

As mentioned in Chapter 1, there are various mechanisms of wear, e.g., abrasion wear, adhesion wear, and diffusion wear. Though all these mechanisms are normally active during a wear process, one or more mechanisms may play a predominant role depending on the various conditions stated in Chapter 1. Thus, abrasion and adhesion are primarily responsible for the flank wear, whereas

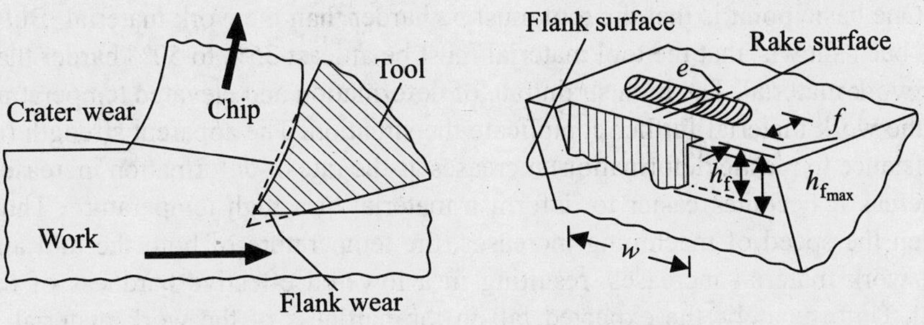

(a) Crater and flank wear (b) Details of crater and flank wear

Fig. 4.19 Wear of cutting tools.

diffusion may play an important role in the development of crater wear at a high speed since then the temperature in the rake face is much higher than that in the flank surface. The typical nature of growth of flank wear is shown in Fig. 4.20a for various cutting speeds. Here, we see that, initially, there is a break in the wear region where the wear grows quickly. Then, the wear rate stabilizes and remains constant for a considerable period; this region is again followed by a rapid growth of wear. The uniform wear zone constitutes the major portion of the period of usability of a cutting tool.

The growth of a crater wear is a more complex phenomenon. The typical nature of variation of *l*, *f*, and *e* is shown in Fig. 4.20b.

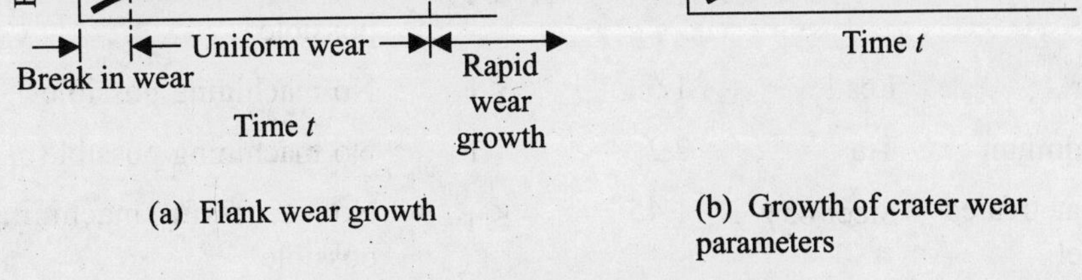

(a) Flank wear growth (b) Growth of crater wear parameters

Fig. 4.20 Growth of tool wear.

4.2.5 CUTTING TOOL MATERIALS

Since the machining operation is basically a deformation process of the work material through the application of a force by the cutting tool, the stability of the geometric form (or form stability) of the tool is a key factor. Obviously, the cutting tool must provide the maximum resistance to any tendency of alteration of its geometric shape. To achieve this, the cutting tool material must be properly selected.

One basic point is that the tool must be harder than the work material. But it has been shown[1] that the tool material must be at least 35% to 50% harder than the work material. The high strain rate of deformation and elevated temperature of the work material further complicate the situation. The apparent strength (or resistance to plastic deformation) increases as the rate of deformation increases, whereas it becomes easier to deform a material at a high temperature. Thus, when the speed of machining increases, the temperature of both the tool and the work material increases, resulting in a lowered effective hardness of the tool. Unfortunately, the expected fall in the hardness of the work material is neutralized by the higher rate of deformation. Moreover, the rate of deformation during a hardness test is orders of magnitude lower than that during machining. So, the condition of hardness ratio, proposed by T.N. Loladze, has to be applied after finding out the modified hardness value, taking care of the elevated temperature and high strain rate of the work material. This condition can be stated as

$$1.35 < \left[\frac{H_{\text{tool}}}{H_{\text{work}}}\right]_{\text{modified}} < 1.5. \tag{4.33}$$

Table 4.5 lists a number of cutting operations for various combinations of tool and work material with a cutting speed of 0.5 m/min, an uncut thickness of

Table 4.5 Performance of various tool/work combinations

Tool material	Work material	Static hardness ratio	Modified hardness ratio	Remark
Copper	Zinc	1.98	≈1	No successful machining possible
Zinc	Cadmium	2.2	≈1	No successful machining possible
Tin	Lead	1.5	<1	No machining possible
Cadmium	Tin	2.2	<1	No machining possible
Heat treated steel	Steel 65γ	1.45	≈1	No successful machining possible

0.2 mm, and an average depth of cut of 6 mm (the rake angle of each of the tools is 10°). It is clear from this table that, though the apparent hardness ratio indicates the possibility of successful machining, in all the cases the modified hardness ratio reveals the correct situation.

The properties of an ideal tool material can be summarized as follows:

[1]Loladze, T.N., Theory of Tool Wear (in Russian), Mashgiz, Moscow, 1958.

(i) As far as possible, it should maintain its hardness which is appreciably higher than that of the work at the elevated temperature.

(ii) It should be tough enough to withstand shocks.

(iii) It should provide a large resistance to the wearing action so that excessive wear does not occur.

(iv) The coefficient of friction between the work and the tool should be low.

(v) Its thermal conductivity and specific heat should be high.

The materials commonly used for making the cutting tools are (i) high carbon steel, (ii) high speed steel (HSS), (iii) cemented carbide, and (iv) ceramic. Of course, for grinding and other machining processes, abrasive minerals, e.g., silicon carbide, aluminium oxide, and diamond, are used. Figure 4.21 shows the variation of hardness with temperature for various tool materials.

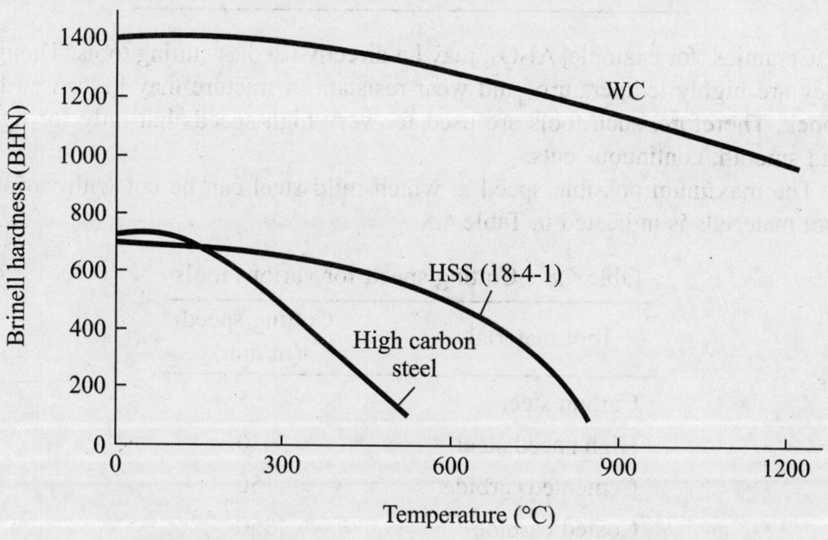

Fig. 4.21 Variation of hardness with temperature for common tool materials.

Carbon tool steel and high speed steel are both ferrous materials and their typical compositions are as given in Table 4.6.

Table 4.6 Per cent composition of tool materials

Material	C	W	Cr	V	Mn	Fe
Carbon tool steel	0.9				0.6	Rest
High speed steel	0.75	18.0	4.0	1.0	0.6	Rest

Cemented carbide is produced by the powder metallurgy technique. For this, powders of several carbide compounds are pressed and bonded together in a matrix to form the cemented material. The matrix is always Co. When machining a steel, TaC and TiC are added to avoid large scale diffusion. Table 4.7 shows some typical carbide compositions. Also, coated carbides have been developed where the cemented carbide is coated with a thin (≈ 5 μm) layer of a ceramic, e.g., TiC, TiN, Al_2O_3, and carbon boron nitride. The coated carbides are used for machining superalloys.

Table 4.7 Per cent composition of carbides

WC	Co	TaC	TiC	Use
94	6			Cast iron
70.7	4.5	12.2	12.6	Steels
72	8.5	11.5	8	Ni alloy steels

Ceramics, for example, Al_2O_3, may be directly used as cutting tools. Though they are highly temperature- and wear-resistant, a fracture may be caused by shock. Therefore, such tools are used for very high speeds but only for light and smooth, continuous cuts.

The maximum possible speed at which mild steel can be cut with various tool materials is indicated in Table 4.8.

Table 4.8 Cutting speed for various tools

Tool material	Cutting speed (m/min)
Carbon steel	5
High speed steel	30
Cemented carbide	150
Coated carbide	350
Ceramic	600

EXAMPLE 4.7 SAE 1040 steel is being machined with an HSS tool with 0° rake angle, $t_1 = 0.25$ mm, and $w = 2$ mm, the coefficient of friction between the tool and the chip being 0.5. The limiting shear stress of the work material is 400×10^6 N/m², $\rho = 7200$ kg/m³, $k = 43.6$ W/m-°C, and $c = 502$ J/kg-°C. The hardness of SAE 1040 steel under the strain rate of the same order as found in machining is 350 HV (Vickers hardness); this may be considered to remain constant with the speed of machining since the effect of increased temperature is neutralized by the increased strain rate. The hardness of the HSS varies with temperature according to the relation

$$H(\theta) = 850[1 - (\frac{\theta}{700})^{3.1}] \text{ HV},$$

where θ is the temperature in °C. The ambient temperature is 40°C. Determine the maximum speed at which cutting is possible.

SOLUTION We have to first find out the average temperature of the tool tip using

$$\theta = \theta_0 + \theta_P + \theta_{S_{av}}$$

for various speeds, where θ_P and $\theta_{S_{av}}$ are given by equations (4.27) and (4.30b).

To make the problem simpler, we shall first find out the average temperature for a speed of 2 m/sec and then, assuming a parabolic variation of temperature with speed [equation (4.31)], determine θ for various speeds. This would make the result approximate, but the calculations would be much simpler. The reader is advised to repeat the problem by determining θ for different speeds, using, for each speed, equations (4.27) and (4.30b).

Now,

$$\theta_{ov} = \frac{(1-\Lambda)W_P + W_S}{\rho c v t_1 w}\ ^\circ\text{C}.$$

With $v = 2$ m/sec, Λ comes out to be 0.1. W_P and W_S are 1304 W and 262 W, respectively. Finally,

$$\theta_{ov} = \frac{(0.9 \times 1304 + 262) \times 10^6}{7200 \times 502 \times 2 \times 0.25 \times 2}\ ^\circ\text{C} = 397^\circ\text{C}.$$

So,

$$\theta_{ov}(v) \approx 309 v^{0.5}\ ^\circ\text{C} \qquad \text{[using equation (4.31)]},$$

where v is the speed in m/sec. Now, in the limiting case, from equation (4.33),

$$\left[\frac{H_{tool}}{H_{work}}\right]_{modified} \approx 1.5$$

or

$$[H_{tool}]_{modified} \approx 1.5 \times 350 \text{ HV}$$
$$\approx 525 \text{ HV}.$$

The corresponding temperature can be found from the given hardness temperature relation. Thus,

$$\theta_{lim} \approx 700\left(1 - \frac{525}{850}\right)^{1/3.1}\ ^\circ\text{C} \approx 513^\circ\text{C}.$$

Finally,

$$513 \approx 309(v_{lim})^{0.5}$$

or

$$v_{lim} \approx \left(\frac{513}{309}\right)^2 \approx 2.76 \text{ m/sec}.$$

4.2.6 TOOL LIFE AND MACHINABILITY

A comparative evaluation of the machining operations has always remained an interesting but a difficult problem. The term "machinability", which loosely means the ease of machining, is used quite commonly. However, the criteria for judging this ease may be different, depending on the point of view or the objective. For example, machining a lead piece with an HSS tool may be quite easy so far as the force requirement is concerned, but a good surface is extremely difficult to achieve. Thus, for a given operation, the machinability may be considered to be good or bad depending on the criterion.

To tackle the problem in a more meaningful manner, we shall first identify the major criteria for judging machinability.

(i) *Machining forces and power consumption* A machining requiring a large force indicates low machinability and vice versa. When the strength of the tool is a matter to worry about, this is the criterion to be considered.

(ii) *Surface finish* In some situations, the major concern can be over the quality of finish and, depending on the severity of this problem, the machinability may be low or high.

(iii) *Tool life* The length of the period for which a tool can be used is defined as the tool life. This criterion is also linked up with the productivity and economics and can be a very good index for an overall judgement of a machining operation.

As long as the available power is enough and does not pose any problem on the machining and we are not concerned with a very special precision machining operation requiring a high quality of finish, the tool life may be considered the index for judging the machinability. In cases where the major objective is to remove a large amount of material quickly and cheaply, the tool life is used as a direct measure of machinability.

A precise definition of tool life is also not a simple problem. As a machining operation progresses, the wear on the flank and rake surfaces keeps on increasing, as mentioned in Section 4.2.4. So, limits on these wears have to be chosen for defining the tool life. The commonly recommended criteria for HSS and carbide tools are given in Table 4.9.

In a very large number of cases, the criterion of the average flank wear h_f serves well. Figure 4.22 shows the growth of flank wear for different speeds. Using $h_f = 0.3$ mm as the tool life criterion, we note that the tool lives are T_1, T_2, and T_3, as shown in the figure. Also, it is obvious that a higher speed of cutting leads to a lower tool life. It has been experimentally established[1] that the tool life equation is

$$vT^n = C, \tag{4.34}$$

where C and n are constants depending on the tool and work material, tool

[1]Taylor, F.W., On the art of cutting metals, *Trans. ASME*, **28**, 31, 1906.

Table 4.9 Limits of tool wear

Tool material	Wear criterion
HSS	Catastrophic failure or $h_f = 0.3$ mm or $h_{f_{max}} = 0.6$ mm (when wear is highly nonuniform)
WC	$h_f = 0.3$ mm or $h_{f_{max}} = 0.6$ mm (when wear is highly nonuniform) or $e = (0.06 + 0.3t_1)$ mm

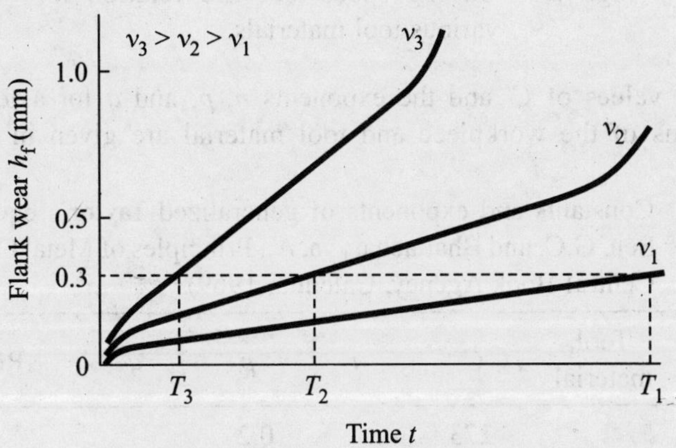

Fig. 4.22 Determination of tool life.

geometry, and cutting conditions (except speed). Figure 4.23 shows the typical variation of tool life with speed for HSS, WC, and ceramic tools, keeping the other conditions the same. It is clear that the tool life for a given speed is normally much higher with WC than that with HSS. A ceramic tool performs better at a high cutting speed. Though cutting speed is the most dominant variable, the other cutting parameters, e.g., the uncut thickness and width of cut, also affect the tool life. When machining is through the production of continuous chips without a built-up edge, the generalized Taylor equation can be written as

$$v = \frac{C'}{T^n t^p w^q}. \tag{4.35}$$

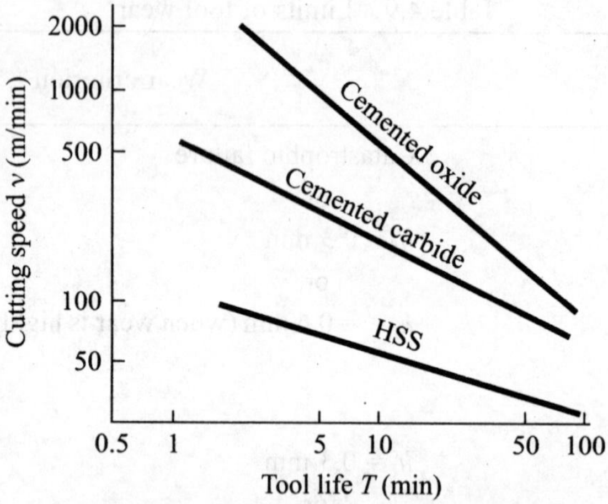

Work material: AISI 1045; hardness, 170 BHN; tool geometry, –10°, –10°, 10°, 10°, 15°, 15°, 1.5 mm; uncut thickness, 0.25 mm; width of cut, 1.6 mm; tool life based on 0.4 mm flank wear

Fig. 4.23 Cutting speed-tool life relation for various tool materials.

The typical values of C' and the exponents n, p, and q for a few common combinations of the workpiece and tool material are given in Table 4.10.

Table 4.10 Constants and exponents of generalized Taylor's equation (after Sen, G.C. and Bhattacharyya, A., Principles of Metal Cutting, New Central Book Agency, Calcutta, 1969)

Work material	Tool material	C'	n	p	q	Restriction
Steel	WTiC 10% Co	273		0.2		$t < 0.3$
		227	0.2	0.35	0.15	$0.3 \leqslant t \leqslant 0.75$
		221		0.45		$t > 0.75$
	WTiC 6% Co	292	0.18	0.3	0.15	
Cast iron	WC	324	0.28	0.4	0.2	

The units of v, T, t, and w are m/min, min, mm, and mm, respectively. It is observed that q is smaller than p, and this indicates that the tool life is more sensitive to the uncut thickness than to the width of cut.

The two most important geometric parameters of a tool, viz., the rake and the clearance angles, also affect the tool life and the typical characteristics are indicated in Fig. 4.24. When the rake angle increases, the tool life starts improving because the cutting force reduces. A further increase in the rake

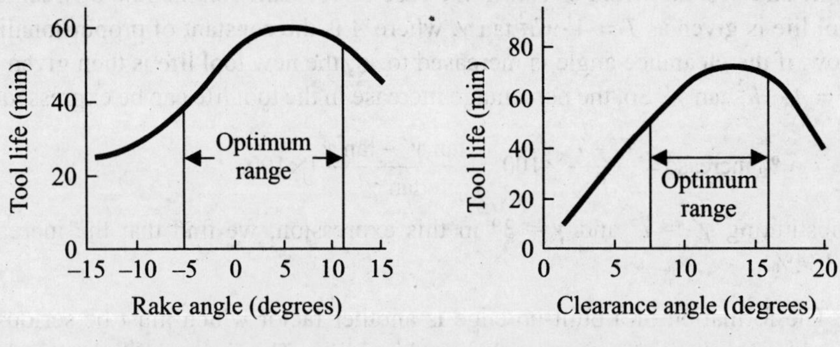

Work material: mild steel; tool material, cemented carbide; cutting speed, 100 m/min; uncut thickness, 0.13 mm; width of cut, 8 mm

Fig. 4.24 Dependence of tool life on tool geometry.

angle results in a larger temperature since the tool becomes thinner and the area available (Fig. 4.25a) for heat conduction reduces. Similarly, when the clearance angle increases, the tool life increases at first. This is due to the fact that for the same volume of flank wear, h_f reduces (Fig. 4.25b). However, with a

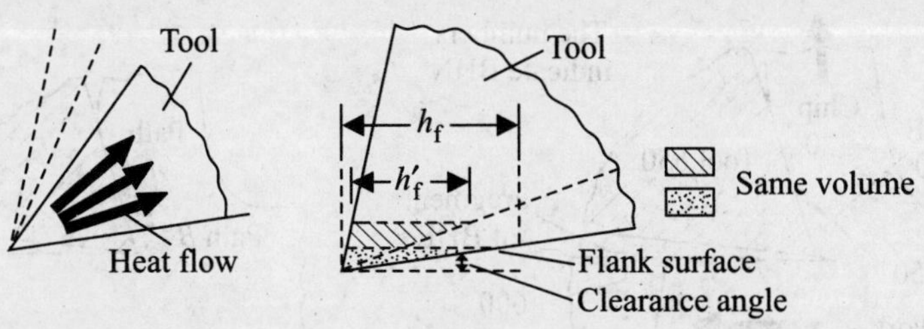

(a) Dependence of heat conduction on tool angle

(b) Dependence of flank wear on clearance angle

Fig. 4.25 Effects of tool angles.

further increase in the clearance angle, the tool becomes thinner and the tool life decreases due to the higher temperature.

EXAMPLE 4.8 A simple cutting tool has a 0° rake angle and a 3° clearance angle. The maximum length of the flank wear allowed before regrinding is 1 mm. Assuming the volume rate of wear from the flank face of the tool to be constant, what will be the percentage increase in the tool life if the clearance angle is increased to 7°?

SOLUTION Since $\alpha = 0°$, the volume of wear corresponding to the development of flank wear h_f can be expressed as $\frac{1}{2}wh_f^2 \tan \gamma$, where γ is the clearance angle and w is the width of cut. In the case of constant volume rate of wear, the tool life is given as $T = A\frac{1}{2}wh_f^2 \tan \gamma$, where A is the constant of proportionality. Now, if the clearance angle is increased to γ', the new tool life is then given by $T' = A\frac{1}{2}wh_f^2 \tan \gamma'$. So, the percentage increase in the tool life can be expressed as

$$\% \text{ increase} = \frac{T' - T}{T} \times 100 = [\frac{\tan \gamma' - \tan \gamma}{\tan \gamma}] \times 100.$$

Substituting $\gamma' = 7°$ and $\gamma = 3°$ in this expression, we find that the increase $\approx 134\%$.

The formation of a built-up edge is another factor which must be seriously considered when investigating the machinability. This is particularly so when the surface finish is of great importance. In Fig. 4.10b, we showed the formation of a built-up edge. When the temperature at the rake surface is below the recrystallization temperature, a seizure of the deformed work material by the tool surface occurs. This is due to the fact that, because of the large plastic deformation (which is maximum at the layers adjacent to the rake surface), the hardness of the chip material increases and the bonding between the tool and the chip breaks along a path not at the interface but inside the chip which is softer. Figure 4.26a shows a typical hardness distribution in the cutting zone[1]. It is clear that such a path (Fig. 4.26b) consumes lesser energy. Thus, the highly

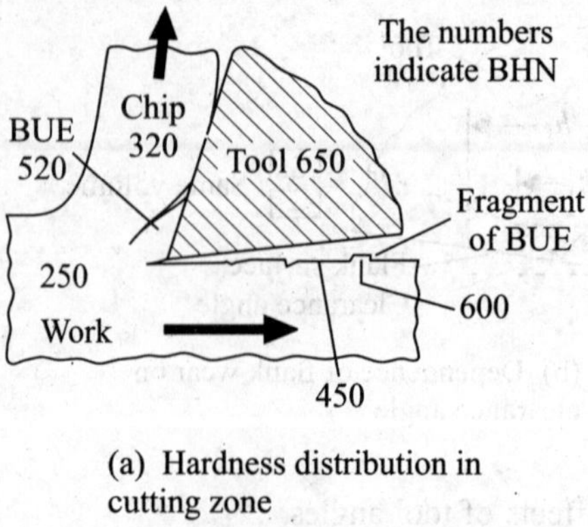

(a) Hardness distribution in cutting zone

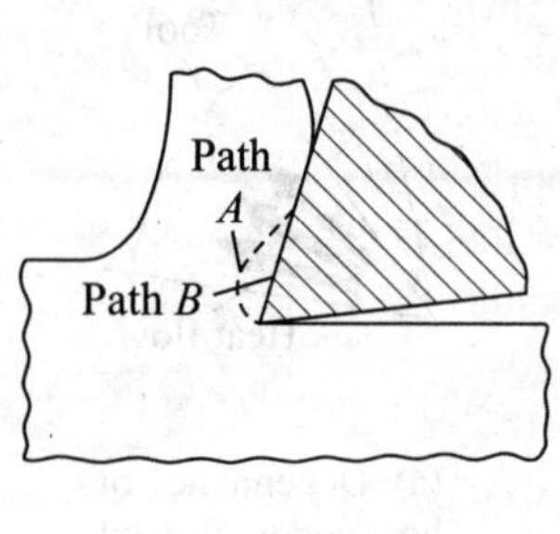

(b) Shear paths

Fig. 4.26 Mechanism of BUE formation.

[1]Shaw, M.C., Metal Cutting Principles, MIT Press, Cambridge, Massachusetts, 1957.

deformed and strain hardened lump sticks to the tool edge and gradually grows in size till the force on the lump is so large that it breaks and the fragments are carried away by the finished surface and the chip. When the temperature is large enough to cross the recrystallization temperature, strain hardening is not very effective and the bonding breaks along the path *B* (Fig. 4.26b), preventing the formation of a built-up edge. Since the built-up edge effectively increases the rake angle, the cutting force and the power consumption reduce. If the built-up edge is stable, it improves the tool life. But when the hard fragments are carried away after the BUE breaks, they can cause a large abrasion wear on the tool, resulting in a lowered tool life. However, for machining a hard material such as cast iron, the formation of a stable built-up edge is welcome since it protects the tool. In general, the effect of BUE on the tool life is erratic. Apart from the effect through abrasion wear, a BUE can lead to tool failure by other mechanisms. For example, when using a carbide tool, a sudden disengagement of the tool from the work may lead to the tearing off of the BUE along with a fragment of the tool material. Again, when the tool cools down after a cut, cracks may develop in the tool due to the higher contraction rate of the welded BUE (Fig. 4.27). The coefficient of thermal expansion of steel is almost twice that of carbide. The effects of the major cutting conditions, i.e., cutting speed and uncut

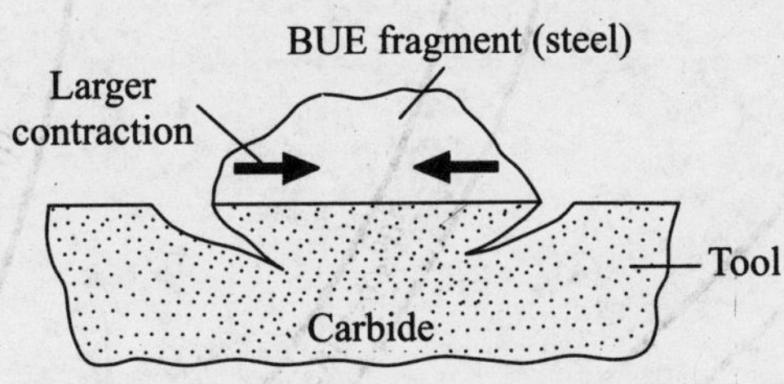

Fig. 4.27 Crack development due to BUE fragment.

thickness, can be represented by the machining chart introduced by E.M. Trent. Figure 4.28 shows such a chart for machining Ni-Cr-Mo steel with cemented carbide. In the case of ferrous metals, the microstructure is one of the most important factors controlling the machinability. Thus, here, a consideration of only the hardness of the work material is not enough. Figure 4.29 shows the generalized behaviour of tool life-cutting speed characteristics for different microstructures when machining steel using a carbide tool. It should be noted that the tool life has been represented not by time but by the volume of material removed. The main structural constituents of cast iron are pearlite and free graphite. The finer the pearlite structure, the lower the machinability. Graphite acts as an internal lubricating agent and prevents the formation of BUE.

4.2.7 CUTTING FLUIDS

Though in a large number of cases machining can be conducted in a dry condition, quite often the use of a cutting fluid is very effective for improving

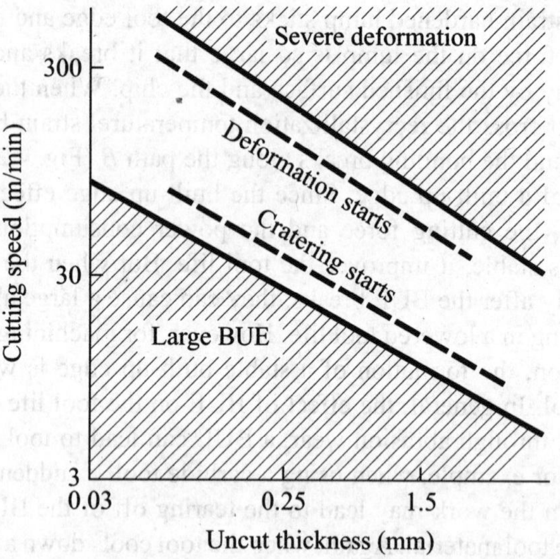

Fig. 4.28 Machining chart for Ni-Cr-Mo steel using cemented carbide tools. (After Trent, E.M., Metal Cutting, Butterworths, London, 1977.)

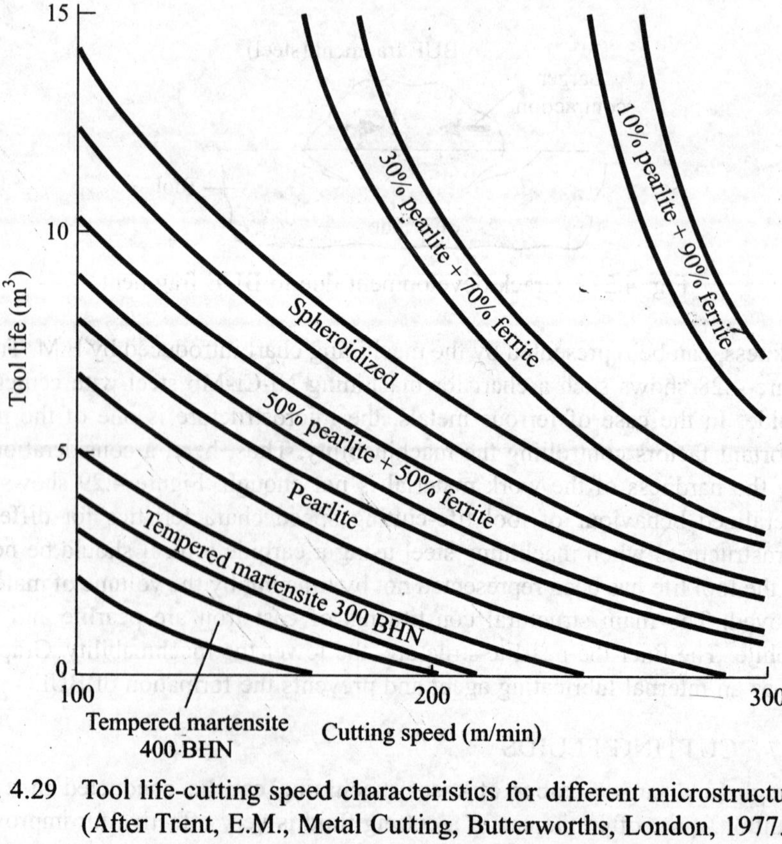

Fig. 4.29 Tool life-cutting speed characteristics for different microstructures. (After Trent, E.M., Metal Cutting, Butterworths, London, 1977.)

the overall machinability. It is not difficult to realize that a suitably chosen fluid may reduce the coefficient of friction at the interfaces. This may be achieved either through lubrication or by lowering the strength of the welded junctions between the tool and the chip by forming a weaker solid (Fig. 4.30). A cutting fluid also prevents the formation of BUE. For example, if a cutting

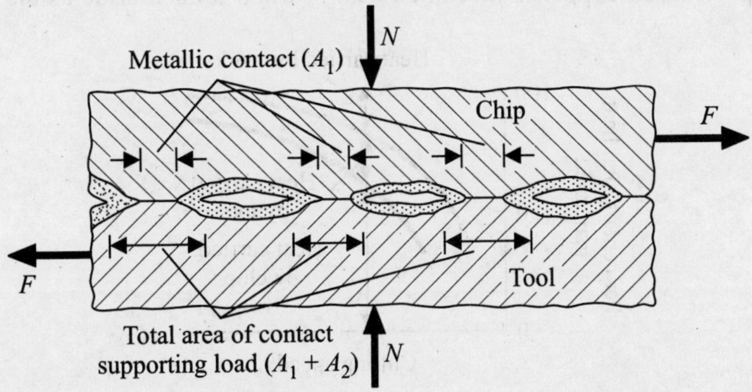

Fig. 4.30 Effect of cutting fluid on chip-tool interface bonding.

fluid containing chlorine is used while machining steel, iron chloride is formed at the interface. It is a weak material and seizure is prevented as shearing takes place only at the interface. The ways in which a cutting fluid affects machining can be summarized and classified as follows:

(i) Cooling down of the chip-tool-work zone by carrying away some of the generated heat.

(ii) Reducing the coefficient of friction at the chip-tool interface due to the formation of a weaker compound at the interface.

(iii) Reducing the thermal distortion caused by temperature gradients generated during machining.

(iv) Washing away the chips and clearing the machining zone.

(v) Protecting the finished surface from corrosion.

Cooling down obviously increases the tool life and reduces the thermal distortion. A reduction in the coefficient of friction lowers the machining force and power consumption. This also prevents the formation of BUE, and the surface finish improves greatly. Washing away of the chips is very helpful in cases where chips are very small as in grinding and milling. An ideal cutting fluid should

(i) have a large specific heat and thermal conductivity,

(ii) have a low viscosity and small molecular size (to help rapid penetration to the chip-tool interface),

(iii) contain a suitable reactive constituent (for forming a low strength compound after reacting with the work material),

(iv) be nonpoisonous and noncorrosive,

(v) be inexpensive and easily available.

Since the action of a cutting fluid takes some minimum time, it is expected that at a high cutting speed the effectiveness of the cutting fluid decreases. Moreover, at a high cutting speed, a large fraction of the generated heat is carried away by the chip (Fig. 4.31a). Figure 4.31b shows how the coefficient of friction changes with speed when copper is machined using carbon tetrachloride as the cutting

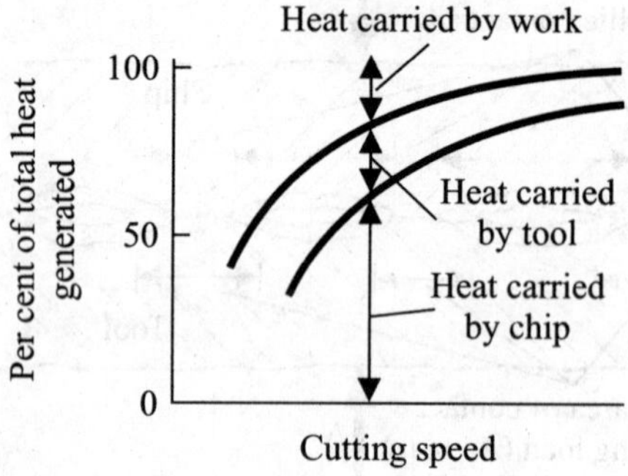

(a) Change in per cent heat distribution with speed

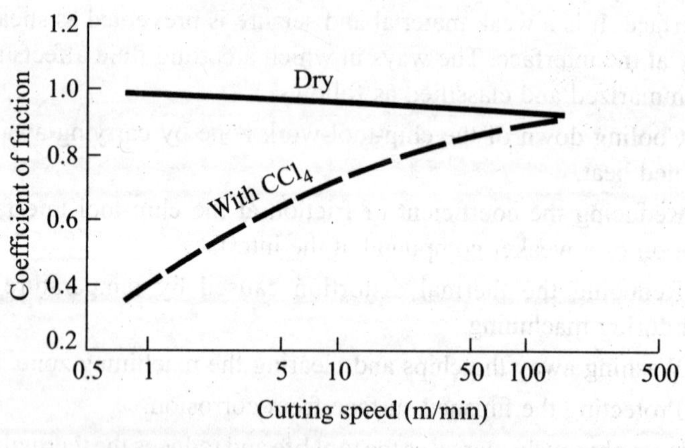

(b) Dependence of coefficient of friction on speed when machining copper

Fig. 4.31 Role of cutting speed.

fluid. Figure 4.32 illustrates a case where 0.15% carbon steel is machined with a tool with +18° rake angle and 0.125 mm uncut thickness, and compares three conditions for cutting.

The cutting fluids are mainly of two types, namely, (i) water based fluids, and (ii) mineral oil based fluids. Additives are used in conjunction with each of these types to accomplish different objectives. Table 4.11 lists the common cutting fluids.

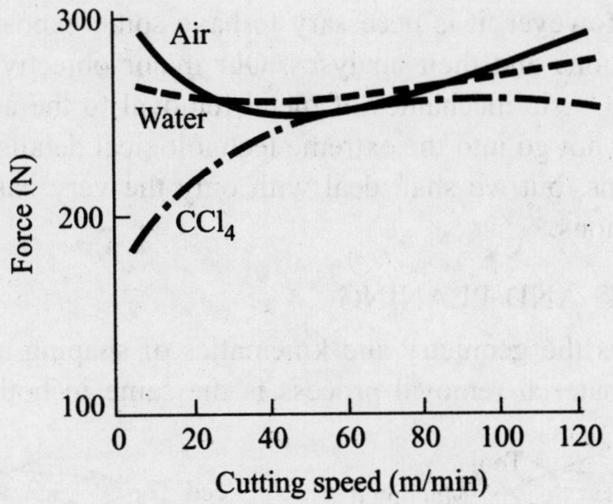

Fig. 4.32 Effect of cutting environment when machining steel.

Table 4.11 Different cutting fluids

- Water based fluids
  - Solutions
  - Dispersions
  - Emulsions
- Mineral oil based fluids
  - Inactive oils
    - Mineral oils
    - Fatty oils
    - Inactive additive oils
  - Active oils
    - Sulphurized mineral oils
    - Sulphochlorinated oils
    - Sulphochlorinated oils with fatty compounds
- Synthetic fluids
- Gases

The effectiveness of a cutting fluid also depends on the technique of application. It has been found that when the cutting zone is flooded, the coolant faces some resistance to enter the chip-tool interface; the effectiveness thus is not high. But if the coolant is forcibly injected into the interface, the fluid acts in a much more efficient manner.

4.3 MACHINING PROCESSES

So far, we have mainly discussed the basic process of material removal and the related aspects. However, it is necessary to have some exposure to the actual machining operations and their analyses. Our major objective in this section is to correlate the basic mechanics of metal removal to the actual process. In this text, we shall not go into the extreme technological details and all possible types of operations, but we shall deal with only the very basic and common machining operations.

4.3.1 SHAPING AND PLANING

Figure 4.33 shows the geometry and kinematics of shaping and planing. The basic nature of material removal process is the same in both the cases. The

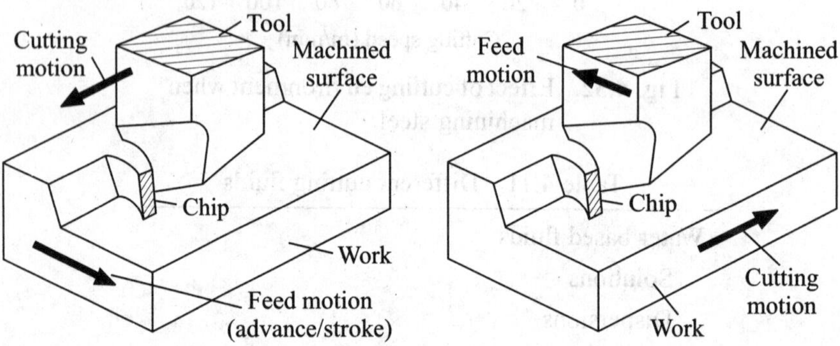

(a) Shaping (b) Planing

Fig. 4.33 Two common intermittent cutting operations.

major difference between the two is that, in shaping, the primary (cutting) motion is provided to the tool and the feed is given to the workpiece, whereas, in planing, it is just the opposite. The cutting operation is intermittent in nature and takes place during the forward stroke. During the return of the tool (or the job, as the case may be), the feed motion is provided when there is no cutting action. Figure 4.34 shows some details of the cutting zone. In an actual cutting operation, the major parameters are the strokes per unit time (N), stroke length (S), quick return ratio (R) (displacement/stroke), depth of cut (d), and the tool angles. To convert these parameters into the basic machining parameters, it would be enough to examine Fig. 4.34 showing a sectional view (section being taken perpendicular to the cutting edge). It should be remembered that, in general, the condition of orthogonal machining is not satisfied but we shall treat the process by assuming that the mechanics of orthogonal machining is applicable. As far as the power consumption is concerned, the results are not very inaccurate. The uncut thickness and the width of cut are given by the relations

$$t_1 = f \cos \psi, \tag{4.36a}$$

$$w = d/\cos \psi, \tag{4.36b}$$

where ψ is the primary principal cutting edge angle. The rake angle is found to

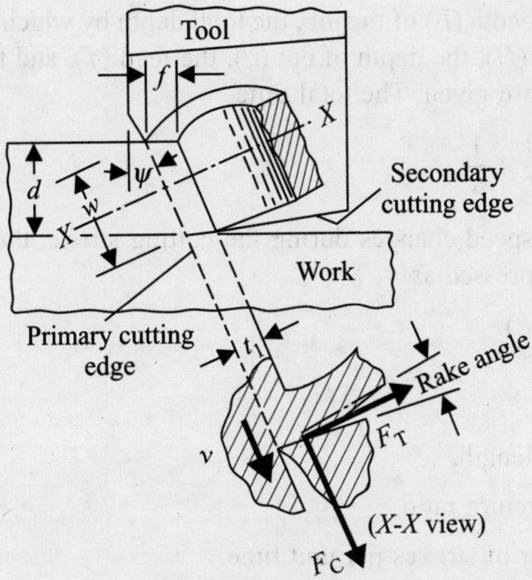

Fig. 4.34 Details of shaping operation.

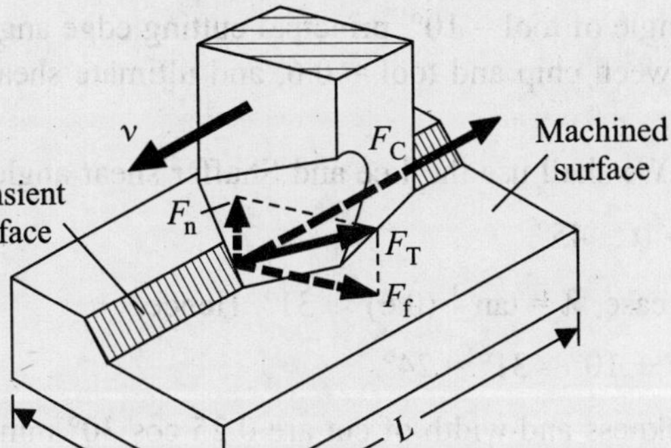

Fig. 4.35 Components of machining force during shaping.

be α (also called normal rake) from the sectional view (Fig. 4.34). Figure 4.35 shows the cutting and thrust components of the force. The cutting component F_C acts against v and F_T acts perpendicular to the transient surface. F_T can be again resolved into two components, namely, F_f (feed component) and F_n (component normal to the machined surface), as

$$F_f = F_T \cos \psi, \tag{4.37a}$$

$$F_n = F_T \sin \psi. \tag{4.37b}$$

The metal removal rate is given by $LdfN$, where L is the length of the job and N is the number of cutting strokes per unit time. The cutting time can also be found out if the breadth (B) of the job, the total depth by which the work surface has to be lowered (H), the depth of cut (d), the feed (f), and the cutting stroke per unit time (N) are given. The total time

$$T_c = \frac{H}{d} \times \frac{B}{f} \times \frac{1}{N}. \tag{4.38}$$

Since the cutting speed changes during the cutting stroke, the average cutting speed v can be expressed as

$$v = \frac{NS(1+R)}{2}, \tag{4.39}$$

where

S = stroke length,

R = quick return ratio,

N = number of strokes per unit time.

EXAMPLE 4.9 Determine the three components of the machining force when shaping a cast iron block with depth of cut = 4 mm, feed = 0.25 mm/stroke, normal rake angle of tool = 10°, principal cutting edge angle = 30°, coefficient of friction between chip and tool = 0.6, and ultimate shear stress of cast iron = 340 N/mm^2.

SOLUTION We shall use the Lee and Shaffer shear angle relationship

$$\phi + \lambda - \alpha = 45°.$$

In the present case, $\lambda = \tan^{-1}(0.6) \approx 31°$. Hence,

$$\phi = 45° + 10° - 31° = 24°.$$

The uncut thickness and width of cut are 0.25 cos 30° mm and 4/cos 30° mm, respectively. Using equation (4.13), we obtain

$$F_C = \frac{0.25 \times 4 \times 340 \times \cos(31° - 10°)}{\sin 24° \times \cos 45°} \text{ N} = 1099 \text{ N}.$$

From equation (4.10), we know

$$F_T = F_C \frac{\sin(\lambda - \alpha)}{\cos(\lambda - \alpha)} = 1099 \times \frac{\sin 21°}{\cos 21°} \text{ N} = 422 \text{ N}.$$

Now, the feed and the normal thrust components are given by equations (4.37a) and (4.37b) as

$$F_f = F_T \cos\psi,$$

$$F_n = F_T \sin\psi$$

or

$$F_f = 422 \cos 30° \text{ N} = 365 \text{ N},$$

$$F_n = 422 \sin 30° \text{ N} = 211 \text{ N}.$$

EXAMPLE 4.10 If the operation in Example 4.9 takes place with 60 strokes/min, what will be the average power consumption if the length of the job is 200 mm? Compare this power consumption with the result obtained by considering the specific power consumption when the hardness of the workpiece is 180 BHN.

SOLUTION Let us assume that the cutting component of F_C remains constant. Thus, the work done during each forward stroke is

$$F_C \times \frac{200}{1000}\ \text{J} = \frac{1099 \times 200}{1000}\ \text{J} = 220\ \text{J}$$

since F_n and F_f do not consume any energy. So, the average power consumption is given by

$$W_{av} = \frac{220 \times 60}{60}\ \text{W} = 220\ \text{W}.$$

Now, let us find out the power consumption using equations (4.21b) and (4.22). In the present case, $U_0 = 0.81$ J/mm^3 from Table 4.4. The value of t_1 is 0.25 cos 30° mm = 0.22 mm. Using equation (4.22), we get

$$U_c = 0.81(0.22)^{-0.4}\ \text{J/mm}^3 = 1.47\ \text{J/mm}^3.$$

The volume rate of material removal is

$$Q = fdLN = \frac{0.25 \times 4 \times 200 \times 60}{60}\ \text{mm}^3/\text{sec} = 200\ \text{mm}^3/\text{sec}.$$

Hence, the power consumption is $W_{av} = U_c Q = 1.47 \times 200$ W = 294 W.

4.3.2 TURNING AND BORING

Turning is one of the most common operations. Surfaces of revolution are generally produced by this operation though the flat surfaces are produced by face turning. All turning operations are done in lathes. The major types of turning operations are (i) turning of cylindrical and stepped cylindrical surfaces, (ii) turning of tapered and curved surfaces of revolution, (iii) turning of screw threads, and (iv) face turning and parting. When an internal surface is machined, the operation is commonly known as boring. The boring operations can also be performed for producing different types of internal surfaces of revolution. Some common turning and boring operations are shown in Fig. 4.36.

We shall discuss here the mechanics of a simple turning operation. This then can be extended to the various other special operations whenever required. Figure 4.37a shows a simple turning operation. The tool used for such an operation is commonly termed as a single point tool. The detailed geometry of this operation is illustrated in Fig. 4.37b. Figure 4.38 shows the different views and angles of a single point turning tool. The parameters in the corresponding basic machining operation can be found out as

$$t_1 = f \cos \psi, \tag{4.40a}$$

$$w = d/\cos \psi, \tag{4.40b}$$

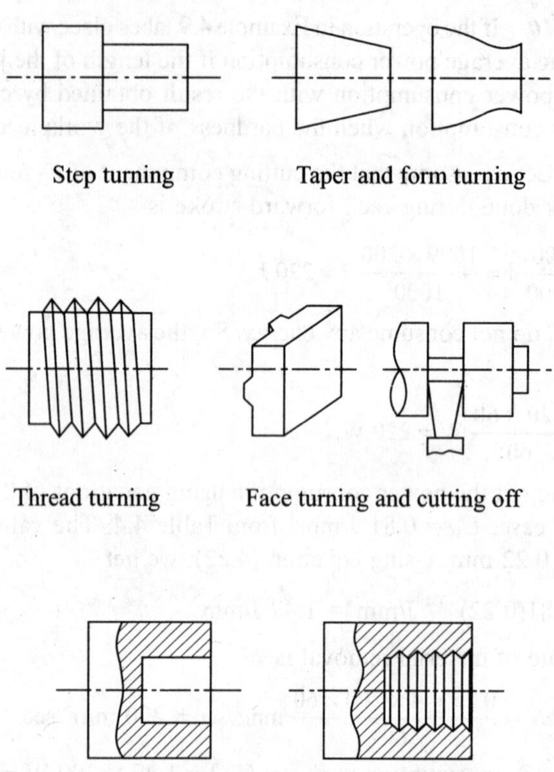

Fig. 4.36 Types of turning operations.

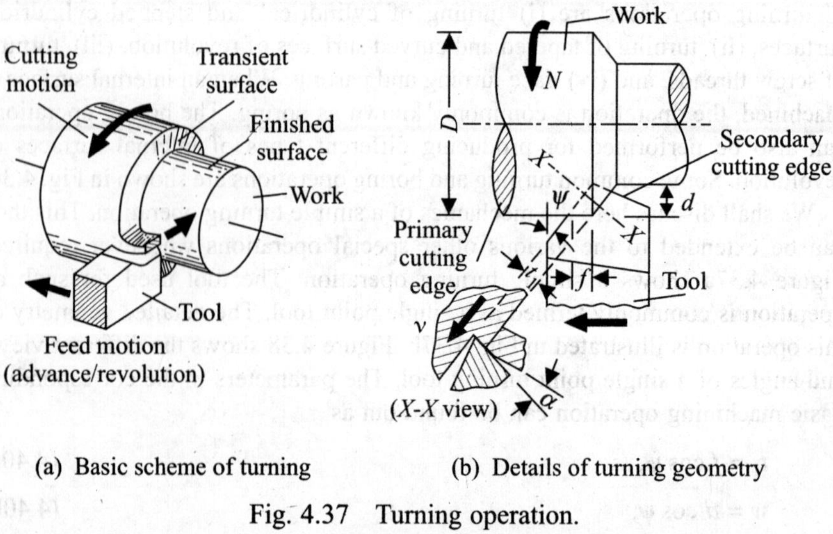

(a) Basic scheme of turning (b) Details of turning geometry

Fig. 4.37 Turning operation.

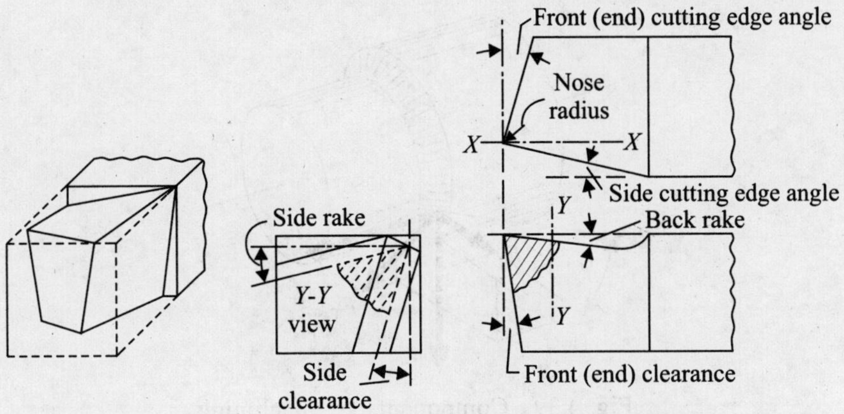

(a) Developing turning tool out of prismatic bar

(b) Tool geometry and angles

Fig. 4.38 Details of single point turning tool.

where ψ is the side cutting edge angle. The normal rake angle α can be found out when the tool angles are specified. In general, the condition of orthogonality is not satisfied, but to keep the discussion within the scope of this text, we shall assume orthogonal machining. The cutting speed is given as

$$v = \pi DN, \tag{4.41}$$

where N is the number of job revolutions per unit time and D is the job diameter. Since the depth of cut d is very small as compared with D, the cutting speed may be assumed to be constant throughout the width of cut and equal to the value given by equation (4.41). To fulfil the condition of orthogonal machining, the cutting edge should be perpendicular to the velocity vector, and it can be easily shown that the condition to be satisfied by the tool angles is

$$\tan \alpha_b \cos \psi = \tan \alpha_s \sin \psi, \tag{4.42}$$

where α_b and α_s are the back and side rake angles. The normal rake angle can be expressed in terms of the other tool angles as

$$\alpha = \tan^{-1}\left[\frac{\tan \alpha_b \sin \psi + \tan \alpha_s \cos \psi}{\sqrt{1 + (\tan \alpha_b \cos \psi - \tan \alpha_s \sin \psi)^2}}\right]. \tag{4.43}$$

The force components on the tool during turning are shown in Fig. 4.39. The cutting component F_C is in the vertical direction and F_T acts in the horizontal plane and perpendicular to the cutting edge. The feed and the radial components F_f and F_R are

$$F_f = F_T \cos \psi, \tag{4.44a}$$

$$F_R = F_T \sin \psi. \tag{4.44b}$$

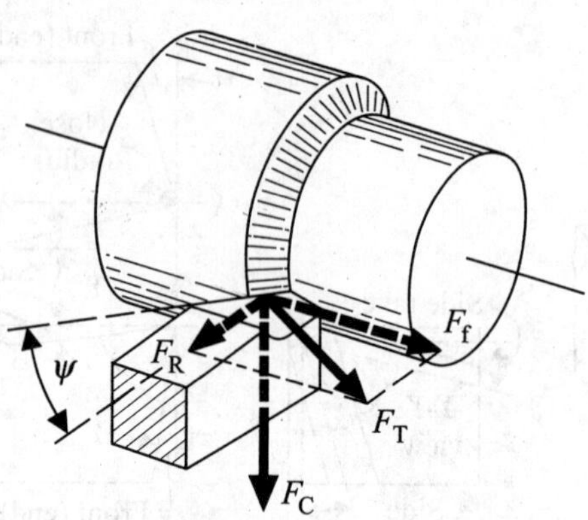

Fig. 4.39 Components of machining force during turning.

The power consumption, neglecting the feed component, is

$$W = F_C v, \tag{4.45}$$

and the material removal rate is given by $f\,dv$. When a cylinder of length L is being turned at a spindle speed N with a feed f (advancement of the tool/revolution of the job), the total machining time is

$$T_c = \frac{L}{fN}. \tag{4.46}$$

EXAMPLE 4.11 The angles of a single point turning tool are as follows (this is the normal order in which the angles are mentioned according to the American Standard System):

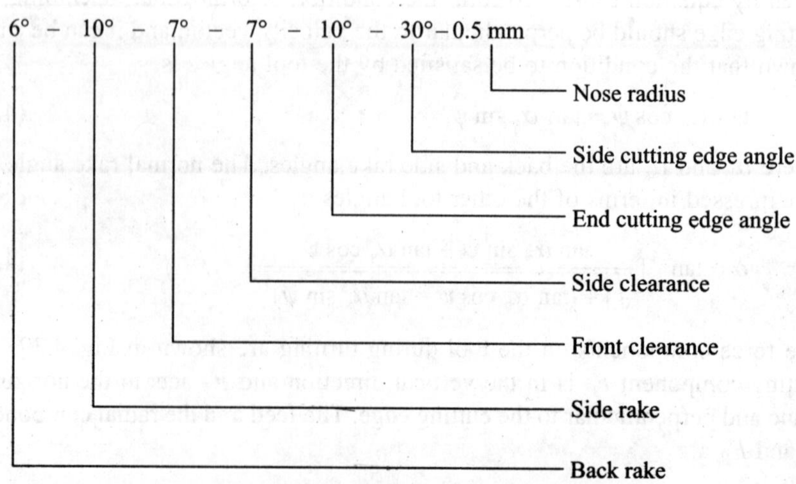

Show that the turning operation with this tool will be close to orthogonal machining, and calculate the normal rake angle.

SOLUTION Strictly speaking, almost all the machining operations are not orthogonal even when the cutting edge is perpendicular to the velocity direction. This is because in any such operation two edges take part in the cutting operation. Of these, one is the principal or primary cutting edge and the other is the secondary cutting edge (Figs. 4.34 and 4.37). Such an operation, where the primary cutting edge is normal to the cutting velocity vector, is called restricted orthogonal machining. However, we shall consider this orthogonal since the major objective of our study is to find out the orders of magnitude and not the exact values. Using equation (4.42), we estimate the value of $\tan \alpha_b \cos \psi - \tan \alpha_s \sin \psi$, and its deviation from zero indicates the degree of nonorthogonality. In our example,

$$\tan \alpha_b \cos \psi - \tan \alpha_s \sin \psi = 0.1051 \times 0.866 - 0.1763 \times 0.5$$
$$= 0.0910166 - 0.08815 = 0.0029.$$

So, the case is close to an orthogonal one. Now, to find out the normal rake α, we use equation (4.43). Thus,

$$\alpha = \tan^{-1}\left[\frac{0.1051 \times 0.5 + 0.1763 \times 0.866}{\sqrt{1 + 0.00292}}\right]$$

$$= \tan^{-1}(0.205225) = 11.6°.$$

EXAMPLE 4.12 A mild steel bar of 100 mm diameter is being turned with a tool with the specification 6°–10°–5°–7°–10°–30°–0.55 mm. The depth of cut is 2.5 mm and the feed is 0.125 mm/revolution. The rpm of the job is 300. Determine the components of the machining force and the power consumption. The ultimate shear stress of the work material is 400 N/mm^2 and the coefficient of friction between the tool and the chip is 0.6.

SOLUTION We have already found out that the normal rake angle $\alpha = 11.6°$ for this tool (see Example 4.11). The friction angle $\lambda = \tan^{-1} 0.6 \approx 31°$. To find out the shear plane angle, let us assume the Lee and Shaffer relation. Thus,

$$\phi = 45° + 11.6° - 31° = 25.6°.$$

From equations (4.13) and (4.40),

$$F_C = \frac{400 \times 0.125 \times 2.5 \times \cos(31° - 11.6°)}{\sin 25.6° \cos 45°} \text{ N} = 386 \text{ N}.$$

The thrust component is

$$F_T = F_C \cdot \frac{\sin 19.4°}{\cos 19.4°} = 136 \text{ N};$$

also,

$$F_f = F_T \cos 30° = 118 \text{ N},$$

$$F_R = F_T \sin 30° = 68 \text{ N}.$$

EXAMPLE 4.13 During a face turning operation, the tool is being fed in a radially outward direction (Fig. 4.40) at a rate of f/revolution. The job is rotating at a speed of N revolutions per unit time. The cutting starts from the inner radius r_0. For a given tool-work feed-depth combination, Taylor's tool life equation is $vT^n = C$ when machining is done at a constant speed v. What will be the expression for tool life during the face turning operation?

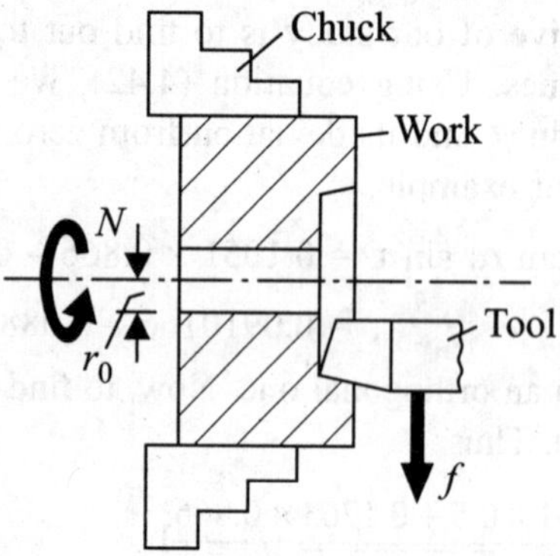

Fig. 4.40 Face turning operation.

SOLUTION To start with, we shall assume that the tool life consists mainly of the contribution of the uniform wear rate zone. Thus, if $h_{f_{\lim}}$ is the limiting value of tool wear and R is the wear rate, then $h_{f_{\lim}} \approx RT$. Substituting T from Taylor's equation, we get

$$h_{f_{\lim}} \approx R\left(\frac{C}{v}\right)^{1/n} \quad \text{or} \quad R \approx \frac{h_{f_{\lim}} v^{1/n}}{C^{1/n}}.$$

This gives the approximate relationship between the wear rate and the speed.

Now, during the face turning operation, the instantaneous radius at time t (considering $t = 0$ when $r = r_0$) is

$$r = r_0 + fNt$$

and the instantaneous cutting speed is

$$v = 2\pi(r_0 + fNt)N.$$

Thus, the instantaneous wear rate can be expressed as

$$R \approx \frac{h_{f_{\lim}}}{C^{1/n}}[2\pi(r_0 + fNt)N]^{1/n}.$$

Hence, during a small interval of time dt, the amount of wear is given as

$$dh = \frac{h_{f_{\lim}}}{C^m}(2\pi N)^m (r_0 + fNt)^m \, dt,$$

where $m = 1/n$. Finally,

$$h_{f_{\lim}} = \int_0^T \frac{h_{f_{\lim}}}{C^m}(2\pi N)^m (r_0 + fNt)^m \, dt$$

since after a time T (tool life) the critical amount of wear is reached. After integration and some algebraic manipulation, the expression for the tool life becomes

$$T = \frac{1}{fN}\{[\frac{(m+1)\,fNC^m}{(2\pi N)^m} + r_0^{m+1}]^{1/(m+1)} - r_0\}.$$

When cutting is started from the centre, the expression takes the form

$$T = [\frac{(m+1)C^m}{(2\pi f N^2)^m}]^{1/(m+1)}.$$

Actually, the face turning method, such as the one in the foregoing example, is used for a quick determination of C and m. Using different speeds N_1 and N_2 if the corresponding tool lives are T_1 and T_2, then it can be shown that

$$n = \frac{2(\log N_2 - \log N_1)}{\log T_1 - \log T_2} - 1.$$

4.3.3 DRILLING

The most common hole making operation is drilling and it is usually performed with the help of a twist drill. Unlike shaping and turning, this involves two principal cutting edges. Figure 4.41 shows a drilling operation. If the total advancement of the drill per revolution (the feed rate) is f, then the share of each cutting edge is $f/2$ because each lip is getting the uncut layer the top surface of which has been finished by the other lip 180° ahead (during 180° rotation, the

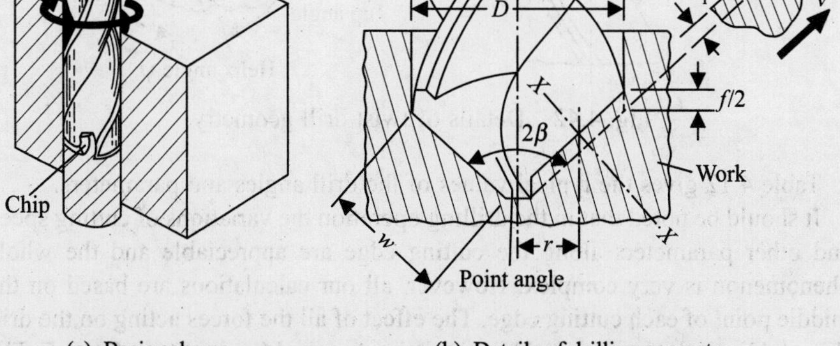

(a) Basic scheme (b) Details of drilling geometry

Fig. 4.41 Drilling operation.

vertical displacement of the drill is $f/2$). The uncut thickness t_1 and the width of cut w are given as

$$t_1 = (f/2)\sin\beta, \tag{4.47a}$$

$$w = (D/2)/\sin\beta, \tag{4.47b}$$

where β is the half point angle (Fig. 4.41b). The rake angle α can be found out from the sectional view (Fig. 4.41b), but it can be easily shown that in the case of a twist drill, α depends on the radial location of the sectioning plane. Figure 4.42 shows three views of a common twist drill. The normal rake angle α can be approximately expressed as

$$\alpha \approx \tan^{-1}\left[\frac{(2r/D)\tan\psi}{\sin\beta}\right], \tag{4.48}$$

r being the radius of the point on the cutting edge where the normal rake is being evaluated, D the nominal diameter of the drill, β the half point angle (Fig. 4.41b), and ψ the helix angle (Fig. 4.42).

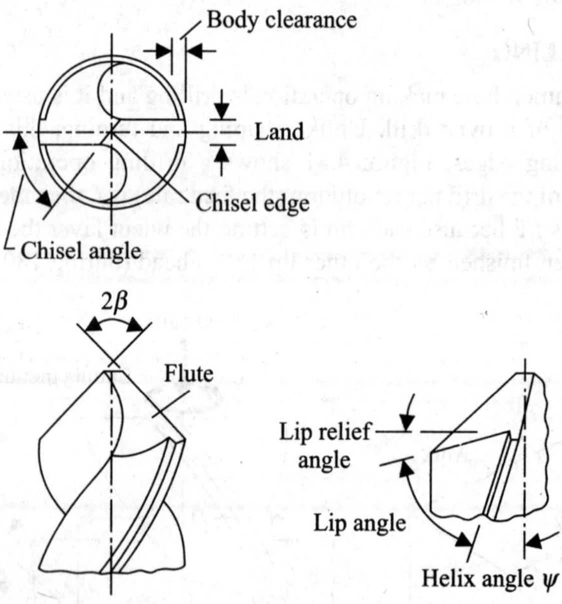

Fig. 4.42 Details of twist drill geometry.

Table 4.12 gives the typical values of the drill angles and parameters.

It should be noted that in the drilling operation the variations of cutting speed and other parameters along the cutting edge are appreciable and the whole phenomenon is very complex. However, all our calculations are based on the middle point of each cutting edge. The effect of all the forces acting on the drill (Fig. 4.43) can be represented by a resisting torque M and a thrust force F. The action at the chisel edge is not truly a cutting action, rather it is one of pushing

Table 4.12 Drill angles used for various materials

Material(s)	Helix angle ψ (degrees)	Point angle 2β (degrees)	Lip relief angle (degrees)	Chisel angle (degrees)
Brittle and hard	22–33 (lower value for smaller size)	80	6–8	55
Steels and cast irons	22–23 (lower value for smaller size)	118	6–8	51
Soft	22–23 (lower value for smaller size)	140	6–8	51

into the material like a wedge. But the effect of the chisel edge on the torque is negligible as it is on the axis of rotation. The contribution of the chisel edge

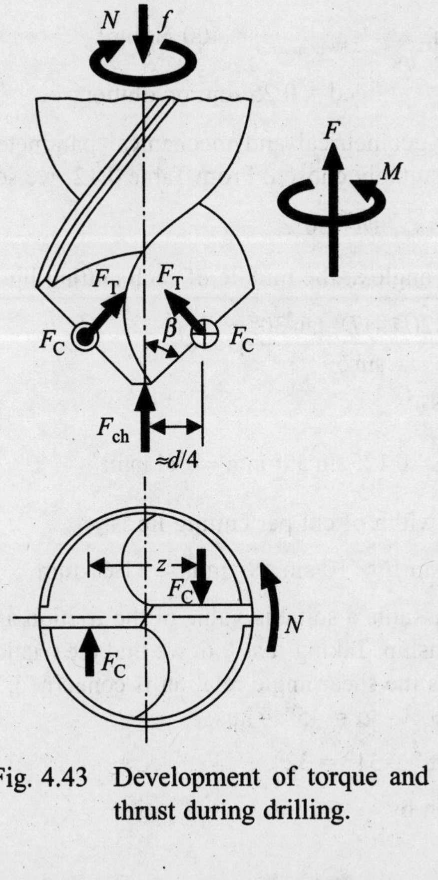

Fig. 4.43 Development of torque and thrust during drilling.

to the development of the thrust force is considerable. The total thrust force F can be expressed as

$$F = 2F_T \sin\beta + F_{ch} + F_{friction}, \tag{4.49}$$

where F_{ch} is the force from the chisel edge and $F_{friction}$ is the friction force. From experimental measurements, it has been found that F_{ch} is almost 60% of the total thrust force and $F_{friction}$ is very small (about 3%). Similarly, the total moment M can be written as (Fig. 4.43)

$$M = F_C z + M_{ch} + M_{friction}, \tag{4.50}$$

where M_{ch} is the moment from the chisel edge and $M_{friction}$ is the moment due to friction. About 80% of M comes from the cutting action of the lips, whereas M_{ch} and $M_{friction}$ are about 8% and 12% of M, respectively. Thus, the final approximate expressions for M and F can be written as

$$M \approx 0.6F_C D, \tag{4.51a}$$

$$F \approx 5F_T \sin\beta. \tag{4.51b}$$

EXAMPLE 4.14 Estimate the torque and the thrust force when drilling a solid block of mild steel with a normal twist drill. Given

$$D = 20 \text{ mm}, \qquad \tau_{s_{\text{work material}}} = 400 \text{ N/mm}^2,$$

$$\text{rpm} = 240, \qquad \text{feed} = 0.25 \text{ mm/revolution}.$$

SOLUTION The geometrical and mechanical parameters, other than those given, have to be suitably chosen. From Table 4.12, we select the values

$$2\beta = 118°, \qquad \psi = 30°.$$

The effective rake angle at the middle of each cutting lip [equation (4.48)] is

$$\alpha \approx \tan^{-1}\left[\frac{\{2(D/4)/D\} \tan 30°}{\sin 59°}\right] \approx 18°.$$

The uncut thickness is

$$t_1 = \frac{f}{2}\sin\beta = 0.125 \sin 59° \text{ mm} = 0.11 \text{ mm};$$

the corresponding width of cut per cutting lip is

$$w \approx (D/2)/\sin\beta \approx 10/\sin 59° \text{ mm} \approx 11.67 \text{ mm}.$$

Next, we have to assume a suitable value of the friction angle λ and a suitable shear angle relationship. Taking $\mu \approx 0.6$, we find the friction angle λ works out to be 31°. So far as the shear angle relation is concerned, we take the Lee and Shaffer relation $\phi + \lambda - \alpha = 45°$. Thus,

$$\phi = 45° + 18° - 31° = 32°.$$

F_C and F_T are given by

$$F_C \approx \frac{0.125 \times 11.67 \times 400 \times \cos(31^\circ - 18^\circ)}{\sin 32^\circ \times 0.707} \text{ N} \approx 1517 \text{ N},$$

$$F_T = \frac{0.125 \times 11.67 \times 400 \times \sin(31^\circ - 18^\circ)}{\sin 32^\circ \times 0.707} \text{ N} \approx 350 \text{ N}.$$

Now, using relations (4.51a) and (4.51b), we see that the drilling torque and thrust work out as

$$M \approx \frac{0.6 \times 157 \times 20}{1000} \text{ N-m} \approx 18.2 \text{ N-m},$$

$$F \approx 5 \times 350 \times \sin 59^\circ \text{ N} \approx 1500 \text{ N}.$$

However, it should be remembered that, since the central portion of the drill (the chisel edge) pushes into the workpiece like a wedge and the cutting speed near the axis of rotation is small, the single shear plane model is not valid for some portion of the cutting edges. Further, the cutting operation near the axis is far from being orthogonal. As a result, the torque and thrust force estimation, using a single shear plane, is not accurate.

4.3.4 MILLING

Milling is perhaps the most versatile machining operation and most of the shapes can be generated by this operation. It is especially more indispensable for machining the parts without rotational symmetry. Unlike turning, shaping, and drilling tools, the milling tool possesses a large number of cutting edges. The shaft on which the cutter is mounted is commonly known as the arbor. The milling operations can be classified into two major groups, namely, (i) horizontal milling, and (ii) vertical milling. In the horizontal milling operation, the cutter axis is horizontal. Figure 4.44 shows some common horizontal milling operations. Horizontal milling can, again, be divided into two groups depending on the relative directions of cutting and feed motion. When the arrangement is like what is shown in Fig. 4.45a, the operation is called up milling. When the cutting and the feed motion are in the same direction (Fig. 4.45b), the operation is called down milling. Since in down milling there is a tendency of the job being dragged into the cutter, up milling is safer and is commonly done. However, down milling results in a better surface finish and longer tool life. When the cutting edges are helical, the cutting operation is smoother and a better finish is obtained. This is due to the gradual engagement of the cutting edge.

The cutter axis is vertical and perpendicular (generally) to the work surface in vertical milling. Some common vertical milling operations are shown in Fig. 4.46.

The scheme of chip formation during plain slab milling using a straight cutter is explained in Fig. 4.47a. The cutter has a diameter D and the depth of cut provided is d. When milling is done with a straight-edged cutter, the operation is orthogonal and the kinematics of chip formation is as shown in Fig. 4.47b. Since all the cutting edges take part in machining, a study of the process is facilitated by considering the action of only a single tooth. If f is the feed velocity of the table in mm/min, the effective feed per tooth in mm will be $f/(NZ)$, where N is

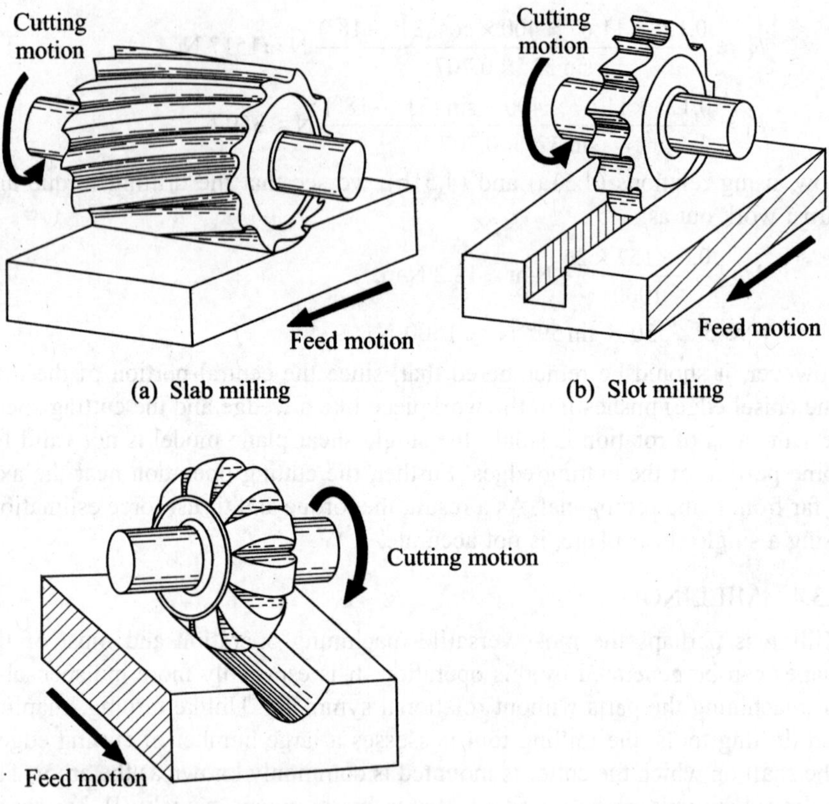

(a) Slab milling (b) Slot milling

(c) Form milling

Fig. 4.44 Types of milling operations.

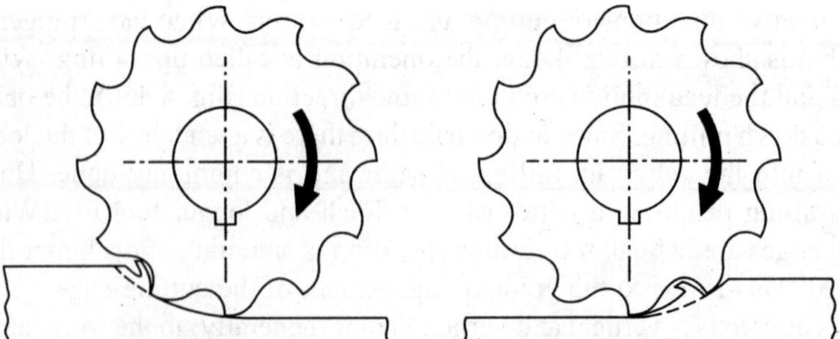

(a) Up milling (b) Down milling

Fig. 4.45 Milling operation.

the cutter rpm and Z is the number of teeth in the cutter. The material removal rate per unit width of the job is given by fd. It is clearly seen from Fig. 4.47b

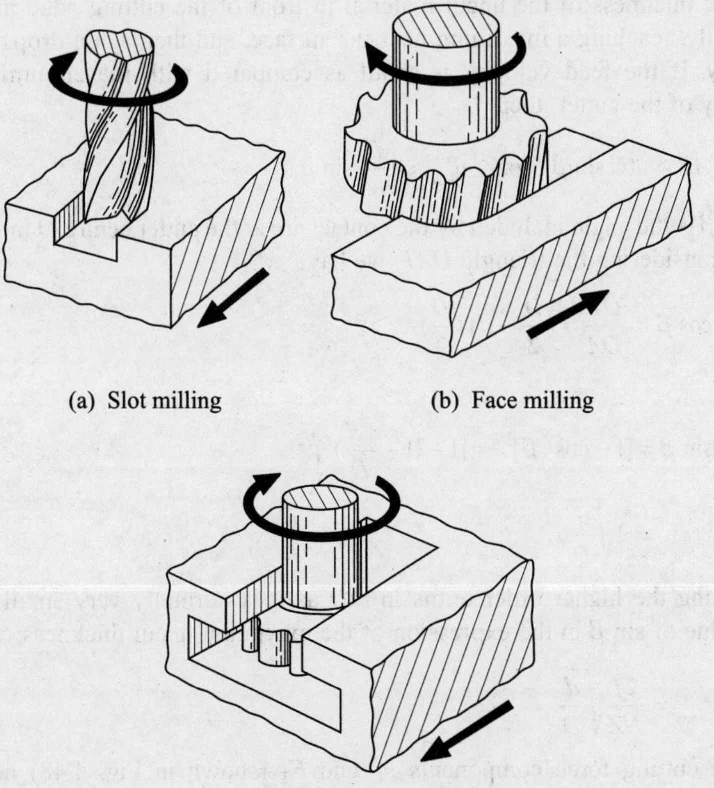

(a) Slot milling (b) Face milling

(c) T-slot milling

Fig. 4.46 Types of vertical milling operations.

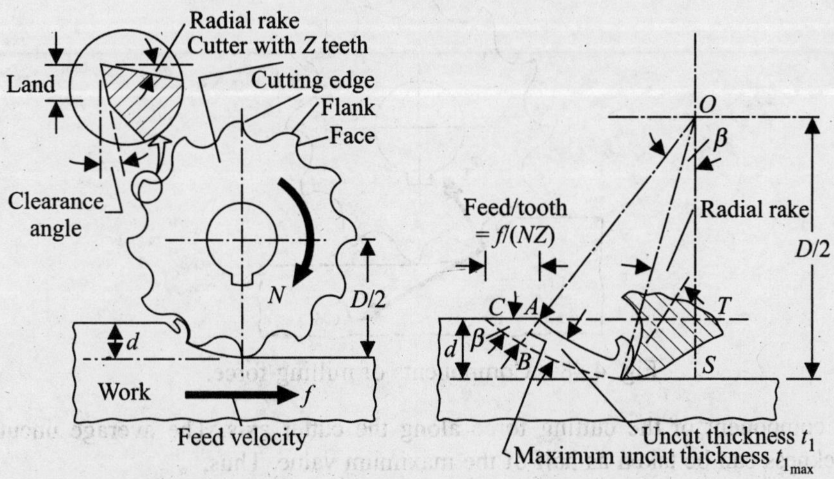

(a) Cutter and operation parameters (b) Details of chip formation

Fig. 4.47 Details of milling operation.

that the thickness of the uncut material in front of the cutting edge increases gradually, reaching a maximum near the surface, and then again drops to zero quickly. If the feed velocity is small as compared with the circumferential velocity of the cutter, then

$$AB \approx AC \sin \beta \quad \text{or} \quad t_{1_{\max}} \approx \frac{f}{NZ} \sin \beta,$$

where β is the angle included by the contact arc at the cutter centre O in radians. Now, considering the triangle OAT, we have

$$\cos \beta \approx \frac{OT}{OA} = (\frac{D}{2} - d)/(\frac{D}{2}).$$

Hence,

$$\sin \beta = [1 - \cos^2 \beta]^{1/2} = [1 - (1 - \frac{2d}{D})^2]^{1/2}$$

$$\approx 2\sqrt{\frac{d}{D}}, \tag{4.52}$$

neglecting the higher order terms in d/D as it is normally very small. Using this value of $\sin \beta$ in the expression of the maximum uncut thickness, we get

$$t_{1_{\max}} = \frac{2f}{NZ}\sqrt{\frac{d}{D}}. \tag{4.53}$$

So, the cutting force components F_C and F_T (shown in Fig. 4.48) not only change in direction but also in magnitude as the cutting edge moves along the cut surface. It is obvious that when cutting with a straight cutter, there is

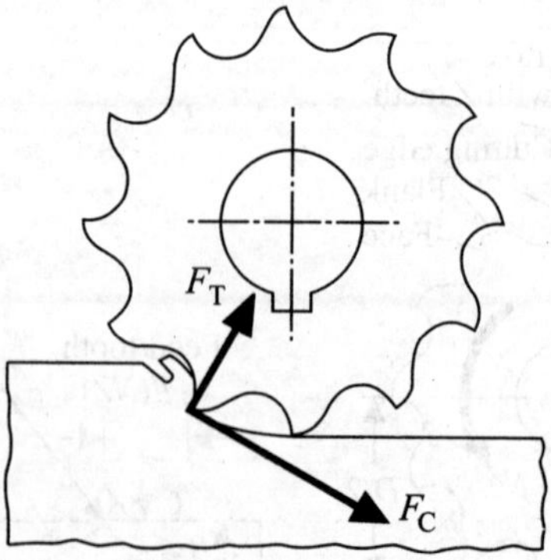

Fig. 4.48 Components of milling force.

no component of the cutting force along the cutter axis. The average uncut thickness can be taken as half of the maximum value. Thus,

$$t_{1_{av}} \approx \frac{f}{NZ}\sqrt{\frac{d}{D}}. \tag{4.54}$$

The average values of F_C and F_T can be approximately found out using this value of uncut thickness. Since F_T acts in the radial direction, it does not produce any torque and the arbor torque is due only to the component F_C. So, the torque M due to one cutting tooth is $F_C(d/2)$ and varies approximately as F_C. Figure 4.49 shows the variation of arbor torque ($\bar{M}$) with arbor rotation

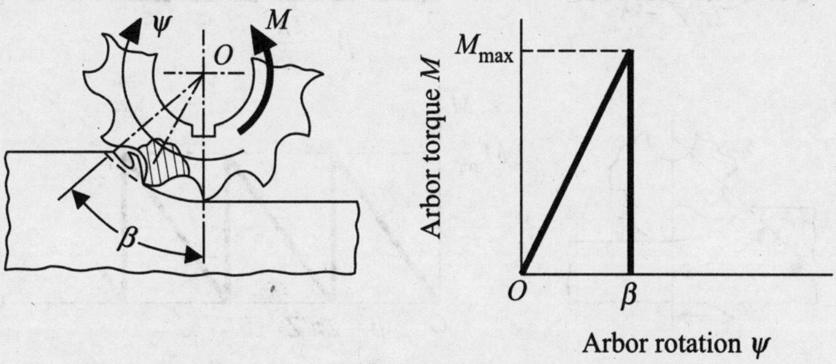

(a) Arbor rotation and resisting torque

(b) Variation of arbor torque due to single cutting tooth

Fig. 4.49 Milling torque.

for the action of only a single tooth. Now, to get the overall torque ($\bar{M}$), the moments due to all the teeth should be properly superimposed. This leads to three different possibilities, namely, (i) $\beta < 2\pi/Z$, (ii) $\beta = 2\pi/Z$, and (iii) $\beta > 2\pi/Z$. Figure 4.50a shows the three different possibilities; the arbor torque corresponding to each of these is shown in Fig. 4.50b. It is apparent from Fig. 4.50 that with a straight-edged cutter the force and arbor torque have sharp variations which can cause vibration problems. When a helical cutter is used, the contact between the cutting edge and the workpiece starts and ends gradually. Here, the arbor torque due to a single tooth and the overall torque are of the type shown in Figs. 4.51a and 4.51b, respectively. The machining power can be calculated by taking the product of the arbor speed and the average overall arbor torque. The average thrust force can be considered to be acting along the midradial line of the work-cutter contact arc.

EXAMPLE 4.15 A mild steel (ms) block of 20 mm width is being milled using a straight slab milling cutter with 20 teeth, 50 mm diameter, and 10° radial rake. The feed velocity of the table is 15 mm/min and the cutter rotates at 60 rpm. If a depth of cut of 1 mm is used, what will be the power consumption?

SOLUTION To start with, let us find out the angle of contact and the angle subtended by two consecutive cutting edges. From relation (4.52),

$$\sin\beta \approx 2\sqrt{\frac{d}{D}} = 2\sqrt{\frac{1}{50}} = 0.28284.$$

So, $\beta = 16.4°$. The angle between two consecutive teeth is $2\pi/20 = 18° > \beta$.

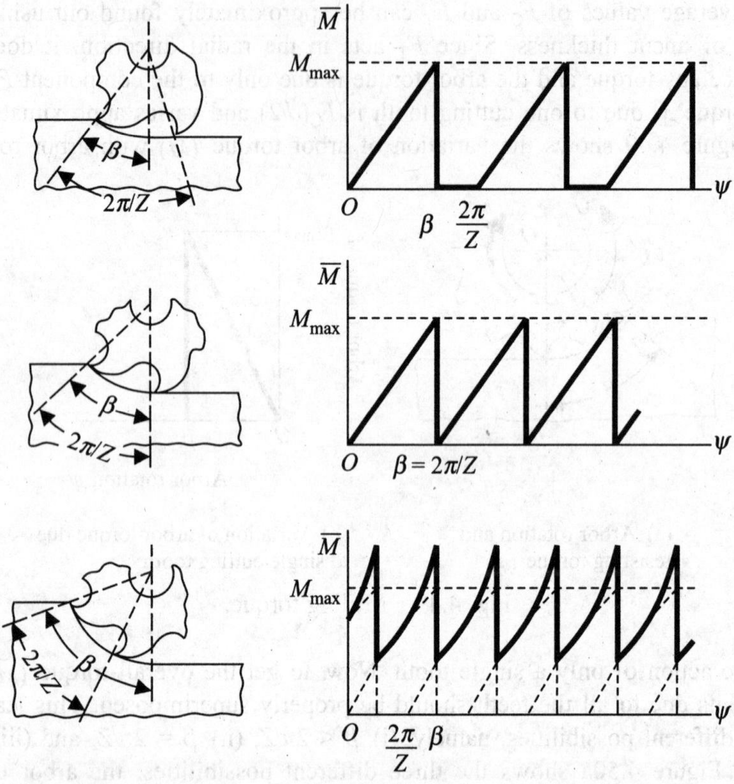

(a) Types of tooth engagements (b) Nature of torque fluctuation

Fig. 4.50 Milling torque fluctuation.

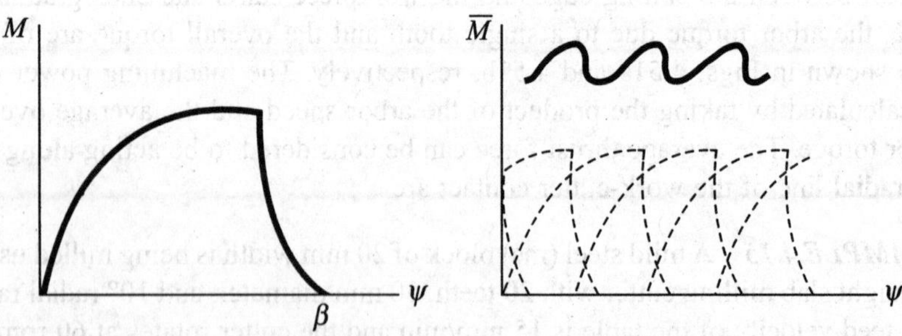

(a) Torque for single tooth (b) Resultant torque

Fig. 4.51 Milling torque for helical cutter.

The maximum uncut thickness is found out, using equation (4.53). Substituting the given data in this equation, we get

$$t_{1_{max}} = \frac{2 \times 15}{60 \times 20}\sqrt{\frac{1}{50}} \text{ mm} = 0.0035 \text{ mm}.$$

The coefficient of friction at the rake face and the shear stress of the work material may be assumed to be 0.5 and 400 N/mm^2, respectively. So,

$$\lambda = \tan^{-1} \mu = \tan^{-1} 0.5 = 26.57°.$$

Following the Lee and Shaffer shear angle relationship (Table 4.3), we get the shear angle $\phi = 28.43°$. The maximum value of the cutting component of the machining force is [using equations (4.9) × (4.10a)]

$$F_{C_{max}} = \frac{wt_{1_{max}} \tau_s \cos(\lambda - \alpha)}{\sin\phi \cos(\phi + \lambda - \alpha)}$$

$$= \frac{20 \times 0.0035 \times 400 \cos(26.57° - 10°)}{\sin 28.43° \cos(\pi/4)} \text{ N}$$

$$= \frac{28.8\sqrt{2} \cos 16.57°}{\sin 28.43°} \text{ N} \approx 81.5 \text{ N}.$$

The variation of torque due to a single tooth with arbor rotation is shown in Fig. 4.52. The maximum value is approximately 2 N-m. Now, to find out the

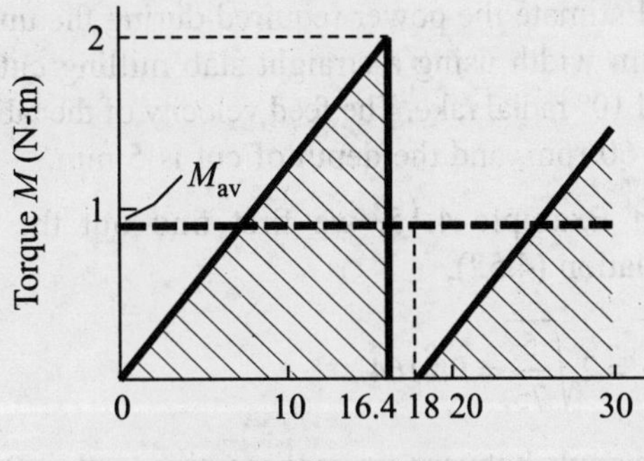

Fig. 4.52 Determination of average torque during milling.

average, we have to divide the area under the M-ψ curve by the angle between two consecutive teeth. Thus,

$$M_{av} = (\tfrac{1}{2} \times 16.4 \times 2)/18 = 0.91 \text{ N-m}.$$

To find out the power consumption, we have to multiply this average torque by the angular speed of the cutter. With 60 rpm, the angular speed is

$$\frac{2\pi \times 60}{60} \text{ rad/sec} = 2\pi \text{ rad/sec}.$$

So, the approximate power requirement will be $2\pi \times 0.91$ W $\approx$ 6 W. We can estimate the power requirement also by using the concept of specific energy. Using equation (4.22) and Table 4.4, we get

$$U_{c_{ms}} \approx 1.4 \times (0.0018)^{-0.4} \text{ J/mm}^3,$$

where 0.0018 is the average uncut thickness in mm. Or

$$U_{c_{ms}} \approx 17.5 \text{ J/mm}^3.$$

The rate of material removal is given by

$$20 \times 1 \times 15 \text{ mm}^3\text{/min} = 5 \text{ mm}^3\text{/sec.}$$

Finally, the power consumption is given by

$$U_{c_{ms}} \times \text{material removal rate,} \quad \text{i.e.,} \quad 17.5 \times 5 \text{ W} = 87.5 \text{ W.}$$

The discrepancy is due to the fact that when calculating the power by mechanics of machining, the ultimate shear stress of ms has been taken as 400 N/mm^2 which is obtained from normal tests. In other words, the size effect has not been taken into account which is quite predominant as the uncut thickness is very small.

It should be further noted that the power required for material removal is only a fraction of the total power supplied, the rest going in overcoming the frictional losses.

EXAMPLE 4.16 Estimate the power required during the up milling of a mild steel block of 20 mm width using a straight slab milling cutter with 10 teeth, 75 mm diameter, and 10° radial rake. The feed velocity of the table is 100 mm/min, the cutter rotates at 60 rpm, and the depth of cut is 5 mm.

SOLUTION As in Example 4.15, we first find out the maximum uncut thickness. From equation (4.52),

$$\sin \beta \approx 2\sqrt{\frac{d}{D}} = 2\sqrt{\frac{5}{75}} = 0.5164.$$

Thus, $\beta = 31.1°$. The angle between two consecutive teeth is $2\pi/10 = 36°$ which is greater than 31.1°.

The maximum uncut thickness is found out, using equation (4.53). Thus,

$$t_{1_{max}} = \frac{2 \times 100}{60 \times 10}\sqrt{\frac{5}{75}} \text{ mm} = 0.086 \text{ mm.}$$

Assuming that the values of the coefficient of friction and ultimate shear stress are the same as in Example 4.15, we have $\lambda = 26.57°$ and $\tau_s = 400$ N/mm^2. The maximum value of the cutting component of force is

$$F_{C_{max}} \approx \frac{20 \times 0.086 \times 400 \cos 16.57°}{\sin 28.43° \cos (\pi/4)} \text{ N}$$

$$\approx 1959 \text{ N.}$$

Therefore, the maximum torque is

$$\frac{1959 \times 75}{2 \times 100} \text{ N-m} = 73.5 \text{ N-m.}$$

The average torque can be found out as

$$M_{av} = \frac{\frac{1}{2}M_{max} \times \beta}{2\pi/N} = \frac{0.5 \times 73.5 \times 31.1}{36} \text{ N-m} = 31.7 \text{ N-m}.$$

The power consumption is

$$\frac{2\pi 60 \times 31.7}{60} \text{ W} \approx 199 \text{ W}.$$

Using the specific power consumption requirement, we have

$$U_{c_{ms}} \approx 1.4 \times (0.043)^{-0.4} \text{ J/mm}^3 = 4.9 \text{ J/mm}^3,$$

where 0.043 mm is the average uncut thickness. The rate of material removal is given by

$$20 \times 5 \times 100 \text{ mm}^3\text{/min} = 166.7 \text{ mm}^3\text{/sec}.$$

Finally, the power consumption is 166.7×4.9 W = 817 W. It is seen that since the uncut thickness is more, the percentage difference between the results obtained by two different approaches in this example is much less than that in Example 4.15.

It has been shown[1] that the average value of the cutting component (which is also the tangential component) F_C can be estimated as

$$F_{C_{av}} = \frac{\Lambda}{(t_{1_{av}})^{\lambda}} a_{av} \text{ N}, \tag{4.55}$$

where Λ, λ are constants depending on the work material and the cutting environment and a_{av} is the average cross-section of uncut thickness in mm considering all the teeth in engagement. To find out a_{av}, we can divide the rate of volume removal by the velocity of the cutter tip. Thus,

$$a_{av} \approx \frac{Bdf}{\pi DN} \text{ mm}, \tag{4.56}$$

where B is the width of cut in mm. Once $F_{C_{av}}$ has been estimated, the power can be found out. The values of Λ and λ for some common engineering materials are given in Table 4.13. Combining relations (4.55) and (4.56), we get

Table 4.13 Values of Λ and λ

Material	Λ	λ
Alloy steel	2100	0.28
Mild steel	1400	0.28
Bronze	400	0.40

[1]Vulf, A.M., Theory of Metal Cutting (in Russian), Mashgiz, Moscow, 1962.

$$\text{power} = \frac{2\pi N}{60} \frac{\Lambda}{(t_{1_{av}})^{\lambda}} \times \frac{Bdf}{\pi DN} \times \frac{D}{2 \times 1000} \text{ W}$$

$$= \frac{\Lambda Bdf}{6(t_{1_{av}})^{\lambda}} \times 10^{-4} \text{ W.} \tag{4.57}$$

EXAMPLE 4.17 Solve Example 4.16 following the method by Vulf.

SOLUTION $t_{1_{max}}$ has already been found to be 0.01 mm. So, $t_{1_{av}}$ can be taken as 0.005 mm. Substituting the values in equation (4.57), we get

$$\text{power} = \frac{1400 \times 20 \times 5 \times 25}{6 \times (0.005)^{0.28}} \times 10^{-4} \text{ W}$$

$$= 257.15 \text{ W.}$$

It has been found that the result obtained by this method is between that computed by the two approaches followed in Examples 4.15 and 4.16 and can be considered to be reasonable.

In the milling operations other than slab milling, the basic kinematics of chip formation is different, and therefore a separate model has to be worked out in each operation. However, an easier approach is to estimate the power consumption, using the method of specific power consumption. To avoid the calculation for uncut thickness, special values of specific energy (generally valid for milling operations with rough cutting conditions), given in Table 4.14, can be used.

Table 4.14 Values of specific energy for milling

Material	U_c (J/mm^3)
Steel (BHN 100)	3.3
Steel (BHN 400)	5.5
Cast iron	2.5
Aluminium	1.1
Brass	1.5
Bronze	2

EXAMPLE 4.18 Solve Example 4.16 using Table 4.14.

SOLUTION The volume rate of material has been found out to be 41.7 mm^3/sec. Taking the value of specific energy to be 3.3 J/mm^3 (from Table 4.14), we find the power consumption works out to be

$$41.7 \times 3.3 \text{ W} = 138 \text{ W}$$

which seems to be a bit low.

4.3.5 MISCELLANEOUS MULTIPOINT MACHINING OPERATIONS

Two other common machining operations, using multipoint tools, are broaching and cutting threads with taps and dies. In broaching, only one motion, i.e., the primary cutting motion, is provided by the machine, whereas the feed is obtained by placing the teeth progressively deeper. Figure 4.53 shows the principle of the broaching operation. Since there is no feed motion, the shape of

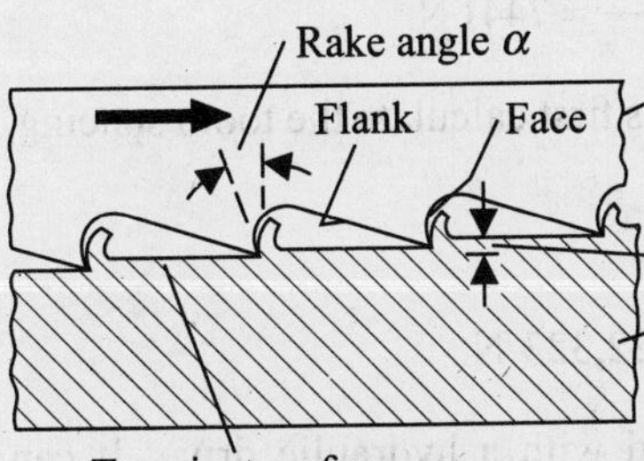

Fig. 4.53 Details of broaching operation.

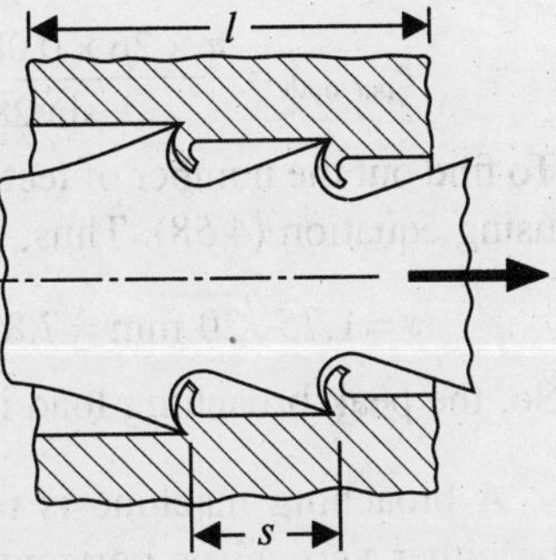

Fig. 4.54 Determination of tooth spacing.

the broach determines the shape of the machined part. Broaching has maximum application in producing internal forms such as splin holes and noncircular holes. Depending on the situation, broaching is done either by pulling or pushing the broach through a hole drilled in the workpiece.

For a smooth operation, it is essential that at least two or three teeth be simultaneously engaged. The thumb rule sometimes followed to determine the tooth spacing (Fig. 4.54) is

$$s = 1.75\sqrt{l} \text{ mm}, \tag{4.58}$$

where s is the tooth spacing and l is the broached length in mm. The cut per tooth f is kept in the range 0.05–0.09 mm. The cutting speed in broaching normally lies in the range 0.1–0.5 m/sec.

The cutting force per tooth can be found out by using the basic mechanics of chip formation and the instantaneous broaching load is obtained as the product of the load per tooth and the number of teeth in contact. If Z is the maximum number of teeth in contact at a time, the load can be found out as

$$F_{\text{max}} = F_{\text{per tooth}} \times Z. \tag{4.59}$$

EXAMPLE 4.19 A circular hole of diameter 25 mm in a 20-mm-thick mild steel plate has to be enlarged to a diameter of 27 mm by broaching. A cut per tooth of 0.08 mm is used. Estimate the peak broaching load. A rake angle of 10° is provided.

SOLUTION First, we find out the load per tooth by using the orthogonal machining model. Let us assume the coefficient of friction and the ultimate shear stress of the work material to be 0.5 and 400 N/mm², respectively. Then,

$$F_{\text{per tooth}} \approx \frac{w \times 0.08 \times 400 \times \cos(\lambda - \alpha)}{\sin\phi \cos(\phi + \lambda - \alpha)} \text{ N.}$$

The width of cut may be taken as the average circumference, i.e., equal to $\pi \times 26$ mm. Using the Lee and Shaffer shear angle relationship, ϕ works out to be 28.43° as $\lambda = \tan^{-1} 0.5 = 26.57°$. Thus,

$$F_{\text{per tooth}} = \frac{\pi \times 26 \times 0.08 \times 400 \times \cos 16.57°}{\sin 28.43° \cos 45°} = 7441 \text{ N.}$$

To find out the number of teeth in contact, let us first calculate the tooth spacing, using equation (4.58). Thus,

$$s = 1.75\sqrt{20} \text{ mm} = 7.83 \text{ mm} \approx 8 \text{ mm.}$$

So, the peak broaching load is 3 × 7441 N = 22,323 N.

A broaching machine is normally provided with a hydraulic drive. It can be either vertical or horizontal. In the normal speed range of broaching, the formation of a built-up edge may be a problem. To avoid this, a copious supply of the cutting fluid is provided.

Often, taps and dies are used for cutting internal and external screw threads. The tapping operation is normally done by hand though machine tapping is also not uncommon.

Another important and common operation, using a multiple cutting edge, is reaming. This operation is used for giving a finish cut to an already drilled hole. A reamer is similar to a drill but has a number of cutting edges and straight flutes. In this operation, the material removal is quite small but the process improves the finish and accuracy. Figure 4.55 shows a typical reamer.

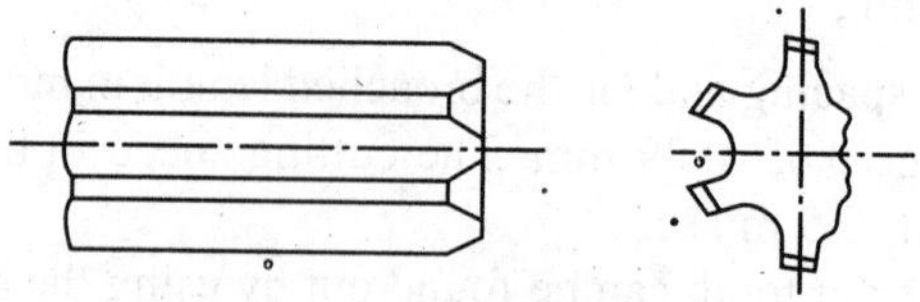

Fig. 4.55 Reamer.

4.4 ABRASIVE MACHINING AND FINISHING OPERATIONS

In our discussion on the basic mechanics of the machining processes, we have emphasized that the material removal is accomplished by plastic deformation of the work material by the tool. This necessitates that the tool material be much harder than the work material, not only to maintain the form but also to avoid excessive wear. Thus, very hard materials are difficult to machine. Table 4.15 shows the hardness of some common materials in the Knoop scale. Obviously,

Table 4.15 Hardness of some materials (Knoop scale)

Material	N/mm^2
Common glass	300–5000
Quartz	8000
Hardened steel	7000–13,000
Emery	14,000
Tungsten carbide	18,000–24,000
Aluminium oxide	20,000–30,000
Titanium carbide	18,000–32,000
Silicon carbide	21,000–30,000
Boron carbide	28,000
Diamond	70,000–80,000

it is possible to make use of hard substances as tool materials. These include Al_2O_3, SiC, and B_4C as they are naturally available and can also be produced synthetically without much problem. Diamond is also quite suitable but its higher cost restricts its application to special cases. However, it is not possible to produce the usual shapes of cutting tools with these materials and the only form in which these can be used is grains. Since the grains of such materials have the capability to abrade the other materials, these grains are commonly known as abrasives and the machining process using such abrasives is called abrasive machining. Abrasives can be used either as powder or in definite geometric forms obtained by bonding these abrasives with some bonding material. The sharp edges of the abrasive grains act as cutting edges which are not only randomly distributed (in the cutting area) but also randomly oriented.

The most common abrasive machining process is grinding where the abrasives are bonded to the shape of a wheel (known as the grinding wheel) which rotates at a high speed (Fig. 4.56). The other common finishing operations using abrasives are honing and lapping.

4.4.1 GRINDING

Though there are various types of grinding operations, to understand the basic mechanics we shall consider the most common grinding operation, namely, surface grinding. Figure 4.57a shows the basic arrangement of surface grinding,

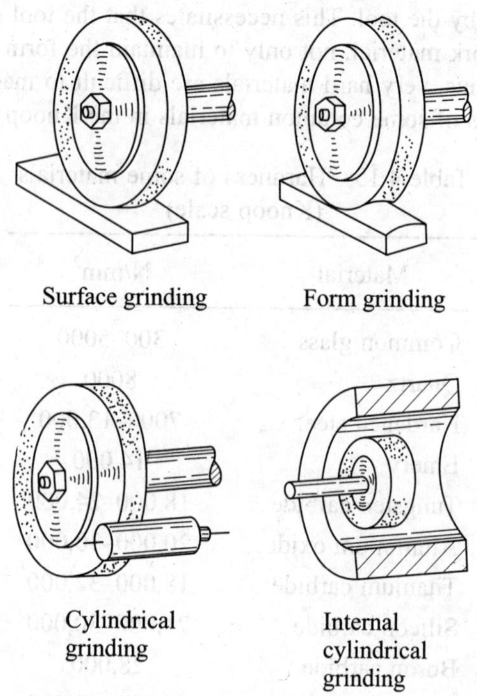

Fig. 4.56 Some common grinding operations.

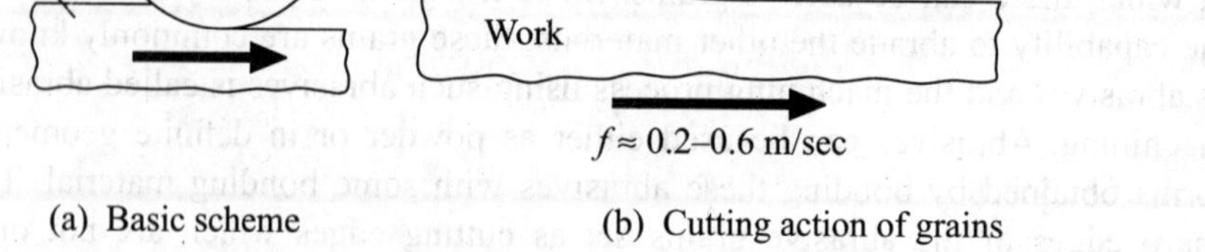

Fig. 4.57 Details of surface grinding.

which has some similarity with the up milling operation except that the cutting points are irregularly shaped and randomly distributed (Fig. 4.57b). The grains actually taking part in the material removal process are called the active grains. Gradually, the sharp edges of the active grains wear out and become blunt. This results in larger forces on the active grains during machining. When the cutting edge is too blunt and the force is sufficiently high, the grain may either get fractured or break away from the wheel. When a fracture takes place, new, sharp cutting edges are generated, and when the whole grain is removed, new grains (below the layer of the active grains) become exposed and active. This gives the grinding wheel self-sharpening characteristics. So, the bonding strength, which dictates the maximum force a grain can withstand, is an important characteristic. The strength of bonding is normally termed as the grade of the wheel. A wheel with a strong bond is called hard and vice versa[1].

Due to the nature of the process, very hot and small chips are produced which may readily get welded to either the grit (abrasive grain) or get welded back on to the workpiece. Moreover, because of random grit orientation, a number of grits may have a very large negative rake angle (Fig. 4.57b) and may rub rather than cut. These two factors make the process of grinding quite inefficient in comparison with the other machining operations from the point of view of specific energy. Apart from this, since the material is removed in the form of exceedingly small chips, the size effect is very prominent.

Mechanics of Grinding

In our analysis of the grinding process, all grains are assumed to be identical. To explain the mechanics, we consider two different types of operations, namely, (i) plunge grinding, and (ii) surface grinding. Figure 4.58a shows a simple

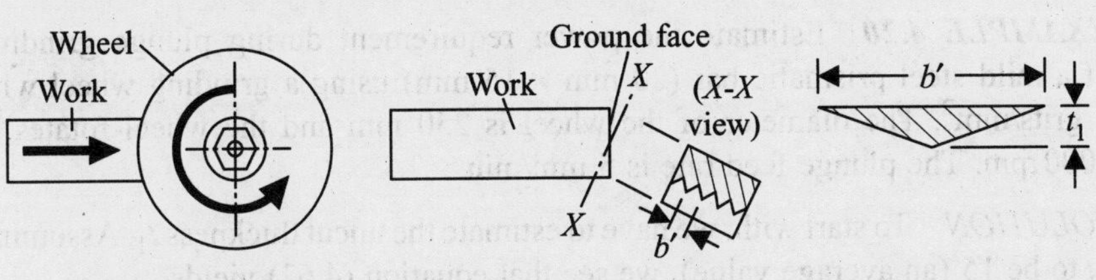

(a) Scheme of plunge grinding (b) Nature of ground face (c) Cross-section of typical chip

Fig. 4.58 Details of plunge grinding.

plunge grinding operation where a job of rectangular cross-section is being fed radially at a rate f (mm/min). The active grains are assumed to be uniformly distributed. The uncut thickness per grit (t_1) can be expressed as

$$t_1 = \frac{f}{ZN} \text{ mm}, \tag{4.60}$$

where Z is the number of active grains per revolution in one line and N is the rpm of the wheel. If the diameter of the wheel is D, the average grain width of cut is b' (mm) as shown in Fig. 4.58b, and the surface density of active grains is C (mm^{-2}), the number of grains/revolution/line (Z) can be found from the expression

$$Z = \pi DCb'. \tag{4.61}$$

[1]It should be remembered that the term *hard* or *soft* in the context of a grinding wheel does not refer to the hardness of the abrasives.

The uncut sections have approximately triangular cross-sections as shown in Fig. 4.58c. The ratio $r_g = b'/t_1$ generally lies between 10 and 20. Since b' can be written as $r_g t_1$, equations (4.60) and (4.61) yield

$$t_1 = \sqrt{\frac{f}{\pi DNCr_g}}. \tag{4.62}$$

Once t_1 is estimated, the value of specific energy U_c can be determined and the power requirement comes out as

$$W = \frac{AfU_c}{60}, \tag{4.63}$$

where A is the cross-sectional area of the job (mm^2). The force per single grit works out to be

$$F_C' = \frac{60{,}000W}{\pi DACN}\ \text{N} \tag{4.64a}$$

$$= \frac{1000 fU_c}{\pi DCN}\ \text{N}. \tag{4.64b}$$

EXAMPLE 4.20 Estimate the power requirement during plunge grinding of a mild steel prismatic bar (20 mm × 15 mm) using a grinding wheel with 3 grits/mm^2. The diameter of the wheel is 250 mm and the wheel rotates at 2000 rpm. The plunge feed rate is 5 mm/min.

SOLUTION To start with, we have to estimate the uncut thickness t_1. Assuming r_g to be 15 (an average value), we see that equation (4.62) yields

$$t_1 = \sqrt{\frac{5}{\pi \times 250 \times 2000 \times 3 \times 15}}\ \text{mm} = 0.00027\ \text{mm}.$$

From equation (4.22) and Table 4.4,

$$U_c \approx 1.4 \times (0.00027)^{-0.4}\ \text{J/mm}^3 = 37.46\ \text{J/mm}^3.$$

Now, the material removal rate is (20 × 15 × 5) mm^3/min. So, the power requirement will be

$$37.46 \times (20 \times 15 \times 5)/60\ \text{W} = 94\ \text{W}.$$

The cutting component of the machining force per grit can also be found out as

$$F_C' = \frac{60{,}000 \times 94}{\pi \times 250 \times 300 \times 3 \times 2000}\ \text{N} \quad \text{[using equation (4.64a)]}$$

$$= 0.04\ \text{N}.$$

Substituting the expression for U_c, i.e.,

$$U_c = U_0(t_1)^{-0.4},$$

in equation (4.64b) and using equation (4.62), we get

$$F_C' = \frac{1000f}{\pi DCN} U_0 [\frac{f}{DNCr}]^{-0.2}$$

$$= \frac{1000 U_0 f^{0.8} r^{0.2}}{(\pi DCN)^{0.8}}. \quad (4.65)$$

Since a larger value of F_C' implies a higher possibility of dislodging the grain from the wheel, a wheel appears to be softer if D, C, or N decreases or f increases, and vice versa.

Figure 4.59 shows the basic scheme of chip formation during surface grinding. Here also, the uncut cross-section is triangular. But the uncut thickness and

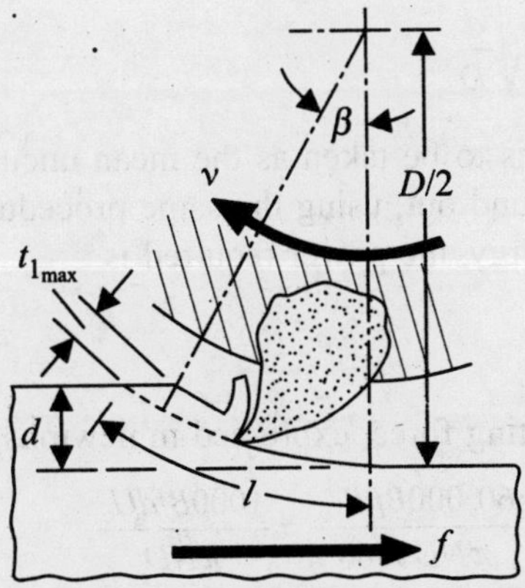

Fig. 4.59 Scheme of chip formation during surface grinding.

width vary (as in milling) and $t_{1_{max}}$ and b'_{max} are their maximum values. The average values may be taken as one-half of these. The average length of chip is given as

$$l \approx \frac{D}{2}\beta.$$

But

$$\cos\beta = (\frac{D}{2} - d)/\frac{D}{2} = 1 - \frac{2d}{D},$$

where d is the depth of cut. $\cos\beta$ can be expanded (keeping only two terms since β is generally small) as

$$\cos\beta \approx 1 - \frac{\beta^2}{2}$$

and, using the foregoing equation, we get $\beta \approx 2\sqrt{d/D}$. Substituting this in the expression for l, we obtain

$$l \approx \sqrt{Dd}. \tag{4.66}$$

The total volume of material removed per unit time is fdB, where B is the width of the cut in mm. The average volume per chip can be approximately taken as $\frac{1}{6}b'_{\max}t_{1_{\max}}l$. The number of chips produced per unit time is clearly $(\pi NDBC)$. Now, taking $r_g = b'_{\max}/t_{1_{\max}}$ as before, we have

$$(\pi NDBC) \times \tfrac{1}{6}b'_{\max}t_{1_{\max}}l = fdB$$

or

$$(\pi NDBC) \times \tfrac{1}{6}r_g t_{1_{\max}}^2 l = fdB$$

or

$$t_{1_{\max}} = \sqrt{\frac{6f}{\pi NDr_g C}\sqrt{\frac{d}{D}}}. \tag{4.67}$$

One-half of this value is to be taken as the mean uncut thickness. The power consumption can be found out, using the same procedure as described before. If U_c is the specific energy, the power required is

$$W = \frac{BfdU_c}{60}\ \text{W}. \tag{4.68a}$$

The total tangential cutting force, expressed in newtons, is

$$F_C = \frac{60{,}000W}{\pi ND} = \frac{60{,}000BfdU_c}{\pi ND \times 60} = \frac{1000BfdU_c}{\pi ND}, \tag{4.68b}$$

where N is the wheel rpm. The number of grits actively engaged at a time is $CBl = CB\sqrt{Dd}$ and the average force per grit is given by

$$F'_C = \frac{60{,}000W}{\pi NDCB\sqrt{Dd}}\ \text{N}.$$

Substituting W from equation (4.68a) and U_c as $U_0(t_{1_{av}})^{-0.4}$ in the foregoing expression and using equation (4.67), we obtain

$$F'_C = \frac{369U_0 f^{0.8} d^{0.4} r_g^{0.2}}{N^{0.8}D^{1.2}C^{0.8}}\ \text{N}. \tag{4.69}$$

Here too, we see that if N, D, or C decreases, or f or d increases, the wheel appears to be softer. This is because F'_C increases, causing a more frequent dislodging of the abrasive grains.

It should be noted that in grinding the radial thrust force F_T (Fig. 4.60) is much more than the tangential cutting component F_C unlike the other machining operations. In surface grinding, the ratio $F_T/F_C \approx 2$.

EXAMPLE 4.21 Estimate the grinding force during surface grinding of a 25-mm-wide mild steel block with a depth of cut of 0.05 mm. The diameter of the wheel is 200 mm and the wheel rotates at 3000 rpm. The number of

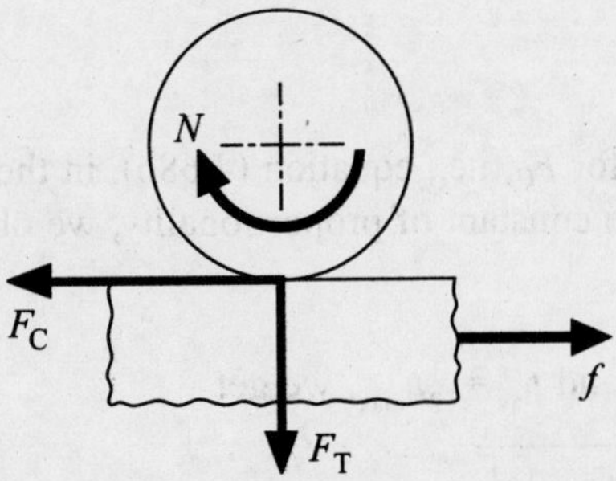

Fig. 4.60 Components of grinding force.

grits/mm^2 is measured and found to be 3. The feed velocity of the table is 100 mm/min.

SOLUTION Assuming r_g to be 15, the maximum uncut thickness per grit is calculated, using equation (4.67), as

$$t_{1_{max}} = \sqrt{\frac{6 \times 100}{\pi \times 3000 \times 200 \times 3 \times 15}\sqrt{\frac{0.05}{200}}} \text{ mm} = 0.00033 \text{ mm.}$$

So,

$$t_{1_{av}} = 0.000165 \text{ mm.}$$

From equation (4.22) and Table 4.4,

$$U_c \approx 1.4 \times (0.000165)^{-0.4} \text{ J/mm}^3 = 45.62 \text{ J/mm}^3.$$

Now, the material removal rate is

$$fdB = 100 \times 0.05 \times 25 \text{ mm}^3\text{/min} = \frac{125}{60} \text{ mm}^3\text{/sec} = 2.083 \text{ mm}^3\text{/sec.}$$

Thus, the power consumption is found [using equation (4.68a)] to be

$$W = 2.083 \times 45.62 \text{ W} \approx 95 \text{ W.}$$

The grinding force F_C is given by

$$F_C = \frac{60{,}000 \times 95}{\pi \times 3000 \times 200} \text{ N} = 3 \text{ N.}$$

The thrust component F_T may be taken as approximately double of F_C, i.e., 6 N, and the total grinding force becomes approximately 6.7 N.

Thermal Aspects

It is extremely difficult to develop a satisfactory analytical model for calculating the surface temperature during grinding. We shall therefore adopt an indirect approach for yielding the approximate results. It is not very illogical to assume that the grinding temperature depends directly on the energy spent per unit surface area ground. Thus (in surface grinding),

$$\theta_s \propto \frac{F_C \pi N D}{Bf}.$$

Using the expression for F_C, i.e., equation (4.68b), in the foregoing expression and taking 1000 as the constant of proportionality, we obtain

$$\theta_s \propto dU_c. \tag{4.70}$$

Since $U_c = U_0(t_{1_{av}})^{-0.4}$ and $t_{1_{av}} = \frac{1}{2}t_{1_{max}}$, we get

$$\theta_s \propto dU_0\left[\frac{1}{2}\sqrt{\frac{6f}{NDCr_g}\sqrt{\frac{d}{D}}}\right]^{-0.4}.$$

Or, for a given material,

$$\theta_s \propto \frac{d^{0.9}D^{0.3}C^{0.2}N^{0.2}}{f^{0.2}}. \tag{4.71}$$

So, the temperature and also the defects caused by higher temperature can be reduced either by decreasing *d, D, C*, or *N*, or by increasing the table feed *f*.

The temperature at the grain-chip interface θ_g during grinding reaches very high values and can easily go beyond 1500°C. Though this temperature may be above the melting temperature, the melting of the chips may or may not take place, depending on whether or not enough time is available. The time during which an individual grit remains in contact with the chip (in surface grinding) is

$$60l/(\pi ND) = \frac{60}{\pi N}\sqrt{\frac{d}{D}} \approx 0.0001 \text{ sec}.$$

The grain-chip interface temperature has been shown to be[1]

$$\theta_g = \Theta U_c \sqrt{\frac{v t_{1_{max}}}{k\rho c}}, \tag{4.72}$$

where Θ is a constant, v is the wheel surface speed, k is the thermal conductivity of the work material, and ρc is the volume specific heat of the work material. Since $U_c \propto t_{1_{max}}^{-0.4}$, we have $\theta_g \propto t_{1_{max}}^{0.1}$ and, using equation (4.67), we can conclude that, in surface grinding, $\theta_g \propto f^{0.05}$. So, the grain-chip interface temperature may slightly increase with *f* but the surface temperature decreases with *f*.

The ground surface may get affected to a depth of about 0.2 mm by thermal and mechanical effects. As a consequence, large residual tensile stresses may develop (Fig. 4.61) which, if sufficiently high, may result in cracks. When the grinding temperature is sufficiently high, microstructural changes may also take place due to heating and rapid quenching; thus, in the case of a steel workpiece, the surface layer may be heated so that it becomes austenite and, due to a quick quenching by the cutting fluid, may be transformed into martensite.

[1] Shaw, M.C. and Outwater, J.O., Surface temperature in grinding, *Trans. ASME*, **74**, 73, 1952.

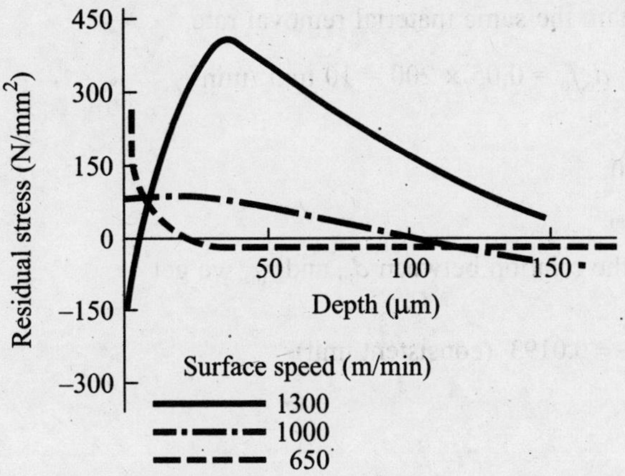

Fig. 4.61 Residual stress in workpiece after surface grinding.

EXAMPLE 4.22 During the surface grinding operation with a depth of cut of 0.05 mm and a feed rate of 200 mm/min, the surface temperature of the workpiece is found to be 850°C. To avoid martensitic transformation of the ferrous workpiece, it is required to maintain the surface temperature below 700°C. However, to maintain the productivity, the material removal rate should be kept the same. Determine the required depth of cut and feed combination to achieve this.

SOLUTION From equation (4.71),

$$\theta_s \propto \frac{d^{0.9}}{f^{0.2}}$$

when the depth of cut and the table feed rate are the only variables. So, if the subscripts o and m are used to indicate, respectively, the original and the modified values of the variables, we then have

$$850 = K\frac{d_o^{0.9}}{f_o^{0.2}}, \qquad 700 = K\frac{d_m^{0.9}}{f_m^{0.2}},$$

where K is the constant of proportionality. Substituting the values of d_o and f_o, we obtain

$$K = \frac{850 \times (200)^{0.2}}{(0.05)^{0.9}} \text{ (consistent unit)}$$

$$= 36{,}354 \text{ (consistent unit).}$$

Using this value of K, we get

$$\frac{d_m^{0.9}}{f_m^{0.2}} = \frac{700}{36{,}354} = 0.0193 \text{ (consistent unit).}$$

Now, to maintain the same material removal rate,

$$d_m f_m = d_o f_o = 0.05 \times 200 = 10 \text{ mm}^2/\text{min}$$

or

$$f_m = \frac{10}{d_m}.$$

Using this in the relation between d_m and f_m, we get

$$\frac{d_m^{0.9}}{(\frac{10}{d_m})^{0.2}} = 0.0193 \text{ (consistent unit)}$$

or

$$d_m^{1.1} = 0.0306 \text{ (consistent unit)}.$$

This finally yields the required value of the new depth of cut as $d_m = 0.042$ mm. Correspondingly, the new feed rate is given as $f_m = 10/d_m = 238$ mm/min.

Grinding Wheel Characteristics

The performance of a grinding wheel depends on the following important factors:

(i) *Abrasive type* The abrasives generally used are aluminium oxide, silicon carbide, and diamond. Diamond is the hardest substance known and is used for very hard work materials such as glass, carbide, and ceramics. Aluminium oxide and silicon carbide are more commonly used for making the grinding wheels. Silicon carbide is harder than aluminium oxide but dulls more rapidly. Generally, aluminium oxide abrasives are selected for the surface grinding of steels and bronzes, whereas silicon carbide is chosen for the surface grinding of cast iron, brass, aluminium, hard alloys, and carbides.

(ii) *Grain size* The size of the grains is generally specified by the grit size. A 60 grit size, for example, is approximately 1/60 inch square. The larger the size of the grains, the more will be the material removal capacity, but the quality of the surface finish deteriorates. Thus, the grain size is determined primarily by the surface quality requirements.

(iii) *Bonding material* The bond materials commonly used are vitrified clay, resinoid materials, silicates, rubber, shellac, and metals. The vitrified bond is strong and rigid. It is the most common type of bond used. The resin bonds are made from synthetic organic materials. Such bonds are strong and fairly flexible. The silicate bonds are essentially the silicates of soda (water glass). These bonds are not as strong as the vitrified bonds, and the grains are dislodged more rapidly. As a result, the operation is cooler. Such bonds are used in grinding tools where the temperature rise should be as small as possible. The rubber bonds are used for making flexible wheels. A high speed operation is possible when the wheel is subjected to a side thrust. A fairly hard vulcanized rubber is used as the bonding material. The shellac bonds are used in making thin but strong wheels possessing some elasticity. Since a smooth finish on a hard surface can be achieved, the shellac bonded wheel is used in grinding parts such as cam shaft and mill roll.

(iv) *Structure* Since the grinding wheel is similar to a milling cutter with a very large number of teeth randomly oriented, it must have voids to allow space for the chips. If the voids are too small for the chips, the chips stay in the wheel, blocking the voids. This is known as *loading* of the wheel. Loading causes inefficient cutting. If the voids are too large, again the cutting action is inefficient since there will be too few cutting edges. In an open structure, the grains are not too densely packed, and in a wheel with a closed structure, the grains are tightly packed. For grinding ductile work materials, larger chips are produced, and to reduce the tendency of loading, an open structure is preferred. In the case of hard and brittle work materials, a closed structure is selected. The structure depends on the required grade and also the nature of cut. For a rough cut, an open structure is more suitable.

(v) *Grade* The grade is determined by the strength of the bonding material. So, a hard wheel means strong bonding and the abrasive grains can withstand large forces without getting dislodged from the wheel. In the case of a soft wheel, the situation is just the opposite. When the work material is hard, the grains wear out easily and the sharpness of the cutting edges is quickly lost. This is known as *glazing* of the wheel. A glazed wheel cuts less and rubs more, making the process inefficient. To avoid this problem, a soft wheel should be used so that the grains which lose the sharpness get easily dislodged as the machining force on the individual grains increases. Thus, the layers of new grains are exposed, maintaining the sharpness of the wheel. When the work material is soft, a hard wheel should be used since the problem of glazing will be absent and a longer wheel life will be achieved. So, for a work material, there exists an optimal grade—too hard a wheel causes glazing, whereas too soft a wheel wears out very fast. Figure 4.62 shows the nature of growth of power requirement as the grinding operation continues for too hard, optimum, and too soft wheels.

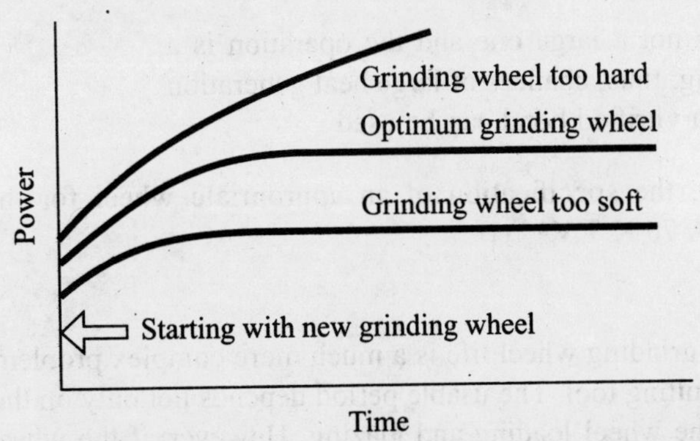

Fig. 4.62 Growth of power requirement for different wheel grades.

Wheel Specification and Selection

Generally, an alphanumeric system is used for a complete specification of a grinding wheel. Figure 4.63 explains the system.

For getting the optimum results, a grinding wheel must be properly selected. The important guidelines that should be observed when choosing a wheel are given in Table 4.16; it should however be noted that such a selection also depends on experience.

EXAMPLE 4.23 The sides of a high speed steel cutting tool bar have to be finish ground on a surface grinder. The machine restricts the diameter of the wheel to 150 mm and no coolant is used. Find out the type of wheel required for this operation.

SOLUTION The problem can be solved in the following manner:

Comments	*Recommendation*
Since the material to be ground is HSS, which is a high strength ferrous alloy, Al_2O_3 abrasives should be used	A
Finish grinding calls for fine grit size, but since no specific finish has been prescribed, there is no point in making the grits finer	70
The material is comparatively harder, so a grade in the softer side of the medium range should be selected	K
For surface grinding, a medium structure can be used; had it been a roughing operation (i.e., large chips), an open-medium structure would have been suitable; in the present case, the operation is a finishing one, so a close-medium structure should be selected; the material is tough, so we should not make the wheel structure too close	7
The wheel is not a large one and the operation is a finish grinding; thus, chances of large heat generation are less and a vitrified bond can be used	V

Thus, finally, the specification of an appropriate wheel for the prescribed operation is A 70 K 7 V.

Wheel Life

The study of grinding wheel life is a much more complex problem than that of an ordinary cutting tool. The usable period depends not only on the wheel wear but also on the wheel loading and glazing. However, if the wheel is properly selected, the loading and glazing may not pose a serious problem, and the wheel wear may be the predominant factor in determining the wheel life. The abrasive

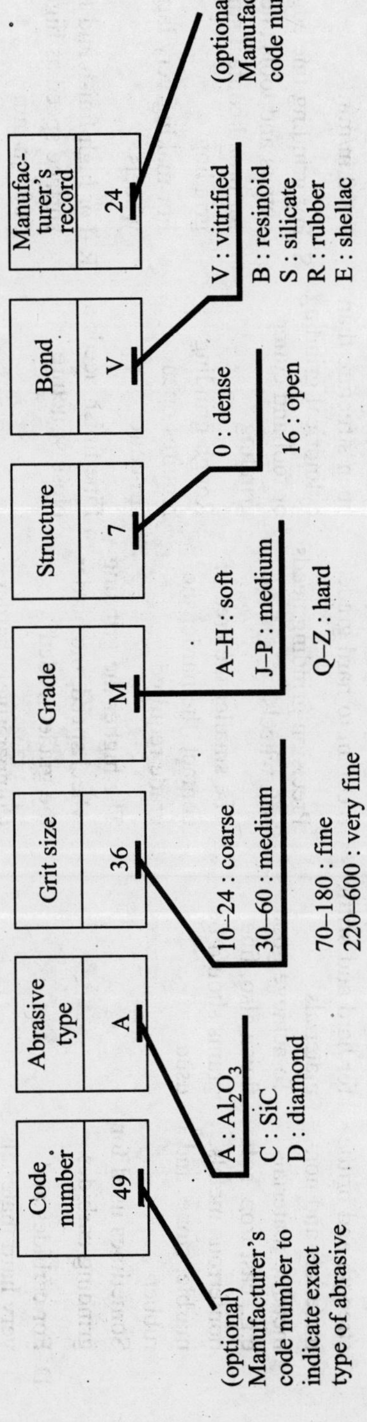

Fig. 4.63 Alphanumeric system for grinding wheel specification.

Table 4.16 Grinding wheel selection

Abrasive type	Grit size	Grade	Structure	Bond material
A–For tensile strength materials, e.g., steel and malleable and wrought iron	Coarse grit for soft or ductile materials when fast grinding is wanted	General rule is to use soft wheels for hard metals and hard wheels for soft metals	Open structure for soft or ductile materials	V–For use with wheel surface speed 2000 m/min
C–For low tensile strength and brittle materials and non-metallic materials, e.g., cast iron, soft nonferrous metals, marble, stone, and rubber Sometimes used for grinding carbides	Fine grains are used for hard and brittle materials To achieve good finish also, fine grains should be used	Rough grinding requires medium to hard grade Precision grinding needs soft wheels The smaller the area of contact, the harder the grade required The higher the grinding wheel speed, the harder the grade needed	Surface grinding requires a more open structure than cylindrical grinding or tool and cutter grinding Rough grinding requires open structure Fine finish needs a close structure	B–For operations requiring very strong wheels and for surface speed up to 3000 m/min S–For grinding fine edges on cutters and tools and also for broad contact surface grinding For making very large wheels
D–For carbides and very hard materials Cost is an important factor and to achieve low cost of operation, SiC is used sometimes instead of diamond		Comparatively harder grades should be used when coolant is applied For old machines, comparatively harder wheels are recommended		R–For high finish and for surface speed as high as 5000 m/min For also making very thin wheels, e.g., cut off wheels E–For thin cut off and high finishing wheels Produces some buffing or polishing effect

grains also have a finite life. An abrasive grain may lose volume in two ways, namely, (i) by a gradual wear of its sharp edges (known as attrition wear), and (ii) by a fracture of a portion.

A grinding wheel wears more rapidly than the tools used in the other conventional machining operations. The commonly-used parameter for evaluating the performance of a wheel is the grinding ratio (G_r) given by the relation

$$G_r = \frac{\text{volume of work material removed}}{\text{volume of wheel worn away}}.$$

If the wheel wear is plotted against the volume of work material removed, the general characteristic would be as shown in Fig. 4.64. We see here that the wear

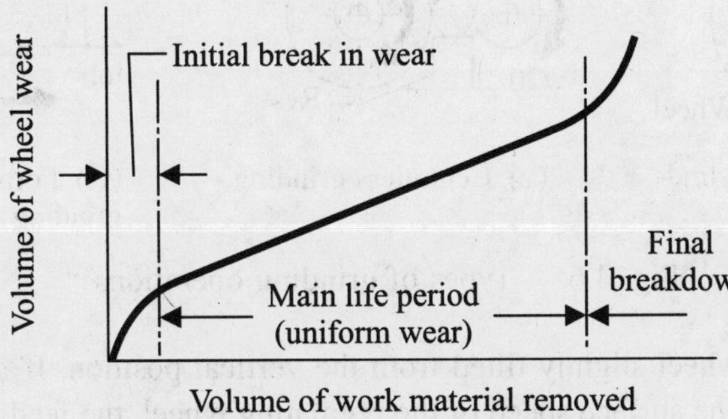

Fig. 4.64 Grinding wheel wear.

curve is quite similar to that of the other cutting tools. Also, there is an initial breakdown period followed by a region of uniform wear rate. Finally, the wheel again starts breaking down rapidly. The middle zone (see Fig. 4.64) is the real usable period which determines the life of a wheel.

Once the period of uniform wear rate ends, the wheel should be reconditioned (as when regrinding a tool) before using it again. The reconditioning of a wheel is done by a process commonly known as *wheel dressing*. This requires holding a hard tipped tool against the rotating wheel and giving the dresser some feed motion to cover the whole width of the wheel. By this process, the dull, used layer is crushed, exposing the fresh surface. Thus, all the problems accumulated by glazing and loading are removed.

Types of Grinding Operations

The common grinding operations are (i) surface grinding with horizontal spindle, (ii) surface grinding with vertical spindle, (iii) external cylindrical grinding, (iv) internal cylindrical grinding, (v) centreless grinding, and (vi) form grinding. The basic principles of these grinding operations are illustrated in Fig. 4.65.

Unlike in cylindrical grinding, in centreless grinding the cylindrical workpiece is supported on a rest. The feed is provided to the long workpiece by keeping

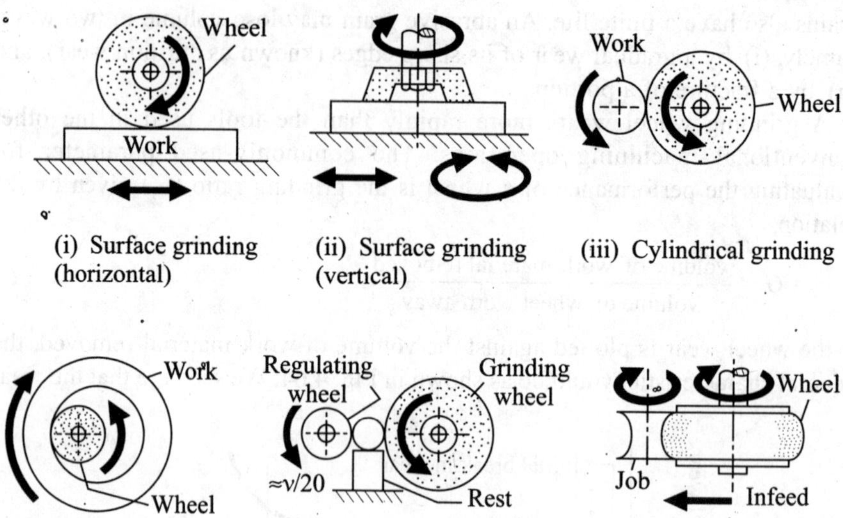

(i) Surface grinding (horizontal)

(ii) Surface grinding (vertical)

(iii) Cylindrical grinding

(iv) Internal cylindrical grinding

(v) Centreless grinding

(vi) Form cylindrical grinding

Fig. 4.65 Types of grinding operations.

the regulating wheel slightly tilted from the vertical position. If ϕ is the angle of tilt and v_r is the surface speed of the regulating wheel, the feed velocity then is $v_r \sin \phi$.

4.4.2 FINISHING OPERATIONS

As already noted, the other important operations using abrasives are (i) honing, and (ii) lapping. However, here the material removal rate is generally very small as compared with that in grinding, and these operations are used only for finishing purposes. Apart from these two operations, there are various finishing operations, e.g., superfinishing and buffing, where abrasives are used.

Honing

The honing operation is used for finishing the inside surface of a hole. Here, abrasives in the form of sticks are mounted on a mandrel which is then given a reciprocating movement (along the hole axis) superimposed on a uniform rotary motion (Fig. 4.66). The grit size normally varies from 80 mesh to 600 mesh. Because of the nature of the path of the abrasive grits on the surface of the work, a random cross-marked surface finish (desirable for lubrication) is obtained. Depending on the work material, the honing speed may vary from 15 m/min to 60 m/min, and the honing pressure lies in the range 1–3 N/mm^2. In special cases, material up to 0.5 mm may be removed by honing. The tolerance and finish achieved in this operation are of the order of 0.0025 mm and 0.25 μm, respectively.

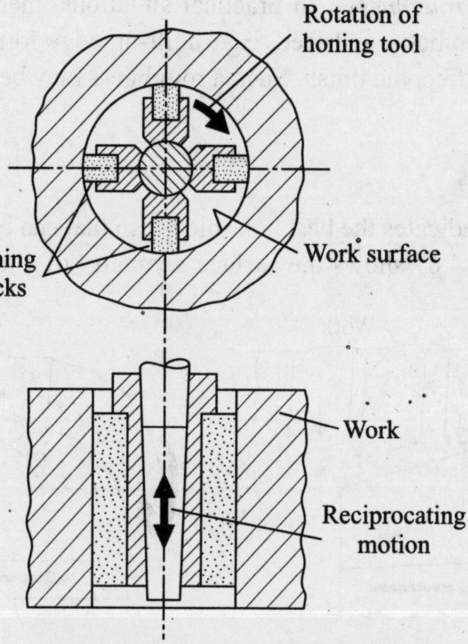

Fig. 4.66 Honing operation.

Lapping

Lapping is another operation for improving the accuracy and finish. It is accomplished by abrasives in the range 120–1200 mesh. A lap is generally made of a material softer than the work material. In this process, straight, narrow grooves are cut at 90° on the lap surface and this surface is charged by sprinkling the abrasive powder. The workpiece is then held against the lap and moved in unrepeated paths. A suitable cutting fluid is applied for lapping. In hand lapping, the work is moved over the lap along a path in the form of 8. Machines are also available for lapping. The material removal is seldom more than 0.0025 mm and the lapping pressure is generally kept in the range 0.01–0.2 N/mm^2, depending on the hardness of the work material.

4.5 SURFACE FINISH

Often, the surface finish requirement may be as important as the desired dimensional accuracy. Therefore, it is important that we know the factors which affect the finish.

4.5.1 SURFACE FINISH IN MACHINING

The resultant roughness produced by a machining operation is the combined effect of two independent quantities. These are:

(i) *Ideal roughness* This is a result of the geometry of the tool and the feed. It is a geometric phenomenon and is the minimum possible magnitude of the unevenness which results from a machining operation.

(ii) *Natural roughness* In practical situations, there are various factors other than tool geometry and feed, e.g., built-up edge formation and vibration, which adversely affect the finish. Such a roughness may be termed as the natural roughness.

Ideal Roughness

Ideal roughness indicates the best possible finish that can be obtained by a given operation. Figure 4.67 shows the surface geometry of a turned surface when a

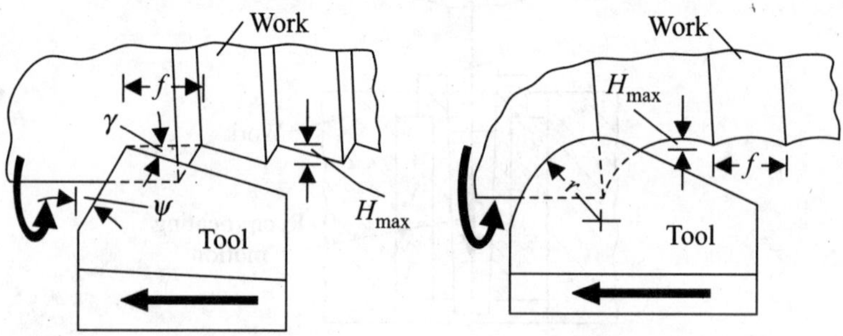

(a) Using tool without nose radius

(b) Using tool with nose radius

Fig. 4.67 Generation of ideal roughness in turning.

tool without and with nose radius is employed. It can be easily shown that (see Fig. 4.67a) the maximum height of unevenness is given as

$$H_{max} = \frac{f}{\tan \psi + \cot \gamma}, \qquad (4.73)$$

where ψ and γ are the side cutting edge angle and the end cutting edge angle, respectively, f being the feed rate.

When a tool with a nose radius r is used, the maximum height of unevenness is given by

$$H_{max} = \frac{f^2}{8r}. \qquad (4.74)$$

So, we see that, for Fig. 4.67a, H_{max} is proportional to f, whereas, for Fig. 4.67b, it is proportional to f^2.

EXAMPLE 4.24 When turning a cylindrical workpiece, two different tools are used. In one tool, no nose radius is provided, the side cutting and the end cutting edge angles being 30° and 7°, respectively. In the other tool, a nose radius of 0.7 mm is provided. The feed used in both the cases is 0.125 mm. Find out the maximum height of unevenness in the generated surfaces.

SOLUTION Using equation (4.73), we find that H_{max} for the first tool is

$$H_{max} = \frac{0.125}{\tan 30° + \cot 7°} \text{ mm} = 0.014 \text{ mm}.$$

For evaluating H_{max} for the other tool, the value of the nose radius has to be substituted in relation (4.74). Thus,

$$H_{max} = \frac{(0.125)^2}{8 \times 0.7} \text{ mm} = 0.0028 \text{ mm}.$$

So, we see that the surface produced by the tool with a nose radius is much smoother.

Ideal roughness during slab milling can also be found out quite easily. Figure 4.68 shows the surface generated by slab milling, using a cutter with

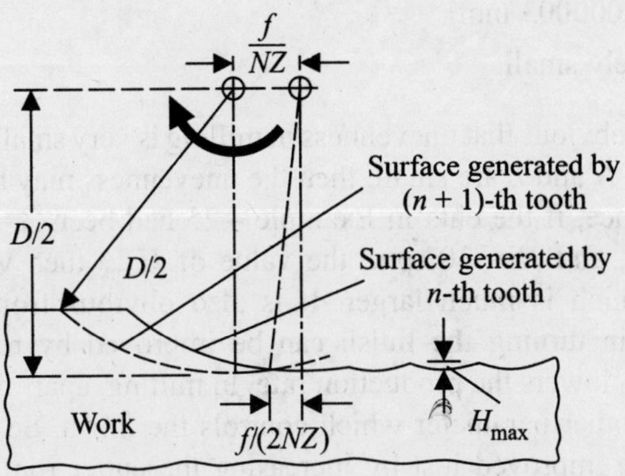

Fig. 4.68 Generation of ideal roughness in slab milling.

straight teeth. Since the feed velocity f of the table is generally much smaller than the speed of the cutter teeth, the surface generated by each tooth may be considered to be cylindrical with the same radius as that of the cutter ($=D/2$). The feed per tooth (or the movement of the table during which the consecutive teeth occupy similar positions) has already been shown to be $f/(NZ)$, N and Z having the usual meaning. So, the centre line of the cylindrical surfaces generated by two consecutive teeth will also be at a distance equal to the feed per tooth (Fig. 4.68). The maximum height of unevenness is easily shown to be

$$H_{max} \approx f^2/(4DN^2Z^2). \tag{4.75}$$

The average value of the unevenness is

$$H_{av} = \frac{f}{4(\tan \psi + \cot \gamma)} \quad \text{(for turning with a tool without nose radius),} \tag{4.76a}$$

$$H_{av} \approx \frac{f^2}{18\sqrt{3r}} \quad \text{(for turning with a tool with nose radius),} \tag{4.76b}$$

$$H_{av} \approx \frac{f^2}{9\sqrt{3}DN^2Z^2} \quad \text{(for slab milling with a straight alter).} \qquad (4.76c)$$

EXAMPLE 4.25 A straight cutter with 12 teeth and 100 mm diameter is used for a plain slab milling operation. The table feed is 25 mm/min and the cutter rotates at 60 rpm. Estimate the maximum height of unevenness.

SOLUTION Relation (4.75) can be directly used and the values of f, D, N, and Z substituted. We thus get

$$H_{max} \approx \frac{(25)^2}{4 \times 100 \times (60)^2 \times (12)^2} \text{ mm}$$

$$= 0.000003 \text{ mm}$$

which is extremely small.

It is therefore obvious that unevenness in milling is very small. However, if the feed is large and N and Z are small, then the unevenness may be comparatively larger. For instance, if the data in Example 4.25 had been $Z = 6$, $D = 75$ mm, $f = 50$ mm/min, and $N = 30$ rpm, the value of H_{max} then would have been 0.00025 mm which is much larger. It is also obvious from our foregoing discussion that in turning the finish can be improved by reducing the feed which, of course, lowers the production rate. In milling, apart from the feed, the cutter rpm is another parameter which controls the finish. So, in slab milling, the finish can be improved just by increasing the cutter rpm which does not affect the production rate. However, it should be remembered that in the case of a multipoint tool (e.g., the milling cutter) the ideal roughness is based on the assumption that the tool or cutter has been manufactured perfectly.

Natural Roughness

As already mentioned, in actual operations, there are various factors which adversely affect the finish. Of these, the most important are (i) the formation of a built-up edge, and (ii) vibration. When the cutting condition is properly chosen, the chatter (vibration) may be avoided. Since the built-up edge formation depends on the cutting condition (i.e., dry or wet) and the cutting speed, it is expected that, for a given cutting condition, the natural roughness will vary with the cutting speed. Except for very low cutting speeds, the intensity of built-up edge formation decreases with the cutting speed, and so the maximum height of surface unevenness is also expected to decrease with the cutting speed. Figure 4.69 shows the experimental results for turning.

4.5.2 SURFACE FINISH IN GRINDING

We have already stated that grinding is similar to milling in some respects. So, the surface roughness generated by the geometric and kinematic factors is similar to that in milling. However, there is one basic difference. The cutting speed being

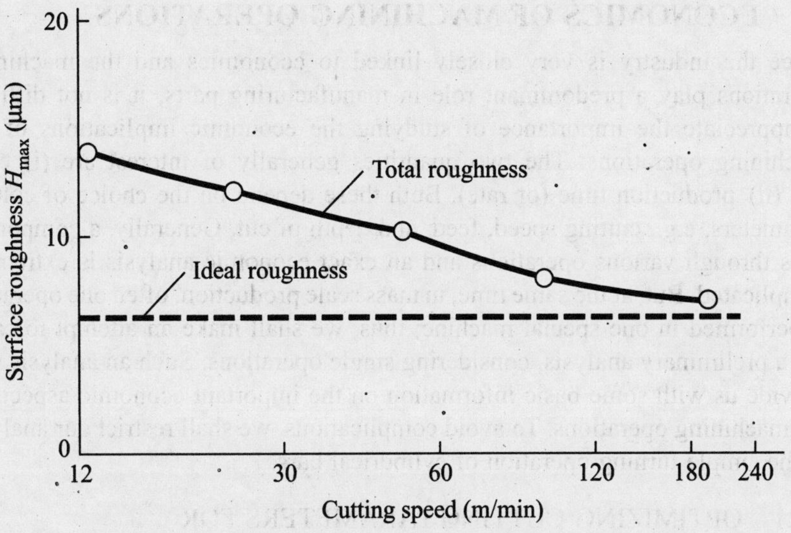

Fig. 4.69 Variation of surface roughness with cutting speed during turning mild steel bar.

very high, the built-up edge formation does not play any role. The total roughness is therefore quite accurately indicated by the ideal roughness. Further, the process of material removal being of a random nature, a statistical analysis is required. But even without going into such detail if we consider the process to be analogous to milling, some indications to this effect can be obtained. Considering an expression for H_{max} equivalent to that given by relation (4.75), we observe that the values of $\dot{N}$ and Z are very high in grinding. So, H_{max} is quite small.

Table 4.17 shows the values of H_{av} for the various machining and abrasive machining operations.

Table 4.17 H_{av} for various machining and abrasive machining processes

Process(es)	H_{av} (μm)
Turning, boring	0.05–25
Milling	0.25–25
Planing, shaping	0.375–25
Drilling	0.75–12.5
Reaming, broaching	0.5–6.25
Grinding	0.025–6.25
Honing	0.025–1.5
Lapping	0.013–0.75

4.6 ECONOMICS OF MACHINING OPERATIONS

Since the industry is very closely linked to economics and the machining operations play a predominant role in manufacturing parts, it is not difficult to appreciate the importance of studying the economic implications of the machining operations. The two quantities generally of interest are (i) cost, and (ii) production time (or rate). Both these depend on the choice of cutting parameters, e.g., cutting speed, feed, and depth of cut. Generally, a component goes through various operations and an exact economic analysis is extremely complicated. But, at the same time, in mass scale production, often one operation is performed in one special machine; thus, we shall make an attempt to carry out a preliminary analysis, considering single operations. Such an analysis will provide us with some basic information on the important economic aspects of the machining operations. To avoid complications, we shall restrict our analysis to the simple turning operation of cylindrical bars.

4.6.1 OPTIMIZING CUTTING PARAMETERS FOR MINIMUM COST

The total cost of a part can be written in the form

$$R = R_1 + R_2 + R_3 + R_4 + R_5, \tag{4.77}$$

where R is the total cost/piece, R_1 is the material cost/piece, R_2 is the set-up and idle time cost/piece, R_3 is the machining cost/piece, R_4 is the tool changing cost/piece, and R_5 is the tool regrinding cost/piece. Let us consider the length and diameter of the cylindrical part to be L and D, respectively, both being in mm. The feed and the cutting speed used are f mm/revolution and v m/min, respectively. Since we are analyzing only one pass, the depth of cut is fixed either by the final diameter required or by the maximum allowable limit without causing chatter (vibration) depending on which one is less. Thus, the depth of cut does not really come into our analysis. For purposes of cost calculation, let us have the rates

λ_1 = cost/min of labour and overheads,

λ_2 = cost of setting a tool for regrinding,

λ_3 = cost/mm of tool ground.

The time spent in the various operations is

t_s = set-up time and idle time/piece, min,

t_m = machining time/piece, min,

tct = tool changing time, min.

Now, the cost of the component can be expressed in terms of the cutting variables.

Material cost The material cost does not depend on the cutting conditions and remains as a constant (R_1).

Set-up and idle time cost It is given by the product of the set-up and idle time and the cost/unit time of labour and overheads. Thus,

$$R_2 = \lambda_1 t_s.$$

This is also independent of the cutting conditions f and v.

Machining cost The machining cost/piece is given by the product of the machining time/piece and the cost/unit time of labour and overheads. So,

$$R_3 = \lambda_1 t_m = \lambda_1 \frac{\pi LD}{1000 fv}$$

since v is in m/min.

Tool changing cost First, we have to find out the number of times the tool has to be changed per pass. If T is the tool life depending on f and v, the tool has to be changed t_m/T times[1]. Thus,

$$R_4 = \lambda_1 \frac{t_m}{T} \cdot tct.$$

Using equation (4.35), we can find out T in terms of the speed v, uncut thickness t_1, and width of cut w. Since t_1 and w are directly proportional (in case of turning) to the feed f [equations (4.40)] and the depth of cut is a constant quantity as already explained, we can write

$$T = \frac{k}{v^{1/n} f^{1/m}},$$

where $m = p/n$ and k is a constant. Thus, the final expression for R_4 becomes

$$R_4 = \lambda_1 tct \frac{\pi LD}{1000k} v^{1/n-1} f^{1/m-1}.$$

Tool regrinding cost If the permitted size of the flank wear is h_f (mm), the minimum length of the tool to be reground (δ) can be found out (see Fig. 4.70) as

$$\delta = h_f \tan \nu_s \text{ mm},$$

where ν_s is the side clearance angle. So, each grinding cost is given as

$$\lambda_2 + \lambda_3 \delta = \lambda_2 + \lambda_3 h_f \tan \nu_s.$$

Since a tool has to be ground t_m/T times, the tool regrinding cost/piece works out to be

$$R_5 = (\lambda_2 + \lambda_3 h_f \tan \nu_s) \frac{t_m}{T}$$

$$= (\lambda_2 + \lambda_3 h_f \tan \nu_s) \frac{\pi LD}{1000k} v^{1/n-1} f^{1/m-1}$$

[1]It may be a fraction, but since a large number of workpieces are considered, no problem is created.

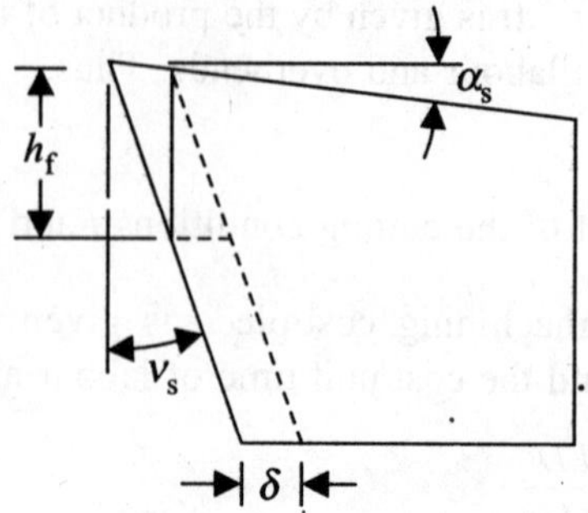

Fig. 4.70 Determination of tool regrinding cost.

If the total cost of a new tool is A and the total length that can be ground off (after which the tool cannot be used) is B mm, then the cost per mm of the tool ground is given by

$$\lambda_3 = \frac{A}{1 + \left(\dfrac{B}{h_f \tan \nu_s}\right)}.$$

Now, the total cost per piece can be expressed as

$$R = R_1 + \lambda_1 t_s + \lambda_1 \frac{\pi LD}{1000 f v} + \lambda_1 tct \frac{\pi LD}{1000k} v^{1/n-1} f^{1/m-1}$$

$$+ (\lambda_2 + \lambda_3 h_f \tan \nu_s) \frac{\pi LD}{1000k} v^{1/n-1} f^{1/m-1}. \quad (4.78)$$

To find out the optimum speed for a given feed, we differentiate R partially with respect to v and equate it to zero. Thus,

$$\left.\frac{\partial R}{\partial v}\right|_{v_{opt}} = -\lambda_1 \frac{\pi LD}{1000 f} v^{-2} + (\lambda_1 tct + \lambda_2 + \lambda_3 h_f \tan \nu_s)$$

$$\left. \times \left(\frac{1}{n} - 1\right) \frac{\pi LD}{1000k} v^{1/n-2} f^{1/m-1} \right|_{v=v_{opt}} = 0$$

or

$$v_{opt} = \left[\frac{nk\lambda_1}{(1-n) f^{1/m} (\lambda_1 tct + \lambda_2 + \lambda_3 h_f \tan \nu_s)}\right]^n. \quad (4.79)$$

Representing the total cost of each regrinding of the tool $(\lambda_2 + \lambda_3 h_f \tan \nu_s)$ by λ_4, we can express the optimum speed for minimum cost (for a given value of feed) as

$$v_{opt} = \left[\frac{nk\lambda_1}{(1-n) f^{1/m} (\lambda_1 tct + \lambda_4)}\right]^n. \quad (4.80)$$

In a similar manner, it can be shown that the optimum feed for a given cutting speed can be expressed as

$$f_{opt} = [\frac{mk\lambda_1}{(1-m)v^{1/n}(\lambda_1 tct + \lambda_4)}]^m. \tag{4.81}$$

The nature of variation of the total cost/piece with speed and feed is shown in Fig. 4.71. From this figure, we see that the minimum value of R becomes smaller as f increases. However, the feed cannot be indefinitely increased for various

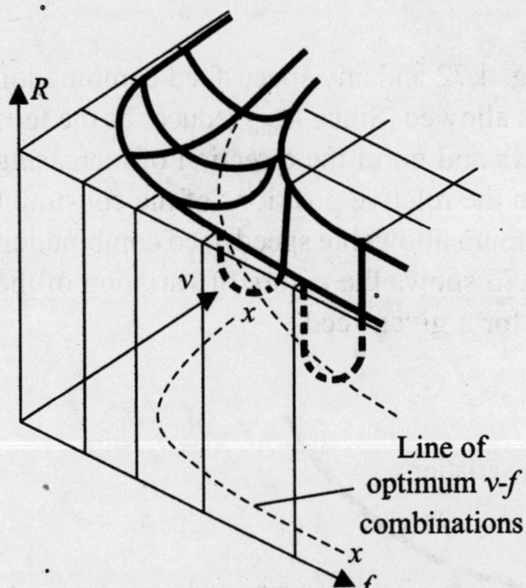

Fig. 4.71 Variation of machining cost with v and f.

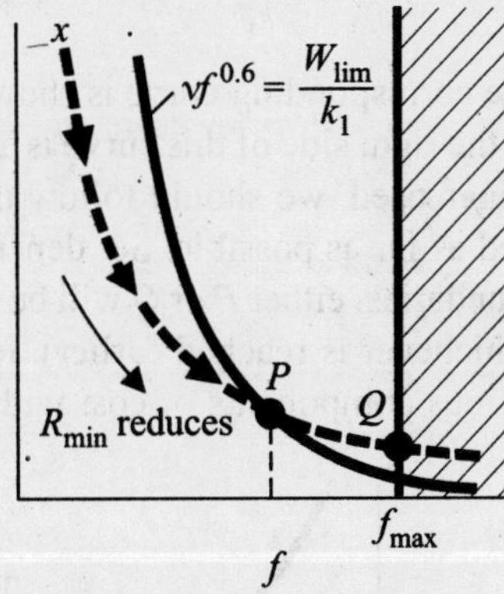

Fig. 4.72 Selection of optimum feed.

reasons, and the most important constraints on feed are the surface finish, cutting force, and power available in the machine tool. Even if we make the tool strong enough to withstand very large forces, the constraints, because of the surface finish requirement and machine tool power, cannot be avoided. From relation (4.74), we know that, for a given nose radius, the maximum unevenness H_{max} varies with f^2. In other words, if $H_{max_{lim}}$ is the limiting value of the unevenness height, then

$$f_{max} = \sqrt{8rH_{max_{lim}}}.$$

In Fig. 4.72, this value is represented by a vertical line which indicates that any point, representing a combination of the cutting speed and feed, in the shaded region is not allowed. The line of optimum v-f combinations xx is shown in Fig. 4.71. From equation (4.23), the cutting component of the machining force is

$$F_C = 1000 U_0 w t_1^{0.6},$$

where t_1 is the uncut thickness in mm. Thus, in terms of the feed f (in mm), we have

$$F_C = k_1 f^{0.6},$$

where k_1 is a constant. The power consumption will then be

$$W = k_1 v f^{0.6}.$$

So, if W_{lim} is the maximum available power in the machine, then the limiting cutting speed-feed combination is given by

$$v f^{0.6} = \frac{W_{lim}}{k_1}.$$

The corresponding curve is shown in Fig. 4.72 and any speed-feed combination on the right side of this curve is again not allowed. Since R_{min} reduces as the feed is increased, we should follow the line *xx* and go in the direction of increasing feed as far as possible. So, depending on the relative positions of the constraint boundaries, either *P* or *Q* will be the optimum allowable speed-feed combination (whichever is reached earlier). Figure 4.73 shows the nature of variation of the various components of cost with speed for a given feed.

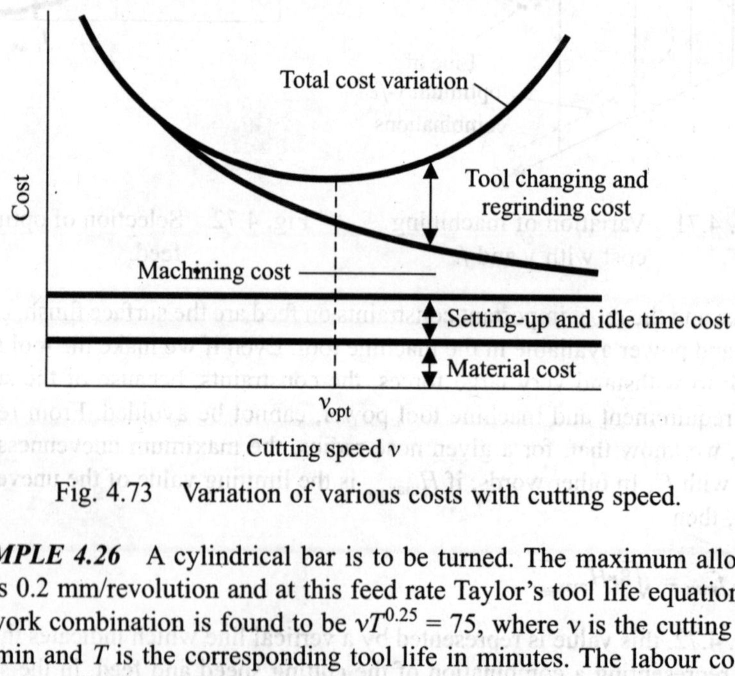

Fig. 4.73 Variation of various costs with cutting speed.

EXAMPLE 4.26 A cylindrical bar is to be turned. The maximum allowable feed is 0.2 mm/revolution and at this feed rate Taylor's tool life equation for a tool-work combination is found to be $vT^{0.25} = 75$, where v is the cutting speed in m/min and T is the corresponding tool life in minutes. The labour cost and overheads is \$0.15 per minute and the total cost involved in each regrinding of the tool is \$2.50. On the average, it takes about 2 minutes to change the tool. Estimate the cutting speed that will lead to the minimum cost.

SOLUTION Since the tool life equation is $vT^{0.25} = 75$, it implies that $n = 0.25$ and

$$\frac{k}{f^{1/m}} = (75)^{1/0.25} = 31{,}640{,}625 \text{ (consistent unit).}$$

Now, using equation (4.80), we have

$$v_{opt} = [\frac{0.25 \times (75)^{1/0.25} \times 0.15}{(1-0.25) \times (0.15 \times 2 + 2.50)}]^{0.25} \text{ m/min}$$

$$= 27.42 \text{ m/min.}$$

EXAMPLE 4.27 A 300-mm-long bar with 30 mm diameter is to be turned on a lathe. The maximum allowable feed is 0.25 mm/revolution. The cost of labour and overheads/min is \$0.25 and each regrinding of the tool involves an expense of \$2. The time required for every tool change is one minute. Two alternative materials *X* and *Y* can be used. Their cost and tool life equation (for a feed of 0.25 mm/revolution) are as given here:

Material	*Cost/piece*	*Tool life equation* (f = 0.25 mm/revolution)
X	\$2.50	$vT^{0.1} = 30$
Y	\$3.00	$vT^{0.16} = 76$

Determine which material should be used from the cost point of view. The setting and idle time involved in each piece is one minute.

SOLUTION We have to find out the minimum total cost for each material. Let us first take up material *X*. To find out the minimum cost, we have to first calculate the optimum speed. Since $k/f^{1/m} = (67)^{1/0.1}$,

$$v_{opt}^{X} = [\frac{0.1 \times 30^{10} \times 0.25}{(1-0.1) \times (0.25 \times 1 + 2)}]^{0.1} \text{ m/min}$$

$$= 19.33 \text{ m/min.}$$

Using equation (4.78), we see that the minimum cost/piece when *X* is used is given by

$$R_{min}^{X} = 2.5 + 0.25 \times 1 + 0.25 \times \frac{\pi \times 300 \times 30}{0.25 \times 19.33 \times 1000}$$

$$+ 0.25 \times 1 \times \frac{\pi \times 300 \times 30}{1000 \times 30^{10}} \times 19.33 \times 19.33^{-1} \times 0.25^{-1}$$

$$+ 2 \times \frac{\pi \times 300 \times 30}{1000 \times 30^{10}} \times 19.33^{10} \times 19.33^{-1} \times 0.25^{-1} \text{ dollars}$$

$$= \$4.38.$$

When material *Y* is used,

$$v_{opt}^{Y} = [\frac{0.16 \times (76)^{1/0.16} \times 0.25}{(1-0.16) \times (0.25 \times 1 + 2)}]^{0.16} \text{ m/min}$$

$$= 41 \text{ m/min.}$$

The corresponding minimum cost/piece becomes

$$R^{Y}_{\min} = 3.75 + 0.25 \times 1 + 0.25 \times \frac{\pi \times 300 \times 30}{0.25 \times 41 \times 1000}$$

$$+ 0.25 \times 1 \times \frac{\pi \times 300 \times 30}{1000 \times (76)^{1/16}} \times (41)^{1/0.16} \times (41)^{-1} \times (0.25)^{-1}$$

$$+ 2 \times \frac{\pi \times 300 \times 30}{1000 \times (76)^{1/0.16}} \times (41)^{1/0.16} (41)^{-1} \times (0.25)^{-1} \text{ dollars}$$

$$= \$4.07.$$

Thus, comparing we conclude that the material Y will be more economical. This is due to the fact that though Y is more expensive, its machinability is much better.

4.6.2 OPTIMIZING CUTTING PARAMETERS FOR MAXIMUM PRODUCTION

Sometimes, the rate of production is more important than the cost per piece. In fact, from the economic point of view, the best situation is when the profit is maximum. A large rate of production may result in a better return, and therefore it is also useful to investigate the conditions leading to the highest possible rate of production.

The maximum production rate can be achieved if the total time required per piece is reduced to a minimum. For this, we shall, for reasons already explained, keep the feed at the highest possible value and search for the optimum velocity. The total time required per piece is given by

t_t = setting and idle time/piece + machining time/piece + tool changing time/piece

or

$$t_t = t_s + t_m + \frac{t_m}{T} tct \text{ min}$$

$$= t_s + \frac{\pi LD}{1000 f v} + \frac{\pi LD}{1000k} v^{1/n-1} f^{1/m-1} tct \text{ min.}$$

For optimum speed to minimize t_t,

$$\left.\frac{\partial t_t}{\partial v}\right|_{v=v_{opt}} = -\frac{\pi LD}{1000 f} v^{-2} + \left(\frac{1}{n} - 1\right) \frac{\pi LD}{1000\text{k}} v^{1/n-2} f^{1/m-1} tct \bigg|_{v=v_{opt}} = 0.$$

Finally, we get

$$v_{opt} = \left[\frac{nk}{(1-n) f^{1/m} tct}\right]^n. \tag{4.82}$$

It is clearly seen that if the value of λ_4 in equation (4.80) is put equal to zero, the result we get is identical with that given by equation (4.82). The reader can easily justify this. Thus, it is clear that the optimum speed for minimum time (or maximum production rate) is always more than that for minimum cost.

EXAMPLE 4.28 **Find out the optimum cutting speed for maximum production rate for the job described in Example 4.26.**

SOLUTION As has already been found out,

$$\frac{k}{f^{1/m}} = (75)^{1/0.25} \text{ (consistent unit).}$$

Substituting the values in equation (4.82), we obtain

$$v_{opt} = \left[\frac{0.25(75)^{1/0.25}}{(1-0.25)\times 2}\right]^{0.25} \text{ m/min}$$

$$= 47.92 \text{ m/min.}$$

So, we see that this optimum speed is much higher than that for minimum cost.

4.6.3 OPTIMUM CUTTING SPEED FOR MAXIMUM EFFICIENCY

The optimum velocity to achieve the maximum efficiency, i.e., the maximum profit rate, can be also found out without much difficulty. Of course, for doing this, a closed form expression for v_{opt} will not be possible, and the numerical or graphical methods have to be employed. If S is the amount received per piece, then the expression for the profit rate is

$$P_r = \frac{S-R}{t_t}.$$

R and t_t can be expressed in terms of v as before, and v_{opt} is found out from the equation

$$\left.\frac{\partial P_r}{\partial v}\right|_{v=v_{opt}} = 0.$$

4.7 EXERCISE PROBLEMS

4.1 The chips from an orthogonal cutting operation with an uncut thickness of 0.2 mm for various rake angles are:

α	15°	10°	5°	0°
t_2 (mm)	0.45	0.5	0.63	1.13

Calculate, for each chip, the corresponding shear angle and shear strain and plot them against α.

4.2 The cutting and the thrust components of the machining force during orthogonal machining of aluminium with a rake angle of 10° are found to be 312 N and 185 N, respectively.

(i) Estimate the coefficient of friction between the tool and the chip.

(ii) If the rake angle is reduced to 0°, keeping all the other parameters the same, and if the coefficient of friction also remains unchanged, estimate the new values of F_C and F_T, using Merchant's first solution.

4.3 During orthogonal machining with a rake angle 10° and an uncut thickness 0.125 mm, the values of F_C and F_T are found to be 517 N and 217 N, respectively. The average chip thickness is also measured and found to be 0.43 mm. Evaluate the machining constant for the work material.

4.4 When the rake angle is zero during orthogonal cutting, show that

$$\frac{\tau_s}{U_c} = \frac{(1 - \mu r)r}{1 + r^2}.$$

4.5 During an orthogonal cutting test, the observations made are:

$$t_1 = 0.25 \text{ mm}, \quad t_2 = 1.2 \text{ mm}, \quad w = 2.5 \text{ mm},$$

$$\alpha = 0°, \quad F_C = 900 \text{ N}, \quad F_T = 810 \text{ N}.$$

Calculate the mean shear strength of the work material.

4.6 Estimate the cutting component of the machining force during the orthogonal machining of an aluminium alloy with an uncut thickness of 0.15 mm, the width of cut being 2.5 mm.

4.7 A metal is being cut orthogonally with a tool with zero rake angle. Show that the rate of heat generation in the shear plane can be expressed as $F_C v(1 - \mu r)$.

4.8 Calculate the mean shear plane temperature rise during orthogonal machining with zero rake. Given

$$U_c = 1.5 \text{ J/mm}^3, \quad \mu = 0.8, \quad t_1 = 0.2 \text{ mm}, \quad r = 0.2,$$

$$\rho = 7000 \text{ kg/m}^3, \quad c = 500 \text{ J/kg-°C}, \quad v = 2 \text{ m/sec}.$$

Assume that 15% of the heat generated goes into the workpiece.

4.9 If the thermal conductivity of the material is 50 W/m-°C and the ambient temperature is 30°C in Exercise 4.8, estimate the peak rake face temperature.

4.10 Estimate the three components of the machining force during shaping of a mild steel block, using the following data: depth of cut = 2.5 mm, feed = 0.125 mm/stroke, normal rake angle of tool = 7°, side cutting edge angle = 20°, coefficient of friction between chip and tool = 0.8, and ultimate shear stress of work material = 450 N/mm^2.

4.11 Prove that the normal rake angle during general turning can be expressed as

$$\tan^{-1}\left[\frac{\tan\alpha_b \sin\psi + \tan\alpha_s \cos\psi}{\sqrt{1 + (\tan\alpha_b \cos\psi - \tan\alpha_s \sin\psi)^2}}\right],$$

where α_b, α_s, and ψ are the back rake, side rake, and side cutting edge angles of the turning tool.

4.12 During the conventional turning of a mild steel bar of 75 mm diameter, the observations made are

turning tool = 5°–11°–6°–8°–25°–0.5 mm,

depth of cut = 2 mm,

feed = 0.15 mm/revolution,

job rpm = 300,

$\tau_s = 450 \text{ N/mm}^2$,

$\mu = 0.8$.

Estimate the three components of the turning force and power consumption following the Lee and Shaffer shear angle relation.

4.13 In a drilling operation using a twist drill, the rotational speed of the drill spindle is 300 rpm, the feed is 0.2 mm/revolution, the point angle is 120°, and the drill diameter is 15 mm. If the specific energy of the work material is 2 J/mm^3, estimate the torque.

4.14 A mild steel block is being drilled with a drill of 10 mm diameter. Given

helix angle = 30°, point angle = 118°,

feed = 0.2 mm/revolution, $\tau_s = 450 \text{ N/mm}^2$.

Estimate the drilling torque and thrust, using the Lee and Shaffer shear angle relation. Assume the coefficient of friction between the chip and the drill to be 0.75.

4.15 In a slab milling operation with a straight teeth cutter, the cutter has 15 teeth with 10° rake angle and rotates at 200 rpm. The diameter of the cutter is 80 mm and the table feed is 75 mm/min, the depth of cut being 5 mm. The width of the mild steel job is 50 mm and the ultimate shear stress of the work material is 420 N/mm^2. Assuming the coefficient of friction between the chip and the cutter to be 0.7 and using the Lee and Shaffer relation, plot the variation of the resultant torque with cutter rotation, and estimate (i) the average power consumption, and (ii) the horizontal and the vertical components of the average milling force. Modify the ultimate shear stress for considering the size effect according to the relation

$$\tau_s = \tau_{so}(t_{1_{av}})^{-0.4},$$

$t_{1_{av}}$ being in mm.

4.16 Solve Exercise 4.15 to estimate the power consumption following the principle of specific energy. Compare the result so arrived at with that obtained in Exercise 4.15.

4.17 The maximum power capacity of a broaching machine is 1.5 kW. Using this machine, a circular hole of diameter 20 mm in a 25-mm-thick mild steel plate has to be enlarged to a diameter of 26 mm. A cut per tooth of 0.075 mm is used and the rake angle provided to the broach teeth is 10°. Assuming the coefficient of friction and shear stress to be 0.6 and 400 N/mm^2, respectively, find out the minimum possible time in which one part can be machined. Use the Lee and Shaffer shear angle relation.

4.18 During the plunge grinding operation of a prismatic bar with cross-section 25 mm × 10 mm, a grinding wheel of 250 mm diameter rotating at 2500 rpm is used. The plunge feed rate is 5 mm/min. What will be the percentage change in the power consumption if the original wheel having 3 grits/mm^2 is replaced by a similar wheel with 9 grits/mm^2?

4.19 When surface grinding a 20-mm-wide mild steel block with a depth of cut 0.08 mm, a feed of 125 mm/min is used. The wheel diameter and rpm are 150 mm and 2000, respectively. The number of grits/mm^2 is found to be 4. It is observed that the wheel is wearing too fast. Reducing the depth of cut to 0.04 mm, the operation seems to be acceptable. To maintain the productivity, it is proposed to keep the material removal rate constant and to increase the wheel speed. Thus, the apparent hardness of the wheel can be increased. Estimate the required wheel rpm.

4.20 Calculate the power consumption during the grinding operation, described in Exercise 4.19, with the original and the modified data.

4.21 The base of a brass bracket has to be rough ground to remove the unevennesses. The four wheels available in the store are (i) A 30 K 12 V, (ii) C 90 M 4 B, (iii) C 30 Q 12 V, (iv) C 50 G 8 V. Select the most suitable wheel.

4.22 Mild steel straight cylindrical pieces (20 mm diameter × 80 mm length) are being turned on a lathe. The machine capacity is such that a power more than 600 W cannot be supplied for the actual machining operation. The diameter has to be reduced to 17 mm in one pass. The maximum unevenness allowed is 15 μm. The turning tool has a nose radius of 0.5 mm. The tool life equation for this work-tool combination is

$$vf^{0.2}T^{0.25} = 25,$$

where v is in m/min, f in mm/revolution, and T in minutes. The cost for labour and overheads is \$0.50 per minute and the total cost involved in each regrinding of the tool is \$3.00. On the average, it takes about three minutes to change the tool. Estimate the most productive cutting speed.

4.23 If the price at which each piece, described in Exercise 4.22, is sold is \$17.50 and the idle and setting time involved per piece is two minutes, find out the most efficient cutting speed.

5 Joining Processes

5.1 INTRODUCTION

Unlike the manufacturing processes employed to produce a single component, the joining processes are used to assemble different members to yield the desired complex configuration. Such a complex geometry is either too difficult or impossible to obtain by using only the manufacturing processes. The joining processes are so intimately related to the overall production system that these are also considered to form a class of manufacturing techniques. The joining of different elements can be either temporary or permanent in nature. Also, the mechanism of bonding may be either mechanical or atomic. All joining processes involving atomic bonding are of a permanent nature. In this chapter, we shall discuss the basic principles and mechanisms of all such processes. In doing so, we shall also consider the different physical, mechanical, and metallurgical factors which govern the quality of the joints. In mechanical bonding, the strength of the joint is less than the combined strength of the original members. In atomic bonding, however, the situation is not necessarily so. A general basic classification of the joining processes is given in Table 5.1. In our discussion, we shall follow this classification.

Another criterion used for a classification of the joining processes is based on the composition of the joint. According to this scheme, all joining processes can be grouped into three different categories, namely, (i) autogeneous, (ii) homogeneous, and (iii) heterogeneous. In the processes belonging to (i), no filler material is added during joining. All types of solid phase welding and resistance welding are examples of this category. In the homogeneous joining processes, the filler material used to provide the joint is the same as the parent material. Arc, gas, and thermit welding belong to this category. In the processes of type (iii), a filler material different from the parent material is used. Soldering and brazing are two such joining processes. It may be noted that two materials which are insoluble in each other, such as iron and silver, can be joined by a heterogeneous process. This may be achieved by using a filler material (e.g., copper and tin) which is soluble in both the parent materials (i.e., iron and silver).

As already shown in Chapter 1, the bonding force between two metallic atoms decreases very sharply with the interatomic distance. When the distance is more than a few atomic spacings (i.e., a few angstroms), the interacting attractive force reduces to almost zero. But the force increases sharply and attains a very large value when the distance is reduced. Thus, if it is possible

Table 5.1 Classification of joining processes

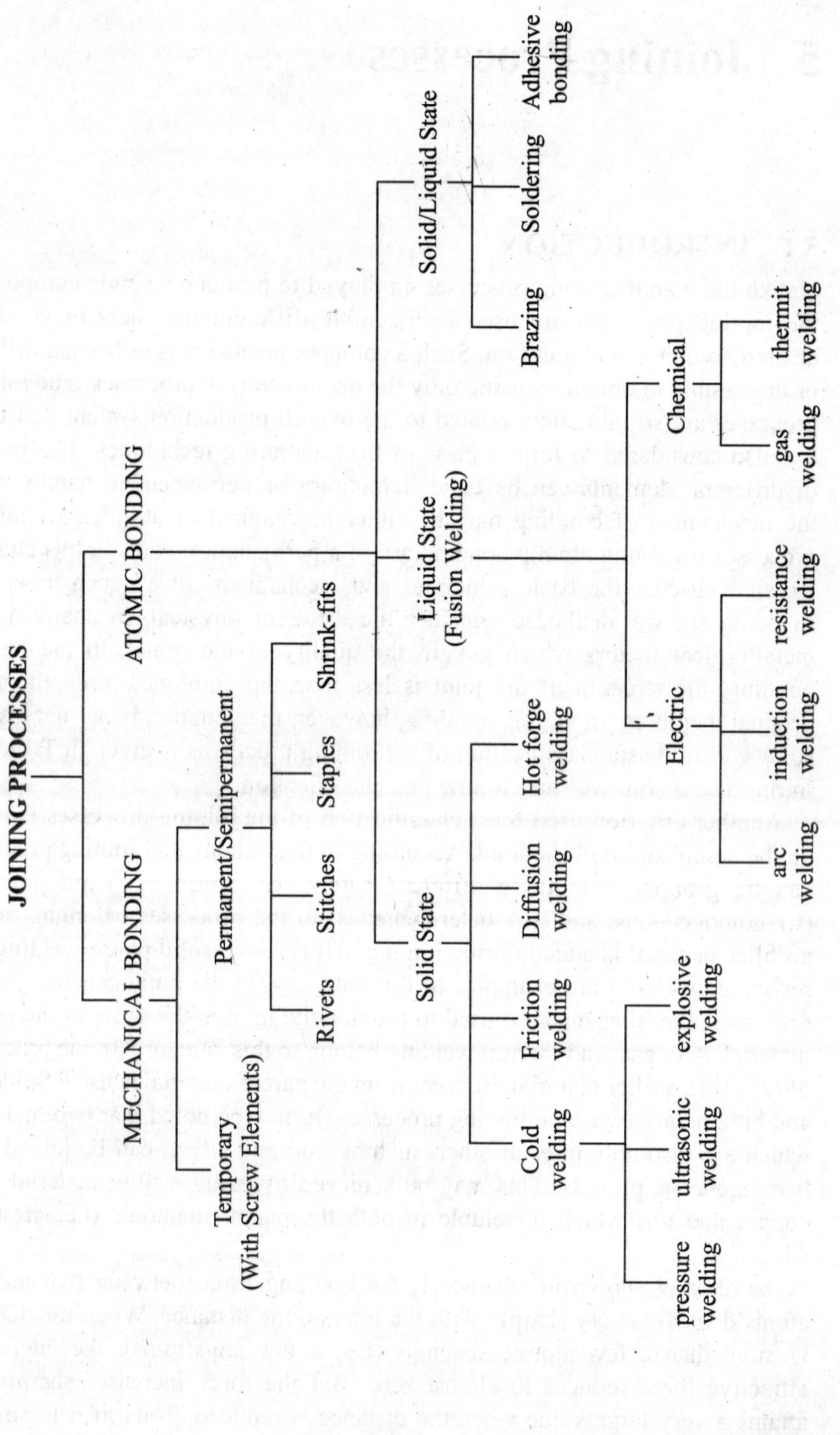

to bring together two metallic surfaces so that nothing but the grain boundaries separate them, the two bodies will adhere with a very large force, resulting in what we call welding. However, in normal atmosphere, the metal surfaces are contaminated with layers of oxides and adsorbed gases. These layers are normally a few hundred angstroms thick. So, it is not possible to generate a strong attractive force when two metal surfaces are brought in contact. But this difficulty can be eliminated when the contaminating layers are removed from the surfaces. Though this may appear as a problem for welding, it is rather fortunate to have these contaminating layers. For example, in the outer space, a major problem is not the welding but the unwelding of surfaces.

5.2 PRINCIPLES OF SOLID PHASE WELDING

The solid state welding processes may be carried out both at the room temperature and at an elevated temperature without, of course, melting any part of the joining surfaces. For a better understanding of the quality of a solid phase joint, it is worthwhile to recapitulate the strength and cohesion of metals. A defect-free crystal fails by a cleavage along a crystallographic plane where the interatomic force is the weakest. As a result, two new surfaces are produced, and the surface energy γ is defined as the work done in order to create these surfaces. The strength of a single crystal (σ_c) is found to be[1]

$$\sigma_c = \left(\frac{E\gamma}{d}\right)^{1/2}, \tag{5.1}$$

where E is the modulus of elasticity of the material and d is the lattice spacing in the cleavage plane. However, in a brittle solid, the failure takes place by the extension of the cracks already present, and the bulk strength is much reduced from that given by equation (5.1). In this case, the bulk strength (σ_b) is expressed as

$$\sigma_b = \left(\frac{E\gamma}{l}\right)^{1/2}, \tag{5.2}$$

where l ($\geqslant d$) is the length of the crack. In Chapter 1, we have already seen that the failure of a polycrystalline ductile material is due to the movement of dislocations, resulting in plastic instability, and this takes place at a stress much lower than that given by equation (5.1).

The foregoing discussion indicates that the bulk strength of a material is much lower than the bonding forces of the constituent atoms. So, a good welding does not require to achieve a strength equal to that between the adjacent lattice planes. Moreover, it should be remembered that at the room temperature, i.e., with negligible creep, even a plane of lattice misfit is not weaker than the bulk material. This information is important because a cold weld junction is essentially a plane of lattice misfit.

[1]Lancaster, J.F., Metallurgy of Welding, Brazing and Soldering, Allen and Unwin, London, 1980.

When two metal surfaces are brought into contact, the real contact takes place through a small area of asperities. This metallic bridging occurs even in the presence of adsorbed surface layers. The bridges so formed have the property of a true grain boundary, and hence are stronger than the bulk material. Some work hardening also takes place in the layers, just beneath the mating surfaces. If the yield strength (or flow pressure) of the material is σ_y with the applied force as p_e, then the fraction of the total area coming in contact, and thereby forming a weld, is simply (p_e/σ_y). However, around the welding zones, there will be some areas which come in contact (without actual flow) where the stresses are still within the elastic range. The experimental results suggest that including this area, the total area of physical contact, with a moderate external pressure, can be taken as $2p_e/\sigma_y$. When the applied load is removed, the two surfaces separate out only when the elastic forces trapped in the regions around the bridges are strong enough to break apart these metallic bridges. It is seen that the softer the material, the better the permanent adherence.

In the solid phase welding processes, the four important factors are (i) surface deformation, (ii) surface films, (iii) recrystallization, and (iv) diffusion.

The surface deformation that takes place during welding is difficult to measure. As such, in pressure welding, the bulk deformation is used as an index of the surface deformation and is expressed as

$$\frac{t_i - t_f}{t_i} \times 100\% \qquad \text{(for a sheet of original and final thicknesses } t_i \text{ and } t_f\text{, respectively),}$$

$$\frac{d_f^2 - d_i^2}{d_f^2} \times 100\% \qquad \text{(for a circular specimen of original and final diameters } d_i \text{ and } d_f\text{, respectively).}$$

The strength of a welded junction increases with increasing bulk deformation. Moreover, no weldment takes place below a certain critical deformation. The amount of deformation necessary for obtaining a specific strength decreases with increasing temperature. A strong weld may be made with only 10% deformation if the working temperature is quite close to the melting point of the material. The ratio of the oxide hardness and the parent metal hardness also effectively governs the amount of necessary deformation.

The greatest hurdle in solid phase welding is posed by the surface oxide layers and oil films. The liquid films can be removed by heating in hot welding, and by means of scratch brushing in cold welding. The oxide films can also be reduced to a certain extent by scratch brushing. Moreover, these oxide layers (being hard and brittle) fracture when the pressure is applied. A lateral movement is very useful (as in ultrasonic welding) since this tends to roll together the fragmented oxide layer into a relatively thick agglomerate. This results in a more metal-to-metal contact area. An excessive oxide contamination is always harmful, resulting in a poor joint efficiency.

A solid phase welding done at the room temperature does not allow recrystallization and grain growth at the interface. This reduces the ductility of the joint to some extent. An increase in working temperature not only increases the ductility but also eliminates some other defects. The phenomenon of diffusion has an important bearing on the performance of a solid phase weld. The shape and the size of the voids at the interface are modified considerably depending on the amount of diffusion.

5.3 PRINCIPLES OF FUSION (LIQUID STATE) WELDING

In a fusion welding process, the material around the joint is melted in both the parts to be joined. If necessary, a molten filler material is also added. Thus, a fusion welding process may be either autogeneous or homogeneous. Metallurgically, there are three distinct zones in a welded part, namely, (i) the fusion zone, (ii) the heat affected unmelted zone around the fusion zone, and (iii) the unaffected original part. The most important factors governing a fusion welding process are

(i) the characteristics of the heat source,

(ii) the nature of deposition of the filler material in the fusion zone, known as the *weld pool*,

(iii) the heat flow characteristics in the joint,

(iv) the gas metal or slag metal reactions in the fusion zone, and

(v) the cooling of the fusion zone with the associated contraction, residual stresses, and metallurgical changes.

5.3.1 HEAT SOURCE

A heat source, suitable for welding, should release the heat in a sharply defined, isolated zone. Moreover, the heat should be produced at a high temperature and at a high rate. The most common sources of heat include (i) the electric arc (as in various arc weldings), (ii) the chemical flame (as in gas welding), (iii) an exothermic chemical reaction (as in thermit welding), and (iv) an electric resistance heating (as in electroslag and other resistance welding processes). The general characteristics of these heat sources are now discussed.

Emission and Ionization of Electric Arc

First of all, let us see how an electric arc is created and maintained between two electrodes of opposite polarity. Figure 5.1 schematically shows an electric circuit used for arc welding where the work is the positive electrode (called the anode) and the electrode rod is the negative electrode (called the cathode). Initially, a good contact is made between the electrode and the work. Thereafter, the electrode is withdrawn. As a result, the metallic bridges start breaking, thus increasing the current density per bridge. Finally, the current density rises to such a high value that the bridges start boiling. Under such conditions, the electrons come out of both the surfaces by a process known as *thermionic emission*. Obviously, the electrons (having the negative charge) coming out of the anode

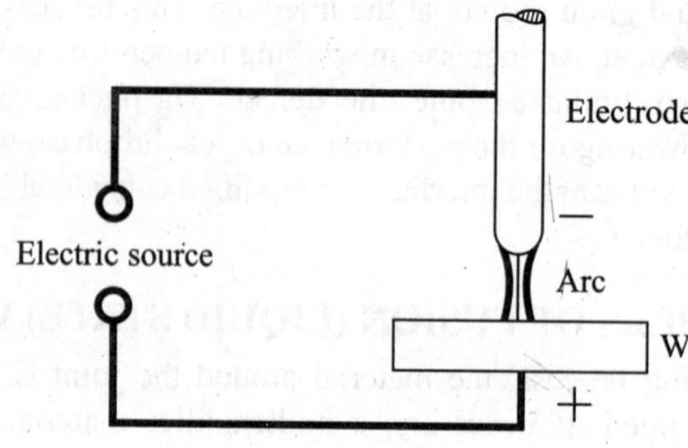

Fig. 5.1 Arc welding scheme.

are pulled back, whereas those coming out of the cathode are also attracted towards the anode.

The rate at which the electrons are emitted from a hot surface is given by

$$I = C\theta^2 \exp(-\beta/\theta), \tag{5.3}$$

where I is in amp/cm^2, θ is the absolute temperature, C is a constant, and β is given by

$$\beta = \phi e/(k\theta) \tag{5.4}$$

with e = charge of an electron, k = Boltzmann's constant, and ϕ (when measured in electron volts) as the *thermionic work function*. ϕ, in fact, represents the kinetic energy necessary to 'boil' out an electron. The values of ϕ for some common metals are shown in Table 5.2. It is obvious from equation (5.3) that a low value of ϕ, together with a high value of θ, makes the emission of electrons easier. Once started, the arc itself becomes a source of ions through a process of ionization (as explained in the next paragraph). These ions are attracted by the cathode and the resulting collisions keep the cathode hot. The total current in the arc is carried by two sets of electrons. The first set, emitted by the cathode, is called primary electrons, and the second set, known as secondary electrons, is produced as a result of the ionization of the arc gap. With tungsten and carbon electrodes, the primary electrons carry most of the current, whereas with copper or aluminium electrodes, the secondary electrons carry most of the current.

An electron of charge e, moving in an electric field of gradient E (volt/distance), experiences a force of magnitude eE. In other words, it accelerates at a rate of eE/m, where m is its mass. So, if it travels through a distance d before colliding with another particle (a neutral atom or another electron), it has a kinetic energy eEd. This kinetic energy is nothing but heat and manifests itself through increased temperature. The interparticle collisions, taking place in the gap between the electrodes, give rise to a process called *thermal ionization*. Normally, these collisions are elastic and both the momentum and kinetic energy are conserved. However, occasionally a collision is such that an electron may be completely knocked out from a neutral atom, producing a free electron and a positively-charged ion. Such a collision is, of course, not elastic in nature.

Table 5.2 Ionization potential and thermionic work function of some common metals

Metal	Ionization potential (V)	ϕ (eV)
Aluminium	6.0	4.1
Copper	7.9	4.4
Iron	7.83	4.4
Tungsten	8.1	4.5
Sodium	5.1	2.3
Potassium	4.3	2.2
Nickel	7.61	5.0

The ions thus produced are attracted towards the cathode, as already explained in the foregoing paragraph. The free electrons (earlier termed as the secondary electrons) help the arc to remain electrically conductive. A definite amount of energy is required to produce ionization in a given atom or molecule. This energy (in electron volts) is numerically equal to the ionization potential (in volts). The ionization potentials for different metal vapours are also shown in Table 5.2.

Most of the ion-producing collisions in an arc are between hot, neutral atoms and molecules. To maintain the conductivity of the arc, only a small fraction of the atoms need to be hot enough to ionize, whereas the rest of the arc should be hot enough to supply the fast atoms. For most common gases and vapours at the atmospheric pressure, the arc temperature is of the order of 6000°C.

Arc structure, characteristics, and power Structurally, we can distinguish five different zones in an electric arc. These are as follows.

(i) *Cathode spot* This is a relatively very small area on the cathode surface, emitting the electrons.

(ii) *Cathode space* It is a gaseous region adjacent to the cathode and has a thickness of the order of 10^{-3} cm. This region has the positive space charge, so a voltage drop is necessary as the electrons are to be pulled across this region.

(iii) *Arc column* This is the visible portion of the arc consisting of plasma (hot ionized gas) where the voltage drop is not sharp.

(iv) *Anode space* This, again, is a gaseous region (thickness $\approx 10^{-3}$ cm) and is adjacent to the anode surface where a sharp drop in the voltage takes place. This is because the electrons have to penetrate the anode surface after overcoming the repulsion of the thermionically-emitted electrons from the anode surface.

(v) *Anode spot* This is the area on the anode surface where the electrons are absorbed. This area is larger than the cathode spot.

The potential drop across an arc is schematically shown in Fig. 5.2. The voltage drop shown in this figure is for given spacing, current, and electrode materials. A change in the materials alters all the values. However, a change in the spacing and the current essentially changes only the drop in the arc column.

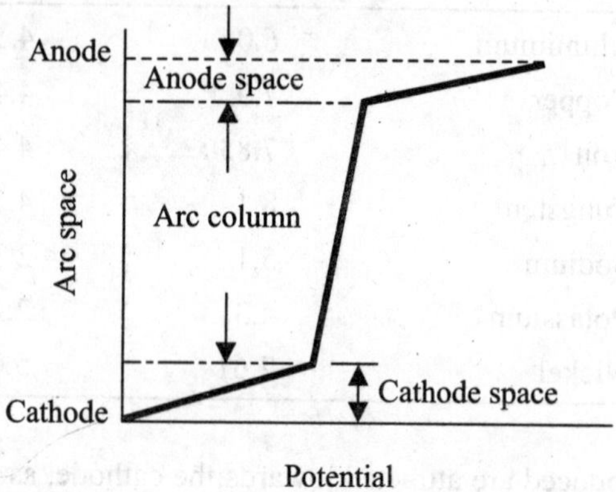

Fig. 5.2 Potential across arc space.

It has been experimentally found that, for given spacing (and, of course, electrode materials), the voltage reduces up to a current value of 50 amp (against the ohmic law of constant resistance) and increases thereafter, as shown in Fig. 5.3. This can be explained as follows. Up to 50 amp of current, the shape of the arc is almost cylindrical and the surface to volume ratio of a

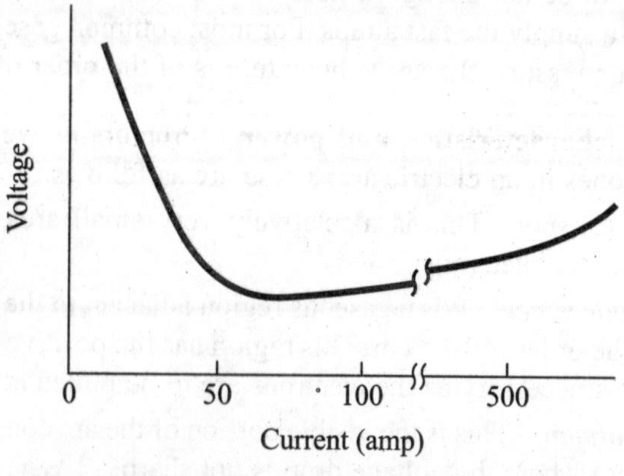

Fig. 5.3 Current-voltage characteristic of arc.

cylinder decreases with increasing radius. Thus, a thick, high current arc loses less heat and essentially burns hotter. This results in a higher conductivity (and consequently lower resistance) as compared with a thin, low current arc. However, beyond 50 amp of current, the arc bulges out and the current path becomes more than the arc gap which again increases the resistance of the arc. Due to these two opposite effects, i.e., higher temperature and longer current path, the voltage drop remains constant over a wide range of the current values.

As a first approximation, we can assume the conductivity of the arc column to be independent of the arc length l. The electrode drops are also independent of the arc length. Hence, we can write the voltage drop across the entire arc as

$$V = A + Bl, \tag{5.5}$$

where A is the electrode drop and Bl represents the column drop.

The voltage-current relationship of an arc (Fig. 5.3) determines the required characteristics of the power source. In Fig. 5.4, we consider two different characteristics of the power source. The curve AB represents a flat characteristic, whereas the curve CD represents a sharply drooping one. In this figure, two

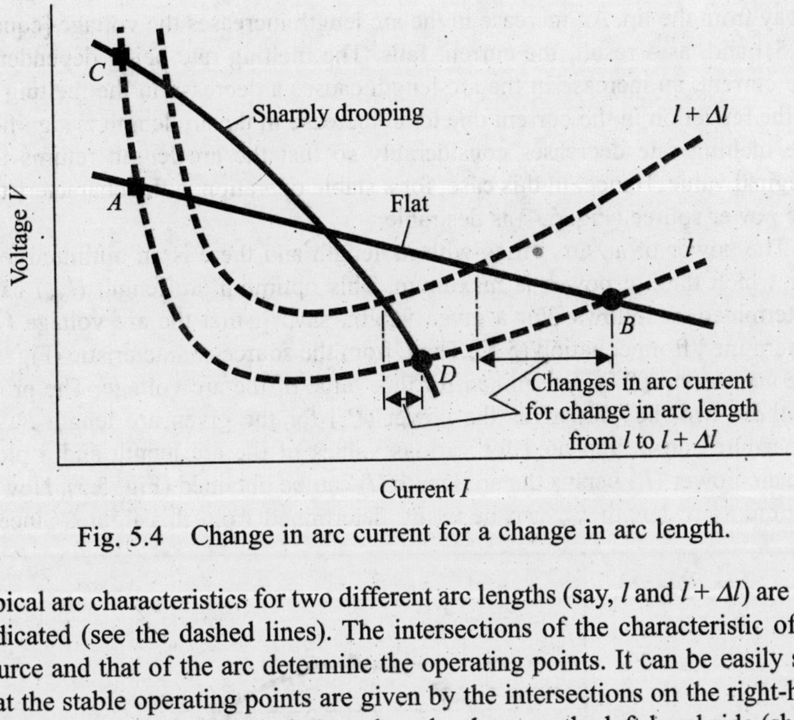

Fig. 5.4 Change in arc current for a change in arc length.

typical arc characteristics for two different arc lengths (say, l and $l + \Delta l$) are also indicated (see the dashed lines). The intersections of the characteristic of the source and that of the arc determine the operating points. It can be easily seen that the stable operating points are given by the intersections on the right-hand side (shown by the solid circles), and not by those on the left-hand side (shown by the solid squares), of the figure. This can be verified by considering a change, say, an increase, in the arc current. At the points shown by the solid circles, such an increase causes an increase in the voltage which, in turn, decreases the generator (source) current. Thus, any disturbance is automatically opposed, and the operating points return to their original values. At the points shown by the solid squares, just the opposite phenomenon takes place, i.e., any disturbance moves the operating points continuously away from their original locations. The changes in the arc current for the two power sources for a given change in the arc length (from l to $l + \Delta l$) are also indicated in Fig. 5.4.

In manual arc welding, an inadvertent change in the arc length is inevitable.

However, this should not cause a large change in the welding current. This obviously makes the sharply drooping characteristic desirable for manual arc welding. With a flat characteristic, for a big change in the arc length, there may not be any point of intersection between the arc and the source characteristics and the arc may blow out. For an efficient striking of the arc, it is necessary that the open circuit voltage of the source be high above the operating voltage. Moreover, it is necessary to have a quick response of the source (low time constant) since the welding process itself is unsteady. The power source should be such that it is not damaged by short circuiting for an appreciable length of time.

In a semiautomatic arc welding process, the arc is maintained between the workpiece and a wire which is driven forward at a constant speed as it melts away from the tip. An increase in the arc length increases the voltage [equation (5.5)] and, as a result, the current falls. The melting rate being dependent on the current, an increase in the arc length causes a decrease in the melting rate. If the reduction in the current due to an increase in the arc length is significant, the melting rate decreases considerably so that the arc length returns to its original value. Hence, in this case, for a stable operation, a flat characteristic of the power source (Fig. 5.4) is desirable.

The power of an arc varies with its length and there is an optimum length for which the arc power is maximum. This optimum arc length (l_{opt}) can be determined as follows. For a given length, say, l_1, first the arc voltage (V) is determined from equation (5.5). Then, from the source characteristic (Fig. 5.4), the arc current (I) is determined for this value of the arc voltage. The product of these two, i.e., VI, gives the power (P_1) for the given arc length l_1. This procedure can be repeated for various values of the arc length and a plot of the arc power (P) versus the arc length (l) can be obtained (Fig. 5.5). Now, the optimum arc length l_{opt} can be easily determined from this figure. Since the

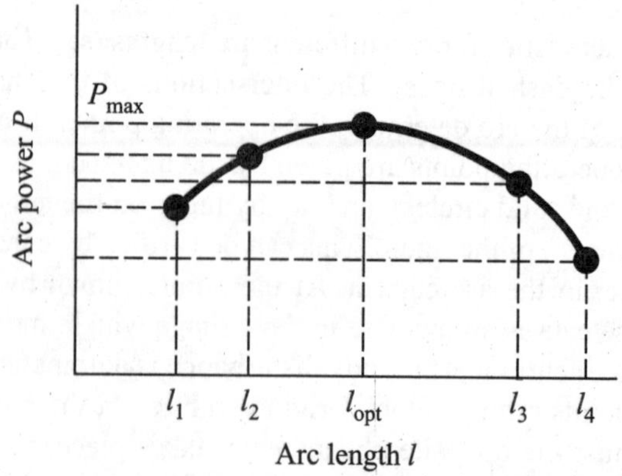

Fig. 5.5 Arc power versus arc length.

electrode drop is utilized with a higher efficiency than the column drop, the actual optimum length is a bit shorter than the optimum obtained from Fig. 5.5.

EXAMPLE 5.1 The voltage-length characteristic of a direct current (dc) arc is given by

$$V = (20 + 40l) \text{ volts,}$$

where l is the length of the arc in cm, The power source characteristic is approximated by a straight line with an open circuit voltage = 80 V and a short circuit current = 1000 amp. Determine the optimum arc length and the corresponding arc power.

SOLUTION The power source characteristic can be written analytically as

$$V = (80 - \frac{80}{1000} I) \text{ volts.} \qquad \text{(a)}$$

The arc characteristic is given as

$$V = (20 + 40l) \text{ volts.} \qquad \text{(b)}$$

Equating (a) and (b), we obtain

$$80 - \frac{80}{1000} I = 20 + 40l$$

or

$$I = (60 - 40l)\frac{1000}{80} \text{ amp.} \qquad \text{(c)}$$

Hence, the power P is obtained from (b) and (c) as

$$P = VI = (20 + 40l)(60 - 40l)\frac{1000}{80} \text{ volt-amp.}$$

For maximum power,

$$\frac{dP}{dl} = 0$$

or

$$40(60 - 40l) - 40(20 + 40l) = 0$$

or

$$1600 = 3200l \quad \text{or} \quad l = 0.5 \text{ cm.}$$

So, the optimum arc length l_{opt} = 0.5 cm when the maximum power of the arc is

$$P_{max} = (20 + 40 \cdot \tfrac{1}{2})(60 - 40 \cdot \tfrac{1}{2})\frac{1000}{80} \text{ volt-amp}$$

$$= 20 \text{ kVA.}$$

In our discussion so far, we have talked of a dc arc. Every half cycle of a 50-Hz alternating current (ac) takes 0.01 sec, whereas an arc takes only about 0.001 sec to reach the equilibrium state. Due to this quick response, both Fig. 5.3 and equation (5.5) are equally applicable for every half cycle of an ac arc as well. It should be remembered, however, that an ac arc must reignite itself after every crossing of the zero current instant. Reignition requires a voltage higher than the normal arc voltage. The process of reignition of an arc is facilitated by the presence of ions having a low ionization potential. So, the electrodes for an ac arc welding are coated with potassium silicate binders, whereas those used for a dc arc welding are normally coated with sodium silicate. From Table 5.2, it is readily seen that potassium has a lower ionization potential as compared with sodium.

Chemical Heat Source

Acetylene (C_2H_2) is the most common chemical heat source and is used in a chemical gas flame. In the presence of excess oxygen, it burns according to the reaction

$$C_2H_2 + \tfrac{5}{2}O_2 = 2CO_2 + H_2O + \text{heat } (\Delta H_1). \tag{5.6}$$

The amount of heat liberated (ΔH_1) is 1.275×10^6 kJ/kg-mole of acetylene. If oxygen is premixed with acetylene (one-to-one mole ratio), then the combustion reaction is

$$C_2H_2 + O_2 = 2CO + H_2 + \text{heat } (\Delta H_2) \tag{5.7}$$

with $\Delta H_2 = 0.448 \times 10^6$ kJ. The carbon monoxide and the hydrogen produced later burn, producing

$$\Delta H_1 - \Delta H_2 = 0.827 \times 10^6 \text{ kJ/kg-mole}$$

of acetylene. However, this heat, since it is generated over a large volume and at a low temperature, does not add much to the welding process.

Once the amount of heat liberated (ΔH) is known, we can roughly estimate the maximum flame temperature with the assumption of an adiabatic flame. This means that the entire ΔH leaves the flame only through the heating of the reaction products. Care must be taken to subtract the latent heat from ΔH if any of the reaction products undergoes a change of phase. The entire reaction is assumed to be completed at the room temperature (say, θ_R). Then, the flame temperature (θ_f) can be computed from the equation

$$\overline{\Delta H} = \Delta H - (\text{latent heat}) = \sum_m \int_{\theta_R}^{\theta_f} nC_p(\theta)\, d\theta. \tag{5.8}$$

In equation (5.8), the summation is taken over n moles of each of the m reaction products and $C_p(\theta)$ is the molar specific heat capacity of these products which vary with the temperature. For reaction (5.7), the flame temperature obtained is around 3560°C, whereas the measured temperature of the flame varies from 1280°C (minimum at the tip) to 3250°C (maximum at the core).

Another chemical source of heat, commonly used for welding, is the reaction

$$8Al + 3Fe_3O_4 = 2Fe + 4Al_2O_3 + \Delta H \tag{5.9}$$

where $\Delta H = 0.242 \times 10^6$ kJ/kg of atomic weight of the contained oxygen. The adiabatic temperature is calculated to be of the order of 3000°C. Reaction (5.9) is utilized in what is known as *thermit welding*.

Contact Resistance Heat Source

The electrical resistance heating, as already stated, too is a heat source. This may be done either by utilizing the contact resistance of the interfaces (as in various resistance welding processes) or by utilizing the resistance of a molten flux and slag (as in electroslag welding).

We have already noted in Section 5.2 that when two metallic surfaces are brought into contact, only a small fraction of the apparent area is in actual metal-to-metal contact. When a current is sent through such an interface, all of it is carried by these tiny metallic bridges. The oxide layers in contact carry no current. As a result, the current flow is constricted, as indicated in Fig. 5.6. Due to this constriction, the resistance to the flow of current increases, and this increment is termed as contact resistance. An estimate of the order of magnitude

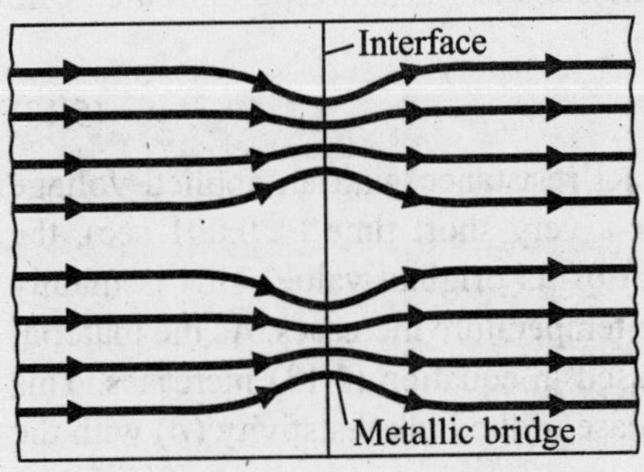

Fig. 5.6 Formation of metallic bridge.

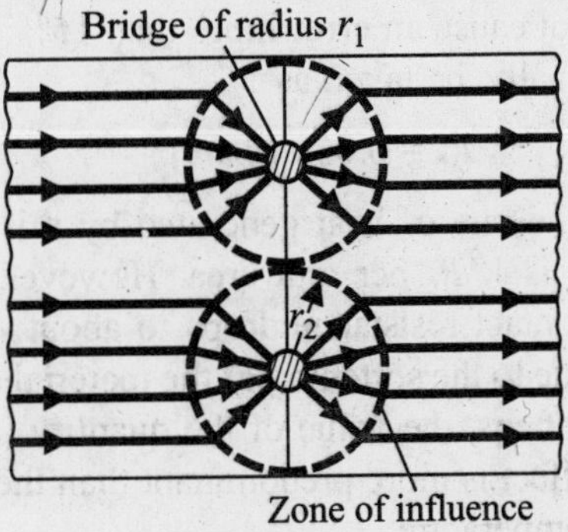

Fig. 5.7 Idealization of constricted current flow.

of this resistance can be obtained by idealizing the constricted current flow (as shown in Fig. 5.7) with the following assumptions:

(i) All bridges are of uniform size and spherical shape with radius r_1.

(ii) All bridges are uniformly spaced at a distance $2r_2$ apart.

(iii) The constriction effect due to each bridge is restricted within a concentric sphere of radius r_2.

(iv) Each bridge is of zero resistance.

Now, if there are n bridges per unit area, then the contact resistance per unit area can be calculated from the resistance of each spherical constriction and considering n such paths in parallel. Each spherical constriction consists of two identical hemispheres in series. The resistance of each hemispherical constriction is given by

$$R = \rho(r_2 - r_1)/S, \tag{5.10}$$

where ρ is the resistivity of the material, $(r_2 - r_1)$ is the length of the current path, and S is the geometric mean area of the two hemispheres of radii r_2 and r_1, respectively. Thus,

$$S = \sqrt{(2\pi r_2^2)(2\pi r_1^2)} = 2\pi r_2 r_1.$$

Substituting this value of S in equation (5.10), we obtain

$$R = \frac{\rho(r_2 - r_1)}{2\pi r_2 r_1} \approx \frac{\rho}{2\pi r_1} \tag{5.11}$$

since $r_2 \gg r_1$. Hence, the total constriction resistance per unit area is given by

$$R_c = \frac{2}{n}\frac{\rho}{2\pi r_1} = \frac{\rho}{n\pi r_1}. \tag{5.12}$$

In the absence of the interface, the resistance of the same path is negligible. Thus, the contact resistance per unit area can be taken as that given by equation (5.12). Experiments show that the assumptions leading to equation (5.12) do not cause an error more than 15%. So, the contact resistance per unit area can, finally, be taken as

$$R_c = 0.85\rho/(n\pi r_1). \tag{5.13}$$

The rate of heat generated by this contact resistance with an applied voltage V is V^2/R_c per unit area. However, after a very short time (≈ 0.001 sec), the contact resistance drops to about $(\frac{1}{10})$-th of its original value. This is mainly due to the softening of the material as the temperature increases. As the material softens, the value of the quantity (nr_1) used in equation (5.13) increases. This effect is more predominant than the increase of the bulk resistivity (ρ) with the temperature.

EXAMPLE 5.2 In a resistance welding process, the applied voltage is 5 V. Determine the rate of heat generated per unit area with 25 bridges/cm^2, each bridge having a radius of 0.1 mm. The resistivity of the material is given to be 2×10^{-5} ohm-cm.

SOLUTION The contact resistance per unit area (R_c) from equation (5.13), with $n = 25/\text{cm}^2$, is

$$R_c = \frac{0.85 \times 20 \times 10^{-5}}{25 \times \pi \times 0.01} = 0.00022 \text{ ohm-cm}^2.$$

Hence, the rate of heat generated per unit area is given as

$$Q = \frac{5 \times 5}{R_c} = \frac{5 \times 5}{0.00022} \text{ W/cm}^2$$

$$= 1.136 \times 10^5 \text{ W/cm}^2.$$

5.3.2 MODES OF METAL TRANSFER IN ARC WELDING

The depth of penetration, the stability of the weld pool, and the amount of spatter loss depend, to a large extent, on the mode of metal transfer from the consumable electrodes. Various forces cause the transfer of metal into the weld pool. The mode of transfer depends on the intersection of these forces and governs the ability of welding in various positions. The major forces which take part in this process are those due to (i) gravity, (ii) surface tension, (iii) electromagnetic interaction, and (iv) hydrodynamic action of plasma.

The force of gravity may be a retaining or a detaching force, depending on whether the electrode is pointing upward or downward. But the surface tension always tends to retain the liquid drop at the tip of the electrode. This force depends on the radius of the electrode, the capillarity constant, and the density of the liquid metal. The electromagnetic force, known as the Lorentz force, is set up due to the interaction of the electric current with its own magnetic field. This force acts in the direction of the current when the cross-section of the conductor is increasing in the direction of the current. Similarly, the force acts in the direction opposite to that of the current if the cross-section of the conductor is reducing in the direction of the current. Figure 5.8 explains how this force accelerates the process of separation of a droplet which has started to separate out. The hydrostatic pressure is created due to the magnetic force. At a high current density, this pressure elongates the liquid drop and also adds to its stiffness. As a result, the liquid drop is projected along the line of the electrode, independent of gravity.

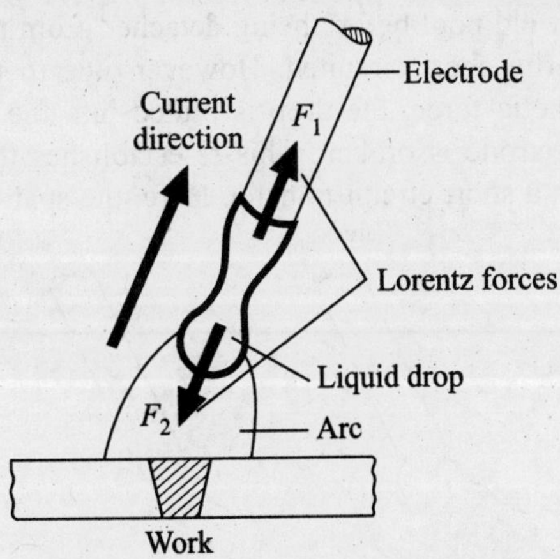

Fig. 5.8 Separation of droplet.

The plasma of the arc also causes the drop to be projected towards the workpiece, whereas a high evaporation rate from the surface of the drop tends to move it in the opposite direction.

All these forces interact in a complicated manner and give rise to two broad classes of metal transfer, namely, free flight transfer and short circuit transfer. In the former, the liquid drop travels freely in the arc space, i.e., gets completely detached from the electrode before contacting the workpiece. The free flight transfer may be (see Fig. 5.9) (i) gravitational, (ii) projected, and

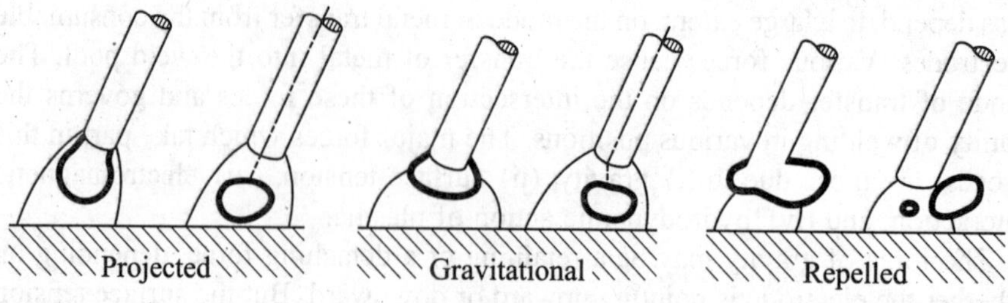

Fig. 5.9 Types of metal transfers.

(iii) repelled. When the transfer is gravitational, the predominant force is that of gravity and the molten drop falls almost vertically from the electrode into the weld pool. If the electromagnetic force, the gas jet, and the hydrostatic pressure are predominant, then the drop is given an initial acceleration towards the weld pool, and thus projected into it independent of gravity. If the resulting force directs the drop away from the weld pool, then the repelled transfer occurs. This situation is encountered when CO_2 is used as the shielding gas, particularly at low and moderate currents. Obviously, the gravitational transfer is not very reliable, and the repelled transfer is undesirable since it causes too much of a spatter loss. The projected transfer is seen in oxide coated carbon steel electrodes where a strong gas jet is set up.

In the short circuit transfer, the liquid drop at the tip of the electrode gets in contact with the weld pool before being detached from the electrode. Thus, the arc is momentarily short circuited. However, due to the surface tension and the electromagnetic force, the drop is pulled into the weld pool and the contact with the electrode is broken. This re-establishes the arc. Figure 5.10 schematically shows a short circuit transfer. Here, the spatter loss is minimum

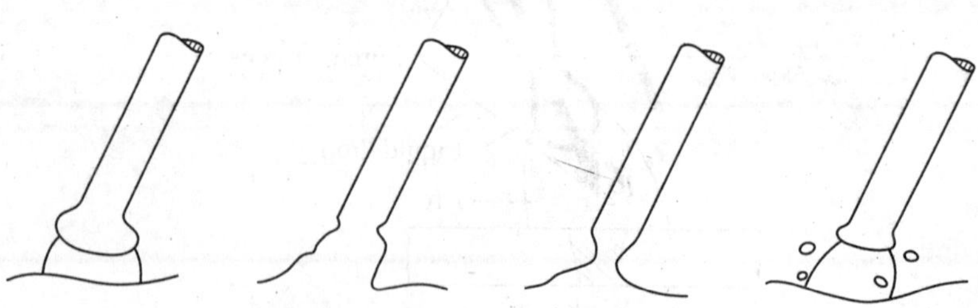

Fig. 5.10 Short circuiting transfer.

and can be achieved by controlling the gap and the other welding variables. This type of transfer, being independent of gravity, is suitable for overhead welding purposes.

5.3.3 HEAT FLOW CHARACTERISTICS

A study of heat flow characteristics can provide an estimate of the minimum heat input rate required to form a weld of a given width. Moreover, a recognition of the major variables controlling the thermal cycle (i.e., the heating and the cooling rate of the heat affected zone) is essential for a successful fusion welding. In the fusion welding processes, the heat source is moving, except in spot welding where the source is stationary. Once the steady state is reached, even with a moving heat source, the temperature distribution relative to the source becomes stationary. The most convenient way of analyzing such a problem with a moving source is to assume the source as stationary and the workpiece as moving in the opposite direction with the same velocity as that of the moving source. This speed is called the welding speed. Two different types of heat sources can be considered. In most cases, the heat is liberated in a small zone which is idealized as a point source, and the heat flow from the source is three-dimensional. In a few cases, e.g., in butt welding of relatively thin plates, the heat is liberated along a line and the heat source is idealized as a line source. In such situations, the heat flow is two-dimensional. These two types of heat sources are explained in Fig. 5.11. For an elaborate analysis of

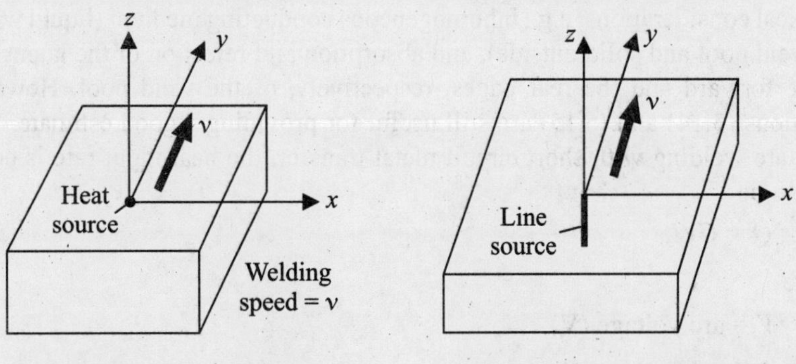

(a) Point source; temperature $= f(x, y, z)$

(b) Line source; temperature $= f(x, y)$

Fig. 5.11 Point and line heat sources.

the temperature distribution in the workpiece under various situations, see the literature. The available results include those of infinite, semi-infinite, and finite medium, each with point and line sources. Of these results, the most useful is the one which gives the minimum heat input rate necessary for maintaining a given width of the weld. For a three-dimensional heat source, this is given as[1]

$$Q = \tfrac{5}{4}\pi wk\theta_{\mathrm{m}}\left(\tfrac{2}{5} + \frac{vw}{4\alpha}\right), \tag{5.14}$$

where

Q = rate of heat input (W),

w = width of the weld (m),

k = thermal conductivity of the work material (W/m-°C),

[1]Lancaster, J.F., Metallurgy of Welding, Brazing and Soldering, Allen and Unwin, London, 1980.

θ_m = melting point of the work material above the initial temperature (°C),

v = speed of welding (m/sec),

α = thermal diffusivity of the work material (m^2/sec)

$= \dfrac{k}{\rho c}$ (ρ = density, c = specific heat).

For a two-dimensional heat source, the corresponding equation is given by

$$Q = 8k\theta_m h\left(\tfrac{1}{5} + \frac{vw}{4\alpha}\right), \tag{5.15}$$

where h = plate thickness. It is clearly seen from equations (5.14) and (5.15) that the most important parameter is vw/α.

It should be noted that the theoretical results fail to accommodate many practical considerations, e.g., inhomogeneous conducting medium (liquid within the weld pool and solid outside), and absorption and rejection of the latent heat at the forward and the rear edges, respectively, of the weld pool. However, equations (5.14) and (5.15) are still useful for providing a good estimate.

In arc welding with short circuit metal transfer, the heat input rate is easily seen to be

$$Q = CVI, \tag{5.16}$$

where

V = arc voltage (V),

I = arc current (A),

C = fraction of total time during which the arc is on.

If the heat input rate given by equation (5.16) falls short of that given by equation (5.14) or equation (5.15) (as the case may be), a lack of side fusion occurs.

EXAMPLE 5.3 In a butt welding process using arc welding, the arc power is found to be 2.5 kVA. The process is used to weld two steel plates, each of 3 mm thickness, as shown in Fig. 5.12. Determine the maximum possible

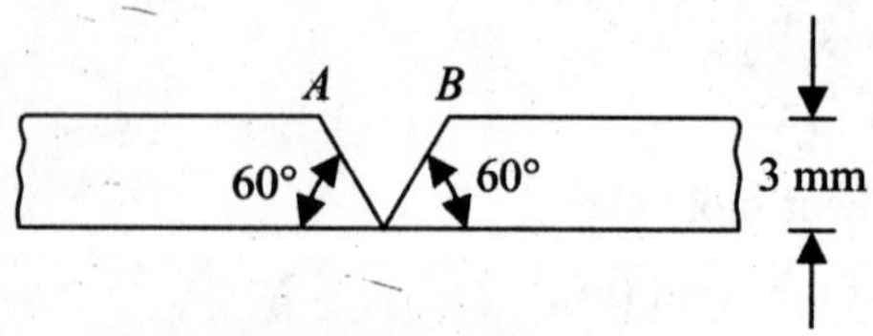

Fig. 5.12 Work for butt welding process.

welding speed. It is assumed that the metal transfer is of short circuit type and the arc is on for 85% of the total time. Given

$$\alpha_{\text{steel}} = 1.2 \times 10^{-5}\ \text{m}^2/\text{sec}, \qquad k_{\text{steel}} = 43.6\ \text{W/m-°C}.$$

The melting point of steel = 1530°C and the ambient temperature = 30°C.

SOLUTION The rate of heat input is given as

$$Q = CVI \qquad \text{[using equation (5.16)]}$$

$$= 0.85 \times 2.5 \times 10^3\ \text{W} = 2.12 \times 10^3\ \text{W}. \qquad \text{(a)}$$

The minimum width of weld to be maintained is given by

$$w = AB = 2\sqrt{3}\ \text{mm} = 2\sqrt{3} \times 10^{-3}\ \text{m}.$$

Also,

$$\theta_{\text{m}} = 1530°\text{C} - 30°\text{C} = 1500°\text{C}, \qquad h = 3 \times 10^{-3}\ \text{mm}.$$

As in the butt welding of thin plates, so too here the source can be approximated as a line source. Thus, using equation (5.15),

$$Q = 8 \times 43.6 \times 1500 \times 3 \times 10^{-3}(0.2 + \frac{vw}{4\alpha}). \qquad \text{(b)}$$

Equating Q from equations (a) and (b), we obtain

$$2.12 \times 10^3 = 8 \times 43.6 \times 15 \times 3(0.2 + \frac{vw}{4\alpha})$$

or

$$0.2 + \frac{vw}{4\alpha} = 1.25 \quad \text{or} \quad \frac{vw}{4\alpha} = 1.05.$$

Since

$$w_{\text{min}} = 2\sqrt{3} \times 10^{-3}\ \text{m},$$

we have

$$v_{\text{max}} = \frac{4 \times 1.2 \times 1.05}{2\sqrt{3} \times 10^{-3}} \times 10^{-5}\ \text{m/sec} = 0.0146\ \text{m/sec}.$$

In fusion welding, the other important heat flow variables are the cooling rate and the thermal cycle. A mathematical analysis leads to the following conclusions which are in accordance with the practical experience.

The cooling rate increases with increasing weld speed, and for a given weld speed, the cooling rate increases with decreasing size of the weld pool. For example, in electroslag welding, since the weld pool is large and the welding speed is very slow, the cooling rate is seen to be very low. On the other hand, in automatic tungsten-inert gas welding, the process is operated at a very high speed with a small weld pool, and this results in a very fast rate of cooling. The thermal cycle at any point in the medium is mainly governed by its distance from the heat source. Obviously, with increasing distance from the source, the maximum temperature is lower and the temperature lags further behind the source. Figure 5.13 shows the variation of temperature with time at different distances from the source.

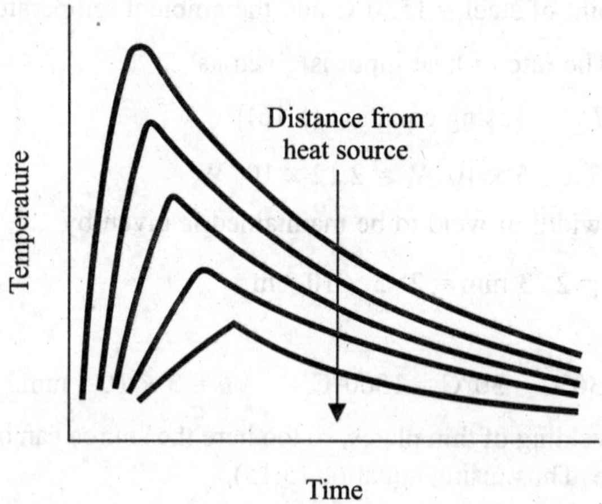

Fig. 5.13 Variation of temperature with time at different distances from source.

5.3.4 GAS METAL REACTION

The absorption of gas in the weld pool from the arc or the flame plays an important role in most fusion welding processes. This is due to the possibility of a reaction between the gas and the liquid metal in the weld pool. The chances of such a reaction are enhanced by the high temperature of the gas and the metal.

There can be two different types of reactions. In one type, the gas may just get dissolved in the liquid metal. In the second type, on the other hand, the gas and liquid metal may react chemically to form a stable compound. In such a case, the situation may be considerably different, depending on the degree of solubility of the reaction product in the weld pool.

As long as the reaction product is soluble, it does not prevent the formation of a weld pool. However, it may result in an embrittlement of the welded joint. An insoluble reaction product produces either surface scales or slags, and thus physically interferes with the formation of the weld pool. In this case, either the excess gas to the weld pool is prevented or a flux is used to dissolve and disperse the reaction product.

When the gas gets dissolved in the liquid weld pool, there is obviously no hindrance towards the formation of the weld pool. However, as the solubility decreases on cooling, degassification starts and, with suitable nuclei, bubbles may form. If these bubbles are trapped, then the quality of the weld is very poor. Even otherwise, this degassification makes the joint porous. This defect is very common in a metal whose oxides are easily reducible by hydrogen, and can be avoided by the addition of a suitable deoxidant in the filler metal.

Another important gas metal reaction is the diffusion of the gas into the parent metal from the weld pool. When the temperature of the thermal cycle is high, this diffusion process may be quite fast. The diffusion of hydrogen into the heat affected zone may, again, cause an embrittlement of the welded joint.

5.3.5 COOLING OF FUSION WELD

The three important effects intimately connected with the cooling of a fusion weld are (i) contraction, (ii) residual stress, and (iii) metallurgical phase transformation. All these effects significantly control the quality of a weld.

(i) *Contraction* During the freezing of the weld pool, a decrease in the volume takes place. Moreover, the direction of freezing, and thus the effect of contraction, depends on the type of joint, as explained in Fig. 5.14. Figure 5.14a shows the solidification of a groove weld. Here, the solidification

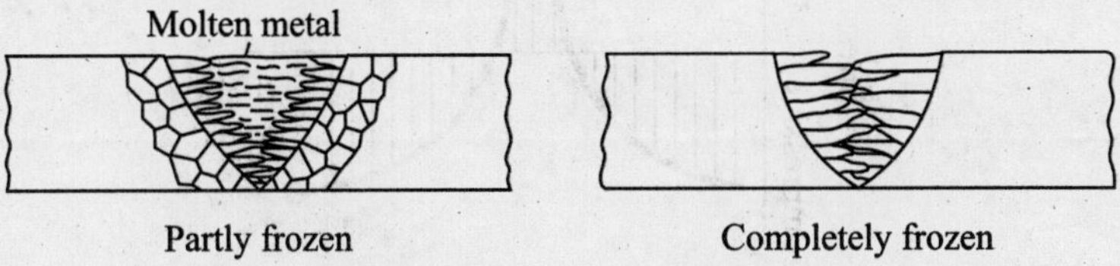

(a) Outside corner (groove)

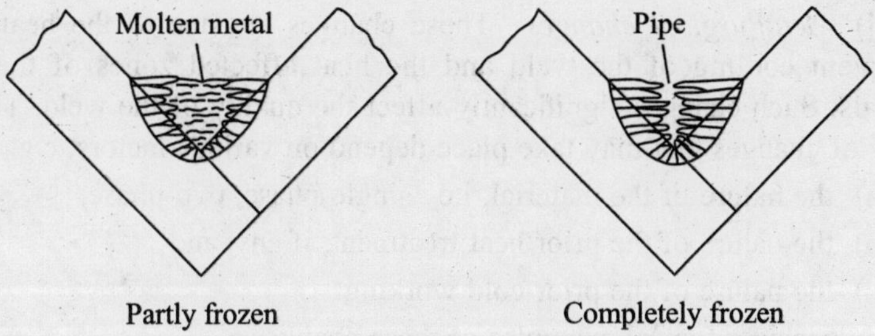

(b) Inside corner (tree)

Fig. 5.14 Solidification patterns for groove welds.

front moves simultaneously from the bottom upwards and from the sides inwards. Further, the molten top portion always makes up for the contraction in the inner layers and *piping* occurs only in the surface layers. Figure 5.14b shows the solidification in a corner joint where more cold metal is near the surface of the weld pool. Thus, the top of the weld pool freezes faster and a long *piping* throughout the joint may occur.

(ii) *Residual stress* During the fusion welding of plates, as the weld pool contracts on cooling, this contraction is resisted by the rest of the plates (which have not melted). As such, a tensile stress is generated in the weld, and this is balanced by the compressive stress in the parent metal. Figure 5.15 shows a typical distribution of these stresses in a plate weld. This residual stress may result in the cracking of a brittle material and is not important as far as a ductile material is concerned.

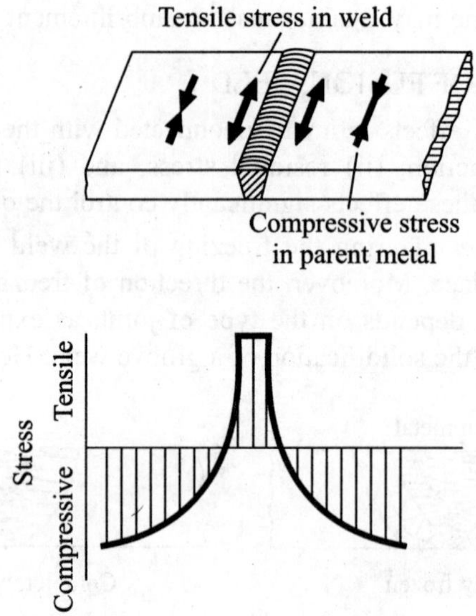

Fig. 5.15 Stress distribution in plate weld.

(iii) *Metallurgical changes* These changes are due to the heating and subsequent cooling of the weld and the heat affected zones of the parent materials. Such changes significantly affect the quality of the weld. The wide variety of changes that may take place depend on various factors, e.g.,

(a) the nature of the material, i.e., single-phase, two-phase,

(b) the nature of the prior heat treatment, if any, and

(c) the nature of the prior cold working.

We now consider typical examples of these changes.

Let us consider the fusion welding of two pieces of a single-phase material, which have been cold worked to yield a desired grain orientation. These cold worked grains result in a high strength and low ductility. However, on fusion welding, a random grain growth again takes place within the melt boundary, which, in turn, results in a low strength. Within the heat affected zone, the grains become coarse due to heat input (annealing), and a partial recrystallization also occurs. In either case, the strength falls much below that of the parent material. With increasing distance from the melt boundary, the grains become finer until the heat unaffected zone with elongated grains is reached. All these changes are shown in Fig. 5.16.

Let us now consider a two-phase material which derives its strength mostly from precipitation hardening. In this case, the strength within the melt boundary is again too low. But, in the immediately adjacent heat affected zone, the thermal cycle results in heating and quenching followed by further aging. This aging process recovers some of the strength. The material beyond this zone is only overaged due to the heat of welding and becomes harder with the loss of

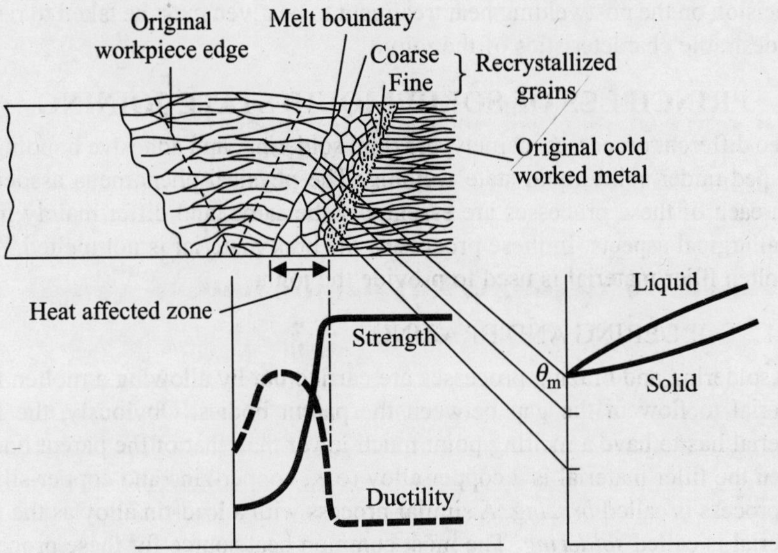

Fig. 5.16 Characteristics of welded joint in pure metal.

strength. Hence, the strength and ductility variation near the joint is as shown in Fig. 5.17.

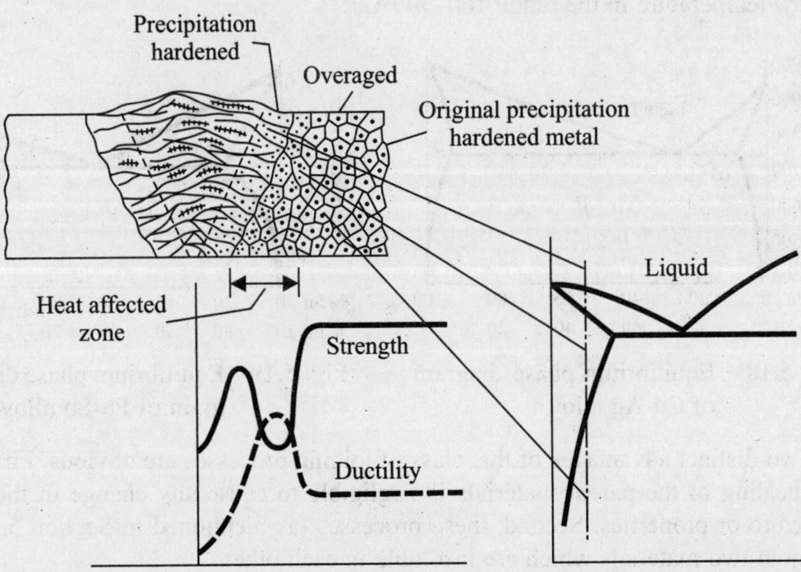

Fig. 5.17 Characteristics of welded joint in precipitation hardened alloy.

The two examples we have considered clearly demonstrate that various types of metallurgical changes are possible during welding, particularly for complex alloys. These changes are governed by the non-equilibrium metallurgy of such alloys, and must be clearly understood to yield a satisfactory fusion weld. Also, a decision on the postwelding heat treatment to be given must be taken to restore the desirable characteristics of the joint.

5.4 PRINCIPLES OF SOLID/LIQUID STATE JOINING

Three different processes, namely, brazing, soldering, and adhesive bonding are grouped under solid/liquid state welding. The physical phenomena associated with each of these processes are essentially the same, and differ mainly in the metallurgical aspects. In these processes, the bulk material is not melted. Also, a molten filler material is used to provide the joint.

5.4.1 SOLDERING AND BRAZING

The soldering and brazing processes are carried out by allowing a molten filler material to flow in the gap between the parent bodies. Obviously, the filler material has to have a melting point much lower than that of the parent bodies. When the filler material is a copper alloy (e.g., copper-zinc and copper-silver), the process is called *brazing*. A similar process with a lead-tin alloy as the filler material is called *soldering*. The most common heat source for these processes is electrical resistance heating.

The copper-silver phase diagram is shown in Fig. 5.18. Here, we see that by varying the composition of the alloy the flow temperature can be controlled in the range 750–980°C. Similarly, the lead-tin phase diagram (Fig. 5.19) indicates a flow temperature in the range 160–300°C.

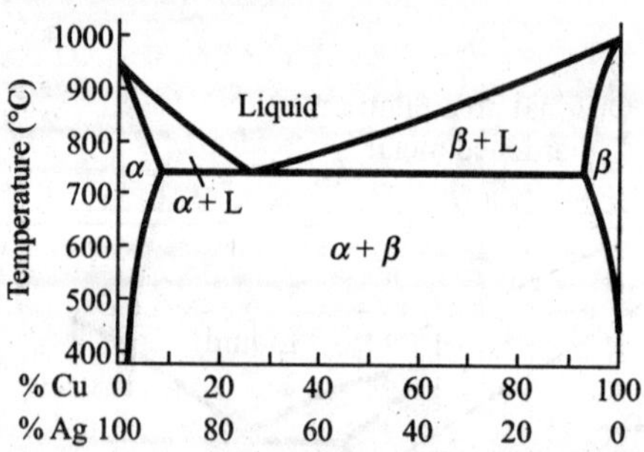

Fig. 5.18 Equilibrium phase diagram of Cu-Ag alloy.

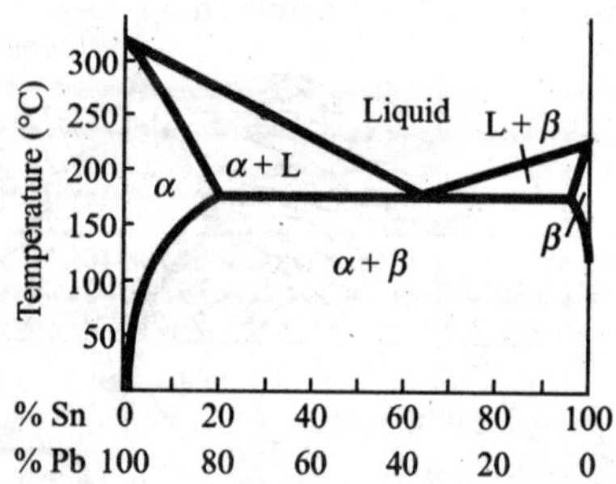

Fig. 5.19 Equilibrium phase diagram of Pb-Sn alloy.

Two distinct advantages of this class of joining processes are obvious. First, the heating of the parent materials is negligible to cause any change in their structure or properties. Second, these processes (as mentioned in Section 5.1) can join two materials which are insoluble in each other.

To produce a perfect joint, the entire gap between the parent bodies must be filled up by the filler material. This is achieved essentially through a capillary action. Thus, the spreading and the wetting capacities of the filler liquid play a predominant role towards producing a satisfactory joint. In some cases, these properties even dictate the composition of the alloy to be used. In what follows, we shall briefly discuss the adhesion between a liquid and a solid surface.

The work of adhesion between a liquid and a solid is governed by the surface free energy according to the relation

$$W_{adh} = \gamma_{S/V} + \gamma_{L/V} - \gamma_{S/L}, \tag{5.17}$$

where $\gamma_{S/V}$ is the surface free energy of the solid/vapour interface, $\gamma_{S/L}$ is the surface free energy of the solid/liquid interface, and $\gamma_{L/V}$ is the surface free energy of the liquid in equilibrium with its vapour. The surface free energy, if expressed in J/m^2, is numerically equal to the surface tension force expressed as N/m. Now, for a liquid drop in contact with a solid surface (Fig. 5.20) having a contact angle α, we see that

$$\gamma_{S/V} = \gamma_{S/L} + \gamma_{L/V} \cos \alpha. \tag{5.18}$$

The condition for wetting is $\alpha = 0$ and this can be written as

$$\gamma_{S/V^0} = \gamma_{S/L} + \gamma_{L/V^0}, \tag{5.19}$$

where V^0 represents a saturated vapour. Thus, for a good wetting, $\gamma_{S/V}$ should be large. This explains the difficulty of soldering or brazing grey cast iron whose surface is contaminated with graphite having a very low surface energy.

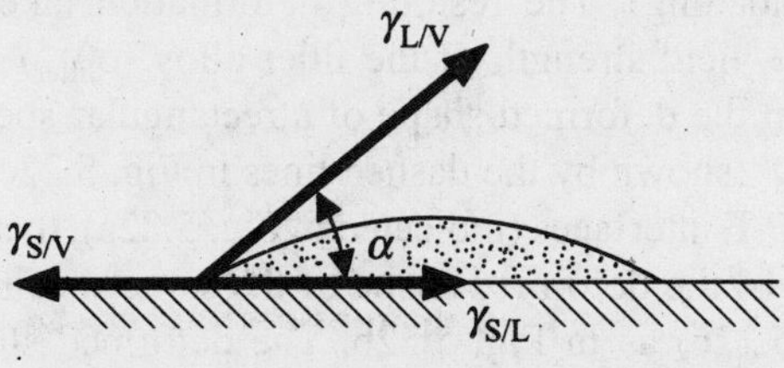

Fig. 5.20 Liquid drop in contact with solid surface.

The theory of capillary flow tells us two important facts, namely,

(i) the liquid drop rises to a greater height with a reduced gap, and

(ii) the rate of rise is slower with a reduced gap.

In fact, the optimum gap maintained, between the parts to be joined, is of the order of 0.1 mm.

The strength of a typically brazed joint varies with the thickness of the joint in a manner shown in Fig. 5.21. Here, we note that an optimum joint thickness (t_o) exists at which the strength is maximum. For a very large thickness of the joint, the strength of the joint approaches that of the brazing alloy (σ_{filler}). Below the optimum thickness, the entire joint is not filled up due to the strong resistance against the capillary flow. Hence, the strength is low due to the lack of a perfect fit.

It can be seen that the maximum joint strength is higher than the strength of the filler material; this can be explained as follows. Usually, the yield stress of the filler material is lower than that of the parent materials. Let the joint be subjected to a tensile loading of stress σ_1 (Fig. 5.22a). With increasing value of σ_1, the filler material tends to yield and the P (parent material)-F (filler

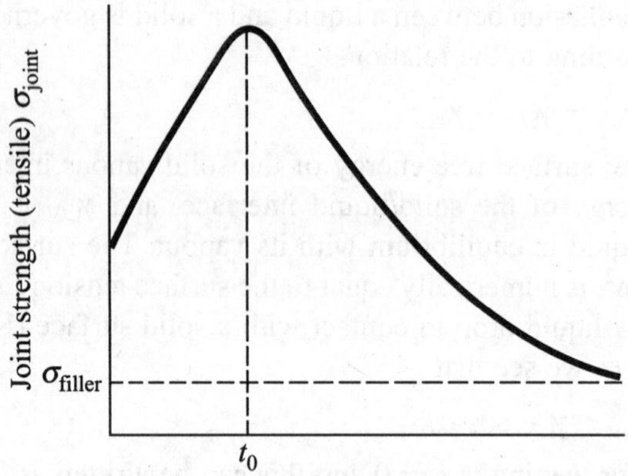

Fig. 5.21 Variation of strength of brazed joint with joint thickness.

material) interfaces tend to resist the yielding (because the parent material does not yield at this loading). The resulting deformation takes the shape shown in Fig. 5.22b. The yield strength of the filler alloy (σ_{filler}) is determined, by a uniaxial test, when the deformed shape of a rectangular specimen *ABCD* takes the shape *A′B′C′D′* (shown by the dashed lines in Fig. 5.22c). If *AD* and *BC* are considered as the P-F interfaces (indicated in Fig. 5.22a), then, in the presence of the parent materials, the deformed shape of the same specimen takes the shape *A″B″C″D″* (Fig. 5.22c) as in Fig. 5.22b. The deformed shape *A″B″C″D″* can

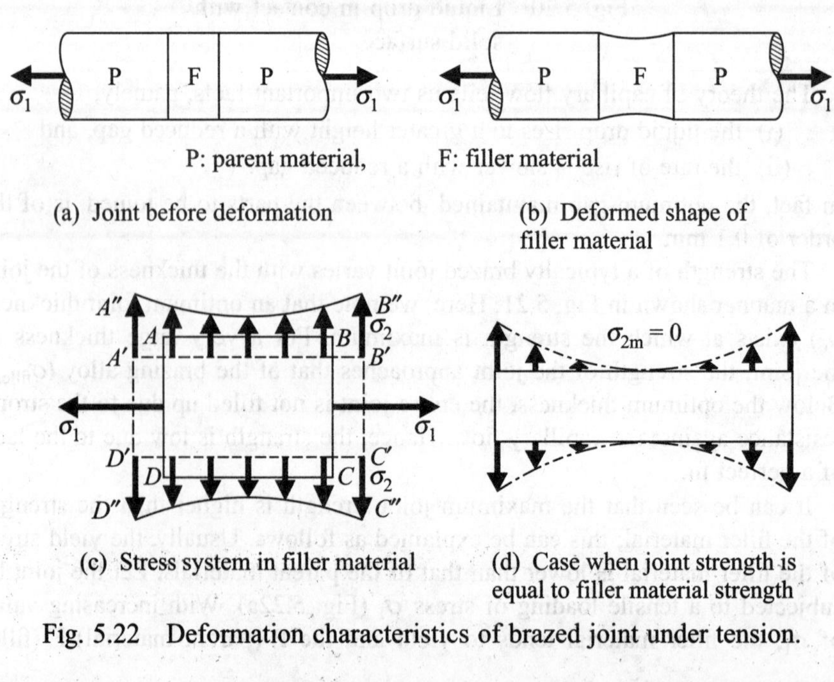

Fig. 5.22 Deformation characteristics of brazed joint under tension.

be obtained from *ABCD* by the simultaneous application of a constant tensile stress (σ_1) on the faces *AD* and *BC*, together with the application of a varying lateral tensile stress (σ_2) on the faces *AB* and *CD*. The magnitude of the lateral stress σ_2 is maximum at *A* and *B* with the minimum (σ_{2m}) occurring at the midpoint of *AB*.

Now, as σ_1 is increased, the joint fails at the midpoint of *AB*. Using the maximum shear stress theory, we find the joint fails if

$$\sigma_1 - \sigma_{2m} = \sigma_{filler}$$

or the strength of the joint

$$\sigma_{joint} = \sigma_1 = \sigma_{filler} + \sigma_{2m}. \tag{5.20}$$

Equation (5.20) clearly shows that σ_{joint} is greater than σ_{filler}. As the joint thickness increases, σ_{2m} tends to zero (Fig. 5.22d) and σ_{joint} approaches the value of σ_{filler}.

5.4.2 ADHESIVE BONDING

The adhesive bonding process is most commonly used in the aircraft and automobile industries where sheet metals are joined in various configurations. In it, unlike in brazing and soldering, there is no uncertainty about the flow of the adhesive in the joint. This is so because the adhesive is first applied on the surfaces and the joint is made with subsequent application of heat and pressure. However, this advantage should not be overemphasized because of the inherent low strength of the resulting joint.

Two types of bonding forces take part in an adhesive joint. Of these, one is the van der Waals force due to the constant movement of the positive and the negative charges of molecules (see Chapter 1). The other is the polar force, between the adhesive and the relatively brittle oxide film, due to the dipole adhesive molecules. This force is normally several orders of magnitude higher than the van der Waals force.

The coefficient of thermal expansion and the elastic properties of the adhesive normally differ very widely from those of the metals to be joined. Therefore, the mechanical properties of an adhesive joint need a careful examination. The factors that affect the strength and the other mechanical properties of such a joint are:

(i) The nature and dimensions of the joint. The strength of a lap joint increases with the overlapped area and reduces with the joint thickness.

(ii) The contact angle at the solid-liquid interface. In the joint, there is always a possibility of an air bubble getting trapped in the metal cavity. The radius *R* of the bubble is determined by the size and shape of the cavity together with the contact angle ϕ (Fig. 5.23). It can be shown that the tensile stress necessary to propagate a crack from the cavity is given by[1]

[1]Lancaster, J.F., Metallurgy of Welding, Brazing and Soldering, Allen and Unwin, London, 1980.

$$\sigma_t = [\frac{\pi E_a \gamma_a}{2R(1-\nu_a^2)}]^{1/2}, \tag{5.21}$$

where E_a, ν_a, and γ_a are Young's modulus, Poisson's ratio, and the surface energy, respectively, of the adhesive.

Moreover, the stress concentration factor due to the formation of a meniscus, shown in Fig. 5.24, depends on the contact angle. The maximum stress concentration factor increases continuously from about 1 to 2.3 as ϕ increases

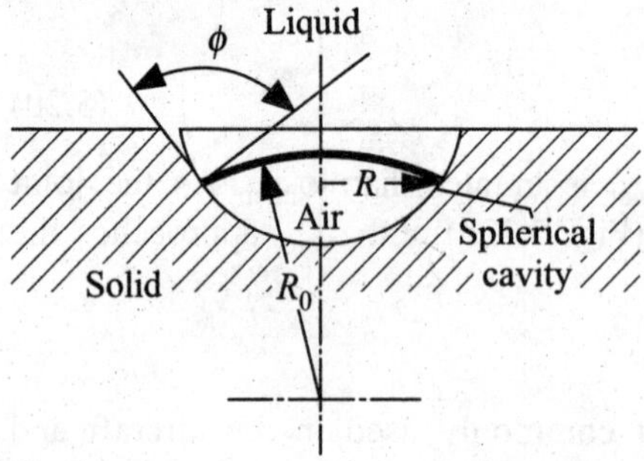

Fig. 5.23 Trapped air bubble in metal cavity.

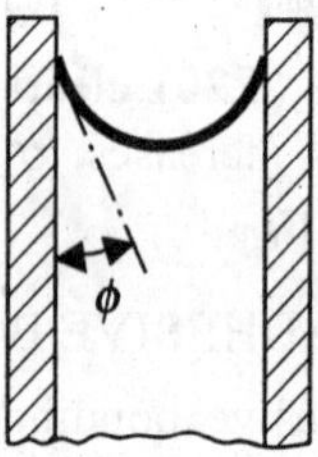

Fig. 5.24 Meniscus formation.

from 0 to 90°. Up to a value $\phi < 30°$, the maximum stress concentration factor and the failure both occur in the body of the adhesive. Beyond $\phi = 30°$, the maximum stress concentration factor, and consequently the failure, shift to the interface. The latter situation normally results in lower strength. In actual practice, the adhesive is extruded (by means of externai pressure) from the joint to prevent the formation of any meniscus.

(iii) The residual stress and the stress concentration. The adhesive in the joint shrinks due to both cooling and polymerization. This shrinkage results in residual tensile stress in the adhesive, and consequently reduces the strength of the joint.

If a lap joint of two metal sheets of equal thickness t (see Fig. 5.25) is considered, it can be shown that the stress concentration factor in shear due to dissimilar materials is given by[1]

$$K = (\frac{c^2 G_a}{2Ett_a})^{1/2}, \tag{5.22}$$

where G_a is the shear modulus of the adhesive and E is Young's modulus of the metal. Other factors, e.g., shape and bending of the adhesive, also give rise to stress concentration.

EXAMPLE 5.4 Determine the maximum shear the lap joint shown in Fig. 5.25

[1]Lancaster, J.F., Metallurgy of Welding, Brazing and Soldering, Allen and Unwin, London, 1980.

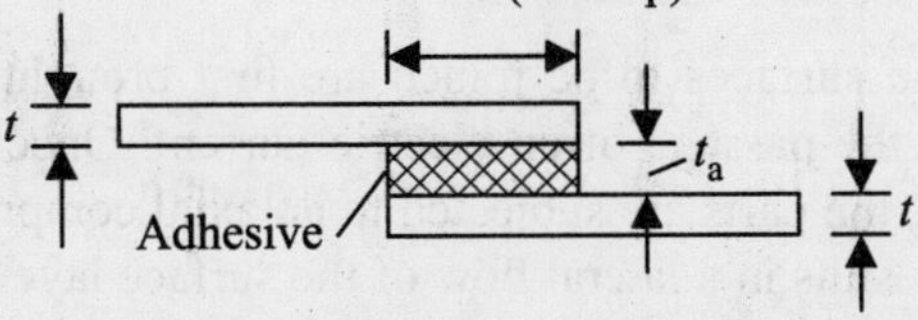

Fig. 5.25 Lap joint using adhesive.

can withstand. The joint is made between two aluminium sheets of 1.2 mm thickness with an adhesive thickness of 0.25 mm. The overlapped length is 12 mm. Given $E_{Al} = 703$ N/mm^2, $G_a = 11.9$ N/mm^2, and the ultimate shear stress of the adhesive $\tau_a = 0.6$ N/mm^2.

SOLUTION From the given data, we know $t = 1.2$ mm, $t_a = 0.25$ mm, and $c = 12$ mm. Substituting these values in equation (5.22), we obtain

$$K = \left(\frac{12 \times 12 \times 11.9}{2 \times 703 \times 1.2 \times 0.25}\right)^{1/2} = 2.19.$$

Hence, the maximum shear stress the joint can withstand is

$$\tau_j = \frac{\tau_a}{K} = \frac{0.6}{2.19} = 0.274 \text{ N/mm}^2.$$

The adhesives used in the bonding processes include epoxy, polyvinyl butyrate, and nitrile rubber. One successful way of adhesive bonding is to first apply a coating of phenolic formaldehyde (in a suitable solvent) and then sprinkle a powder of polyvinyl formaldehyde on this coating. The surfaces can then be clamped together under pressure and cured to a specific temperature.

5.5 VARIOUS JOINING PROCESSES

In this section, we shall briefly consider the various joining processes. In so doing, we shall not go into the minute details of the technology of each process. Instead, we shall attempt to correlate the science and mechanisms (so far discussed) with the actual joining operations.

5.5.1 SOLID PHASE WELDING AT ELEVATED TEMPERATURE

Depending on the nature of the heat source and the method of application of the pressure, the solid phase welding processes can be classified as (i) forge welding, (ii) butt welding, (iii) oxyacetylene pressure welding, (iv) flash butt welding, and (v) friction welding.

Forge Welding

In forge welding, the parts to be welded are heated in a furnace and then hammered together to form the weld. In the case of tubes, the welding pressure is applied by the forging rolls. The amount of bulk deformation varies in a wide range.

Butt Welding

In butt welding, the surfaces to be joined are first brought in contact. Then, they are heated by the passage of an electric current. Once the required temperature is attained, the parts are subjected to an axial compression (Fig. 5.26). This compression results in a lateral flow of the surface layers (e.g., oxide) and brings the clean metal surfaces in contact. The applied pressure is controlled accurately; it is held almost constant with a sharp rise near the end of the operation.

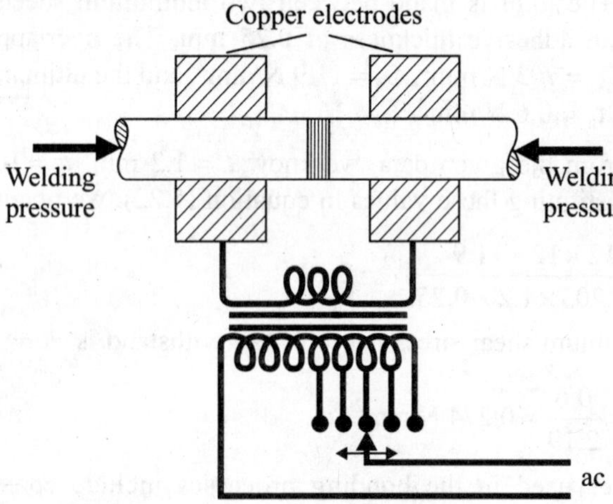

Fig. 5.26 Scheme of electric butt welding.

Oxyacetylene Pressure Welding

The oxyacetylene pressure welding process is similar to butt welding. The only difference is that here the joint is heated by an oxyacetylene ring burner. In both the processes, a bulk deformation of the order of 50% may take place.

Flash Butt Welding

In flash butt welding, the parts are brought in contact with a light pressure. The interface is heated by the passage of an electric current, as in butt welding. However, the heating is continued till the interface melts when the dies (Fig. 5.27) are brought closer. Thus, the liquid metal, along with the oxide film, is driven into the die cavity. This brings the *solid*, clean metal faces in contact under high pressure to make the weld. It should be noted that the form of the die cavity eases both the flow of the liquid metal into it and the final trimming off of the flash.

Friction Welding

In the friction welding process, the parts to be welded are kept in contact and rotated relative to each other. The interface is heated up due to friction. After the desired temperature is attained, an axial pressure is applied to complete the welding. After the welding is complete, the welded parts rotate together as one piece until stopped. One obvious limitation of this process is that the parts to

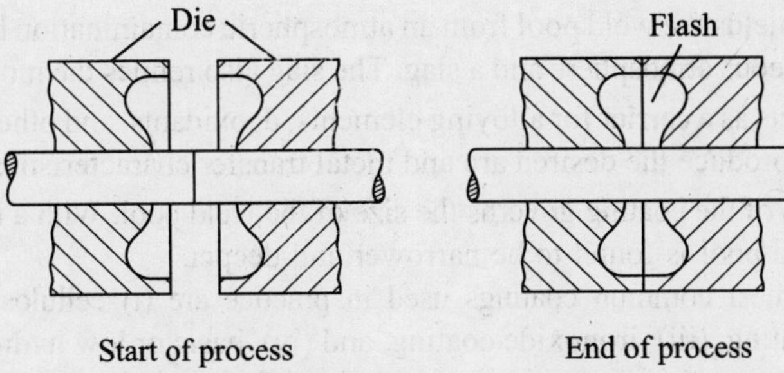

Fig. 5.27 Flash butt welding.

be welded must have a rotational symmetry. It should also be noted that the interface never melts as the softening of the material reduces the friction, and consequently the heat input, making the process self-regulating.

5.5.2 ARC WELDING

Each of the following fusion welding processes uses an electric arc as the source of heat:

(i) Arc welding with coated electrodes.
(ii) Tungsten-inert gas welding.
(iii) Consumable metal-inert gas welding.
(iv) Submerged arc welding.

Though all these processes use an electric arc as the source of heat, they differ in many aspects, including the power source and the arc characteristics. This will be apparent from the discussion that follows.

Arc Welding with Coated Electrodes

Manual arc welding with coated electrodes is the most common welding process (Fig. 5.28). The process is applicable for almost all metals and alloys with the

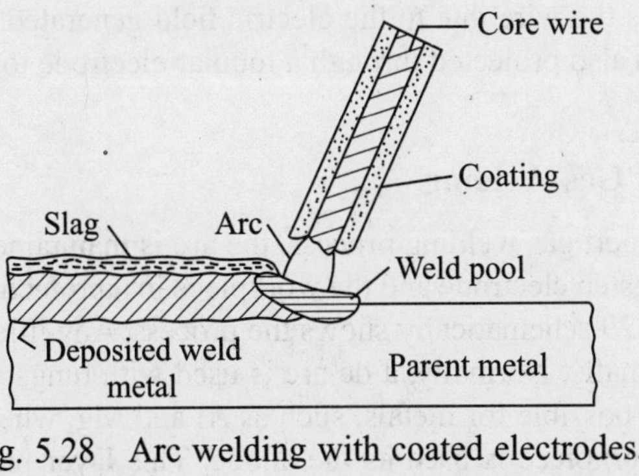

Fig. 5.28 Arc welding with coated electrodes.

exception of pure copper, pure aluminium, and some low melting point and reactive metals. The coating serves the following purposes:

(i) It shields the weld pool from an atmospheric contamination by creating a suitable gaseous atmosphere and a slag. The slag also refines the molten metal.

(ii) It acts as a carrier for alloying elements, deoxidants, and other elements necessary to produce the desired arc and metal transfer characteristics.

The thickness of the coating governs the size of the weld pool. With a high thickness, the weld pool is found to be narrower and deeper.

The four most common coatings used in practice are (i) cellulosic coating, (ii) rutile coating, (iii) ironoxide coating, and (iv) basic or low hydrogen coating. The first two coatings are similar, the only difference being that in the rutile coating the content of TiO_2 (rutile) is somewhat more. For the same thickness, the cellulosic coating penetrates more than the rutile coating. On decomposition, the organic (cellulose) material generates hydrogen and carbon monoxide at a sufficient rate to form the protective gaseous atmosphere. Since the rate of hydrogen evolution is high, the chances of an embrittlement of the joint due to hydrogen are more.

In the ironoxide coating, the gas evolved is less than that in the cellulosic and rutile coatings, and the protection of the weld pool depends heavily on the slag.

The basic or low hydrogen coating avoids an embrittlement due to hydrogen. Here, the coating contains mainly calcium carbonate and calcium fluoride. On decomposition, these generate a CO-CO_2 mixture which serves as the protective atmosphere. The rate of gas evolution is substantially lower in this case. Moreover, since CO-CO_2 reduces less (than H_2), a short arc length is maintained for full protection.

The nature of coating also controls the mode of metal transfer. In general, the cellulosic, rutile, and basic coatings give rise to a short circuit transfer, whereas the ironoxide coating results in a projected transfer.

The coating on the electrode can be applied in various ways. The wire electrode can be dipped in the paste of flux which, on drying, results in the coating. The coating can also be applied by extrusion (as in the cladding operation—see Chapter 3). Sometimes, the base wire is fed through a magnetic flux which is attracted towards the wire due to the electric field generated on the passage of current. A flux is also projected through a tubular electrode to serve the purpose of the coating.

Tungsten-Inert Gas Welding

In the tungsten-inert gas welding process, the arc is maintained between a non-consumable tungsten electrode and the workpiece in a protective inert gas atmosphere. Figure 5.29 schematically shows the process. Any filler material needed is supplied externally. Normally, a dc arc is used with tungsten as the negative pole. This is not possible for metals, such as Al and Mg, where the oxide layer persists if the workpiece is used as the anode. This layer prevents the formation of the weld pool. The mobile cathode spot can disperse the oxide layer but excessive heat is generated at the tungsten electrode if this is used as the anode. Hence, an ac arc is used for such materials. To avoid the melting of the electrode, thorium or zirconium is added to the tungsten (to increase the melting point).

Argon is most commonly used to provide the inert atmosphere. Nitrogen is sometimes used for welding copper. This is a special type, costly welding process

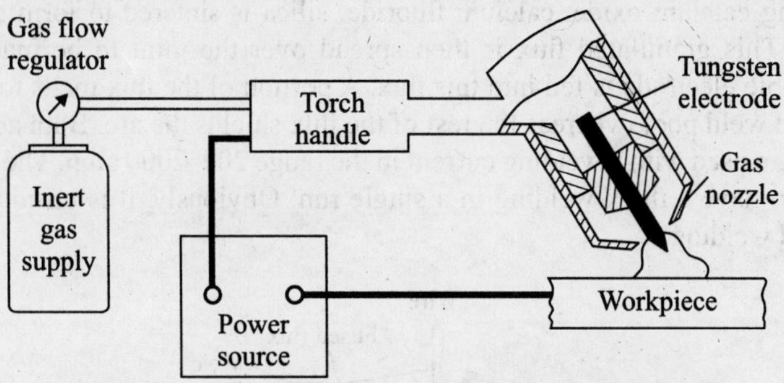

Fig. 5.29 Tungsten-inert gas welding equipment.

used only for aluminium, magnesium, and other reactive metals. To prevent the possible little contamination, an argon deoxidant is added to the filler rod.

Metal-Inert Gas Welding

Figure 5.30 schematically represents a metal-inert gas welding process. Here, the arc is maintained between a consumable electrode and the workpiece in an inert gas atmosphere. The coiled electrode wire is fed by drive rolls as it melts away at the tip. Except for aluminium, a dc source is used with the consumable electrode as the positive terminal. The difference in this respect with the tungsten-inert gas welding should be noted. For welding steel, a shielding is provided by CO_2 for lower cost. Normally, a high current density in the electrode (of the order of 10,000 amp/cm^2) is used so that a projected type of metal transfer results. The welding current is in the range 100–300 amp. The process is primarily meant for thick plates and fillet welds.

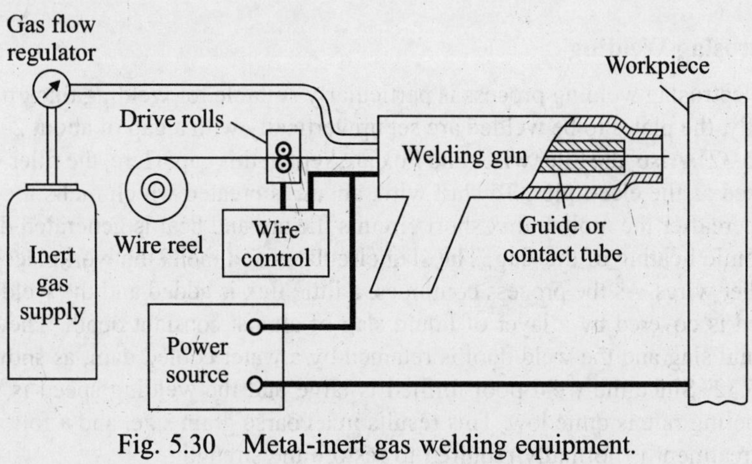

Fig. 5.30 Metal-inert gas welding equipment.

Submerged Arc Welding

In submerged arc welding, the arc is maintained underneath a mass of fusible, granular flux. The process is schematically shown in Fig. 5.31. First, the flux containing calcium oxide, calcium fluoride, silica is sintered to form a coarse powder. This granulated flux is then spread over the joint to be made. The consumable electrode is fed into this flux. A portion of the flux melts to protect the liquid weld pool, whereas the rest of the flux shields the arc. Both ac and dc sources are used with a welding current in the range 200–2000 amp. The process is used to have a thick welding in a single run. Obviously, it is unsuitable for overhead welding.

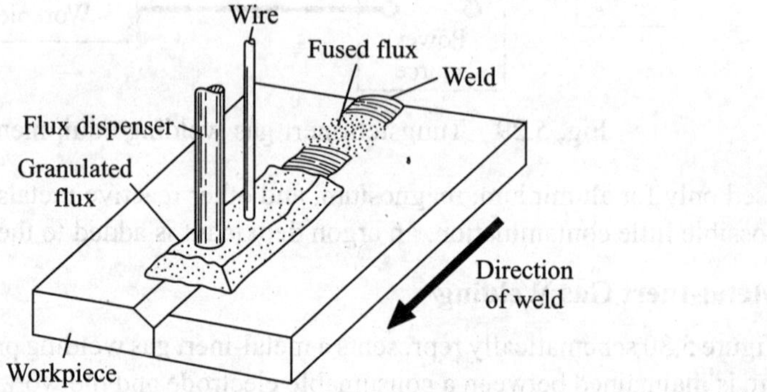

Fig. 5.31 Submerged arc welding.

5.5.3 RESISTANCE WELDING

The following processes use an ohmic resistance heating as the source of heat:

(i) Electroslag welding.
(ii) Spot welding.
(iii) Projection welding.
(iv) Seam welding.

Electroslag Welding

The electroslag welding process is particularly suitable for welding thick plates. Initially, the plates to be welded are set up vertically with a gap of about 2–3 cm (Fig. 5.32). Also, the filler wires and flux are kept in this gap. Here, the filler wires are used as the electrodes. To start with, an arc is created which melts the flux, and thereafter the molten flux short circuits the arc and heat is generated due to the ohmic heating of the slag. The slag circulates and melts the workpieces and the filler wires. As the process continues, a little flux is added and the weld pool formed is covered by a layer of liquid slag of almost constant depth. The layer of liquid slag and the weld pool is retained by a water cooled dam, as shown in Fig. 5.32. Since the weld pool formed is large and the welding speed is slow, the cooling rate is quite low. This results in a coarse grain size, and a follow up heat treatment is normally required to restore the strength.

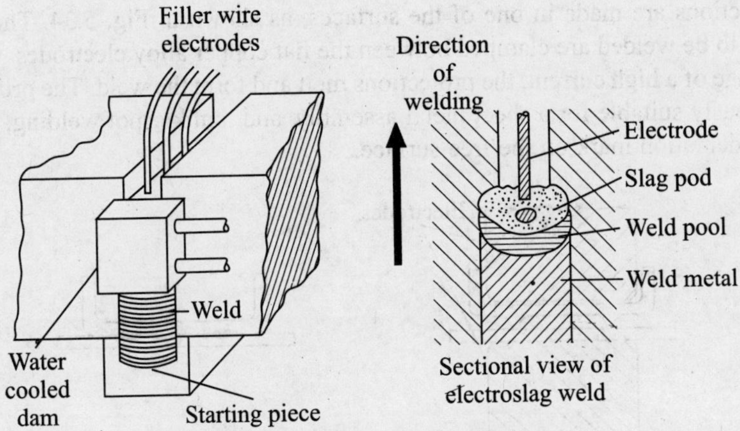

Fig. 5.32 Electroslag welding.

Spot Welding

In spot welding, the parts to be joined are normally overlapped (Fig. 5.33). The workpieces are clamped between two water cooled copper electrodes. On the passage of a high transient current, the interface melts over a spot and forms the weld. The cooling at the electrode limits the size of the spot. A very high current (40 amp or more) is needed for a very short duration (of the order of a fraction of a second) to complete the welding. The interfaces to be joined are initially cleaned by various methods, including scratch brushing and vapour degreasing. A spot weld normally contains some porosity at the weld centre, which, unless excessive, is harmless. The spot welding process is difficult to use for highly conductive materials such as aluminium and magnesium. If a series of spot welds are to be made, obviously then a higher current is necessary for each subsequent spot in view of the short circuiting provided by the preceding welds.

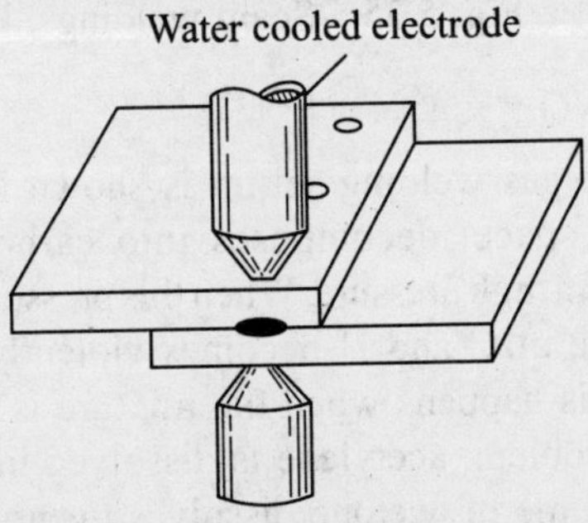

Fig. 5.33 Resistance spot welding.

Projection Welding

Projection welding is a variation of the spot welding process in which small projections are made in one of the surfaces, as shown in Fig. 5.34. Then, the parts to be welded are clamped between the flat copper alloy electrodes. On the passage of a high current, the projections melt and form the weld. The process is obviously suitable for a sheet metal assembly, and, unlike spot welding, leaves no indentation mark on the free surface.

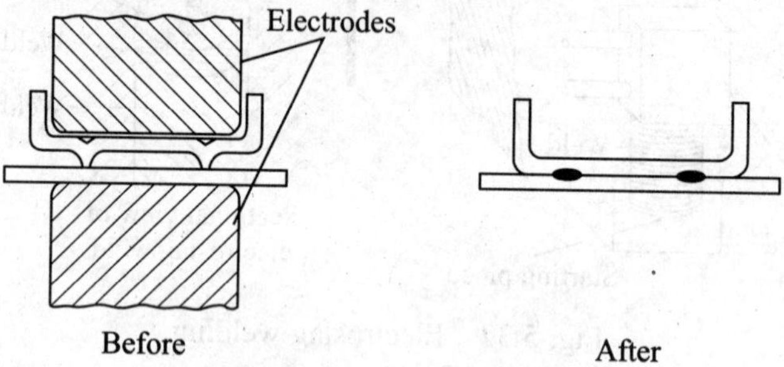

Fig. 5.34 Projection welding.

Seam Welding

Seam welding is a continuous spot welding process where the overlapped parts to be welded are fed through a pair of copper alloy electrodes to form a continuous seam (Fig. 5.35).

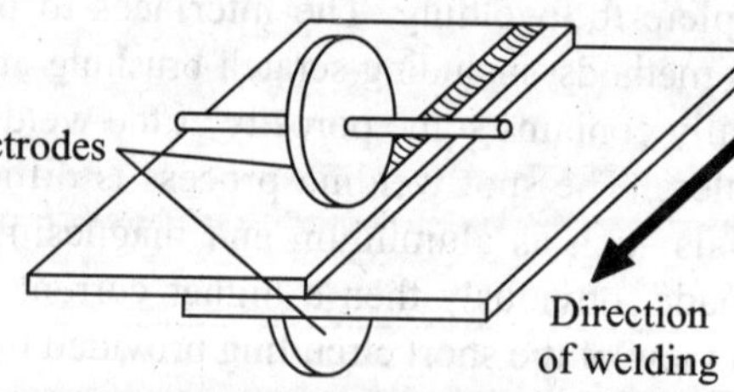

Fig. 5.35 Seam welding.

5.5.4 GAS WELDING

A typical oxyacetylene gas welding set-up is shown in Fig. 5.36. Acetylene, if kept in a confined space, decomposes into carbon and hydrogen. This decomposition results in a high pressure. When this pressure reaches a value around 0.2 N/mm^2, the mixture of C and H becomes violently explosive even in the absence of oxygen. This happens when the mixture is subjected to a spark or shock. To avoid this problem, acetylene is dissolved in acetone. At a pressure of 0.1 N/mm^2, one volume of acetone dissolves twenty volumes of acetylene, and the solubility increases almost linearly to 300 volumes at a pressure of 1.2 N/mm^2. An excess of oxygen or acetylene is used, depending on whether a decarburizing or a carburizing flame is desired. In the welding of brass, bronze, and copper-zinc-tin alloys, a decarburizing flame is used, whereas a carburizing flame is used for the welding of high carbon steel.

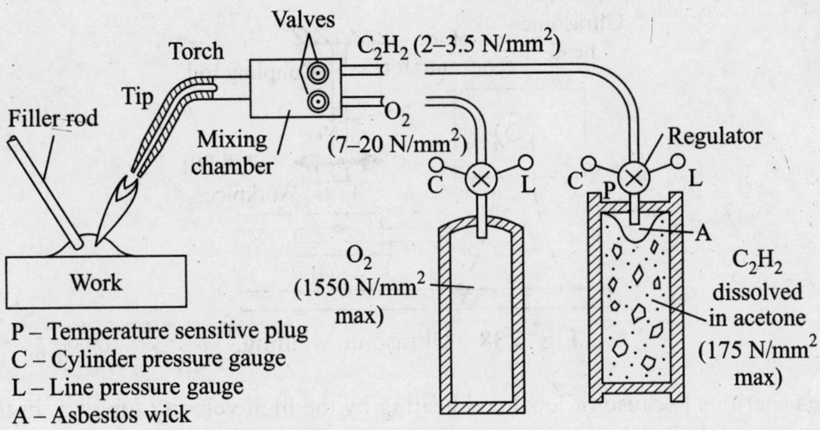

Fig. 5.36 Scheme of gas welding.

5.5.5 THERMIT WELDING

The thermit welding process utilizes a chemical heat source and is normally used for an on-site welding of rails (Fig. 5.37). The chemical reaction, already discussed in Section 5.3.1, indicates that the temperature obtained is much higher than is necessary and the reaction product is iron and not steel. To avoid these, the reaction mix is added with carbon, ferromanganese, and ferrosilicon to cool off the reaction and to produce steel at the end of the reaction. The reaction is started by burning a magnesium ribbon dipped in the reaction mix.

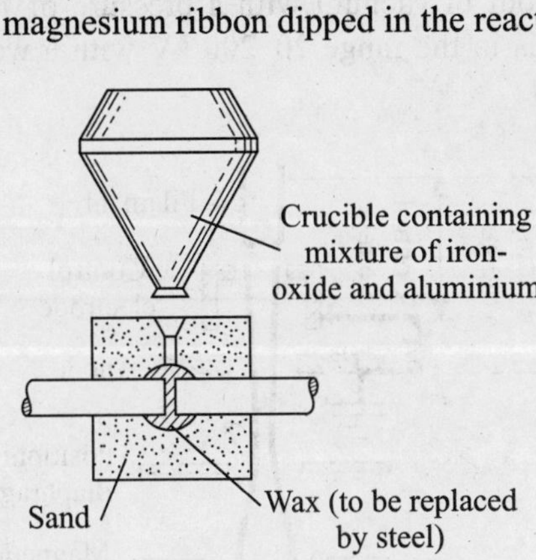

Fig. 5.37 Thermit welding.

5.5.6 ULTRASONIC WELDING

The ultrasonic welding process is used only for the welding of thin strips and foils. The core of a magnetostriction ultrasonic generator is coupled to the work through a bar having a suitably-shaped tip (Fig. 5.38). The tip applies a transverse pressure between the workpieces and the simultaneous application of ultrasonic vibration to the tip results in a spot weld. The welding takes place due to a combination of fracturing of the brittle oxide layers and softening of

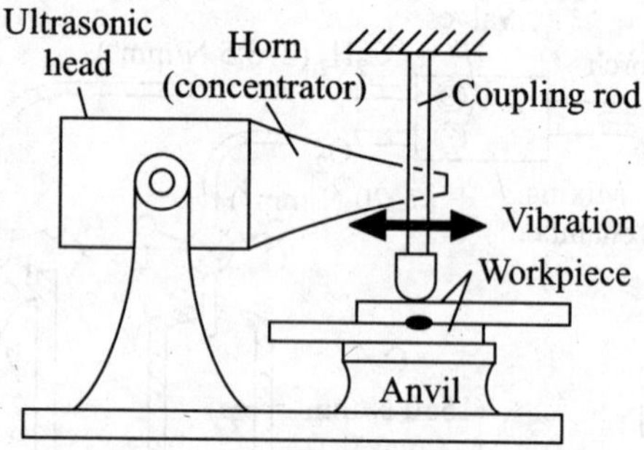

Fig. 5.38 Ultrasonic welding.

the asperities because of localized heating by the high velocity rubbing. In this process, no bulk heating, with the consequent bad effects (e.g., metallurgical changes and mechanical deformation), takes place.

5.5.7 ELECTRON BEAM WELDING

In electron beam welding, the heat for fusion is obtained from the kinetic energy of a dense beam of high velocity electrons (see Chapter 6). The electrons are emitted by a cathode and thereafter accelerated by a ring-shaped anode, and focussed by means of an electromagnetic field to finally impinge upon a very narrow area (a few microns in diameter) of the workpiece (Fig. 5.39). The entire operation is carried out in vacuum with a pressure of 10^{-3} mm of Hg. The accelerating voltage is in the range 20–200 kV with a welding current of the order of milliamperes.

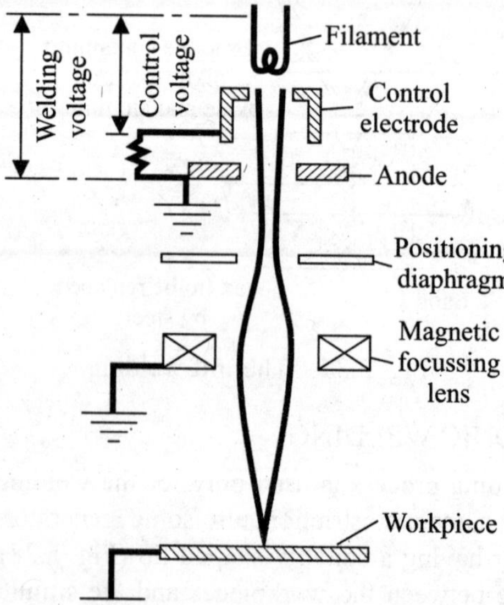

Fig. 5.39 Electron beam welding.

When the applied voltage crosses the critical value, the electron beam penetrates into the metal. Under such a condition, if the workpiece is traversed relative to the beam, an extremely narrow bead of weld is formed. The process is not suitable for a metal which vaporizes excessively or emits a lot of gas when melted. The process can be used to join dissimilar materials and reactive metals, and also for joints requiring a precise control of the weld profile and penetration (as in the fabrication of gas turbine parts).

5.5.8 LASER BEAM WELDING

A laser beam, instead of a highly focussed electron beam, can also be used for welding. A major advantage in this process is that the operation need not be carried out in vacuum. A laser beam is capable of producing a power density as high as 10^7 W/cm^2 (see Chapter 6).

5.5.9 EXPLOSIVE WELDING

The explosive welding process is used to join two plates, face to face. One of the workpieces, called the *target plate*, is held fixed (Fig. 5.40). The other

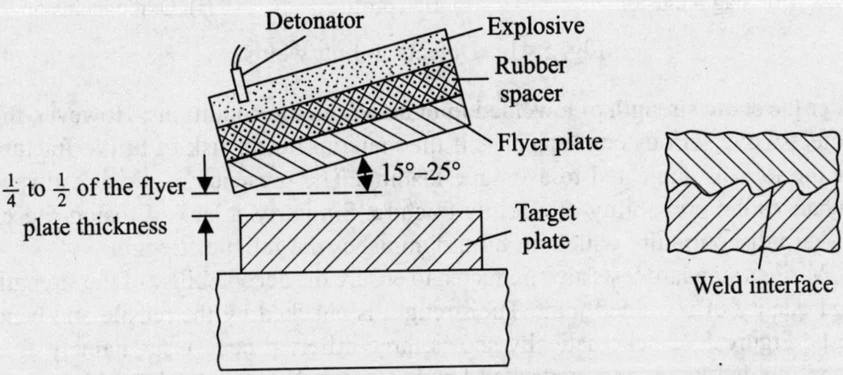

Fig. 5.40 Explosive welding.

one, called the *flyer plate*, is kept at an angle to the target plate. The minimum distance between the two plates is of the order of $\frac{1}{4}$ to $\frac{1}{2}$ of the flyer plate thickness. An explosive charge is kept on the top of the flyer plate with an intervening layer of rubber spacers. When the explosive charge is detonated, the flyer plate comes and hits the target plate with a huge velocity and the two plates are welded face to face. This process can be used to join dissimilar materials and the weld interface is seen to be wavy, as indicated in Fig. 5.40.

5.6 WELD DEFECTS AND INSPECTION

In a fusion weld, the defects often found include the lack of fusion, lack of penetration, inclusion of slag or oxide, presence of cracks, porosity, and undercut or excessive penetration (i.e., bad profile). These defects are shown diagrammatically for butt welds in Fig. 5.41. Such defects do not significantly

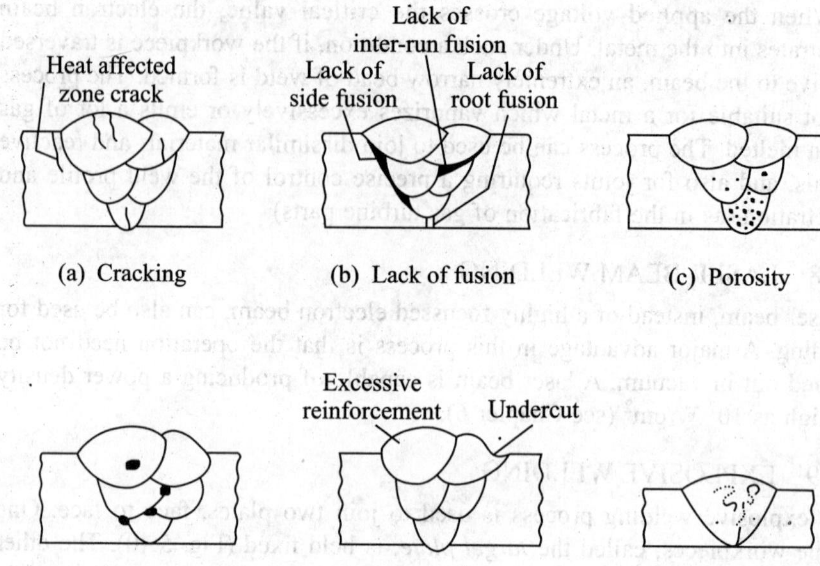

Fig. 5.41 Defects in butt welds.

alter the static strength of a welded joint under ductile conditions. However, the defects have serious consequences if the material runs a risk of brittle fracture or the joint is subjected to a fatigue loading. The presence of a crack always enhances the probability of a brittle fracture. Similarly, a lack of fusion causes a sharp discontinuity which, in turn, diminishes the fatigue strength.

Various standard tests are conducted to ensure the acceptability of the strength and quality of a welded joint. The strength is checked by the tensile and bend tests. Figure 5.42 schematically shows three different bend tests, namely, free bend, guided bend, and controlled bend. The bend test is conducted with two

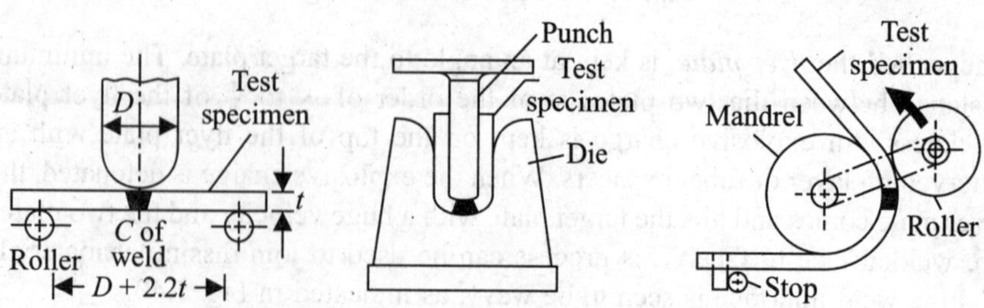

Fig. 5.42 Various bend tests for welded specimens.

specimens; in one specimen, the outside face of the weld is put to tension, and in the other specimen, the root is subjected to tension. The bend test is difficult to conduct on very thick plates. In such a situation, the side bend test is performed with a slice, 3–6 mm thick, cut at right angles to the plate surface and to the weld axis. The side bend test reveals a lack of side fusion very clearly but is less sensitive to a root defect than the normal bend test. The impact strength of a joint is tested by the usual methods after providing a notch in the joint.

Nondestructive inspection techniques are used to locate the internal defects and minute cracks in a welded joint. For the details on such techniques, viz., radiographic examination, ultrasonic testing, magnetic particle test, dye-penetrant testing, see Chapter 2 (Section 2.8).

5.7 EXERCISE PROBLEMS

5.1 The voltage-length characteristic of a dc arc is given by

$$V = (20 + 4l) \text{ volts},$$

where l = length of the arc in mm. During a welding operation, it is expected that the arc length will vary between 4 mm and 6 mm. It is desired that the welding current be limited to the range 450–550 amp. Assuming a linear power source characteristic, determine the open circuit voltage and the short circuit current of the power source.

5.2 Consider the voltage-length arc characteristic described in Exercise 5.1. Let the power source characteristic be given by

$$\left(\frac{V}{V_0}\right)^2 + 2\left(\frac{I}{I_s}\right)^2 = 1,$$

where V_0 is the open circuit voltage and I_s is the short circuit current. If V_0 and I_s remain the same as obtained in Exercise 5.1, determine the range of welding current for the arc length variation mentioned in Exercise 5.1.

5.3 Refer to the power source characteristics stated in Exercises 5.1 and 5.2. Of these, which is suitable for (i) manual operation? (ii) semiautomatic operation? Justify your answer.

5.4 Two different pairs of sheets of the same material have to be spot welded. In one pair, there are 25 bridges per cm^2 and the average radius of each bridge is 0.1 mm. The other pair of sheets contains 50 bridges per cm^2 with the same average radius of each bridge. Determine the ratio of the voltages to be applied in these two cases in order to generate the same rate of heating per unit area.

5.5 Solve Example 5.3 by assuming the heat flow to be three-dimensional (i.e., with a point source).

5.6 While making a lap joint of two thin sheets, using the same amount of adhesive, two designs are proposed. In one design, a joint of thickness 0.3 mm with an overlap length of 12 mm is made. In the other design, a joint of thickness 0.2 mm with an overlap length of 18 mm is made. Estimate the ratio of the maximum shear stresses the two joints can withstand.

6 Unconventional Machining Processes

6.1 INTRODUCTION

With the development of technology, more and more challenging problems are faced by the scientists and technologists in the field of manufacturing. The difficulty in adopting the traditional manufacturing processes can be attributed mainly to the following three basic sources:

(i) New materials with a low machinability.

(ii) Dimensional and accuracy requirements.

(iii) A higher production rate and economy.

The many new materials and alloys that have been developed for specific uses possess a very low machinability. Producing complicated geometries in such materials becomes extremely difficult with the usual methods. Also, sometimes the combination of the material properties and the job dimensions is such that the use of the traditional processes becomes impossible. Examples of these types of jobs are machining a complicated turbine blade made of superalloys, and producing holes and slots (both through and blind) in materials such as glass and semiconductors. At times, the job becomes difficult because of the dimensional complications. So, drilling a noncircular hole or a microhole becomes problematic (and sometimes impossible) if the traditional processes are used. Apart from the situations cited, higher production rate and economic requirements may demand the use of nontraditional (or unconventional) machining processes.

To tackle such difficult jobs, two approaches are possible, viz., (i) a modification of the traditional processes (e.g., hot machining), and (ii) the development of new processes. In this chapter, we shall discuss the common unconventional machining processes. Such processes are becoming increasingly unavoidable and popular; therefore, a knowledge of these is essential for a mechanical engineer.

The basic objective of all machining operations is to remove the excess material to obtain the desired shape and size. These operations use various types of energies. Table 6.1 shows the possible machining processes using the different types of energies and various methods of material removal.

Since the unconventional machining processes differ widely in nature and characteristics, we shall not attempt a generalized introduction. Instead, in what follows, we shall take up and discuss each such process separately.

Table 6.1 Unconventional machining processes

Energy type	Mechanics of material removal	Energy source	Process
Mechanical	Plastic shear	Mechanical motion of tool/ job	Conventional machining
	Erosion	Mechanical/fluid motion	Abrasive jet machining (AJM)
			Ultrasonic machining (USM)
Electrochemical	Ion displacement	Electric current	Electrochemical machining (ECM)
Mechanical and electrochemical	Plastic shear and ion displacement	Electric current and mechanical motion	Electrochemical grinding (ECG)
Chemical	Corrosive reaction	Corrosive agent	Chemical machining (CHM)
	Fusion and vaporization	Electric spark	Electric discharge machining (EDM)
		High speed electrons	Electron beam machining (EBM)
Thermal		Powerful radiation	Laser beam machining (LBM)
		Ionized substance	Ion beam machining (IBM)
			Plasma arc machining (PAM)

6.2 ABRASIVE JET MACHINING (AJM)

In AJM, the material removal takes place due to the impingement of the fine abrasive particles. These particles move with a high speed air (or gas) stream. Figure 6.1 shows the process along with some typical parameters of the process. The abrasive particles are typically of 0.025 mm diameter and the air discharges at a pressure of several atmospheres.

6.2.1 MECHANICS OF AJM

When an abrasive particle impinges on the work surface at a high velocity,

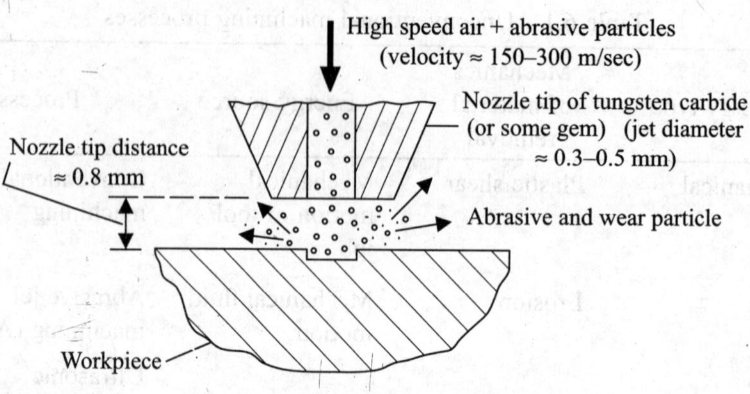

Fig. 6.1 Abrasive jet machining.

the impact causes a tiny brittle fracture and the following air (or gas) carries away the dislodged small workpiece particle (wear particle). This is shown in Figs. 6.2a and 6.2b. Thus, it is obvious that the process is more suitable when the work material is brittle and fragile. A model for estimating the material removal rate (mrr) is available[1]. The mrr due to the chipping of the work surface by the impacting abrasive particles is expressed as

$$Q = \chi Z d^3 v^{3/2} \left(\frac{\rho}{12 H_w}\right)^{3/4},$$

where Z is the number of abrasive particles impacting per unit time, d is the mean diameter of the abrasive grains, v is the velocity of the abrasive grains, ρ is the density of the abrasive material, H_w is the hardness of the work material (the flow stress), and χ is a constant.

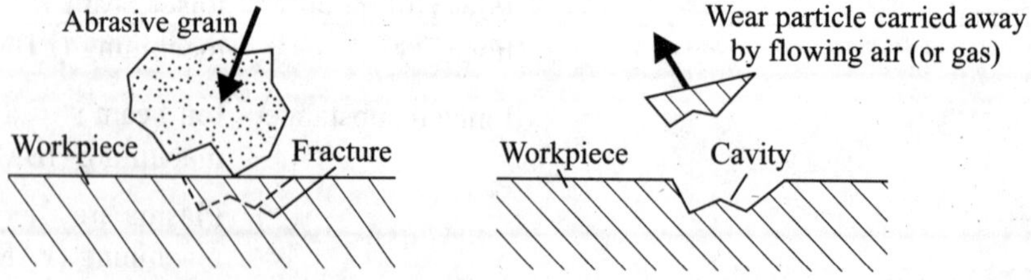

Fig. 6.2 Scheme of material removal in AJM.

6.2.2 PROCESS PARAMETERS

The process characteristics can be evaluated by judging (i) the mrr, (ii) the geometry of the cut, (iii) the roughness of the surface produced, and (iv) the rate of nozzle wear. The major parameters which control these quantities are

[1]Pandey, P.C. and Shan, H.S., Modern Machining Processes, Tata McGraw-Hill, New Delhi, 1980.

(i) the abrasive (composition, strength, size, and mass flow rate),
(ii) the gas (composition, pressure, and velocity),
(iii) the nozzle (geometry, material, distance from and inclination to the work surface).

We shall now discuss each of these parameters as also their effects.

The Abrasive

Mainly two types of abrasives are used, viz., (i) aluminium oxide and (ii) silicon carbide. However, generally aluminium oxide abrasives are preferred in most applications. The shape of these grains is not very important, but, for a satisfactory wear action on the work surface, these should have sharp edges. Al_2O_3 and SiC powders with a nominal grain diameter of 10–50 μm are available. The best cutting is achieved when the nominal diameter is between 15 μm and 20 μm. A reuse of the abrasive powder is not recommended as the (i) cutting capacity decreases after the first application, and (ii) contamination clogs the small orifices in the nozzle. The mass flow rate of the abrasive particles depends on the pressure and the flow rate of the gas. When the mass fraction of the abrasives in the jet (mixing ratio) increases, the mrr initially increases, but with a further increase in the mixing ratio, it reaches a maximum and then drops (Fig. 6.3a). When the mass flow rate of the abrasive increases, the mrr also increases (Fig. 6.3b).

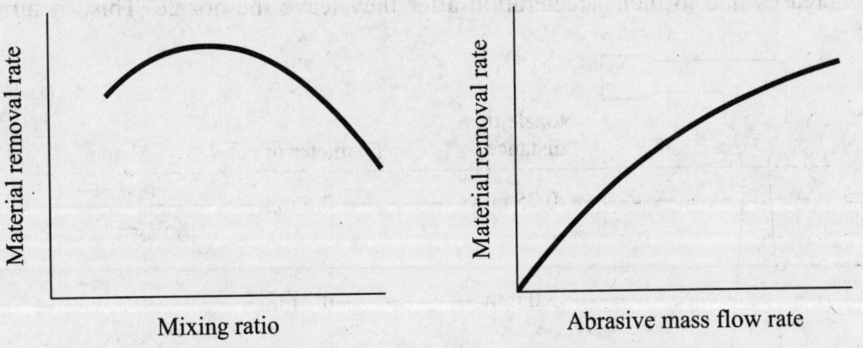

(a) Variation with mixing ratio

(b) Variation with abrasive mass flow rate

Fig. 6.3 Material removal rate characteristics in AJM.

The Gas

The AJM units normally operate at a pressure of 0.2 N/mm^2 to 1 N/mm^2. The composition of gas affects the mrr in an indirect manner as the velocity-pressure relation depends on this composition. A high velocity obviously causes a high mrr even if the mass flow rate of the abrasive is kept constant.

The Nozzle

The nozzle is one of the most vital elements controlling the process characteristics. Since it is continuously in contact with the abrasive grains flowing at a high speed, the material must be very hard to avoid any significant wear. Normally, WC or sapphire is used. For a normal operation, the cross-sectional area of the orifice is between 0.05 mm^2 and 0.2 mm^2. The shape of the orifice can be either circular or rectangular. Two typical nozzle tips and heads are shown in Fig. 6.4. The average life of a nozzle is very difficult to ascertain. A WC nozzle lasts between 12 hr and 30 hr, whereas a sapphire nozzle lasts for 300 hr approximately.

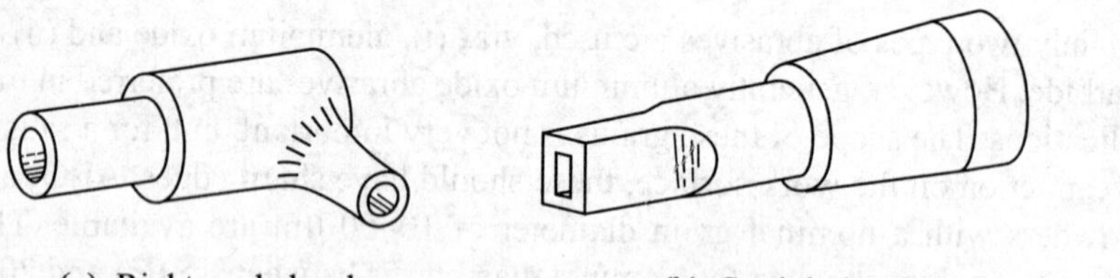

(a) Right angle head (b) Straight head

Fig. 6.4 Types of nozzle heads.

One of the most important factors in AJM is the distance between the work surface and the tip of the nozzle, normally called the *nozzle tip distance* (NTD). The NTD affects not only the mrr from the work surface but also the shape and size of the cavity produced. Figure 6.5 shows the effect of NTD. When the NTD increases, the velocity of the abrasive particles impinging on the work surface increases due to their acceleration after they leave the nozzle. This, in turn,

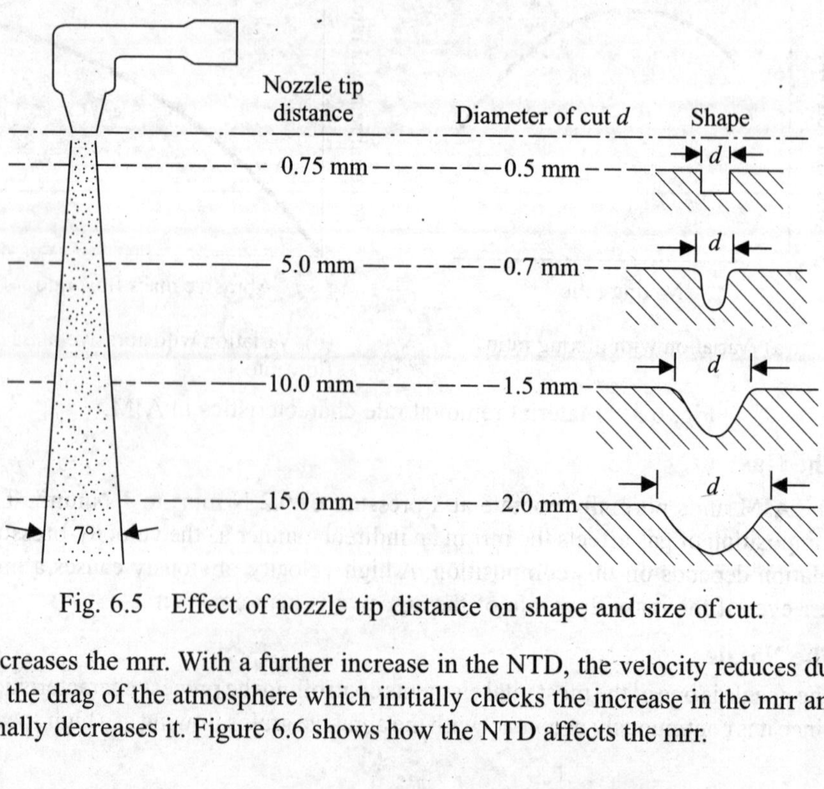

Fig. 6.5 Effect of nozzle tip distance on shape and size of cut.

increases the mrr. With a further increase in the NTD, the velocity reduces due to the drag of the atmosphere which initially checks the increase in the mrr and finally decreases it. Figure 6.6 shows how the NTD affects the mrr.

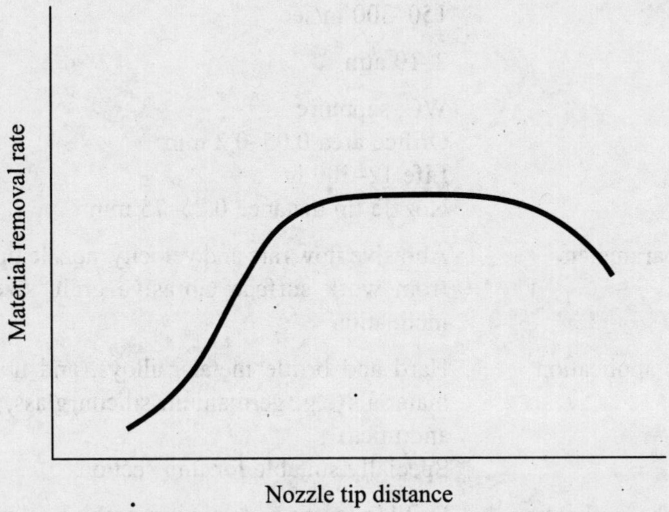

Fig. 6.6 Effect of nozzle tip distance on material removal rate.

6.2.3 ABRASIVE JET MACHINES

The abrasive jet machines are manufactured and marketed by a single manufacturer (namely, S.S. White Co., New York) under the name "Airbrasive". Figure 6.7 shows the principal features of an abrasive jet machine.

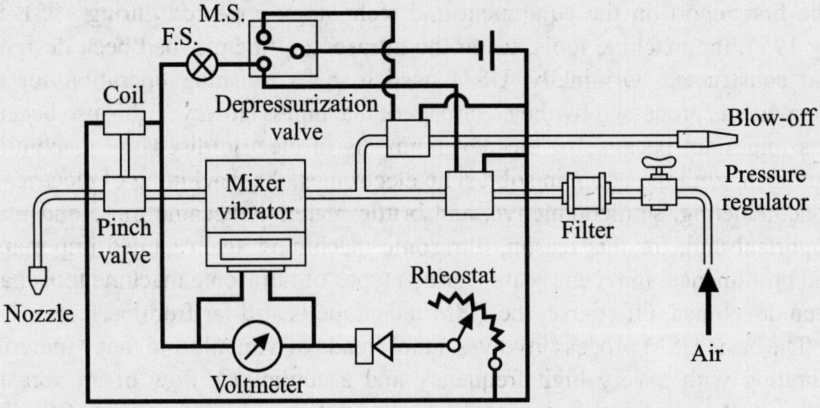

Fig. 6.7 Schematic diagram of abrasive jet machining equipment.

6.2.4 SUMMARY OF AJM CHARACTERISTICS

Mechanics of material removal	Brittle fracture by impinging abrasive grains at high speed
Media	Air, CO_2
Abrasives	Al_2O_3, SiC 0.025 mm diameter, 2–20 g/min, nonrecirculating

Velocity	150–300 m/sec
Pressure	2–10 atm
Nozzle	WC, sapphire Orifice area 0.05–0.2 mm^2 Life 12–300 hr Nozzle tip distance 0.25–75 mm
Critical parameters	Abrasive flow rate and velocity, nozzle tip distance from work surface, abrasive grain size and jet inclination
Materials application	Hard and brittle metals, alloys, and nonmetallic materials (e.g., germanium, silicon, glass, ceramics, and mica) Specially suitable for thin sections
Shape (job) application	Drilling, cutting, deburring, etching, cleaning
Limitations	Low metal removal rate (40 mg/min, 15 mm^3/min), embedding of abrasive in workpiece, tapering of drilled holes, possibility of stray abrasive action

6.3 ULTRASONIC MACHINING (USM)

The use of ultrasonics in machining was first proposed by L. Balamuth in 1945. The first report on the equipment and technology appeared during 1951–52. By 1954, the machine tools, using the ultrasonic principle, had been designed and constructed. Originally, USM used to be a finishing operation for the components processed by the electrospark machines. However, this use became less important because of the developments in electric discharge machining. But, then, with the boom in solid state electronics, the machining of electrically nonconducting, semiconductive, and brittle materials became more and more important and, for this reason, ultrasonic machining again gained importance and prominence. In recent years, various types of ultrasonic machine tools have been developed. Of course, the USM technique is still far from perfect.

The basic USM process involves a tool (made of a ductile and tough material) vibrating with a very high frequency and a continuous flow of an abrasive slurry in the small gap between the tool and the work surface (Fig. 6.8). The tool is gradually fed with a uniform force. The impact of the hard abrasive grains fractures the hard and brittle work surface, resulting in the removal of the work material in the form of small wear particles which are carried away by the abrasive slurry. The tool material, being tough and ductile, wears out at a much slower rate.

6.3.1 MECHANICS OF USM

The physics of ultrasonic machining is neither complete nor uncontroversial. The reasons of material removal during USM are believed to be

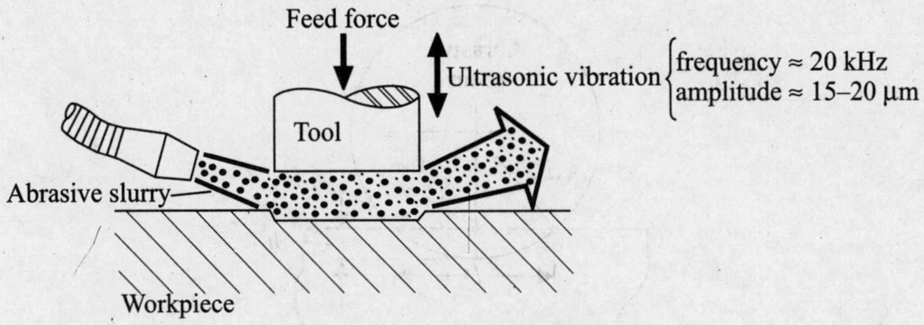

Fig. 6.8 Ultrasonic machining.

(i) the hammering of the abrasive particles on the work surface by the tool,

(ii) the impact of the free abrasive particles on the work surface,

(iii) the erosion due to cavitation, and

(iv) the chemical action associated with the fluid used.

A number of researchers have tried to develop the theories to predict the characteristics of ultrasonic machining. The model proposed by M.C. Shaw is generally well-accepted and, despite its limitations, explains the material removal process reasonably well. In this model, the direct impact of the tool on the grains in contact with the workpiece (which is responsible for the major portion of the material removal) is taken into consideration. Also, the assumptions made are that

(i) the rate of work material removal is proportional to the volume of work material per impact,

(ii) the rate of work material removal is proportional to the number of particles making impact per cycle,

(iii) the rate of work material removal is proportional to the frequency (number of cycles per unit time),

(iv) all impacts are identical,

(v) all abrasive grains are identical and spherical in shape.

Thus,

$$Q \propto vZ\nu, \tag{6.1}$$

where

Q = volume of work material removal rate,

v = volume of work material dislodged/impact,

Z = number of particles making impact/cycle,

ν = frequency.

Let us now consider the impact of a rigid, spherical abrasive grain of diameter d on the work surface. Figure 6.9 shows the indentation caused by such an impact

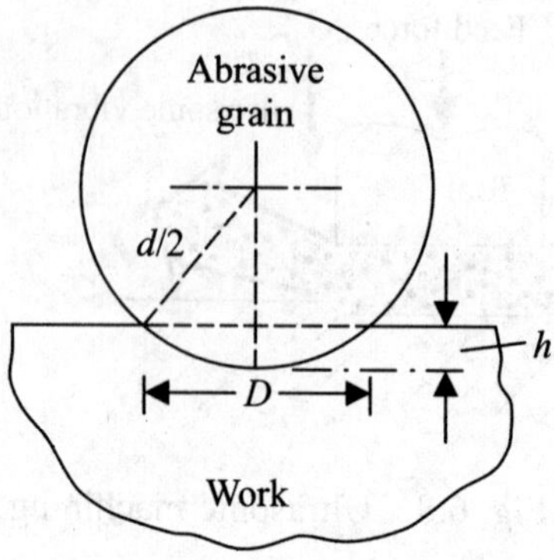

Fig. 6.9 Scheme of idealized grain indentation.

at an instant of time. If D is the diameter of the indentation at any instant and h the corresponding depth of penetration, we get, from Fig. 6.9,

$$\left(\frac{D}{2}\right)^2 = \left(\frac{d}{2}\right)^2 - \left(\frac{d}{2} - h\right)^2$$

or

$$D^2 = d^2 - (d - 2h)^2 \quad \text{or} \quad D^2 = 4dh - 4h^2.$$

Since h is normally very small as compared with the abrasive grain diameter d, the indentation diameter can be approximately expressed as

$$D \approx 2\sqrt{dh}. \tag{6.2}$$

Assuming the volume of material dislodged per impact to be proportional to D^3, we get

$$Q \propto (dh)^{3/2} Zv. \tag{6.3}$$

Since the mean speed of the tool is low, the mean static feed force F applied to the tool must be equal to the mean force of the tool on the grains. When the duration of an impact is Δt and the maximum value of the impact force F_i is $F_{i_{max}}$, the nature of variation of F with time is as shown in Fig. 6.10. Now,

$$F = \frac{1}{T}\int_0^T F_i(t)\, dt,$$

where T is the time period of each cycle. It will not be very much erroneous to assume the nature of variation of F_i to be triangular, yielding

$$F \approx \tfrac{1}{2} F_{i_{max}} \Delta t \frac{1}{T}. \tag{6.4}$$

The various tool positions during a cycle are as shown in Fig. 6.11. The position A indicates the instant the tool face touches the abrasive grain, and the period of movement from A to B represents the impact. The indentations, caused by

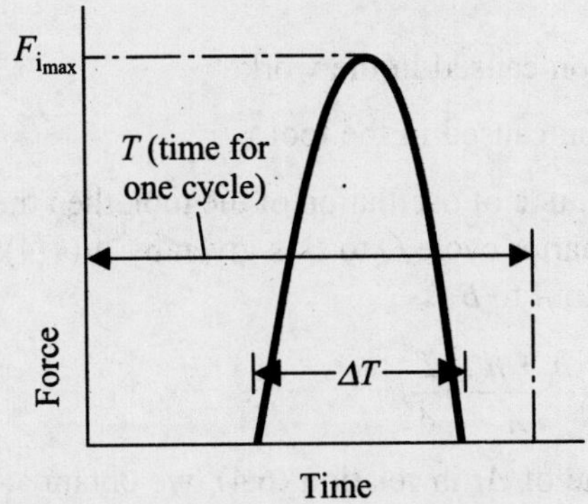

Fig. 6.10 Force during indentation.

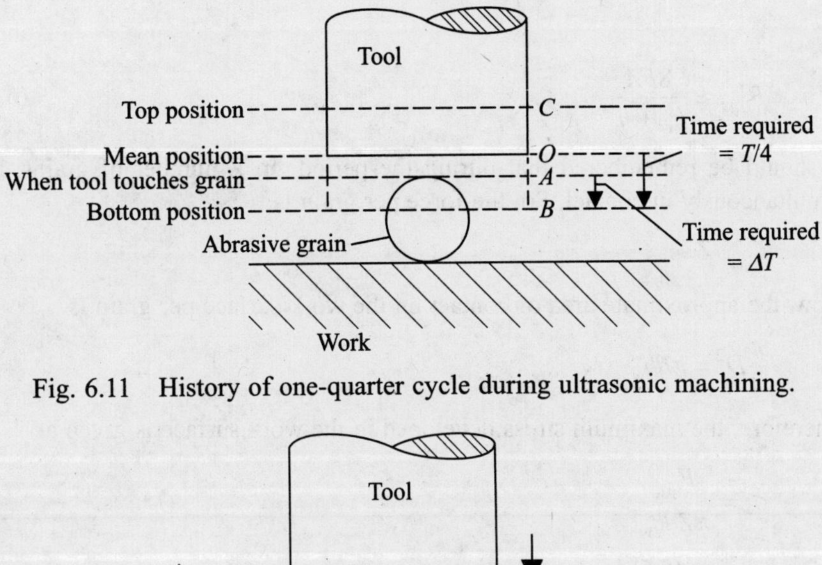

Fig. 6.11 History of one-quarter cycle during ultrasonic machining.

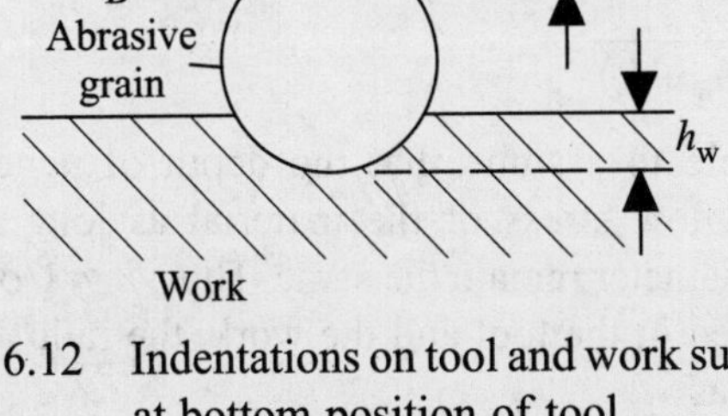

Fig. 6.12 Indentations on tool and work surface at bottom position of tool.

the grain on the tool and the work surface at the extreme bottom position of the tool, are shown in Fig. 6.12. If the distance travelled by the tool from the position A to the position B is h (the total indentation), then

$$h = h_w + h_t,$$

where

h_w = indentation caused in the work,

h_t = indentation caused in the tool.

Now, if A is the amplitude of oscillation of the tool, then the average velocity of the tool during the quarter cycle O to B is given by $A/(T/4)$. Therefore, the time required to travel from A to B is

$$\Delta t \approx \frac{h}{A} \bullet \frac{T}{4} = (\frac{h_t + h_w}{A})\frac{T}{4}.$$

Substituting this value of Δt in relation (6.4), we obtain

$$F \approx \tfrac{1}{2} F_{i_{max}} (\frac{h_t + h_w}{A}) \frac{T}{4} \frac{1}{T}$$

or

$$F_{i_{max}} \approx \frac{8FA}{h_w + h_t}. \tag{6.5}$$

It should be remembered that during the period Δt, Z number of grains are simultaneously in contact. So, the force per grain is

$$F_{i_{max}}/Z.$$

Now, the approximate area of contact on the work surface per grain is

$$\frac{\pi}{4} D^2 = \pi d h_w.$$

Therefore, the maximum stress developed in the work surface is given as

$$\sigma_w = \frac{F_{i_{max}}}{\pi Z d h_w}.$$

Using relation (6.5) in this equation, we get

$$\sigma_w = \frac{8FA}{\pi Z d h_w (h_w + h_t)}. \tag{6.6}$$

It is quite reasonable to assume that the depth of penetration is inversely proportional to the flow stress of the material as long as the load and the indenting sphere's diameter remain the same. Thus, $h \propto 1/\sigma$. So, if σ_t and σ_w are the stresses developed in the tool and the work, the ratio of the corresponding indentation is given as

$$\frac{h_t}{h_w} = \frac{\sigma_w}{\sigma_t} = \lambda. \tag{6.7}$$

Since the flow stress σ and the Brinell hardness H are the same, equations (6.6) and (6.7) yield

$$h_w^2 = \frac{8FA}{\pi Z d H_w (1+\lambda)}. \tag{6.8}$$

Again, it may be assumed that the number of grains acting is inversely proportional to the square of the diameter of each grain for a given area of the tool face. Therefore,

$$Z \propto \frac{C}{d^2}$$

or

$$Z = \chi \frac{C}{d^2}, \tag{6.9}$$

where C represents the concentration of the abrasive grains in the slurry and χ is a constant of proportionality. Substituting Z from equation (6.9) in equation (6.8), we get

$$h_w^2 = \frac{8FAd}{\pi \chi H_w C(1+\lambda)}$$

or

$$h_w = \sqrt{\frac{8FAd}{\pi \chi H_w C(1+\lambda)}}. \tag{6.10}$$

When h_w is substituted in the right-hand side of relation (6.3), we get the expression for the volume rate of material removal from the work surface. Thus,

$$Q \propto [d\{\frac{8FAd}{\pi \chi H_w C(1+\lambda)}\}^{1/2}]^{3/2} \chi \frac{C}{d^2} \nu$$

or

$$Q \propto \frac{A^{3/4} d^{1/4} F^{3/4} C^{1/4}}{H_w^{3/4}} \nu. \tag{6.11}$$

This rate of material removal is through the direct hammering action of the grains due to the vibrating tool. Some grains, reflected by the fast moving tool face, also impinge on the work face, and we can estimate the indentation caused by such freely moving grains. Figure 6.13 shows a grain reflected by the tool. During vibration, the maximum velocity of the tool face is $2\pi\nu A$. Since the original velocity of an abrasive grain is small, its maximum velocity is, obviously, of the order of $2\pi\nu A$. So, the corresponding maximum kinetic energy of the abrasive grain is given by

$$(\text{KE})_{\max} = \tfrac{1}{2}(\frac{\pi}{6} d^3 \rho)(2\pi\nu A)^2 = \tfrac{1}{3}\pi^3 \rho d^3 \nu^2 A^2, \tag{6.12}$$

where ρ is the density of the abrasive material. If we assume that during the

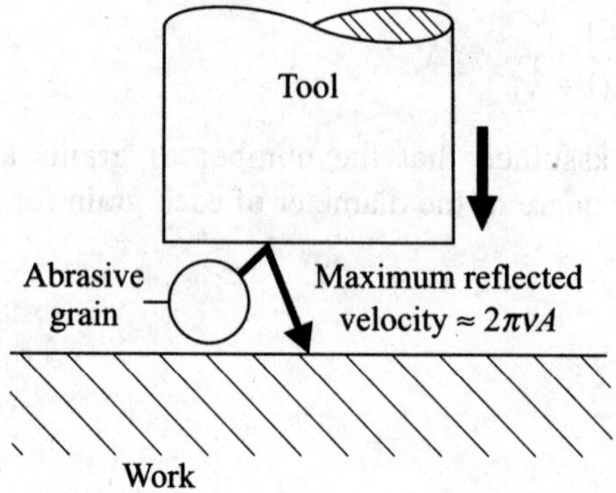

Fig. 6.13 Rebounding abrasive grain during ultrasonic machining.

indentation caused by such an impinging grain the contact force increases linearly with the indentation, then

$$(\text{KE})_{\max} \approx \tfrac{1}{2} F'_{i_{\max}} h'_{w}, \tag{6.13}$$

where h'_w is the depth of indentation. Using the same reasoning as given before, we obtain

$$\sigma_w \equiv H_w = \frac{F'_{i_{\max}}}{\pi d h'_w}.$$

Therefore,

$$F'_{i_{\max}} = \pi d h'_w H_w. \tag{6.14}$$

Using equations (6.12) and (6.14) in relation (6.13), we get

$$\tfrac{1}{3}\pi^3 \rho d^3 v^2 A^2 = \tfrac{1}{2}\pi d h'_w H_w h'_w.$$

Finally, the maximum possible value of the depth of indentation caused by a freely moving grain is found from the foregoing equation as

$$h'_w = \pi \sqrt{\frac{2\rho}{3H_w}}\, dvA. \tag{6.15}$$

Comparing the values of h_w and h'_w under normal conditions, we see that h'_w is very small as compared with h_w, and so it can be concluded that most of the material is removed by the directly impacting abrasive grains.

Relation (6.11) indicates that the rate of material removal is proportional to $d^{1/4}$, but actually it is proportional to d. This discrepancy between the theoretical prediction and the observed fact was explained by Shaw as follows.

The actual shape of an abrasive grain is not spherical, as shown in Fig. 6.14. Instead of having a smooth surface, it has projections of average diameter d_1. The average diameter of the projections is observed to be proportional to the

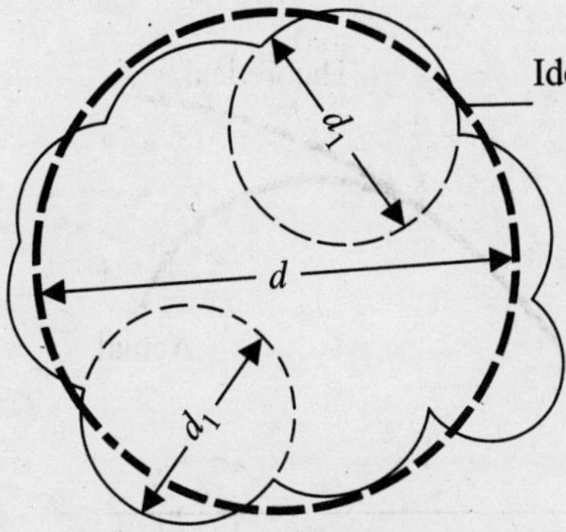

Fig. 6.14 Spherical projections on abrasive grain.

square of the nominal diameter of the grain (d). So,

$$d_1 = \mu d^2. \tag{6.16}$$

Now, for small depths of indentations, the effective diameter will be the diameter of the indenting projections (d_1). Therefore, from relation (6.3), we get

$$Q \propto (d_1 h_w)^{3/2} Z v, \tag{6.17}$$

where

$$h_w^2 = \frac{8FA}{\pi Z d_1 H_w (1+\lambda)},$$

$$Z = \frac{\chi C}{d^2}.$$

Using the expression for d_1, i.e., equation (6.16), in relation (6.17), we finally find the mrr to be

$$Q \propto \frac{dF^{3/4} A^{3/4} C^{1/4}}{H_w^{3/4} (1+\lambda)^{3/4}} v. \tag{6.18}$$

Relation (6.18) shows that the mrr is proportional to d, a fact also experimentally confirmed.

The Shaw theory has a number of limitations. For example, it does not correctly predict the effects of variation of A, F, and v. When F is increased, the mrr increases, as shown in Fig. 6.15. This is also confirmed by relation (6.18). However, in practice, Q starts decreasing after some value of F because the abrasive grains get crushed under heavy load.

6.3.2 PROCESS PARAMETERS

The important parameters which affect the process are the

(i) frequency,
(ii) amplitude,
(iii) static loading (feed force),

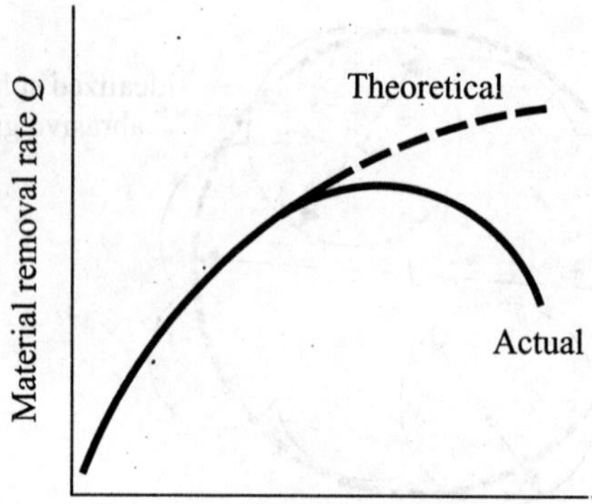

Fig. 6.15 Variation of material removal rate with feed force.

(iv) hardness ratio of the tool and the workpiece,
(v) grain size,
(vi) concentration of abrasive in the slurry.

(i) As can be seen from relation (6.18), the mrr increases linearly with the frequency. In practice also, the mrr increases with the frequency (see Fig. 6.16a) but the actual characteristic is not exactly linear. The mrr tends to be somewhat lower than the theoretically-predicted value.

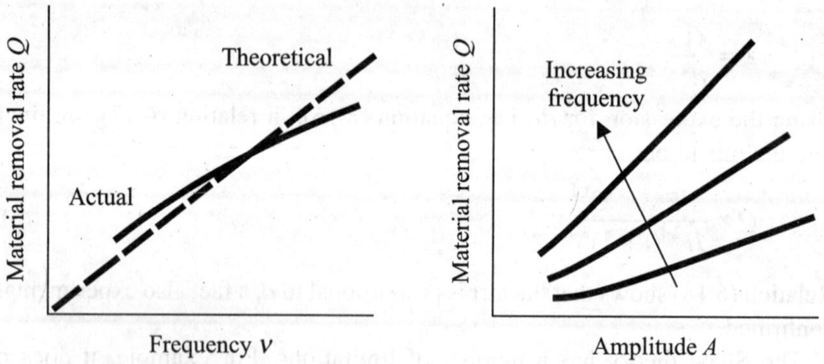

(a) Variation with frequency (b) Variation with amplitude

Fig. 6.16 Material removal rate characteristics in USM.

(ii) When the amplitude of vibration is increased, the mrr is expected to increase, as can be seen from relation (6.18). The actual nature of the variation is as shown in Fig. 6.16b for different values of the frequency. Again, the actual characteristic is somewhat different from the theoretically-predicted one. The main source of discrepancy stems from the fact that we calculated the duration of penetration Δt by considering the average velocity ($=A/(T/4)$). The characteristic of variation of Δt, given by

$$\frac{1}{v}\left[\frac{1}{4} - \frac{1}{2\pi}\sin^{-1}\left(1 - \frac{h}{A}\right)\right],$$

is quite different from that obtained from the approximate expression, i.e., $(h/A)(T/4)$.

(iii) We have already said that with an increase in static loading (i.e., the feed force), the mrr tends to increase. However, in practice, it tends to decrease beyond a certain critical value of the force as the grains start getting crushed. The nature of variation of the mrr with the feed force (for various amplitudes) is shown in Fig. 6.17a.

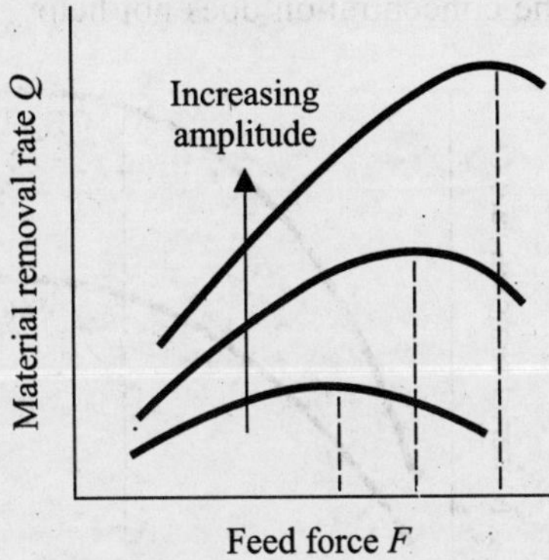

(a) Variation with feed rate

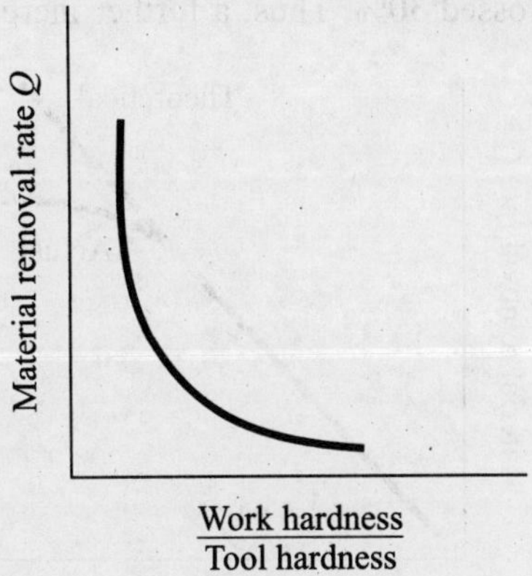

(b) Variation with hardness ratio

Fig. 6.17 Material removal rate characteristics in USM.

(iv) The ratio of the workpiece hardness and the tool hardness affects the mrr quite significantly, and the characteristic is as shown in Fig. 6.17b. Apart from the hardness, the brittleness of the work material plays a very dominant role. Table 6.2 indicates the relative material removal rates for different work materials, keeping the other parameters the same. Clearly, a more brittle material is machined more rapidly.

Table 6.2 Relative material removal rates (v = 16.3 kHz, A = 12.5 μm, grain size = 100 mesh)

Work material	Relative removal rate
Glass	100.0
Brass	6.6
Tungsten	4.8
Titanium	4.0
Steel	3.9
Chrome steel	1.4

(v) Relation (6.18) indicates that the mrr should rise proportionately with the mean grain diameter d. However, when d becomes too large and approaches the magnitude of the amplitude A, the crushing tendency increases, resulting in a fall in the mrr as shown in Fig. 6.18a.

(vi) Since the concentration directly controls the number of grains producing impact per cycle and also the magnitude of each impact, the mrr is expected to depend on C. But relation (6.18) shows that the mrr is expected to be proportional to $C^{1/4}$. The actual variation is shown in Fig. 6.18b for B_4C and SiC abrasives. This is in a fairly good agreement with the theoretical prediction. Since the mrr increases as $C^{1/4}$, the increase in the mrr is quite low after C has crossed 30%. Thus, a further increase in the concentration does not help.

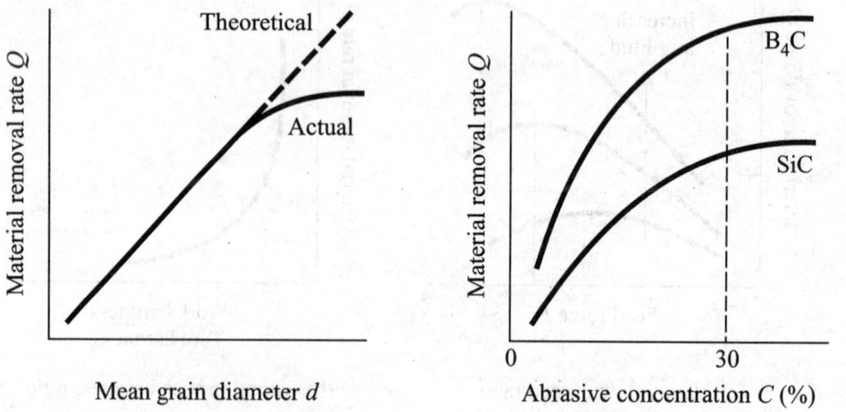

(a) Variation with grain diameter (b) Variation with concentration

Fig. 6.18 Material removal rate characteristics in USM.

Apart from the foregoing process parameters, some physical properties (e.g., viscosity) of the fluid used for the slurry also affect the mrr. Experiments show that the mrr drops as the viscosity increases (Fig. 6.19a).

Though the mrr is a very important consideration for judging the performance of an USM operation, the quality of finish obtained has also to be considered for a proper evaluation. In an USM operation, the surface finish depends mainly on the size of the abrasive grains. Figure 6.19b shows a typical variation of the mean value of the surface unevenness with the mean grain size for both glass and tungsten carbide as the work material. It is clear that the surface finish is much more sensitive to the grain size in the case of glass. This is because of the fact that, for a high hardness, the size of the fragments dislodged through a brittle fracture does not depend much on the size of the impacting particles.

EXAMPLE 6.1 Find out the approximate time required to machine a square hole (5 mm × 5 mm) in a tungsten carbide plate of thickness 4 mm. The abrasive grains are of 0.01 mm diameter. The feeding is done with a constant force of 3.5 N. The amplitude of tool oscillation is about 25 μm, the frequency being 25

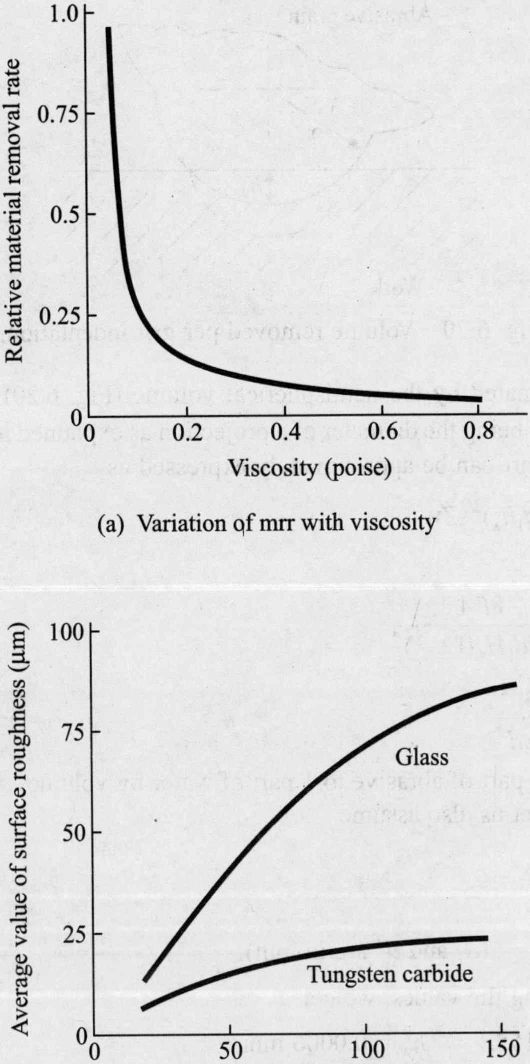

(a) Variation of mrr with viscosity

(b) Dependence of surface finish on grain size

Fig. 6.19 Dependence of material removal rate on viscosity and effect of grain size on surface finish.

kHz. The fracture hardness of WC can be approximately taken as 6900 N/mm^2. The slurry contains 1 part of abrasive to about 1 part of water.

SOLUTION Since relation (6.18) yields only a qualitative result, let us first find out an approximate expression giving the mrr in terms of the other quantities. To do this, let us assume that the volume removed per grit indentation

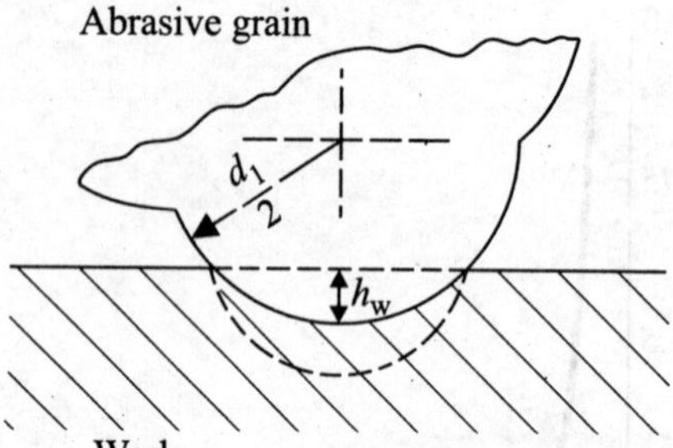

Fig. 6.20 Volume removed per grit indentation.

can be approximated by the hemispherical volume (Fig. 6.20) $\frac{2}{3}\pi(D/2)^3$ with $D = 2\sqrt{d_1 h_w}$, d_1 being the diameter of a projection as explained in Section 6.3.2. Therefore, the mrr can be approximately expressed as

$$Q \approx \tfrac{2}{3}\pi(d_1 h_w)^{3/2} Zv,$$

where

$$h_w^2 \approx \frac{8FA}{\pi Z d_1 H_w (1+\lambda)},$$

$$Z \approx \tfrac{1}{2}\cdot\frac{4a^2}{\pi d^2}$$

since we have 1 part of abrasive to 1 part of water by volume, a being the side of the square. Let us also assume

$$\lambda = \frac{H_{work}}{H_{tool}} = 5,$$

$$d_1 = d^2 \qquad (d_1 \text{ and } d^2 \text{ are in mm}).$$

Then, substituting the values, we get

$$Z = 159{,}155, \qquad h_w = 0.0006 \text{ mm},$$

and ultimately

$$Q = 0.122 \text{ mm}^3/\text{sec}.$$

Since the volume to be removed is 100 mm^3, the approximate time required is 13.66 minutes.

The actual time required is more than that obtained. This is because the process is not 100% efficient, i.e., all the impacts do not produce brittle fractures.

EXAMPLE 6.2 Determine the percentage change in the machining time for an USM operation cutting WC plates when the tool material is changed from copper to stainless steel.

SOLUTION If Q_c and Q_s are the material removal rates with copper and stainless steel as the tool materials, respectively, then, from relation (6.18), we get

$$\frac{Q_c}{Q_s} = (\frac{1+\lambda_s}{1+\lambda_c})^{3/4},$$

where

$$\lambda_s = \frac{\text{hardness of WC}}{\text{hardness of stainless steel}},$$

$$\lambda_c = \frac{\text{hardness of WC}}{\text{hardness of copper}}.$$

Since WC is much harder than both stainless steel and copper, λ_s and λ_c both will be much larger than unity. Thus,

$$\frac{Q_c}{Q_s} \approx (\frac{\lambda_s}{\lambda_c})^{3/4} = (\frac{H_c}{H_s})^{3/4},$$

where H_c and H_s are the hardness values for copper and stainless steel, respectively. Assuming $H_c/H_s \approx \frac{1}{3}$, we get

$$\frac{Q_c}{Q_s} \approx 0.44.$$

If t_c and t_s are the machining time for copper and stainless steel tools, respectively, then

$$\frac{t_c}{t_s} = \frac{Q_s}{Q_c} = 2.27.$$

Now, the percentage change in cutting time when the tool is changed from copper to steel is

$$\frac{t_c - t_s}{t_c} \times 100 = (1 - \frac{t_s}{t_c}) \times 100 = (1 - 0.44) \times 100 = 56 \text{ (reduction)}.$$

6.3.3 ULTRASONIC MACHINING UNIT

The main elements of an ultrasonic machining unit are shown in Fig. 6.21. The important components of the machine are

(i) the acoustic head,
(ii) the feeding unit,
(iii) the tool,
(iv) the abrasive slurry and pump unit,
(v) the body with work table.

Acoustic Head

The acoustic head (Fig. 6.22) is perhaps the most important part of the machine.

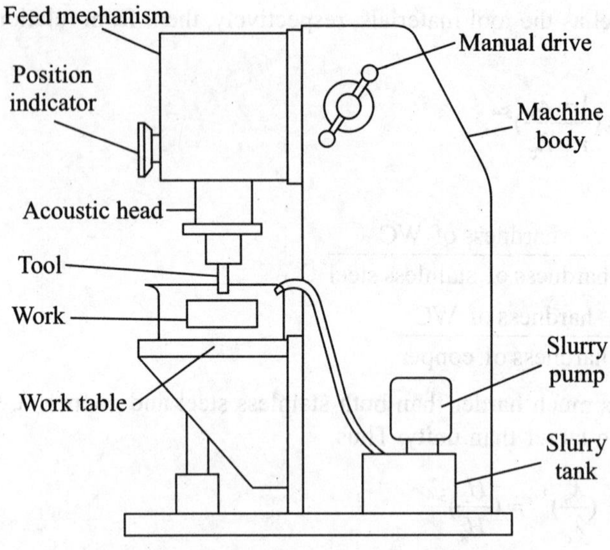

Fig. 6.21 Ultrasonic drilling machine.

Its function is to produce a vibration in the tool. It consists of a generator for supplying a high frequency electric current, a transducer to convert this into a mechanical motion in the form of a high frequency vibration, a holder to hold the head, and a concentrator to mechanically amplify the vibration while transmitting it to the tool.

Most transducers work on the magnetostrictive principle because of the high efficiency, high reliability in the 15–30 kHz range, low supply voltage, and simple cooling arrangement. Stampings are used to reduce loss as in transformers. The dimensions are so chosen that the natural frequency coincides with the electric supply frequency. Almost all the modern machines use the magnetostriction transducers made of nickel (stampings of 0.1–0.2 mm thickness).

The main purpose of the concentrator is to increase the amplitude to the level needed for cutting. Various types of concentrators are used (Fig. 6.23a). Figure 6.23b shows how the amplitude of longitudinal vibration of the transducer-concentrator assembly is amplified. It should be noted that the system has to be held to the main body at a nodal point, as shown.

Feed Mechanism

The objective of the feed mechanism is to apply the working force during the machining operation. An instrument showing the movement of the tool indicates the depth of machining. The basic types of feed mechanisms are the

(i) counterweight type,

(ii) spring type,

(iii) pneumatic and hydraulic type,

(iv) motor type.

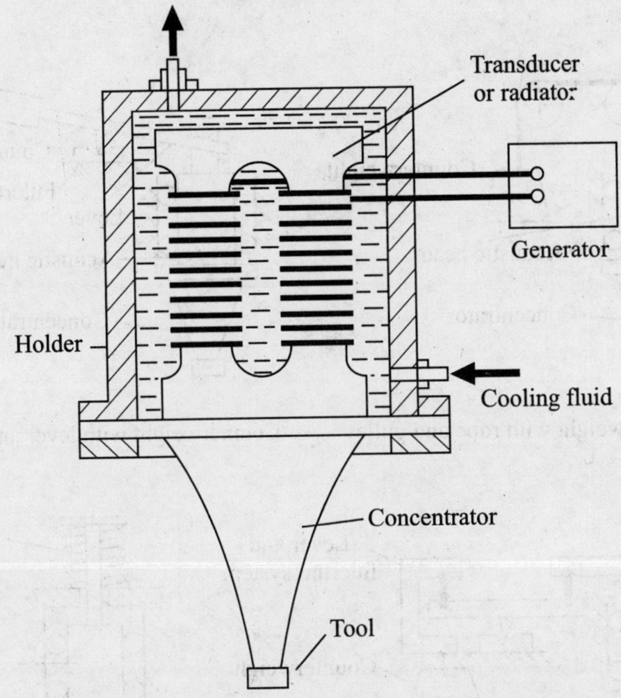

Fig. 6.22 Simplified diagram of acoustic head.

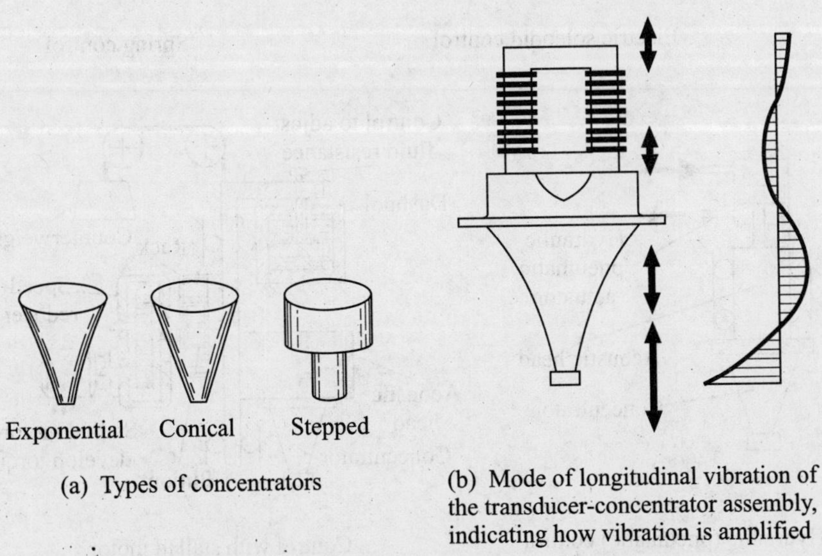

Fig. 6.23 Concentrators and their function.

Figure 6.24 schematically shows these arrangements.

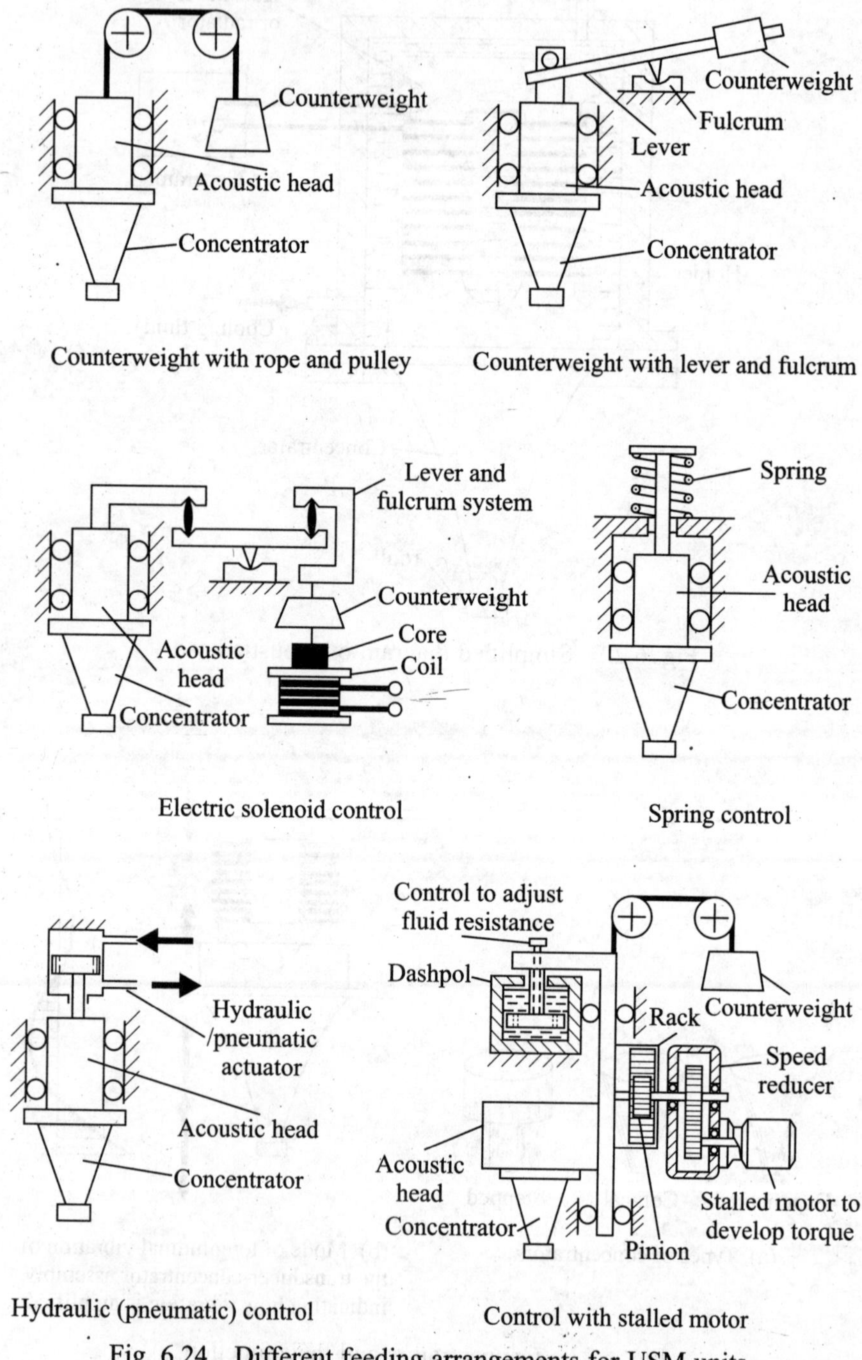

Fig. 6.24 Different feeding arrangements for USM units.

Tool

As already mentioned, the tool is made of a strong, but at the same time ductile, metal. Generally, stainless steels and low carbon steels are used for making the tools. Aluminium and brass tools wear ten and five times faster than steel tools, respectively. The geometrical features are decided by the process. The diameter of the circle circumscribed about the tool should not be more than 1.5 to 2 times the diameter of the end of the concentrator, and the tool should be as short and rigid as possible. When the tool is made hollow, the internal contour should be parallel to the external one to ensure uniform wear. The thickness of any wall or projection should be at least five times the grain size of the abrasive. In a hollow tool, the walls should not be made thinner than 0.5 mm to 0.8 mm. When designing the tool, consideration should be given to the side clearance which is normally of the order of 0.06 mm to 0.36 mm, depending on the grain size of the abrasive.

Abrasive Slurry

The most common abrasives are (i) boron carbide (B_4C), (ii) silicon carbide (SiC), (iii) corundum (Al_2O_3), (iv) diamond, and (v) boron silicarbide (very efficient) whose abrasive power is about 10% more than that of B_4C. B_4C is the best and most efficient among the rest but it is expensive. SiC is used on glass, germanium, and some ceramics. The cutting time with SiC is about 20–40% more than that with B_4C. Corundum is much less efficient and the cutting time is about 3–4 times of that with B_4C. Diamond dust is used only for cutting diamonds and rubies.

Though water is the most commonly used fluid in the slurry, other liquids, such as benzene, glycerol, and oils, are also used. It has been found that the mrr tends to decrease with increasing viscosity.

6.3.4 EFFECTS OF USM ON MATERIALS

Since the cutting force involved is very small, the process produces no appreciable stress and heating. So, the material structure remains unaffected. However, during cutting through a hole, chipping may occur at the exit side of the hole. To avoid this, the workpiece made of a brittle material is fastened to a base usually made of glass.

6.3.5 SUMMARY OF USM CHARACTERISTICS

Mechanics of material removal	Brittle fracture caused by impact of abrasive grains due to tool vibrating at high frequency
Medium	Slurry
Abrasives	B_4C, SiC, Al_2O_3, diamond 100–800 grit size
Vibration	
Frequency	15–30 kHz

Amplitude	25–100 μm
Tool	
Material	Soft steel
$\frac{\text{Material removal rate}}{\text{Tool wear rate}}$	1.5 for WC workpiece, 100 for glass workpiece
Gap	25–40 μm
Critical parameters	Frequency, amplitude, tool material, grit size, abrasive material, feed force, slurry concentration, slurry viscosity
Materials application	Metals and alloys (particularly hard and brittle), semiconductors, nonmetals, e.g., glass and ceramics
Shape application	Round and irregular holes, impressions
Limitations	Very low mrr, tool wear, depth of holes and cavities small

6.4 ELECTROCHEMICAL MACHINING (ECM)

Electrochemical machining is one of the most potential unconventional machining processes. Though it is a new process for metal working, the basic principle had been well-known for a long time. This process may be considered as the reverse of electroplating with some modifications. Further, it is based on the principle of electrolysis. In a metal, electricity is conducted by the free electrons, but it has been established that in an electrolyte the conduction of electricity is achieved through the movement of ions. Thus, the flow of current through an electrolyte is always accompanied by the movement of matter.

The electrolysis principle has been in use for long for electroplating where the objective is to deposit metal on the workpiece. But since in electrochemical machining the objective is to remove metal, the workpiece is connected to the positive, and the tool to the negative, terminal. Figure 6.25 shows a workpiece and a suitably-shaped tool, the gap between the tool and the work being full of a suitable electrolyte. When the current is passed, the dissolution of the anode occurs. However, the dissolution rate is more where the gap is less and vice

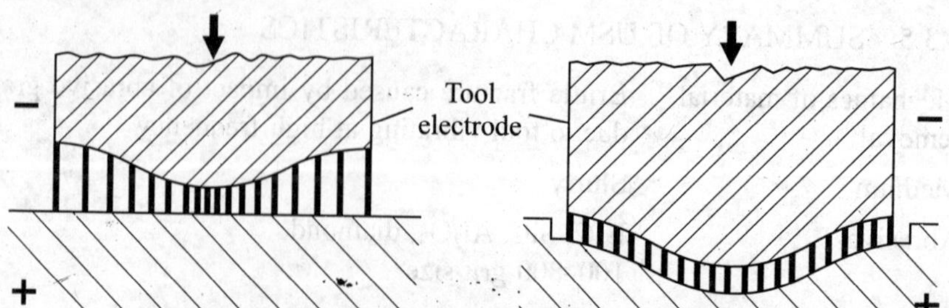

Fig. 6.25 Scheme of electrochemical machining.

versa as the current density is inversely proportional to the gap. Now, if the tool is given a downward motion, the work surface tends to take the same shape as that of the tool, and at a steady state, the gap is uniform, as shown in Fig. 6.25. Thus, the shape of the tool is reproduced in the job.

In an electrochemical machining process, the tool is provided with a constant feed motion. The electrolyte is pumped at a high pressure through the tool and the small gap between the tool and the workpiece (Fig. 6.26). The electrolyte is so chosen that the anode is dissolved but no deposition takes place on the cathode (the tool). The order of the current and voltage are a few thousand amperes and 8–20 volts. The gap is of the order of 0.1–0.2 mm. In a typical machine, the metal removal rate is about 1600 mm^3/min for each 1000 amp. Approximately 3 kWh are needed to remove 16×10^3 mm^3 of metal, which is almost 30 times the energy required in a conventional process (of course, when the metal is readily machinable). But with ECM, the rate of metal removal is independent of the workpiece hardness. So, ECM becomes advantageous when either the work material possesses a very low machinability or the shape to be machined is complicated.

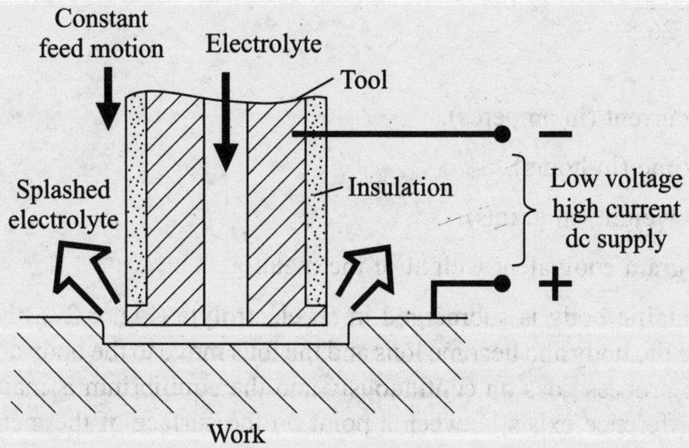

Fig. 6.26 Electrochemical machining.

Unlike most other conventional and unconventional processes, here there is practically no tool wear. Though it appears that, since machining is done electrochemically, the tool experiences no force, the fact is that the tool and work are subjected to very large forces exerted by the high pressure fluid in the gap.

6.4.1 ELECTROCHEMISTRY OF ECM PROCESS

The electrolysis process is governed by the following two laws proposed by Faraday:

(i) The amount of chemical change produced by an electric current, that is, the amount of any material dissolved or deposited, is proportional to the quantity of electricity passed.

(ii) The amounts of different substances dissolved or deposited by the same quantity of electricity are proportional to their chemical equivalent weights. In the quantitative form, Faraday's two laws state that

$$m \propto It\varepsilon, \tag{6.19}$$

where

m = weight (in grams) of a material dissolved or deposited,

I = current (in amperes),

t = time (in seconds),

ε = gram equivalent weight of the material.

Introducing the constant of proportionality F, commonly called Faraday (=96,500 coulombs), we find that relation (6.19) becomes

$$m = \frac{It\varepsilon}{F}. \tag{6.20}$$

Using the numerical value of F, we get

$$m = \frac{It\varepsilon}{26.8}, \tag{6.21}$$

where

I = current (in amperes),

t = time (in hours),

m = weight (in grams),

ε = gram equivalent weight of the metal.

When a metallic body is submerged in an electrolyte (Fig. 6.27), the metallic atoms leave the body and become ions and the ions move to the body and become atoms. The process goes on continuously and the equilibrium is maintained. A potential difference exists between a point on the surface of the metallic body

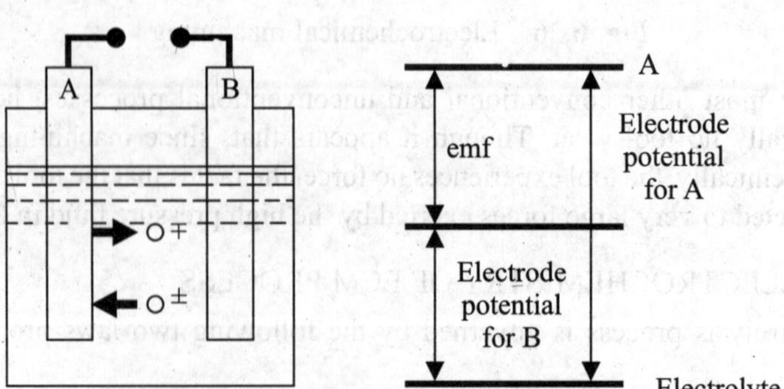

(a) Material dissolution and deposition (b) Cell emf

Fig. 6.27 Electrode dissolution and electrode potential.

(electrode) and an adjacent point in the electrolyte. This potential difference is known as the electrode potential. The electrode potential varies depending on the electrode-electrolyte combination. If two different electrodes (A and B) are immersed, a potential difference between these electrodes will exist since the potentials of A and B with respect to the common electrolyte are different. This potential difference is the electromotive force (emf) of the cell, generated by the electrodes and the electrolyte. This is explained in Fig. 6.27. For example, if Fe and Cu electrodes are dipped in brine (solution of kitchen salt in water) as shown in Fig. 6.28a, the electrode potentials are

$$Fe \rightleftharpoons Fe^{2+} + 2e^- \rightarrow -0.409 \text{ volt}$$

$$Cu \rightleftharpoons Cu^{2+} + 2e^- \rightarrow +0.304 \text{ volt}$$

difference between
electrode potentials = 0.713 volt

The nature of the electrolysis process depends on the electrolyte used. To understand how ECM is realized, let us consider the aqueous solution of sodium

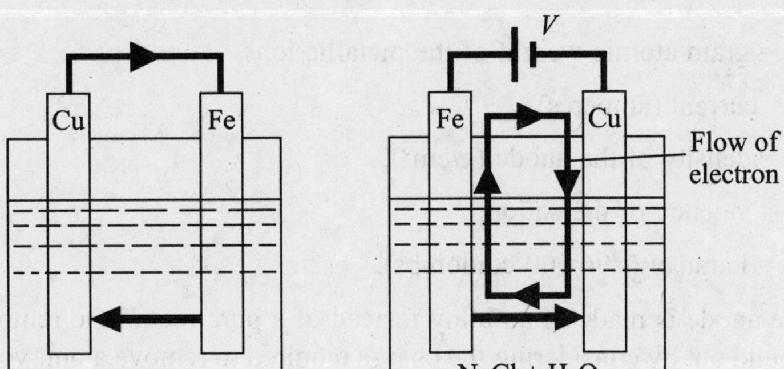

(a) Case without extra source of emf

(b) Case with extra source of emf

Fig. 6.28 Principle of electrolysis.

chloride as the electrolyte. When a voltage difference is applied across the electrodes (Fig. 6.28b), the reactions at the anode and the cathode are

$$Fe \rightarrow Fe^{++} + 2e \qquad \text{(at anode)}$$

[the electrode metal (Fe) dissolves, leaving two electrons]

$$2H_2O + 2e \rightarrow H_2\uparrow + 2(OH)^- \qquad \text{(at cathode)}$$

The water gets two electrons from the electrode and, as a result, the hydrogen gas is evolved and hydroxyl ions are produced. The positive metal ions tend to move towards the cathode and the negative hydroxyl ions are attracted towards the anode. Then, the positive metal ions combine with the negatively-charged hydroxyl ions to form ferrous hydroxide as

$$Fe^{++} + 2(OH)^{--} \rightarrow Fe(OH)_2$$

This ferrous hydroxide forms an insoluble precipitate. So, with this kind of electrode metal-electrolyte combination, the anode dissolves and H_2 generates at the cathode, leaving the cathode shape unchanged. This is the most important characteristic of the electrochemistry of the ECM process. It should be noted that for ECM the choice of electrodes and the electrolyte must be such that no deposition at either electrode can take place.

The gram equivalent weight of the metal is given by $\varepsilon = A/Z$, where A is the atomic weight and Z is the valency of the ions produced. Using this in equation (6.20), we get the rate of mass removal in the form

$$\dot{m} = \frac{AI}{ZF}. \tag{6.22}$$

If the density of the anode material is ρ, the volumetric removal rate is given by

$$Q = \frac{AI}{\rho ZF} \text{ cm}^3/\text{sec}, \tag{6.23}$$

where

A = gram atomic weight of the metallic ions,

I = current (amperes),

ρ = density of the anode (g/cm^3),

Z = valency of the cation,

F = Faraday (=96,500 coulombs).

When the anode is made of an alloy instead of a pure metal, the removal rate can be found out by considering the charge required to remove a unit volume of each element. If the atomic weights and the valencies (of the corresponding ions entering the electrolyte) are $A_1, A_2, A_3, \ldots$ and $Z_1, Z_2, Z_3, \ldots$, respectively, and the composition (by weight) of the alloy is x_1% of element 1, x_2% of element 2, . . . , then a volume v cm^3 of the alloy contains $v\rho x_i/100$ gram of the i-th element, where ρ is the overall density of the alloy in g/cm^3. The charge required to remove all of the i-th element in volume v is given by

$$\frac{v\rho x_i}{100} \cdot \frac{Z_i F}{A_i}.$$

Thus, the volume of the alloy removed per unit charge is

$$Q = \frac{100}{\rho F}\left(\frac{1}{\sum_i (x_i Z_i / A_i)}\right) \text{ cm}^3/\text{amp-sec}$$

or

$$Q = \frac{0.1035 \times 10^{-2}}{\rho}\left(\frac{1}{\sum_i (x_i Z_i / A_i)}\right) \text{ cm}^3/\text{amp-sec}. \tag{6.24}$$

EXAMPLE 6.3 In an electrochemical machining process with a pure iron workpiece, a removal rate of 5 cm^3/min is desired. Determine the current required.

SOLUTION The gram atomic weight, valence at which dissolution takes place, and density of iron are 56 g, 2, and 7.8 g/cm^3, respectively. Now, using equation (6.23) along with the given data, we get

$$\frac{5}{60} = \frac{56 \times I}{7.8 \times 2 \times 96{,}500}.$$

So, the current required is given by

$$I = \frac{5 \times 7.8 \times 2 \times 96{,}500}{60 \times 56} \text{ amp} \quad \text{or} \quad I = 2240 \text{ amp}.$$

In the actual ECM process, there are many other factors which affect the removal rate. Also, the process is seldom as ideal as we have described. As a result, the actual removal rate may differ slightly from that obtained theoretically from equation (6.23). In Fig. 6.29, the theoretical and the actual removal rates with nickel as the work material are shown. The theoretical removal rate has

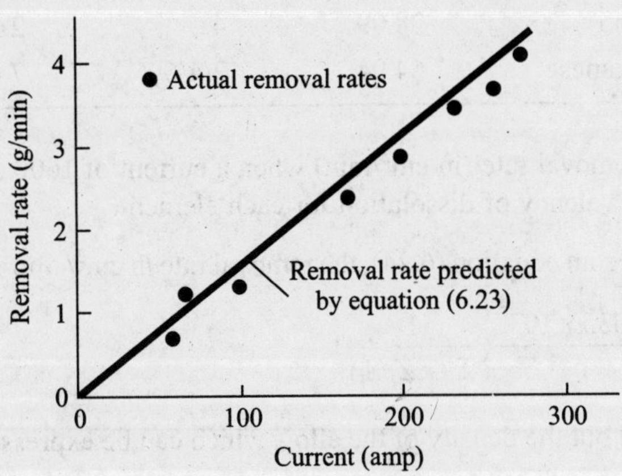

Fig. 6.29 Removal rate versus current for nickel.

been based on divalent dissolution. When the current is more, i.e., the dissolution takes place at a higher potential difference, trivalent dissolution also takes place. Therefore, at larger currents, the theoretical value tends to be more than the actual one. Sometimes, the dissolution valence also depends on the electrolytes. For example, copper dissolves in the monovalent form in chloride solutions, whereas in nitrate solutions, the dissolution takes place in the divalent state. Table 6.3 shows the important data on some elements.

EXAMPLE 6.4 The composition (% by weight) of the Nimonic 75 alloy is as given here:

Ni	Cr	Fe	Ti	Si	Mn	Cu
72.5	19.5	5.0	0.4	1.0	1.0	0.6

Table 6.3 Atomic weight, valency of dissolution, and density of important elements

Metal	Gram atomic weight	Valency of dissolution	Density (g/cm^3)
Aluminium	26.97	3	2.67
Chromium	51.99	2/3/6	7.19
Cobalt	58.93	2/3	8.85
Copper	63.57	1/2	8.96
Iron	55.85	2/3	7.86
Nickel	58.71	2/3	8.90
Tin	118.69	2/4	7.30
Titanium	47.9	3/4	4.51
Tungsten	183.85	6/8	19.3
Zinc	65.37	2	7.13
Silicon	28.09	4	2.33
Manganese	54.94	2/4/6/7	7.43

Calculate the removal rate (in cm^3/min) when a current of 1000 amp is passed. Use the lowest valency of dissolution for each element.

SOLUTION From equation (6.24), the removal rate in cm^3/amp-sec is given by

$$Q = \frac{0.1035 \times 10^{-2}}{\rho}\left(\frac{1}{\sum_i (x_i Z_i / A_i)}\right).$$

First, let us find out the density of the alloy, which can be expressed in the form

$$\rho = \frac{100}{\sum_i (x_i/\rho_i)},$$

where ρ_i is the density of the i-th element. So, using the values given in Table 6.3, we get

$$\rho = \frac{100}{\frac{72.5}{8.9} + \frac{19.5}{7.19} + \frac{5.0}{7.86} + \frac{0.4}{4.51} + \frac{1.0}{2.33} + \frac{1.0}{7.43} + \frac{0.6}{8.96}}$$

$$= \frac{100}{8.15 + 2.71 + 0.64 + 0.09 + 0.43 + 0.13 + 0.07}$$

$$= 8.18 \text{ g/cm}^3.$$

Using this value of ρ in the expression for Q and substituting the values of Z_i

and A_i from Table 6.3, we get

$$Q = \frac{0.1035 \times 10^{-2}}{8.18}$$

$$\times \frac{1}{\dfrac{72.5 \times 2}{58.71} + \dfrac{19.5 \times 2}{51.99} + \dfrac{5 \times 2}{55.85} + \dfrac{0.4 \times 3}{47.9} + \dfrac{1 \times 4}{28.09} + \dfrac{1 \times 2}{54.94} + \dfrac{0.6 \times 1}{63.57}}$$

$$= \frac{0.1035 \times 10^{-2}}{8.18 \times 3.62} = 0.35 \times 10^{-4} \text{ cm}^3\text{/sec-amp.}$$

Thus, when a 1000-ampere current is used, the removal rate in cm^3/min is

$0.35 \times 10^{-4} \times 1000 \times 60 = 2.1.$

The experimentally-observed value is about 2 cm^3/min.

The relationship between the voltage applied across the electrodes and the flow of current is not very simple. The total potential profile (see Fig. 6.30) consists of the following:

(i) Electrode potential.

(ii) Overvoltage due to activation polarization. The electrochemical changes at an electrode are in equilibrium when no current flows. The electrode potential acts as a barrier to a faster rate of reaction. So, an additional energy has to be supplied to get the required mrr.

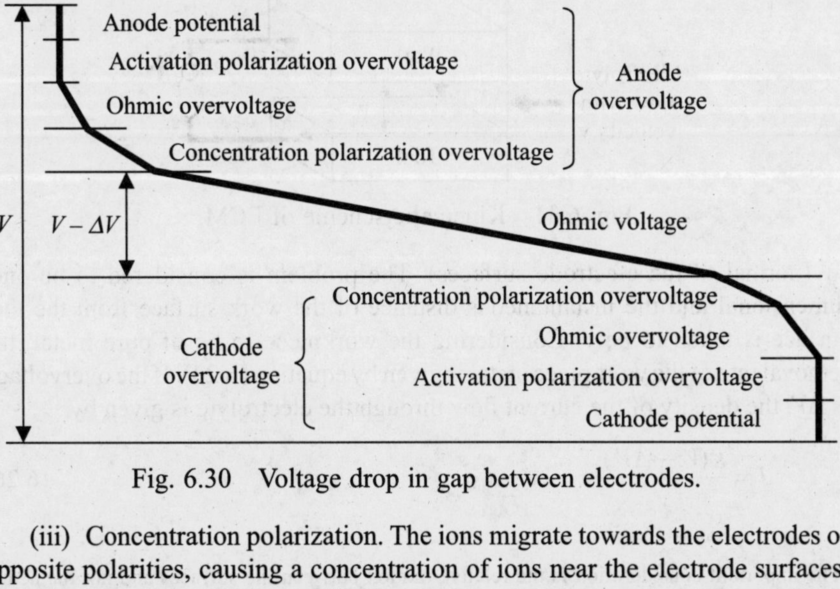

Fig. 6.30 Voltage drop in gap between electrodes.

(iii) Concentration polarization. The ions migrate towards the electrodes of opposite polarities, causing a concentration of ions near the electrode surfaces. Each ion must pass through this concentration barrier to release its charge at the electrode surface. So, an extra voltage is required for the migration of ions through the concentration layers.

(iv) Ohmic overvoltage. The films of solid materials forming on the electrode surface offer an extra resistance to the passage of current.

(v) Ohmic resistance of electrolyte. The ohmic voltage drop occurs across the bulk of the electrolyte. This is the main voltage drop and is the only part of the circuit within the electrolyte which obeys Ohm's law.

If the total overvoltage at the anode and the cathode is ΔV and the applied voltage is V, the current I is then given by

$$I = \frac{V - \Delta V}{R}, \tag{6.25}$$

where R is the ohmic resistance of the electrolyte.

The conductivities of the tool and the workpiece are much larger than the conductivity of the electrolyte. The typical electrolyte conductivity is about 0.1–1.0 Ω^{-1} cm^{-1}, whereas that of iron is 10^5 Ω^{-1} cm^{-1}. Thus, the surfaces of the tool and the workpiece may be considered as equipotential. The conductivity of the electrolyte is not really constant because of the temperature variation and accumulation of bubbles. However, for simple calculations, it may be treated as constant.

6.4.2 KINEMATICS AND DYNAMICS OF ECM

Figure 6.31 shows a set of electrodes with plane and parallel surfaces. The work (the upper electrode) is being fed[1] with a constant velocity f in the direction

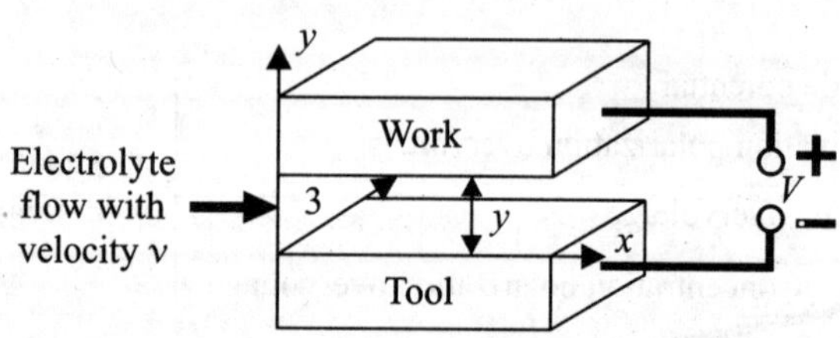

Fig. 6.31 Kinematic scheme of ECM.

$-y$ (normal to the electrode surfaces). The problem is considered to be one-dimensional and the instantaneous distance of the work surface from the tool surface is taken to be y. Considering the workpiece to be of pure metal, the removal rate of the workpiece metal is given by equation (6.23). If the overvoltage is ΔV, the density of the current flow through the electrolyte is given by

$$J = \frac{\kappa(V - \Delta V)}{y}, \tag{6.26}$$

[1]The important consideration is the relative motion between the surfaces and not the actual motion of the individual electrodes. So, even though the feed motion is normally given to the tool, our analysis remains valid.

where κ is the conductivity of the electrolyte. Now, the removal of work material causes the surface of the workpiece to recede (in the y-direction) with respect to the original surface with a velocity given by Q', where Q' is the volume rate of workpiece metal removal per unit area of the workpiece surface. Thus, the rate at which the gap between the work and the tool surface changes is

$$\frac{dy}{dt} = \frac{AJ}{\rho ZF} - f.$$

Using equation (6.26), we find this equation becomes

$$\frac{dy}{dt} = [\frac{\kappa A(V - \Delta V)}{\rho ZF}]\frac{1}{y} - f. \tag{6.27}$$

Replacing the term within the square brackets by a constant parameter λ, we obtain

$$\frac{dy}{dt} = \frac{\lambda}{y} - f. \tag{6.28}$$

This is the basic equation representing the dynamics of ECM process.

We shall now investigate a few basic cases.

Zero Feed

When the feed velocity is zero, i.e., $f = 0$, equation (6.28) reduces to

$$\frac{dy}{dt} = \frac{\lambda}{y}.$$

If the initial gap is y_0, the solution of this equation is then given by

$$y^2 = y_0^2 + 2\lambda t \tag{6.29}$$

or

$$y = \sqrt{y_0^2 + 2\lambda t}.$$

So, the gap increases with time, as indicated in Fig. 6.32a.

Constant Feed

An ever increasing gap is not desirable in an ECM process. So, in practice, the electrode is provided with a constant feed velocity of suitable magnitude. Thus, in equation (6.28), f is constant. Obviously, when the feed rate f equals the velocity of recession of the electrode surface due to metal removal, the gap remains constant. This gap (which depends on the feed velocity) is called the equilibrium gap (y_e). Thus, for the equilibrium gap, equation (6.28) yields

$$0 = \frac{\lambda}{y_e} - f$$

or

$$y_e = \frac{\lambda}{f}. \tag{6.30}$$

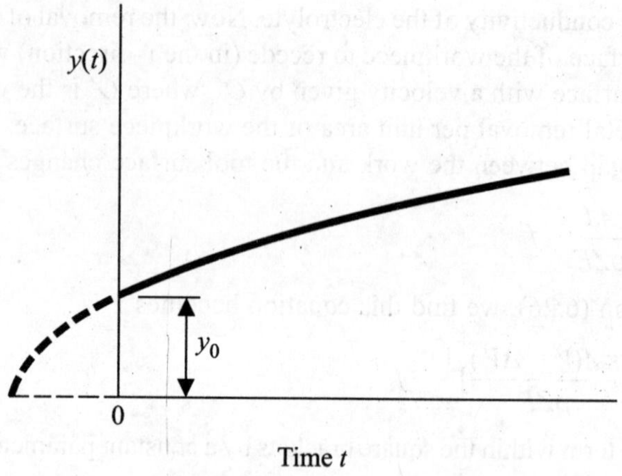

(a) Variation of electrode gap with time for zero feed

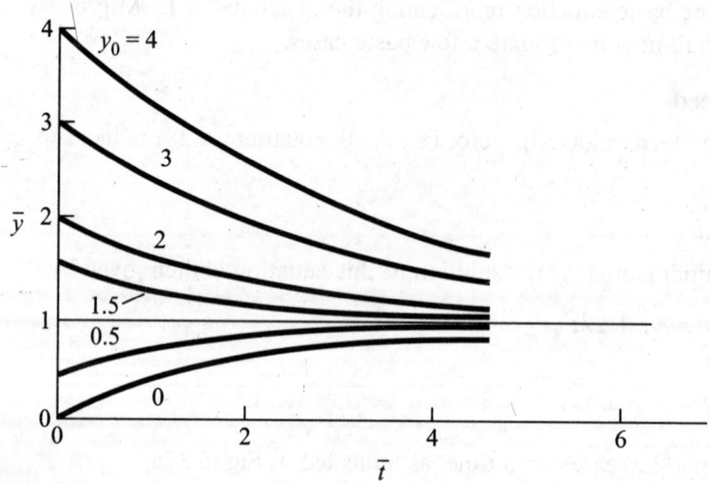

(b) Plot of $\bar{y}$ versus $\bar{t}$

Fig. 6.32 Electrode gap characteristics in ECM.

Let us now use the nondimensional quantities

$$\bar{y} = \frac{y}{y_e} = \frac{fy}{\lambda}, \qquad \bar{t} = \frac{t}{y_e/f} = \frac{f^2 t}{\lambda}. \tag{6.31}$$

Then, equation (6.28) takes the form

$$\frac{d\bar{y}}{d\bar{t}} = \frac{1}{\bar{y}} - 1.$$

With the initial condition $\bar{y} = \bar{y}_0$, the solution of this equation yields

$$\bar{t} = \bar{y}_0 - \bar{y} + \ln\left(\frac{\bar{y}_0 - 1}{\bar{y} - 1}\right). \tag{6.32}$$

Figure 6.32b shows the plot of $\bar{y}$ versus $\bar{t}$ for different values of the initial gap. It is seen that the gap always approaches the equilibrium value irrespective of the initial condition.

Feed Motion Inclined to Surface

When the feed velocity vector is inclined to the surface (Fig. 6.33), the component of the feed normal to the surface is $f \cos \theta$. In this case, the equilibrium gap is given by $\lambda/(f \cos \theta)$.

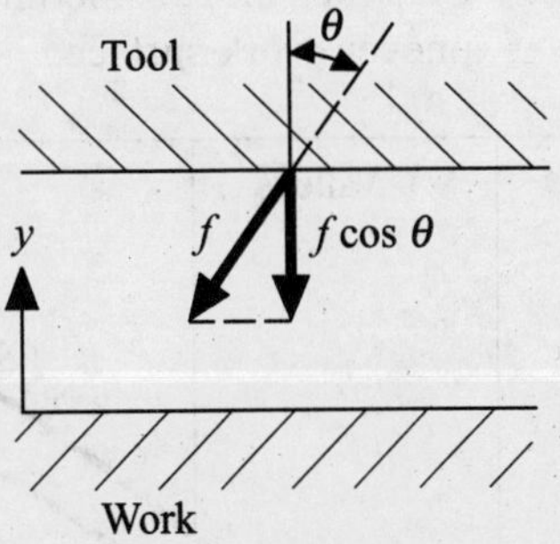

Fig. 6.33 Kinematics of ECM when feed direction is not normal to electrode surfaces.

Machining Uneven Surface

When an uneven work surface is subjected to ECM, the metal is removed from all portions of the surface (unlike other machining operations). The portion projecting outwards (the hills) is nearer the tool surface and gets machined more quickly than that projecting inwards (the cavities). Thus, the ECM process has the effect of smoothening out the unevenness. As shown in Fig. 6.34, the equilibrium work surface position ($\bar{y} = 1$) can be regarded as the desired final workpiece surface. The deviations from this desired surface are the defects characterized by the nondimensional depth or height ($\bar{\delta}$), depending on whether the defect is a valley or a hill. Since $\delta = y - y_e$,

$$\bar{\delta} = \bar{y} - 1.$$

Equation (6.32) can be written in terms of $\bar{\delta}$, using the foregoing relation. Thus,

$$\bar{t} = \bar{\delta}_0 - \bar{\delta} + \ln \frac{\bar{\delta}_0}{\bar{\delta}}. \tag{6.33}$$

Theoretically, it would take an infinite time to remove a defect completely; in practice however, as soon as $\bar{\delta}$ goes below a preassigned allowable value, the process is finished. Figure 6.35 shows how the hills and the valleys are smoothened out.

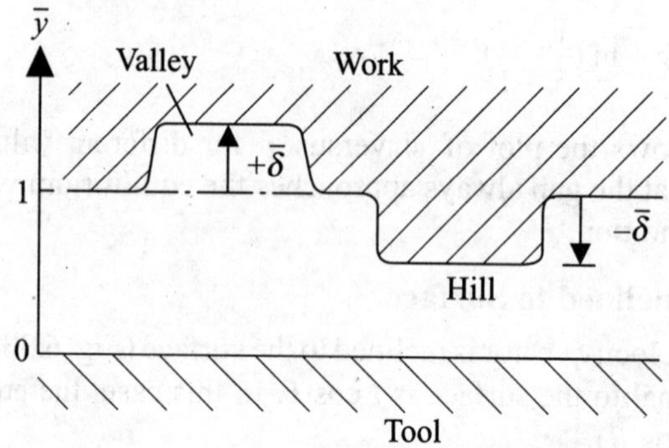

Fig. 6.34 Electrochemical smoothening of uneven work surface.

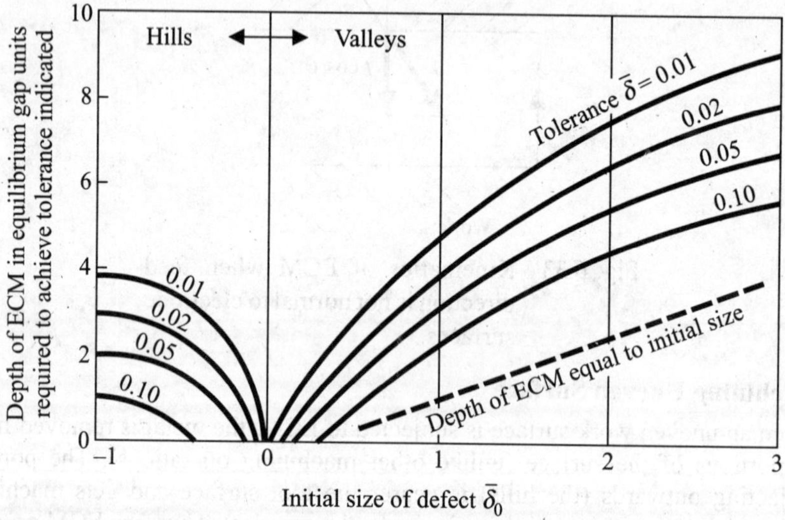

Fig. 6.35 Smoothening of hills and valleys.

EXAMPLE 6.5 In an ECM operation with the flat surfaces, a 10-V dc supply is used. The conductivity of the electrolyte is 0.2 ohm^{-1} cm^{-1}, and a feed rate of 1 mm/min is used. The workpiece is of pure iron. Calculate the equilibrium gap. Consider the total overvoltage to be 1.5 V.

SOLUTION For iron, $A = 55.85$ g, $Z = 2$, and $\rho = 7.86$ g/cm^3. Now, the equilibrium gap is given by

$$y_e = \frac{\lambda}{f} = \frac{\kappa A(V - \Delta V)}{\rho Z F f} = \frac{0.2 \times 55.85 \times (10 - 1.5)}{7.86 \times 2 \times 96{,}500 \times (0.1/60)}$$

$$= 0.04 \text{ cm.}$$

Theoretically, the equilibrium gap can have any value, but, in practice, the tool and the work surfaces are never perfectly flat. So, if the equilibrium gap is too small, the surface irregularities of the electrodes may touch each other, causing a short circuit.

EXAMPLE 6.6 The surface irregularities of the electrodes (with flat surfaces) are 5 μm and 8 μm. These are the heights of the peaks of the asperities as shown in Fig. 6.36. If the work is of pure iron and a dc voltage of 12 V is employed, estimate the largest possible feed rate which can be used. Assume the conductivity and the overvoltage to be the same as in Example 6.5.

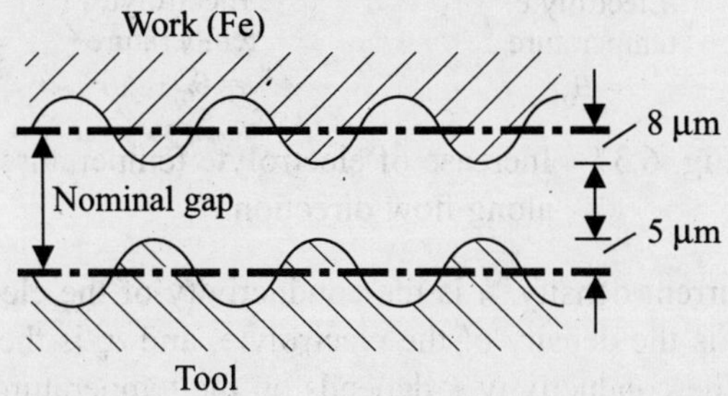

Fig. 6.36 Nominal gap between undulating work and tool surfaces.

SOLUTION The minimum allowable value of the nominal gap so that the electrodes do not touch each other is (0.0005 + 0.0008) = 0.0013 cm. So, when $y_e = 0.0013$ cm, the corresponding feed rate is given by

$$f_{max} = \frac{\kappa A(V - \Delta V)}{ZFy_e} = \frac{0.2 \times 55.85(12 - 1.5)}{7.86 \times 2 \times 96{,}500 \times 0.0013} \text{ cm/sec}$$

$$= 0.059 \text{ cm/sec} = 35.7 \text{ mm/min}.$$

So far as the machining forces are concerned, it may appear at the first glance that the forces acting on the tool and the workpiece are negligible since material removal takes place in the atomic level. But since the electrolyte has to be provided with an adequate rate of flow, normally the pressure is large. This hydrostatic force acting on the tool and the workpiece may be quite considerable in magnitude. The electrolyte flow is necessary

(i) to avoid the ion concentration,
(ii) to avoid the deposition on the tool,
(iii) to remove the precipitation,
(iv) to avoid the overheating of the electrolyte.

Of these four major objectives, the last one is very important, and an estimate of the required flow rate of the electrolyte can be worked out on the basis of this. Assuming all the heat, generated by the flow of electric current, to remain in the electrolyte (i.e., neglecting the conduction through the electrodes), the rise in temperature ($d\theta$) in passing a length dx of the gap (Fig. 6.37) is given by

$$d\theta = \frac{J^2 \, dx}{\kappa v \rho_e c_e}, \tag{6.34}$$

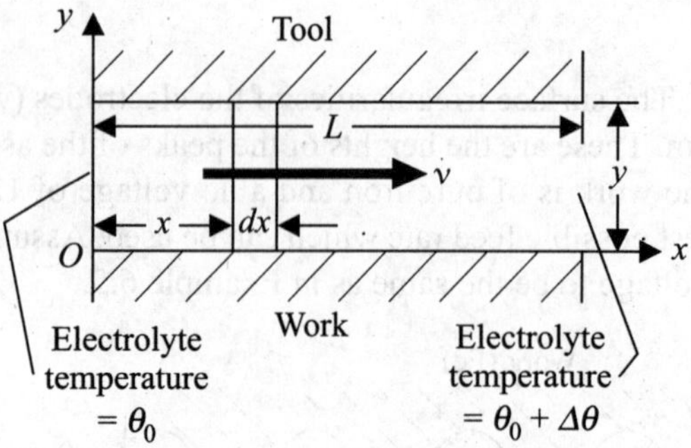

Fig. 6.37 Increase of electrolyte temperature along flow direction.

where J is the current density, κ is the conductivity of the electrolyte, v is the flow velocity, ρ_e is the density of the electrolyte, and c_e is the specific heat of the electrolyte. The conductivity κ depends on the temperature as

$$\kappa = \kappa_0[1 + \alpha(\theta - \theta_0)], \tag{6.35}$$

where κ_0 is the conductivity at the temperature θ_0, κ is the conductivity at the temperature θ, and α is a constant. Using equation (6.35) in equation (6.34), we obtain

$$d\theta = \frac{J^2\,dx}{\kappa_0[1+\alpha(\theta-\theta_0)]\rho_e c_e v}.$$

Rearranging this equation and integrating, we get

$$\int [1+\alpha(\theta-\theta_0)]\,d\theta = \frac{J^2}{\kappa_0 \rho_e c_e v}\,dx + C,$$

C being a constant. If the total temperature rise along the total length L of the gap is θ, this equation, after integration and rearrangement, takes the form

$$v = \frac{J^2 L}{\kappa_0(\Delta\theta + \frac{\alpha}{2}\Delta\theta^2)\rho_e c_e}. \tag{6.36}$$

If the gap is y and the viscosity of the electrolyte is η, the total pressure required (assuming turbulent flow) to maintain a flow velocity v is given by

$$p = [\frac{0.3164\,\rho_e v^2 L}{4y(Re)^{0.25}} + \frac{\rho_e v^2}{2}], \tag{6.37}$$

Re being Reynolds number (with the characteristic dimension $2y$). In equation (6.37), the first term is the pressure required to overcome the viscous friction and the second term is the pressure required to overcome the inertia. However, the pressure, acting on the tool face, is given by the first term.

EXAMPLE 6.7 During an ECM operation on an iron workpiece with a square-face copper tool (using brine as the electrolyte), both having a flat surface, a feed rate of 2 mm/min is used. The dc voltage used is 10 V and the total overvoltage is 1.5 V. The dimension of the tool face is 25.4 mm × 25.4 mm. The boiling temperature of the electrolyte is 95°C. Find out the total force acting on the tool. Use the data

viscosity of electrolyte = 0.876×10^{-3} kg/m-sec,

density of electrolyte = 1.088 g/cm^3,

specific heat of electrolyte = 0.997,

conductivity of electrolyte = 0.2 Ω^{-1} cm^{-1},

ambient temperature (initial temperature of electrolyte) = 35°C.

Neglect the variation in electrolyte conductivity due to the temperature change. The electrolyte is fed from one side of the square-shaped tool.

SOLUTION First, let us find out the equilibrium gap. For iron, A = 55.85 g, Z = 2, and ρ = 7.86 g/cm^3. Using equation (6.30), we find that the equilibrium gap is

$$y_e = \frac{\lambda}{f} = \frac{\kappa A(V - \Delta V)}{\rho Z F f} = \frac{0.2 \times 55.85 \times (10 - 1.5) \times 60}{7.86 \times 2 \times 96{,}500 \times 0.2} \text{ cm}$$

$$= 0.02 \text{ cm.}$$

The current density with this gap is

$$J = \frac{\kappa(V - \Delta V)}{y_e} = \frac{0.2 \times 8.5}{0.02} \text{ amp/cm}^2 = 85 \text{ amp/cm}^2.$$

Now, the allowable rise in temperature $\Delta\theta$ (to avoid boiling) being 95°C – 35°C = 60°C, the required velocity of the electrolyte flow is given by equation (6.36) with $\alpha = 0$. Thus,

$$v = \frac{J^2 L}{\kappa_0 \Delta\theta \rho_e c_e}$$

$$= \frac{85^2 \times 2.54}{0.2 \times 60 \times 1.088 \times 0.997} \text{ cm/sec} = 1410 \text{ cm/sec.}$$

With this velocity, Reynolds number with the characteristic dimension $2y_e$ and viscosity η_e becomes (using consistent units)

$$Re = \frac{\rho_e v 2 y_e}{\eta_e}$$

$$= 7004.$$

Now, from the first term of equation (6.37), the pressure on the tool is found to be

$$p_{\text{tool}} = \left(\frac{0.3164 \times 1.088 \times 1410^2 \times 2.54}{4 \times 0.02 \times 7004^{0.25}}\right) \times 10^{-4}\ \text{kN/m}^2$$

$$= 238\ \text{kN/m}^2.$$

The area of the tool face is 2.54^2 cm^2 = 6.45 cm^2, and the force acting on the tool is given by (assuming a linear drop in the pressure)

$$\tfrac{1}{2} \times 238 \times 10^{-4} \times 6.45\ \text{kN} = 79\ \text{N}.$$

6.4.3 EFFECTS OF HEAT AND H_2 BUBBLE GENERATION

In our analysis in Section 6.4.2, the different parameters and properties were assumed to be uniform throughout the face of the electrodes. But, in practice, it is not true. A variation in these properties affects the machining process. Also, the electrolyte conductivity changes as the electrolyte passes along the gap due to (i) the increase in electrolyte temperature, (ii) the evolution of hydrogen bubbles, and (iii) the formation of precipitates, the last effect being small. Because of the flow of electricity, the electrolyte temperature gradually increases and the conductivity changes, resulting in a nonuniformity in the current density along the direction of electrolyte flow. Apart from this, bubbles are formed since hydrogen is generated during machining. These bubbles are swept by the electrolyte, and the concentration of such bubbles tends to increase along the direction of electrolyte flow. As a result, the overall conductivity and the current density vary along the same direction. The resultant effect of these causes the equilibrium gap between the electrodes to vary.

6.4.4 SURFACE FINISH

Since, in general, a very good surface finish is desired in the parts machined by ECM, a study of the possibilities that may result in a bad finish is important. The surface finish is adversely affected by the (i) selective dissolution, (ii) sporadic breakdown of the anodic film, (iii) flow separation and formation of eddies, and (iv) evolution of H_2 gas.

Selective Dissolution

In alloys, the different constituents have varying electrode potentials. In pure metals too, the dissolution potentials at the grain boundaries are different from those inside the grains. Let us consider the work surface (with two constituents A and B) shown in Fig. 6.38a. In this figure, the voltage profile across the gap has also been shown. Let the dissolution potential of the constituent B (V_{d_B}) be greater than the dissolution potential of the constituent A (V_{d_A}). So, the required potential difference between a point on the surface and the adjacent electrolyte for ECM to start must be either V_{d_A} or V_{d_B}, depending on the local constituent. Since the whole anode surface is equipotential and the electrolyte potential varies across the gap as shown, the surface of a grain of B must project away from the surface of the constituent A (to meet the electrolyte with a lower potential) so that a larger difference, V_{d_B}, is achieved. Thus, in the steady state, the work surface will be uneven and not very smooth.

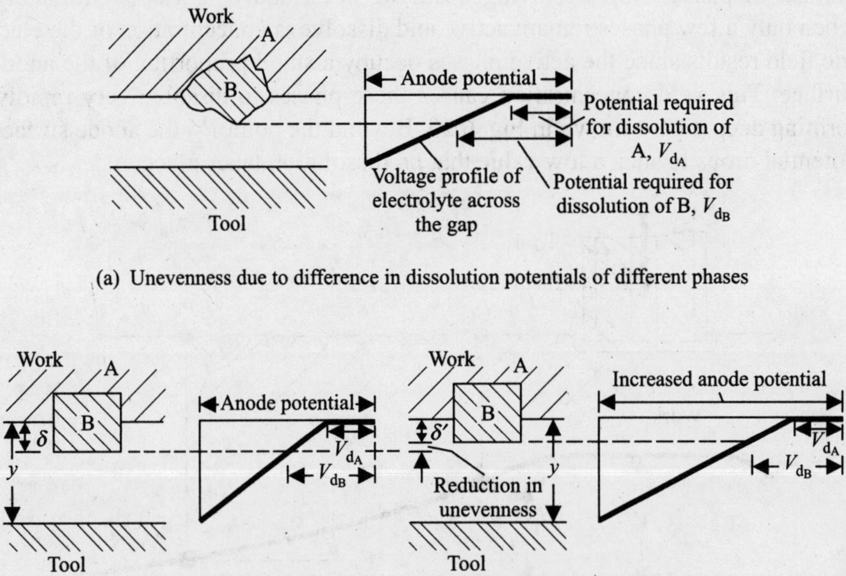

(a) Unevenness due to difference in dissolution potentials of different phases

(b) Reduction in unevenness with increase in anode potential

Fig. 6.38 Surface unevenness in multiphase materials.

When the potential gradient is higher, the unevenness is less. Figure 6.38b shows two situations with different potential gradients, the other parameters remaining the same. It is obvious from this figure that the height of the projection of a grain of the constituent B is less when the potential gradient is higher. An approximate expression of the projection height can also be derived as follows. From Fig. 6.38b,

$$(y-\delta)\frac{V-V_{d_A}}{y} = V - V_{d_B},$$

where V is the anode potential. Rearranging, we obtain

$$\delta = \left(1 - \frac{V-V_{d_B}}{V-V_{d_A}}\right)y. \tag{6.38}$$

Sporadic Breakdown of Anodic Film

The main reason for the sporadic breakdown of the anodic film is the gradual fall in the potential difference between the work surface and the electrolyte in the region away from the machining area. Figure 6.39 shows the variation of the surface potential of the anode in this region. Here, till the point P_1, the potential is enough to cause the dissolution of all the phases. At P_1, the available potential falls below the dissolution potential of one phase, and so the anode stops dissolving. Beyond P_1, the anode surface potential continues to drop and an increasing number of phases stop dissolving, resulting in an uneven surface. Ultimately, when only a few phases remain active and dissolve, a concentration of the electric field results since the active phases occupy a small proportion of the anode surface. This field concentration causes these phases to dissolve very rapidly, forming deep pits as shown in Fig. 6.39. Beyond the point P_2, the anode surface potential drops to such a low value that no dissolution takes place.

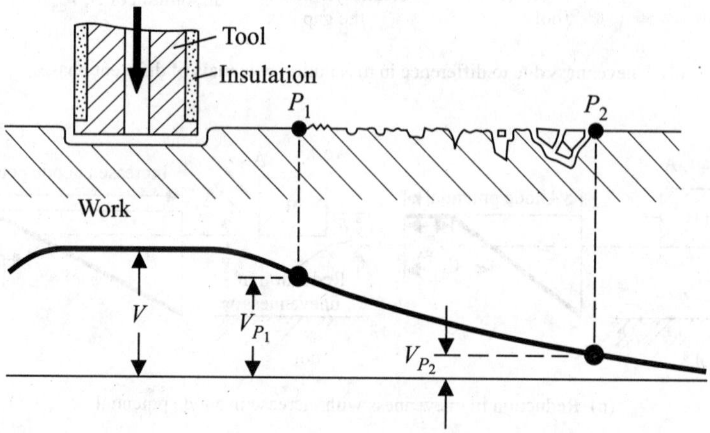

Fig. 6.39 Surface unevenness due to sporadic breakdown of anode.

Flow Separation and Formation of Eddies

The presence of hills and valleys on the anode surface may cause a separation of electrolyte flow and eddy formation. In these eddies, separated from the main stream, a large concentration of the metal ions may build up, resulting in a high concentration overpotential in the eddies. This introduces a localized variation in the removal rates, and consequently an uneven finished surface. Apart from the presence of hills and valleys, the flow separation may be caused by an improper design of the tool and the electrolyte flow path. So, a great care has to be taken in designing the electrolyte flow path in a tool.

Evolution of H_2 Gas

The flowing electrolyte collects the evolving hydrogen gas generated at the cathode. The presence of H_2 in the electrolyte reduces the specific conductivity of the solution. This effect increases as the H_2 concentration goes on increasing downstream, and the overall effect is a deterioration of the surface finish.

Apart from the foregoing four mechanisms, there are some other sources of surface deterioration. But since their importance is of a lower magnitude, we shall not discuss them.

6.4.5 TOOL DESIGN

There are two major aspects of tool design. These are:

(i) Determining the tool shape so that the desired shape of the job is achieved for the given machining conditions.

(ii) Designing the tool for considerations other than (i), e.g., electrolyte flow, insulation, strength, and fixing arrangements.

Theoretical Determination of Tool Shape

When the desired shape of the machined workpiece surface is known, it is possible to theoretically determine the required geometry of the tool surface for a given set of machining conditions.

Let the applied potential, the overvoltage, and the feed rate be V, ΔV, and f, respectively. The equilibrium gap between the anode and the cathode surfaces can be expressed as

$$g_e^1 = \frac{\kappa A(V - \Delta V)}{\rho ZFf \cos\theta}, \tag{6.39}$$

where θ is the inclination of the feed direction to the surface normal. The coordinates x and y are selected so that the y-axis and the feed direction are parallel. Let us consider a two-dimensional case where there is no variation in the z-direction. The work surface geometry is prescribed to be

$$y = \phi(x) \tag{6.40}$$

as shown in Fig. 6.40. When the steady state is reached, any point $P_w(x_w, y_w)$

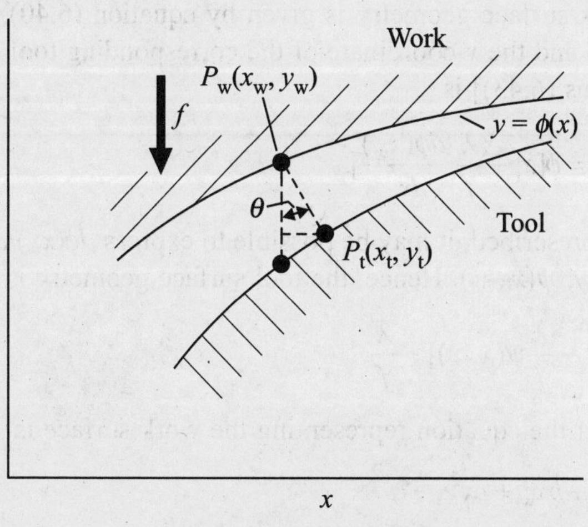

Fig. 6.40 Generation of tool surface for given work surface.

[1]In our analysis here, the gap is represented by g instead of y to avoid confusion with the coordinate y.

on the work surface can be transformed into the corresponding point $P_t(x_t, y_t)$ on the tool surface so that

$$P_w P_t = g_e. \tag{6.41}$$

Then,

$$\begin{aligned} y_w - y_t &= PP_t \cos\theta = g_e \cos\theta, \\ x_t - x_w &= P_w P_t \sin\theta = g_e \sin\theta. \end{aligned} \tag{6.42}$$

Using equation (6.39) in equations (6.42), we get

$$y_t = y_w - \frac{\kappa A(V - \Delta V)}{\rho Z F f} = y_w - \frac{\lambda}{f},$$

$$x_t = x_w + \frac{\kappa A(V - \Delta V)}{\rho Z F f} \tan\theta = x_w + \frac{\lambda}{f} \tan\theta.$$

With

$$\tan\theta = \frac{dy_w}{dx_w},$$

these equations become

$$\begin{aligned} y_t &= y_w - \frac{\lambda}{f}, \\ x_t &= x_w + \frac{\lambda}{f}\left(\frac{dy_w}{dx_w}\right). \end{aligned} \tag{6.43}$$

Since the work surface geometry is given by equation (6.40), the relationship between the y- and the x-coordinate of the corresponding tool surface obtained [using equations (6.43)] is

$$y_t + \frac{\lambda}{f} = \phi\left[x_t - \frac{\lambda}{f} \frac{d\phi(x_w)}{dx_w}\right].$$

When $\phi(x)$ is prescribed, it may be possible to express $d\phi(x_w)/dx_w$ as a function of x_t and y_t, say, $\psi(x_t, y_t)$. Hence, the tool surface geometry is represented by

$$y = \phi\left[x - \frac{\lambda}{f}\psi(x, y)\right] - \frac{\lambda}{f}. \tag{6.44a}$$

For example, if the equation representing the work surface is

$$y_w = a + bx_w + cx_w^2,$$

then

$$\frac{d\phi(x_w)}{dx_w} = b + 2cx_w.$$

Using equations (6.43) in this equation, we get

$$\frac{d\phi(x_w)}{dx_w} = b + 2c[x_t - \frac{\lambda}{f}\frac{d\phi(x_w)}{dx_w}]$$

or

$$\frac{d\phi(x_w)}{dx_w} = \frac{b + 2cx_t}{1 + 2c\dfrac{\lambda}{f}} \equiv \psi(x_t, y_t).$$

Substituting the foregoing expression of $\psi(x_t, y_t)$ in equation (6.44a), we find that the required tool surface geometry becomes

$$y = a + bx + cx^2 - \frac{\lambda}{f} - \frac{\lambda}{f}[\frac{(b + 2cx)^2}{1 + 2c\lambda/f}]. \tag{6.44b}$$

For a similar case in three dimensions, i.e., when the work geometry is represented by

$$y = a + bx + cx^2 + dz + ez^2 + gxz,$$

the required tool geometry is given by

$$y = a + bx + cx^2 + dz + ez^2 + gxz - \frac{\lambda}{f}$$

$$- \frac{\lambda}{f}[\frac{(b + 2cx + gz)^2 + (d + 2ez + gx)^2}{1 + 2(c + e)\lambda/f}]. \tag{6.44c}$$

EXAMPLE 6.8 The geometry of a workpiece surface with single curvature is given by the equation

$$y = 10 + 0.3x - 0.05x^2,$$

where x and y are in cm. The process data are

applied potential = 15 V,

overvoltage = 0.67 V,

feed velocity = 0.75 mm/min (given to the work in the $-y$-direction),

work material = copper,

electrolyte conductivity = 0.2 Ω^{-1} cm^{-1}.

Determine the equation of the required tool surface geometry.

SOLUTION From Table 6.3, we find that for copper $Z = 1$, $A = 63.57$, and $\rho = 8.96$ g/cm^3. The value of F is 96,500 coulombs. Converting the feed in cm/sec, we get $f = 0.00125$ cm/sec.

First, we find λ as

$$\lambda = \frac{\kappa A(V - \Delta V)}{\rho ZF} = \frac{0.2 \times 63.57 \times (15 - 0.67)}{8.96 \times 1 \times 96{,}500} \text{ cm}^2\text{/sec}$$

$$= 2.11 \times 10^{-4} \text{ cm}^2/\text{sec}.$$

Thus,

$$\frac{\lambda}{f} = \frac{2.11 \times 10^{-4}}{12.5 \times 10^{-4}} \text{ cm} = 0.169 \text{ cm}.$$

Now,

$$\phi(x) = 10 + 0.3x - 0.05x^2.$$

Finally, the equation of the tool surface geometry is found by using equation (6.44b). Thus,

$$y = 10 + 0.3[x - 0.169\frac{(0.3 - 0.1x)}{1 - 0.1 \times 0.169}]$$

$$- 0.05[x - 0.169\frac{(0.3 - 0.1x)}{1 - 0.1 \times 0.169}]^2 - 0.169,$$

where x and y are in cm. Simplifying this equation, we finally get the tool surface equation as

$$y = 9.8154 + 0.3157x - 0.0517x^2,$$

where both x and y are in cm.

However, it should be remembered that the method in Example 6.8 is applicable for smooth surfaces with gentle variations, i.e., where the current flow lines are more or less parallel to one another. For more complex shapes and for surfaces involving sharp curves and sudden changes, a solution of the electric field is necessary. For a description of the various methods of solution for such problems, the interested reader should consult the standard reference books. When the closed form expression of the workpiece surface is not available, the surface can be divided into small, straight or curved segments of known geometry, and then the method we have described can be applied.

Design for Electrolyte Flow

A sufficient electrolyte flow between the tool and the workpiece is necessary to carry away the heat and the products of machining and to assist the machining process at the required feed rate, producing a satisfactory surface finish. Cavitation, stagnation, and vortex formation should be avoided since these lead to a bad surface finish. One basic rule is that there should be no sharp corners in the flow path. All corners in the flow path should have a radius of at least 0.7–0.8 mm as shown in Fig. 6.41.

The initial shape of a component generally does not comply with the tool shape and only a small fraction of the area is close to the tool surface at the beginning. The problem of supplying the electrolyte over such an area is usually solved by the flow restriction techniques.

In many situations, when the initial work shape conforms to the tool shape,

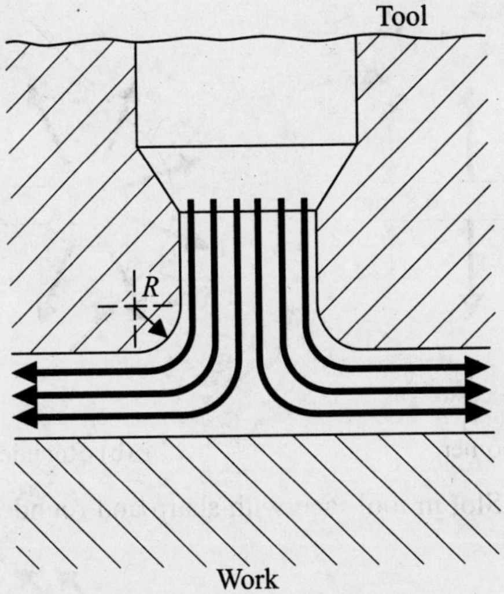

Fig. 6.41 Avoiding sharp corners in flow path.

the machining process itself causes the formation of boss or ridge in the workpiece; this helps in a proper distribution of the electrolyte flow (Fig. 6.42).

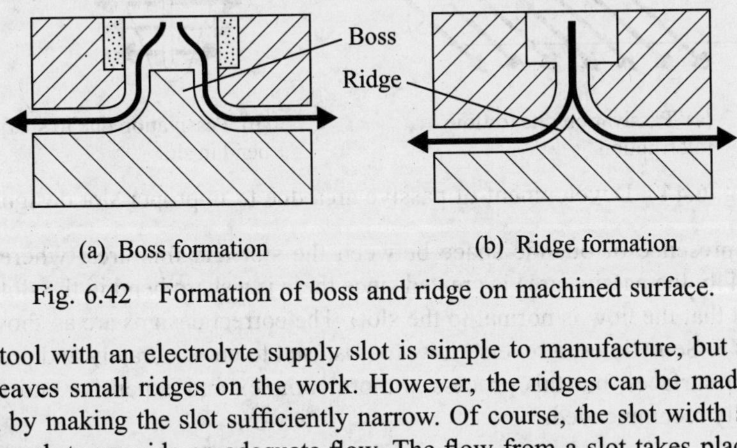

Fig. 6.42 Formation of boss and ridge on machined surface.

A tool with an electrolyte supply slot is simple to manufacture, but such a slot leaves small ridges on the work. However, the ridges can be made very small by making the slot sufficiently narrow. Of course, the slot width should be enough to provide an adequate flow. The flow from a slot takes place in a direction perpendicular to the slot and the flow at the end is poor. Therefore, the slot should be terminated near the corners of the workpiece surface as shown in Fig. 6.43a. The distance between the tip of the slot and the corners should be at least 1.5 mm, whereas a slot with a width 0.7–0.8 mm is recommended. When a workpiece corner is rounded, the slot end should be made larger as shown in Fig. 6.43b. The shape and location of the slot should be such that every portion of the surface is supplied with electrolyte flow and no passive area exists. Figure 6.44 shows two situations where the passive areas exist since the slot design is faulty. In Fig. 6.44a, the passive area is not getting the supply because

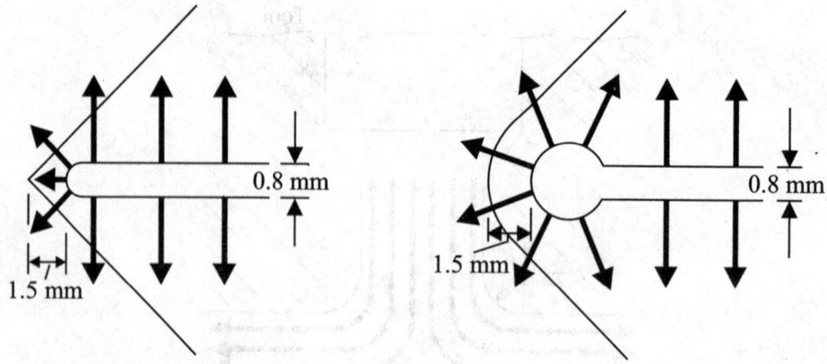

(a) Sharp corner (b) Rounded corner

Fig. 6.43 Slot in tool face with sharp and rounded corners.

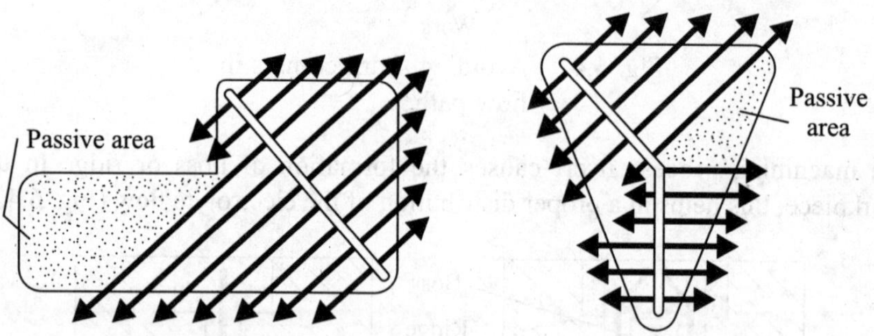

(a) Passivation due to flow interruption (b) Passivation due to sharp bend in slot

Fig. 6.44 Development of passive area due to improper slot design.

of the presence of outside space between the slot and this area, whereas in Fig. 6.44b, the passive area is created since there is a sharp bend in the slot (and the fact that the flow is normal to the slot). The correct designs are as shown in Fig. 6.45. Sometimes, a reverse flow tool is used to cut accurately and produce superior surfaces, but this process is more complex and expensive and is not generally recommended.

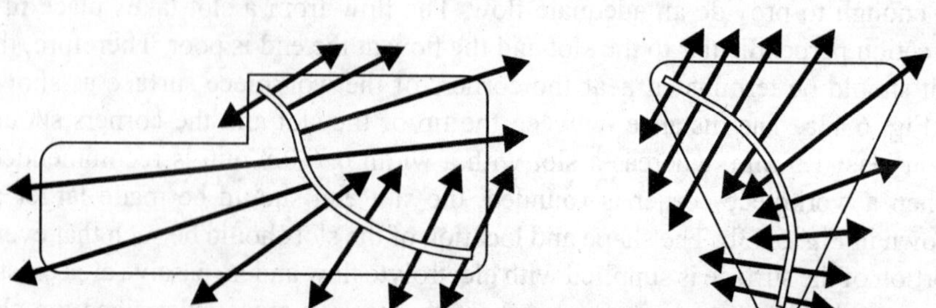

Fig. 6.45 Slot design to avoid development of passive area.

The techniques for controlling the electrolyte flow when the initial work surface does not conform to the tool shape are illustrated in Fig. 6.46. The general rules for putting a flow restrictor can be stated as follows. The flow restrictor must be adjacent to the area of initial close proximity (between the tool and the work surface) and should not increase the flow path appreciably. Also, it must be at the electrolyte entry or exit position.

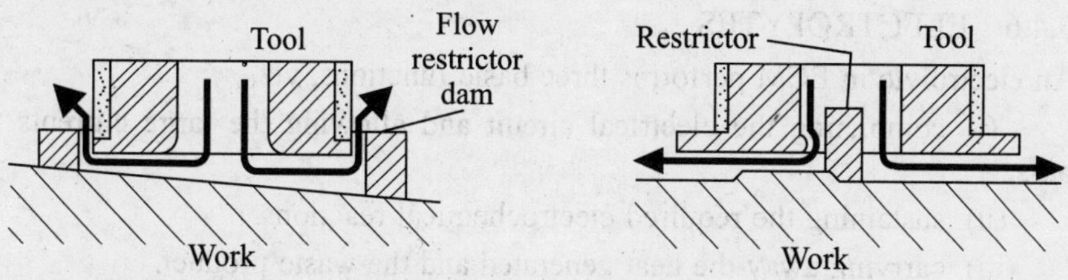

Fig. 6.46 Control of electrolyte flow by restrictor dams.

Design for Insulation

The areas on a tool where electrochemical machining is not desirable have to be insulated. In die sinking also, the tool should be properly insulated to minimize stray machining. Figure 6.47 shows the ECM process without and with a proper insulation. Figure 6.48 illustrates die sinking without and with a proper insulation.

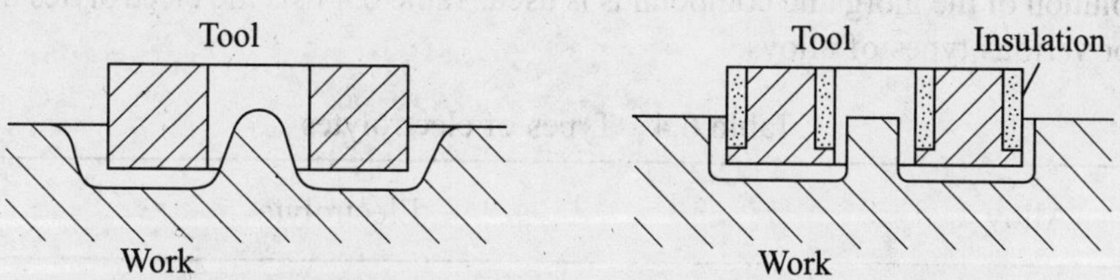

(a) Tool without insulation (b) Insulated tool

Fig. 6.47 Effect of insulation.

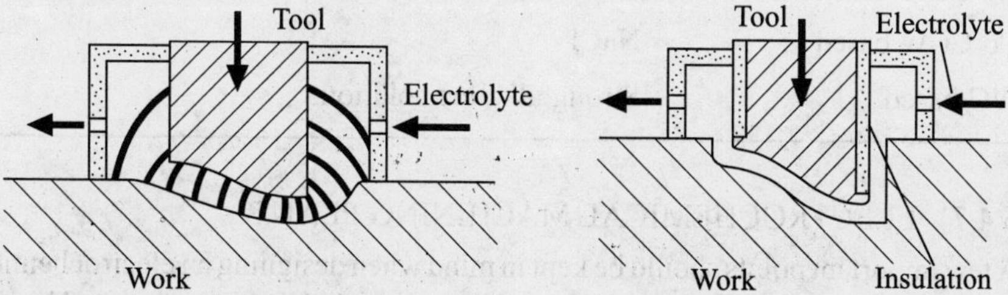

(a) Tool without insulation (b) Insulated tool

Fig. 6.48 Stray machining with uninsulated tool.

The insulation must be tough and securely bonded to the tool surface. It can be provided by securing the reinforced solid plastic material to the toll with epoxy resin cement and plastic screws. Sometimes, the insulation can also be done by applying a synthetic rubber coating on the artificially oxidized copper tool surface. For this, a hot chemical oxidizing solution is used. The boundaries of the insulation layer should not be exposed to a high velocity electrolyte flow as this may tend to tear up the glued layer.

6.4.6 ELECTROLYTES

An electrolyte in ECM performs three basic functions, viz.,

(i) completing the electrical circuit and allowing the large currents to pass,

(ii) sustaining the required electrochemical reactions,

(iii) carrying away the heat generated and the waste product.

The first function requires the electrolyte, ideally, to have a large electrical conductivity. The second function requires the electrolyte to be such that at the anode the workpiece material is continuously dissolved, and a discharge of the metal ion on the cathode should not occur. Generally, the cationic constituent of the electrolyte is hydrogen, ammonia, or alkali metals. The dissolution of the anode should be sustained at a high level of efficiency. Also, the electrolyte must have a good chemical stability. Apart from all these, the electrolyte should be inexpensive, safe, and as noncorrosive as possible. Generally, an aqueous solution of the inorganic compounds is used. Table 6.4 lists the electrolytes used for various types of alloys.

Table 6.4 Types of electrolytes

Alloy	Electrolyte
Iron based	Chloride solutions in water (mostly 20% NaCl)
Ni based	HCl or mixture of brine and H_2SO_4
Ti based	10% hydrofluoric acid + 10% HCl + 10% HNO_3
Co-Cr-W based	NaCl
WC based	Strong alkaline solutions

6.4.7 ELECTROCHEMICAL MACHINING PLANT

A few important points should be kept in mind when designing an electrochemical machine. These include the stiffness and the material of the components. Though, at a first glance, it appears that the machining force is negligible as there is no physical contact between the tool and the workpiece surface, very large forces may develop between them due to the high pressure of the electrolyte required to maintain an adequate flow velocity through the narrow gap. So, the machine must possess enough rigidity to avoid any significant deflection of the tool which may destroy the accuracy of the parts being machined. A change of temperature may also cause some relative displacement between the tool and the workpiece, and the design should take care of it.

To avoid corrosion, wherever possible, the nonmetallic materials should be used. When strength and stiffness are required, the plastic coated metals should be used. The material used to hold the workpiece is exposed to anodic attack, and Ti appears to be most suitable because of its passivity. When different metals are in contact in the presence of the electrolyte, especially when the machine is idle, corrosion may occur. To minimize this, the metals in contact should be so chosen that they do not differ much in their electrochemical behaviour. The slideways cannot be protected permanently, and so they are heavily coated with grease. Sometimes, a corrosion protection may be provided by applying a small electrical potential in such a direction that the whole structure becomes more noble electrochemically. This is commonly known as cathodic protection.

The pump is the most important element of the ancillary plant. Generally, the positive displacement pumps (similar to gear pumps) made of stainless steel are used. The tank for the electrolyte, the pipeline, and the valves are normally made of PVC. The schematic diagram of an ECM plant is shown in Fig. 6.49.

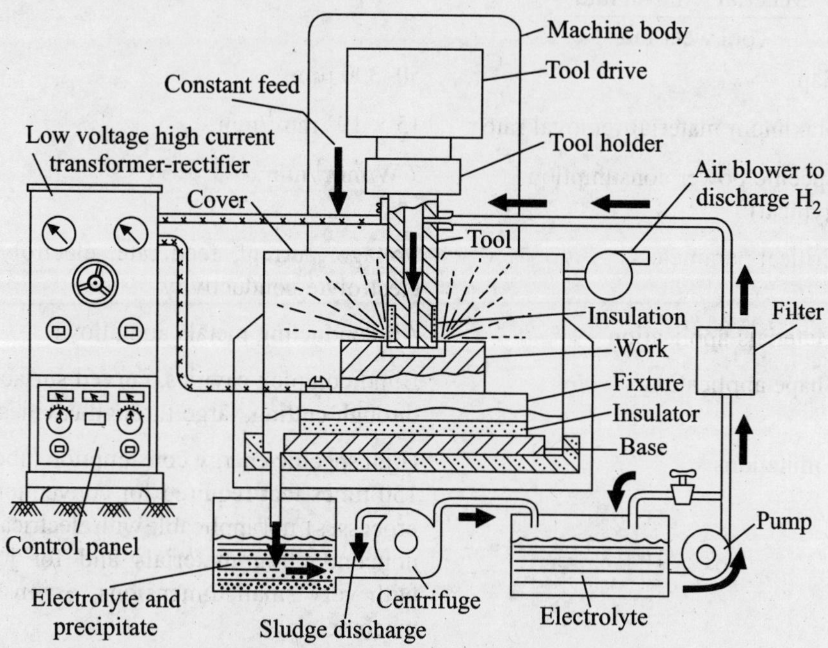

Fig. 6.49 Electrochemical machining plant.

6.4.8 EFFECTS OF ECM ON MATERIALS

In contrast with the conventional machining processes, the material removal during ECM is smooth and gentle. As a result, the maximum residual compressive stress is very low in the workpiece surface. Moreover, the depth of the work hardened surface layer is negligible. When the depth of the work hardened surface layer is about 0.5 mm and 1.5 mm for turning and milling, respectively, that in ECM is only about 0.001 mm. Similarly, the order of magnitude of the residual stress in a surface machined by a conventional process is about 50 kg/mm^2, whereas that with ECM is almost zero. This results in a 10–25% lower fatigue strength of the parts produced by ECM. This is because the micro crack tips are exposed at the surface produced by ECM and also because the process leaves a stress-free surface. For increasing the fatigue strength, some mechanical processes (e.g., mechanical polishing, glass bead blasting, and vapour blasting) can be used.

6.4.9 SUMMARY OF ECM CHARACTERISTICS

Mechanics of material removal	Electrolysis
Medium	Conducting electrolyte
Tool	
Materials	Cu, brass, steel
$\dfrac{\text{Material removal rate}}{\text{Tool wear rate}}$	∞
Gap	50–300 μm
Maximum material removal rate	15×10^3 mm^3/min
Specific power consumption (typical)	7 W/mm^3/min
Critical parameters	Voltage, current, feed rate, electrolyte, electrolyte conductivity
Materials application	All conducting metals and alloys
Shape application	Blind complex cavities, curved surfaces, through cutting, large through cavities
Limitations	High specific energy consumption (about 150 times that required for conventional processes), not applicable with electrically nonconducting materials and for jobs with very small dimensions, expensive machine

6.5 ELECTRIC DISCHARGE MACHINING (EDM)

The use of a thermoelectric source of energy in developing the nontraditional techniques has greatly helped in achieving an economic machining of the extremely low machinability materials and difficult jobs. The process of material removal by a controlled erosion through a series of electric sparks, commonly known as electric discharge machining, was first started in the USSR around 1943. Then onwards, research and development have brought this process to its present level.

The basic scheme of electric discharge machining is illustrated in Fig. 6.50. When a discharge takes place between two points of the anode and the cathode,

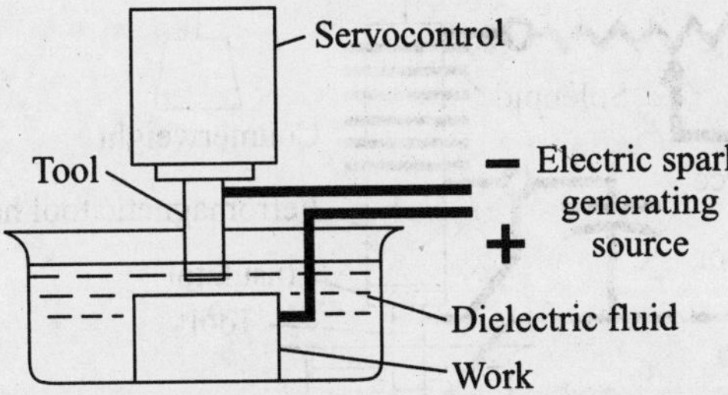

Fig. 6.50 Basic scheme of electric discharge machining.

the intense heat generated near the zone melts and evaporates the materials in the sparking zone. For improving the effectiveness, the workpiece and the tool are submerged in a dielectric fluid (hydrocarbon or mineral oils). It has been observed that if both the electrodes are made of the same material, the electrode connected to the positive terminal generally erodes at a faster rate. For this reason, the workpiece is normally made the anode. A suitable gap, known as the spark gap, is maintained between the tool and the workpiece surfaces. The sparks are made to discharge at a high frequency with a suitable source. Since the spark occurs at the spot where the tool and the workpiece surfaces are the closest and since the spot changes after each spark (because of the material removal after each spark), the sparks travel all over the surface. This results in a uniform material removal all over the surface, and finally the work face conforms to the tool surface. Thus, the tool produces the required impression in the workpiece. For maintaining the predetermined spark gap, a servocontrol unit is generally used. The gap is sensed through the average voltage across it and this voltage is compared with a preset value. The difference is used to control the servomotor. Sometimes, a stepper motor is used instead of a servomotor. Of course, for very primitive operations, a solenoid control is also possible, and with this, the machine becomes extremely inexpensive and simple to construct[1]. Figure 6.51 schematically shows the arrangement of a solenoid controlled electric discharge machine. The spark frequency is normally in the range 200–500,000 Hz, the spark gap being of the order of 0.025–0.05 mm. The peak voltage across the gap is kept in the range 30–250 volts. An mrr up to 300 mm^3/min can be obtained with this process, the specific power being of

[1]Shankar, Kripa and Ghosh, A., A study of electrospark machining characteristics with electromagnetic spark-gap controlling mechanism, *Int. J. Mach. Tool Des. Res.*, **15**, 209, 1975.

the order of 10 W/mm^3/min. The efficiency and the accuracy of performance have been found to improve when a forced circulation of the dielectric fluid is provided. The most commonly used dielectric fluid is kerosene. The tool is generally made of brass or a copper alloy.

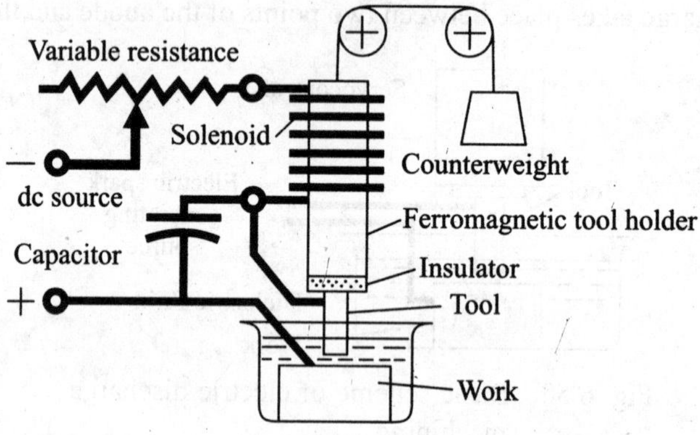

Fig. 6.51 Solenoid controlled electric discharge machine.

6.5.1 MECHANICS OF EDM

Figure 6.52 shows the details of the electrode surfaces. Though the surfaces may appear smooth, asperities and irregularities are always present, as indicated

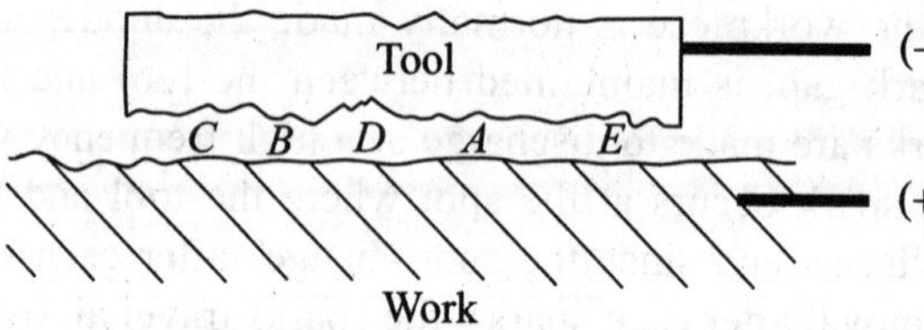

Fig. 6.52 Details of electrode surface characteristics.

(in an exaggerated manner, of course). As a result, the local gap varies, and at a given instant, it is minimum at one point (say, A). When a suitable voltage is built up across the tool and the workpiece (the cathode and the anode, respectively), an electrostatic field of sufficient strength is established, causing cold emission of electrons from the cathode at A. These liberated electrons accelerate towards the anode. After gaining a sufficient velocity, the electrons collide with the molecules of the dielectric fluid, breaking them into electrons and positive ions. The electrons so produced also accelerate and may ultimately dislodge the other electrons from the dielectric fluid molecules. Ultimately, a narrow column of ionized dielectric fluid molecules is established at A connecting the two electrodes (causing an avalanche of electrons, since the conductivity of the ionized column is very large, which is normally seen as a spark). As a result of this spark, a compression shock wave is generated and a very high temperature is developed on the electrodes (10,000–12,000°C). This high temperature causes the melting and vaporization of the electrode materials, and the molten metals are evacuated by a mechanical blast, resulting in small craters on both the electrodes at A. As soon as this happens, the gap between the electrodes at A increases and the next location of the shortest gap is somewhere else (say, B). Therefore, when the cycle is repeated, the next spark takes place at B. In this way, the sparks wander all over the electrode surface and, ultimately, the process results in a uniform gap. So, depending on the negative electrode shape, an impression is created on the other electrode. Generally, the rate of material removal from the cathode is comparatively less than that from the anode due to the following reasons:

(i) The momentum with which the stream of electrons strikes the anode is much more than that due to the stream of the positive ions impinging on the cathode though the mass of an individual electron is less than that of the positive ions.

(ii) The pyrolysis of the dielectric fluid (normally a hydrocarbon) creates a thin film of carbon on the cathode.

(iii) A compressive force is developed on the cathode surface.

Therefore, normally, the tool is connected to the negative terminal of the dc source.

If the tool is stationary relative to the workpiece, the gap increases as the material removal progresses, necessitating an increased voltage to initiate the sparks. To avoid this problem, the tool is fed with the help of a servodrive which senses the magnitude of the average gap and keeps it constant.

In what follows, we shall attempt a theoretical determination of the material removal rate during electric discharge machining. In so doing, though the quantitative results will not be obtained, many important features will become evident. For now, it would be sufficient to understand the effect of only one spark.

The quantity of material removal due to a single discharge can be determined by considering the diameter of the crater and the depth to which the melting temperature is reached. To do this, we shall make the following assumptions:

(i) The spark is a uniform circular heat source on the electrode surface and the diameter ($=2a$) of this circular source remains constant.

(ii) The electrode surface is a semi-infinite region.

(iii) Except for the portion of the heat source, the electrode surface is insulated.

(iv) The rate of heat input remains constant throughout the discharge duration.

(v) The properties of the electrode material do not change with the temperature.

(vi) The vaporization of the electrode material is neglected.

Figure 6.53 shows the details of the idealized heat source. In our analysis, $H =$ amount of heat input (cal), $\theta =$ temperature (°C), $t =$ time (sec), $k =$ thermal conductivity (cal/cm-sec-°C), $\alpha =$ thermal diffusivity (cm^2/sec), $t_d =$ discharge

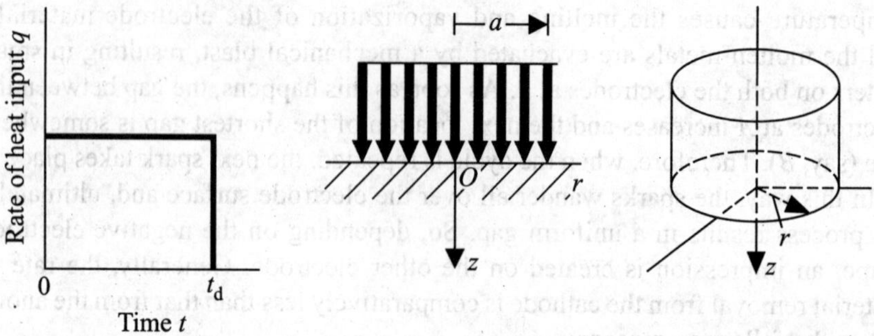

(a) Constant rate of heat input (b) Uniform heat flux (c) Circular heat source

Fig. 6.53 Description of idealized heat source during EDM.

duration (sec), and θ_m = melting temperature (°C).

Because of circular symmetry, the temperature at any point depends on r and z. The equation for heat conduction is

$$\frac{\partial \theta}{\partial t} = \alpha\left(\frac{\partial^2 \theta}{\partial r^2} + \frac{1}{r}\frac{\partial \theta}{\partial r} + \frac{\partial^2 \theta}{\partial z^2}\right). \tag{6.45}$$

The initial and the boundary conditions are

$$t \leqslant 0, \qquad \theta(r, z, t) = 0,$$

$$t > 0, \qquad r > a, \qquad \frac{\partial \theta}{\partial z} = 0,$$

$$0 < r \leqslant a, \qquad -k\frac{\partial \theta}{\partial z} = \frac{H}{\pi a^2 t_d}.$$

Since, intuitively, it can be seen that the depth to which the melting temperature is reached is maximum at the centre, our interest lies in the solution at $r = 0$. The temperature at a point on the axis at the end of the discharge (assuming that the maximum temperature is reached at $t = t_d$ as the heat input stops at this instant) is given by

$$\theta(0, z, t_d) = \frac{1}{2}\frac{H}{\pi k a t_d}\int_0^\infty J_0(\xi a)J_1(\xi a)\left[e^{-\xi z}\,\text{erfc}\left\{\frac{z}{2\sqrt{\alpha t_d}} - \xi\sqrt{\alpha t_d}\right\}\right]\frac{d\xi}{\xi}, \tag{6.46}$$

where ξ is a dummy variable. If Z is the depth to which the melting temperature is reached, the equation obtained is

$$\theta_m = \frac{2H\sqrt{\alpha t_d}}{\pi k a^2 t_d}\left[\text{ierfc}\,\frac{Z}{2\sqrt{\alpha t_d}} - \text{ierfc}\,\frac{\sqrt{Z^2 + a^2}}{2\sqrt{\alpha t_d}}\right], \tag{6.47}$$

where

$$\text{ierfc}(\zeta) = \frac{1}{\sqrt{\pi}} e^{-\zeta^2} - \zeta \,\text{erfc}\,\zeta,$$

$$\text{erfc}(\zeta) = 1 - \text{erf}(\zeta),$$

$$\text{erf}(\zeta) = \frac{2}{\sqrt{\pi}} \int_0^{\zeta} e^{-x^2}\, dx.$$

To take care of the latent heat of the molten material, the actual heat input rate can be found out by subtracting the heat used to melt the material from the total heat supplied by the spark. Thus, the rate of heat input is given by

$$\frac{H_{\text{total}} - H_{\text{m}}\rho\pi a^2 Z}{\pi a^2 t_{\text{d}}} \text{ cal/cm}^2\text{-sec,} \tag{6.48}$$

where

H_{total} = total amount of heat released (cal),

H_{m} = latent heat (cal/g),

ρ = density of the material (g/cm^3).

The diameter of the crater has been assumed to be equal to $2a$, i.e., the spark diameter, which, under the idealized condition of uniform strength, is given by

$$2a = KW^{n_1} t_{\text{d}}^{n_2} \text{ cm,} \tag{6.49}$$

where W is the total pulse energy (in joules), and n_1, n_2, K are constants characterizing the properties of the electrodes and the dielectric medium. The melting temperature depth Z is related to the crater volume as

$$v_{\text{c}} = \frac{\pi}{6} h_{\text{c}}(3a^2 + h_{\text{c}}^2) \text{ cm}^3, \tag{6.50}$$

where

v_{c} = crater volume (cm^3),

h_{c} ($\propto Z$) = crater depth (cm).

So, it is clear that Z gives an indication of the volume of material removed by each spark. Figure 6.54a shows the theoretical values of Z for a given spark energy and a constant spark diameter for Cu, Al, and Zn as the electrode materials. Figure 6.54b depicts the actual nature of variation of the crater volume with t_{d} for different spark energies. The trends are quite similar. One important feature which becomes evident from these results is that the material removal is very low for a small discharge time and increases with t_{d}. Then, reaching a peak value, it suddenly drops to zero. Also, it has been established that the material removed per discharge strongly depends on the melting point of the material.

The effect of cavitation in the mechanical removal process is also important. The mrr during a single spark plotted against time is as shown in Fig. 6.55. Clearly,

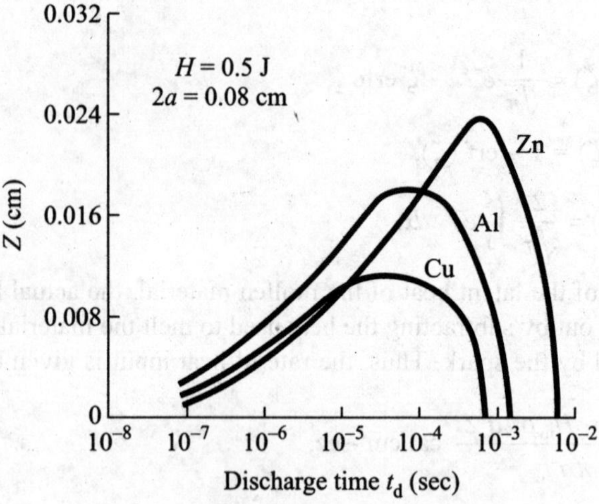

(a) Variation of Z

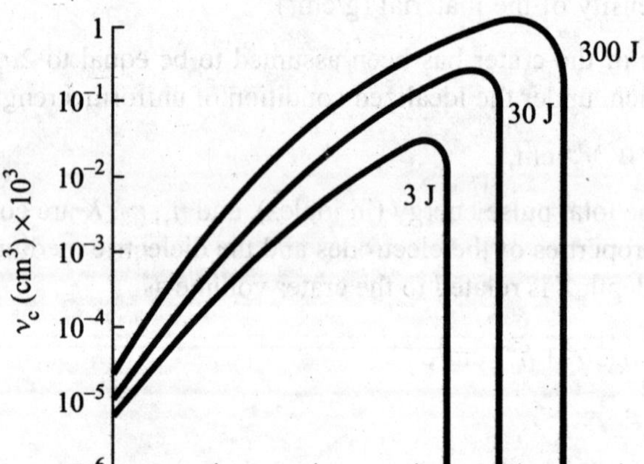

(b) Variation of V_c

Fig. 6.54 Variation of melting temperature depth Z and crater volume v_c with discharge time t_d.

the mrr is maximum when the pressure is below atmospheric, showing the importance of cavitation.

For arriving at a rough estimate, empirical relationships have been developed for the material removal rate during EDM. Since the size of the crater depends on the spark energy (assuming all other conditions remain unchanged), the depth and diameter of the crater are given by

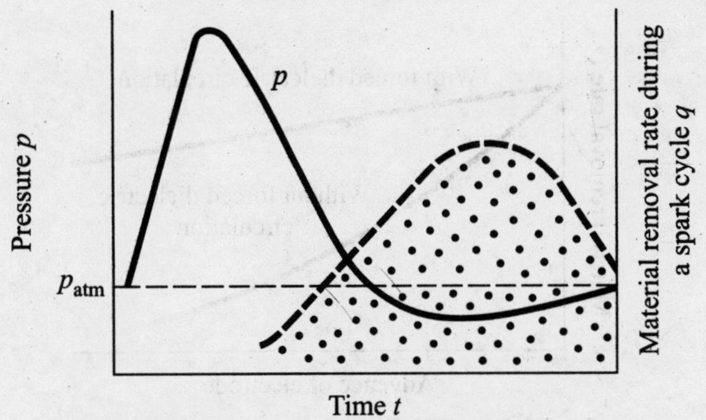

Fig. 6.55 Role of cavitation in material removal.

$$h_c = (K_1 W^{1/3}) \text{ cm}, \tag{6.51a}$$

$$2a = (K_2 W^{1/3}) \text{ cm}, \tag{6.51b}$$

where W is the spark energy (joules), K_1 and K_2 being the constants. For Cu electrodes and kerosene as the dielectric medium, $K_1 \approx 0.4$ and $K_2 \approx 0.045$. Using equations (6.51) in equation (6.50), we obtain

$$v_c = \frac{\pi}{6} K_1 (\tfrac{3}{4} K_2^2 + K_1^2) W \text{ cm}^3. \tag{6.52}$$

Another approximate relationship, for getting a rough estimate of the mrr (under normal working conditions), is in terms of the melting point of the workpiece material and is given by

$$Q \approx 4 \times 10^4 \theta_m^{-1.23} \text{ mm}^3\text{/amp-min}, \tag{6.53}$$

where

Q = metal removal rate (mm^3/amp-min),

θ_m = melting temperature (°C).

In this relation, we have assumed average sparking condition.

The mrr also depends strongly on the circulation of the dielectric fluid. Without a forced circulation, the wear particles repeatedly melt and reunite with the electrode. Figure 6.56 shows the nature of the mrr characteristics without and with a forced circulation of the dielectric.

After the discharge is completed, the dielectric medium around the last spark should be allowed to deionize. For this, the voltage across the gap must be kept below the discharge voltage until deionization is complete, otherwise the current again starts flowing through the gap at the location of the preceding discharge. The time required for a complete deionization depends on the energy released by the preceding discharge. A larger energy release results in a longer deionization time.

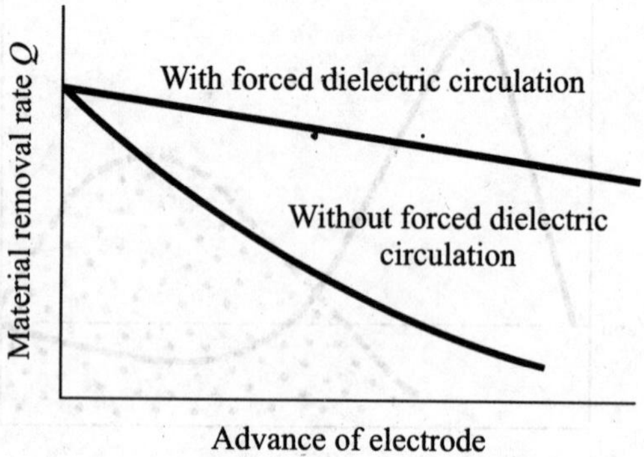

Fig. 6.56 Effect of forced circulation of dielectric fluid.

6.5.2 EDM CIRCUITS AND OPERATING PRINCIPLES

Several basically different electric circuits are available to provide the pulsating dc across the work-tool gap. Though the operational characteristics are different, in almost all such circuits a capacitor is used for storing the electric charge before the discharge takes place across the gap. The suitability of a circuit depends on the machining conditions and requirements. The commonly-used principles for supplying the pulsating dc can be classified into the following three groups:

(i) Resistance-capacitance relaxation circuit with a constant dc source.
(ii) Rotary impulse generator.
(iii) Controlled pulse circuit.

Resistance-Capacitance Relaxation Circuit

The resistance-capacitance relaxation circuit was used when the electric discharge machines were first developed. Figure 6.57a shows a simple *RC* circuit. As is clear from this figure, the capacitor *C* (which can be varied) is charged through a variable resistance *R* by the dc source of voltage V_0. The voltage across the

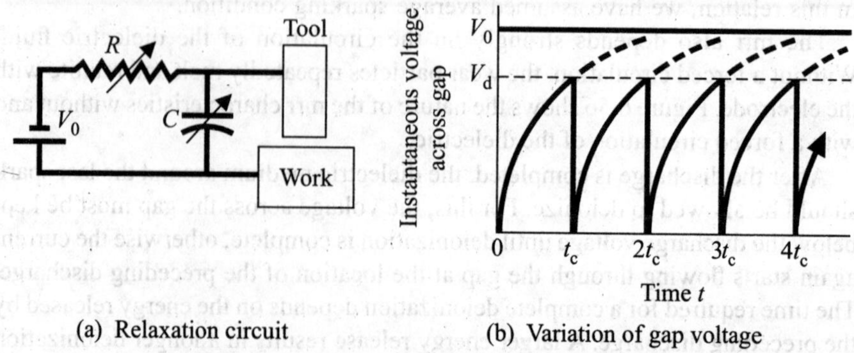

Fig. 6.57 Principle of *RC* relaxation circuit.

gap (which is almost the same as that across the capacitor) V varies with time according to the relation

$$V = V_0(1 - e^{-t/(RC)}), \tag{6.54}$$

where t denotes the time starting at the instant V_0 is applied. So, V will approach V_0 asymptotically, as shown in Fig. 6.57b, if allowed to do so. If the tool-work gap and the dielectric fluid are such that a spark can take place when the voltage across the gap reaches a value V_d (commonly known as the discharge voltage), a spark will occur, discharging the capacitor completely whenever the voltage across the tool-work gap (V) reaches V_d. The discharge time is much smaller (about 10%) than the charging time and the frequency of sparking (ν) is approximately given by the following equation (since the time required for deionization is also very small under normal circumstances):

$$\nu \approx \frac{1}{t_c} = \frac{1}{RC \log_e \left(\dfrac{V_0}{V_0 - V_d}\right)}, \tag{6.55}$$

where t_c denotes the charging time which is equal to the time required for the gap voltage to reach a value V_d. The energy released per spark is given by

$$E = \tfrac{1}{2} C V_d^2. \tag{6.56}$$

If t_c $(= 1/\nu)$ is the cycle time, then the average value of the power delivered W_{av} is given by

$$W_{av} = \frac{E}{t_c + t_s}.$$

Since, obviously,

$$V_d = V_0[1 - \exp\{-t_c/(RC)\}], \tag{6.57}$$

the average power can be expressed, after using equation (6.56), in the form

$$W_{av} = \tfrac{1}{2}\frac{C}{t_c} V_0^2 [1 - \exp\{-t_c/(RC)\}]^2.$$

Rearranging, we obtain

$$W_{av} = \frac{V_0^2}{2R}\frac{1}{\zeta}(1 - e^{-\zeta})^2, \tag{6.58}$$

where $\zeta = t_c/(RC)$. For maximum power delivery,

$$\left.\frac{\partial W_{av}}{\partial \zeta}\right|_{\zeta=\zeta_{opt}} = 0.$$

Using this condition in equation (6.58), we get

$$(2\zeta_{opt} + 1)\exp(-\zeta_{opt}) = 1$$

which yields $\zeta_{opt} = 1.26$. Once ζ $[= t_c/(RC)]$ is known, the optimum value of (V_d/V_0) is fixed as

$$\left(\frac{V_d}{V_0}\right)_{opt} = 1 - \exp(-\zeta_{opt}) = 1 - e^{-1.26} = 0.72. \tag{6.59}$$

Thus, for maximum power delivery, the discharge voltage should be 72% of the supply voltage V_0.

If we assume the material removed per spark to be proportional to the energy released per spark, then the mrr can be expressed as

$$Q \approx K(\tfrac{1}{2}V_d^2 C)\nu,$$

where K is the constant of proportionality denoting the fraction of power effectively used in material removal. Using relation (6.55) in the foregoing expression, we get

$$Q \approx K\tfrac{1}{2}V_d^2 C \frac{1}{RC \log_e [V_0/(V_0 - V_d)]}$$

or

$$Q \approx K \frac{V_d^2}{2R \log_e [V_0/(V_0 - V_d)]}. \tag{6.60}$$

So, it is evident from equation (6.60) that, for a given circuit, the mrr increases as R is decreased. However, R cannot be decreased below a critical value as, otherwise, arcing, instead of sparking, will take place.

The critical value of the resistance depends on the inductance L of the discharging circuit. It can be shown that, for a purely inductive discharging circuit, the critical value is

$$R_{min} = \sqrt{\frac{L}{C}}. \tag{6.61}$$

However, the discharging circuit is seldom purely inductive and R should not be lowered below a value $30\sqrt{L/C}$. In the case of machining steels under normal conditions, the removal rate can be approximately expressed as

$$Q \approx 27.4W^{1.54}, \tag{6.62}$$

where Q is the removal rate of steel in mm^3/min and W is the power input in kW.

EXAMPLE 6.9 During an electric discharge drilling of a 10-mm square hole in a low carbon steel plate of 5 mm thickness, brass tool and kerosene are used. The resistance and the capacitance in the relaxation circuit are 50 Ω and 10 μF, respectively. The supply voltage is 200 volts and the gap is maintained at such a value that the discharge (sparking) takes place at 150 volts. Estimate the time required to complete the drilling operation.

SOLUTION Since the work material is steel, we can use relation (6.62) for estimating the removal rate. So, the power input (average) has to be first found out. The amount of energy released per spark, according to equation (6.56), is

$$\tfrac{1}{2} \times 10 \times 10^{-6} \times 150^2 \text{ J} = 0.113 \text{ J}.$$

The cycle time can be approximately found out, using relation (6.55). Thus,

$$t_c \approx RC \log_e \left(\frac{V_0}{V_0 - V_d}\right)$$

or

$$t_c \approx 50 \times 10 \times 10^{-6} \log_e \left(\frac{200}{50}\right) \text{ sec} = 50 \times 10^{-5} \times 1.39 \text{ sec}$$

$$= 7 \times 10^{-4} \text{ sec.}$$

So, the average power input is

$$W = \frac{0.113}{7 \times 10^{-4}} \times 10^{-3} \text{ kW} = 0.16 \text{ kW}.$$

Now, finally the mrr may be found out by using relation (6.62) as

$$Q = 27.4 \times 0.16^{1.54} \text{ mm}^3/\text{min} = 1.633 \text{ mm}^3/\text{min}.$$

The total amount of material to be removed is given by

$$10^2 \times 5 = 500 \text{ mm}^3.$$

Therefore, the time required to accomplish the job is

$$\frac{500}{1.633} \approx 306 \text{ min.}$$

The general characteristics of the mrr are indicated in Fig. 6.58. Here, all the variations shown directly follow from relation (6.60). However, the dependence of mrr (shown in Fig. 6.58d) on the spark gap is not so direct. When the gap is too small, the discharge voltage is small, and, though the frequency is high, the mrr is small. Similarly, when the gap is too large, though the discharge voltage is large, the frequency falls down drastically, resulting in a drop in the mrr. So, there exists an optimum spark gap.

Rotary Impulse Generator

The relaxation circuit for spark generation, though simple, has certain disadvantages. Of these, an important disadvantage is that the mrr is not high. For increasing the removal rate, an impulse generator is used for spark generation. Figure 6.59 shows the schematic diagram of such a system. The capacitor is charged through the diode during the first half cycle. During the following half cycle, the sum of the voltages generated by the generator and the charged capacitor

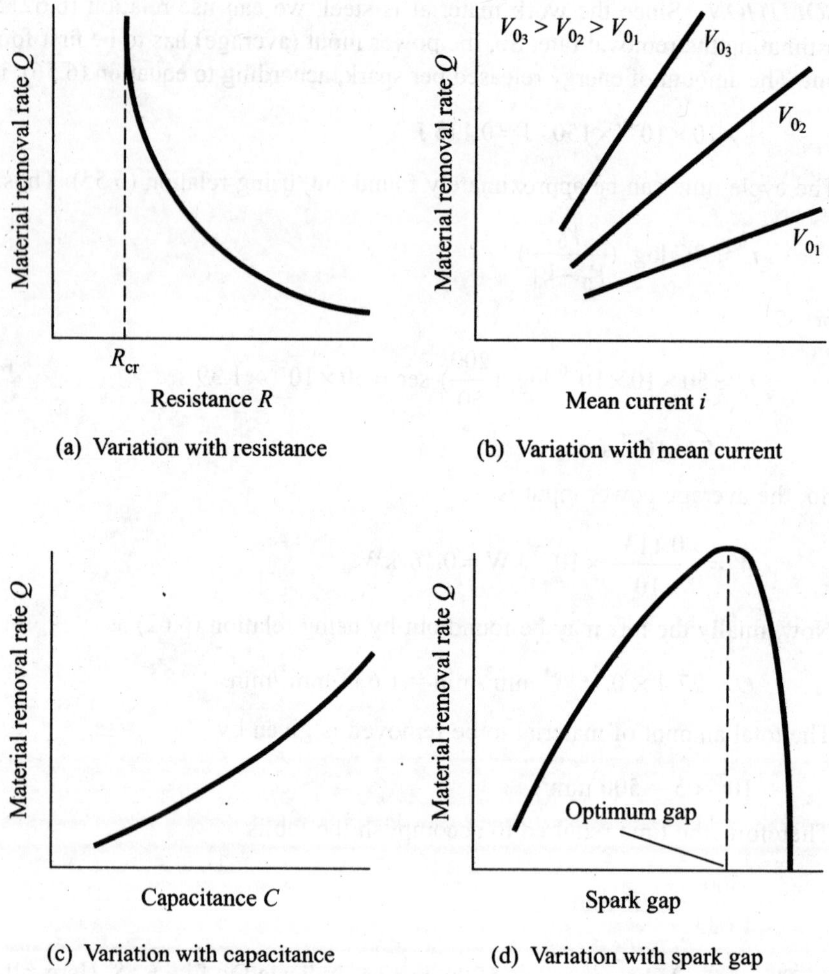

(a) Variation with resistance

(b) Variation with mean current

(c) Variation with capacitance

(d) Variation with spark gap

Fig. 6.58 Material removal rate characteristics in EDM using *RC* relaxation circuit.

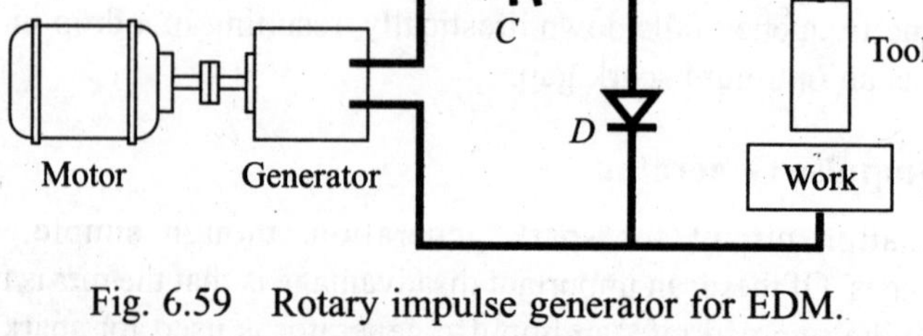

Fig. 6.59 Rotary impulse generator for EDM.

is applied to the work-tool gap. The operating frequency is the frequency of the sine wave generation which depends on the motor speed. Though the mrr is higher, such a system does not produce a good surface finish.

Controlled Pulse Circuits

In the two systems we have discussed, there is no provision for an automatic prevention of the current flow when a short circuit is developed. To achieve such an automatic control, a vacuum tube (or a transistor) is used as the switching device. This system is known as a controlled pulse circuit. Figure 6.60 schematically shows such a system. During sparking, the current which flows through the gap comes from the capacitor. When the current flows through the gap, the valve tube (VT) is biased to cut off and behaves like an infinite resistance. The bias control is done through an electronic control (EC). As soon as the current in the gap ceases, the conductivity of the tube increases, allowing the flow of current to charge the capacitor for the next cycle.

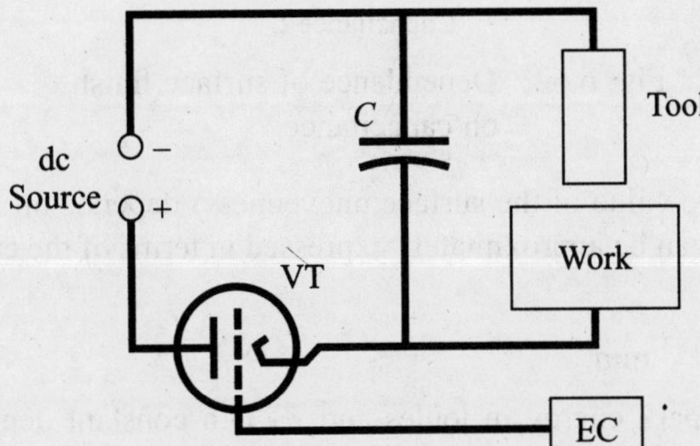

Fig. 6.60 EDM with controlled pulse circuit.

The circuit can be simplified and the operating stability improved if the flow of current is allowed cyclically with an imposed frequency. This can be done by controlling the bias with the help of an oscillator. In this case, the capacitor is not needed. Figure 6.61 shows such a circuit, using a transistor.

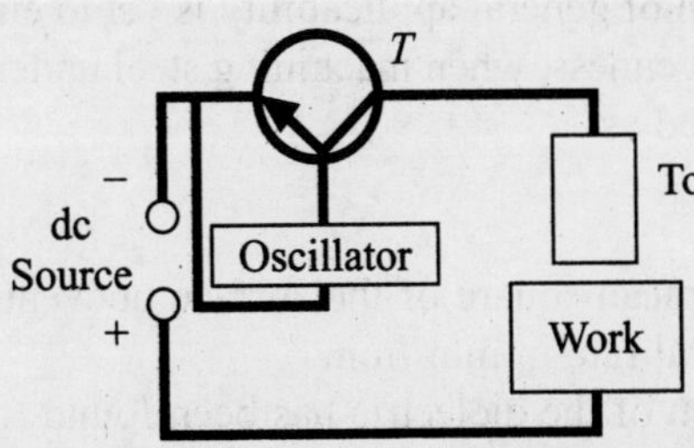

Fig. 6.61 EDM with controlled pulse circuit without capacitor.

6.5.3 SURFACE FINISH AND MACHINING ACCURACY

Since the material removal in EDM is achieved through the formation of craters due to the sparks, it is obvious that large crater sizes (especially the depth) result in a rough surface. So, the crater size, which depends mainly on the energy/spark, controls the quality of the surface. Figure 6.62 shows how H_{rms}

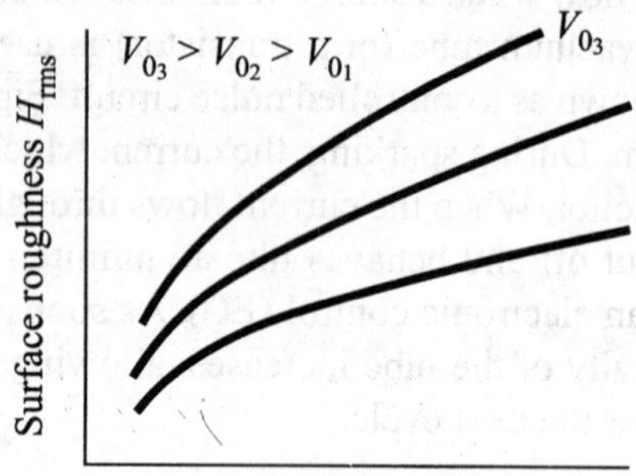

Fig. 6.62 Dependence of surface finish on capacitance.

(root mean square value of the surface unevenness) depends on C and V_0. The crater depth (h_c) can be approximately expressed in terms of the energy released per spark (E) as

$$h_c \approx K_1 E^{0.33} \text{ mm}, \tag{6.63}$$

where E is the spark energy in joules and K_1 is a constant depending on the material. For copper as the work material, K_1 is approximately 4. Using equation (6.56), we find relation (6.63) takes the form

$$h_c \approx 0.78 K C^{0.33} V_d^{0.66}. \tag{6.64}$$

The dependence of surface finish on the pulse energy E and the comparison of surface finish with that obtained by the conventional processes are indicated in Fig. 6.63. A lot of effort has been spent on determining a suitable relationship between the rate of material removal and the quality of surface finish. But a very dependable relation of general applicability is yet to emerge. However, the mrr and the surface unevenness, when machining steel under normal conditions, are approximately related as

$$H_{rms} \approx 1.11 Q^{0.384}, \tag{6.65}$$

where H_{rms} is the root mean square of the surface unevenness in microns and Q is the material removal rate in mm^3/min.

The forced circulation of the dielectric has been found to generally improve the surface finish. The cross-sections of the brass electrode surface produced by EDM with and without the forced circulation (voltage 40 V, current 0.2 A, frequency 1.12 kHz) are shown in Fig. 6.64. It is clear that the forced circulation leads to a significant improvement in the surface finish.

EXAMPLE 6.10 A steel workpiece is being machined with $R = 50$, $C = 10\ \mu$F, $V_0 = 200$ volts, and $V_d = 150$ volts. Estimate the surface roughness.

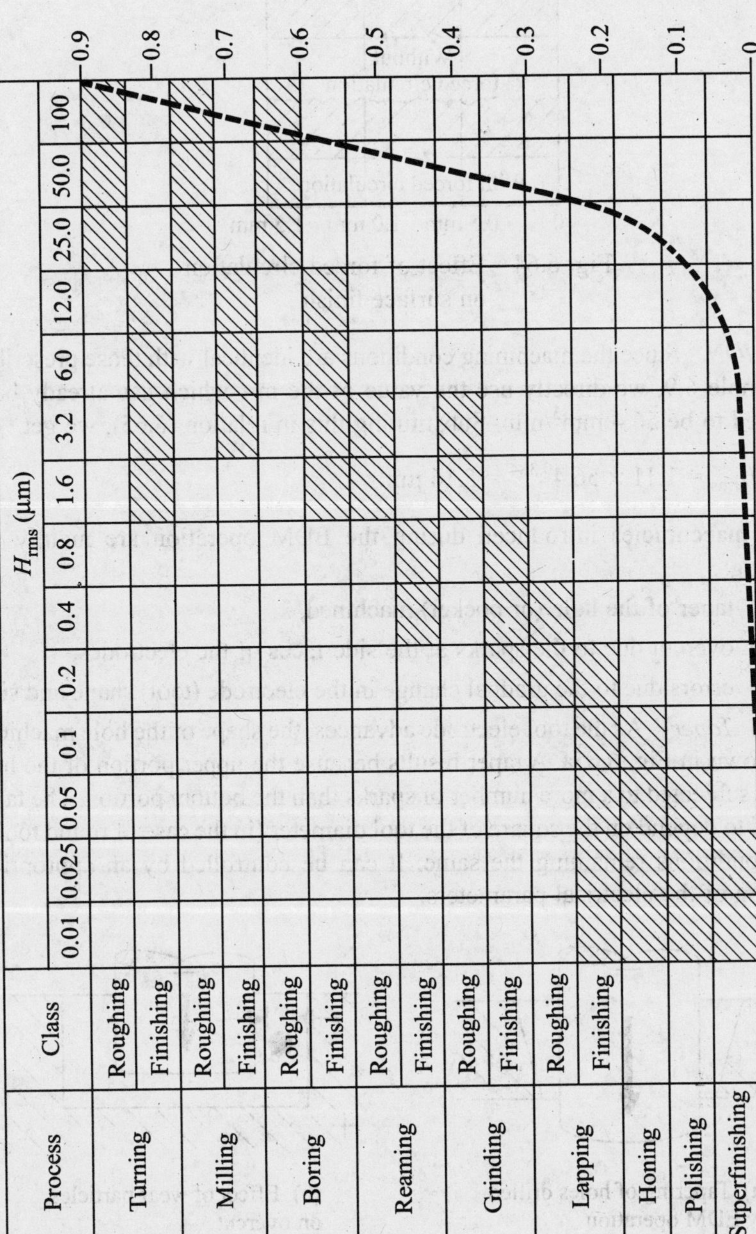

Fig. 6.63 Dependence of surface finish on pulse energy and comparison with conventional processes.

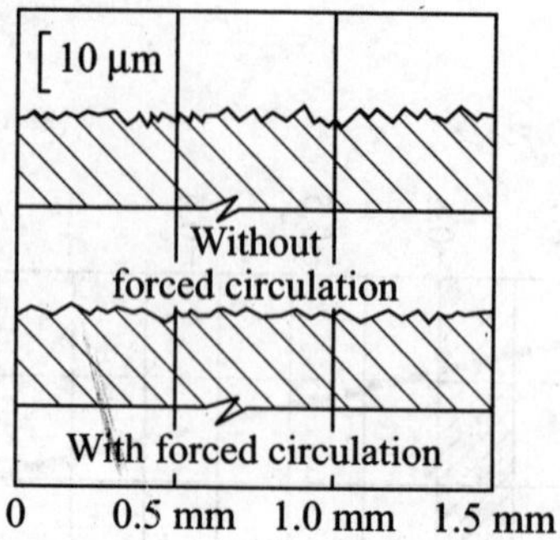

Fig. 6.64 Effect of forced circulation on surface finish.

SOLUTION Since the machining conditions are identical with those prescribed in Example 6.9, we directly use the value of the mrr which has already been estimated to be 56.4 mm^3/min. Substituting this in relation (6.65), we get

$$H_{rms} = 1.11 \times 56.4^{0.384} = 5.16 \ \mu m.$$

The inaccuracies introduced during the EDM operation are mainly the following:

(i) taper of the hole (or pocket) machined,

(ii) overcut due to the sparks at the side faces of the electrodes,

(iii) errors due to the gradual change in the electrode (tool) shape and size.

(i) *Taper* As the tool electrode advances, the shape of the hole machined is as shown in Fig. 6.65a. A taper results because the upper portion of the hole walls is subjected to a more number of sparks than the bottom portion. The taper is found to depend on the square of the tool diameter (in the case of round tools), other conditions remaining the same. It can be controlled by an appropriate alteration of the electrical parameters.

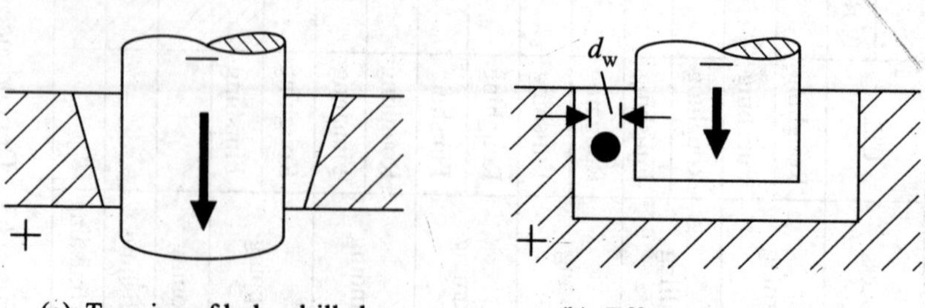

(a) Tapering of holes drilled by EDM operation

(b) Effect of wear particles on overcut

Fig. 6.65 Sources of dimensional inaccuracy in EDM.

(ii) *Overcut* An overcut is that dimension by which the hole in the workpiece exceeds the electrode size. The magnitude of the overcut is dependent on the spark length and, to some extent, on the crater dimensions. When the wear particles are present in the gap, the effective length of the spark (and hence the magnitude of the overcut) is increased by d_w (diameter of the wear particle), as shown in Fig. 6.65b. When a 10-mm-diameter brass tool is used for machining steel (with no forced circulation), the dependence of the overcut and taper on the energy of the individual sparks is as indicated in Fig. 6.66.

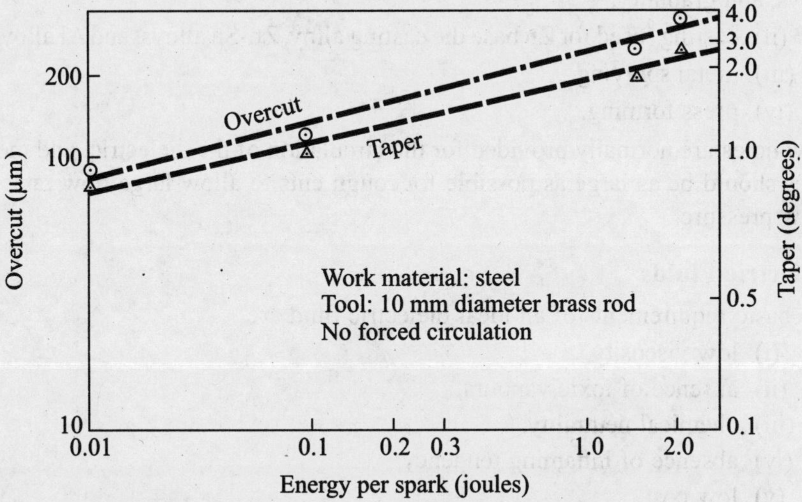

Fig. 6.66 Effect of spark energy on overcut and taper.

6.5.4 TOOL ELECTRODE AND DIELECTRIC FLUIDS

The electrodes play an extremely important role in the EDM operation, and therefore certain aspects of the tool electrode should be kept in mind to achieve better results.

Tool Electrode Wear

During the EDM operation, the electrode (i.e., the tool), as already mentioned, also gets eroded due to the sparking action. The materials having good electrode wear characteristics are the same as those that are generally difficult to machine. One of the principal materials used for the tool is graphite which goes directly to the vapour phase without melting. The wear ratio (r_Q), defined by the ratio of the material removed from the work to the material removed from the tool, is found to be related to r_θ (=melting point of the work/melting point of the tool) as

$$r_Q \approx 2.25 r_\theta^{-2.3}. \tag{6.66}$$

Electrode Material

The selection of the electrode material depends on the

(i) material removal rate,

(ii) wear ratio,

(iii) ease of shaping the electrode,
(iv) cost.

The most commonly used electrode materials are brass, copper, graphite, Al alloys, copper-tungsten alloys, and silver-tungsten alloys.

The methods used for making the electrodes are

(i) conventional machining (used for copper, brass, Cu-W alloys, Ag-W alloys, and graphite),
(ii) casting (used for Zn base die casting alloy, Zn-Sn alloys, and Al alloys),
(iii) metal spraying,
(iv) press forming.

Flow holes are normally provided for the circulation of the dielectric, and these holes should be as large as possible for rough cuts to allow large flow rates at a low pressure.

Dielectric Fluids

The basic requirements of an ideal dielectric fluid are

(i) low viscosity,
(ii) absence of toxic vapours,
(iii) chemical neutrality,
(iv) absence of inflaming tendency,
(v) low cost.

The ordinary water possesses almost all these properties, but since it causes rusting in the work and the machine, it is not used. Another reason why water is not recommended is as follows. The electrodes are constantly under some potential difference, and due to the good conductivity of water, the ECM process starts distorting the workpiece. Also, power is wasted. However, in some cases, deionized water is used.

The most commonly used type of fluid is hydrocarbon (petroleum) oil. Kerosene, liquid paraffin, and silicon oils are also used as dielectric fluids.

6.5.5 EFFECTS OF EDM ON METAL SURFACES

The high temperature generated by the sparks causes the melting and vaporization of the metal and, obviously, this high temperature affects the properties of the shallow layers (2.5–150 μm) of the surface machined.

The outermost layer is rapidly chilled, and it is therefore very hard. The layer directly below this is in a somewhat tempered condition. Figure 6.67 shows the variation of hardness with depth for both rough and finish EDM operations on steel. It is clear that in finish machining such a hardening is not prominent. However, the outer layer is tempered and the hardness is low.

The hardening of the surface layer during the EDM operation imparts a better wear resistance characteristic. However, the fatigue strength reduces due to the micro cracks that develop in the surface layer during chilling. Figure 6.68 shows the comparison between the fatigue strength of identical parts produced

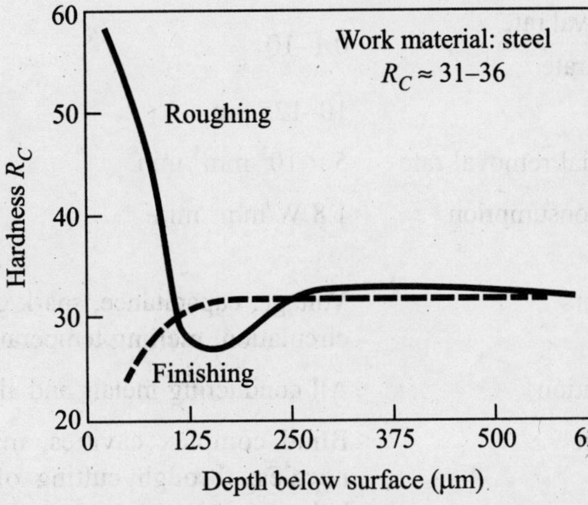

Fig. 6.67 Effect of EDM on subsurface hardness.

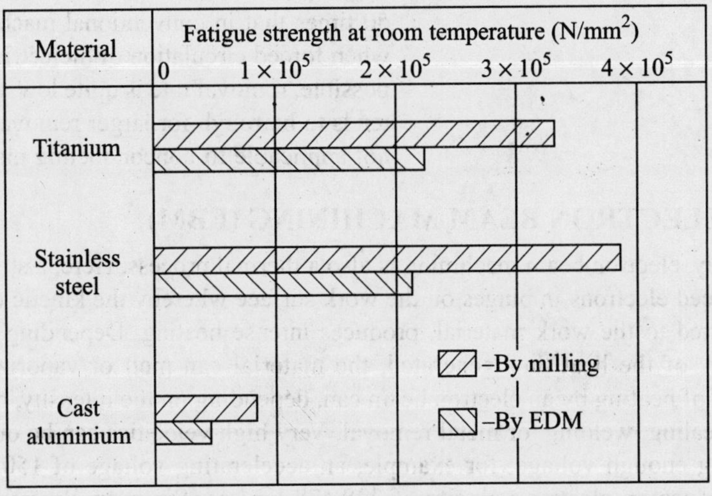

Fig. 6.68 Effect of EDM on fatigue strength of work.

by conventional milling and EDM. The properties of the thin surface layers do not have much effect on the tensile strength. Their structure gets transformed and, due to the sparks, their chemical composition alters to some extent. These generally reduce the erosion resistance.

6.5.6 SUMMARY OF EDM CHARACTERISTICS

Mechanics of material removal	Melting and evaporation aided by cavitation
Medium	Dielectric fluid
Tool	
Materials	Cu, brass, Cu-W alloy, Ag-W alloy, graphite

$\dfrac{\text{Material removal rate}}{\text{Tool wear rate}}$	0.1–10
Gap	10–125 μm
Maximum material removal rate	5×10^3 mm^3/min
Specific power consumption (typical)	1.8 W/mm^3/min
Critical parameters	Voltage, capacitance, spark gap, dielectric circulation, melting temperature
Materials application	All conducting metals and alloys
Shape application	Blind complex cavities, microholes for nozzles, through cutting of noncircular holes, narrow slots
Limitations	High specific energy consumption (about 50 times that in conventional machining); when forced circulation of dielectric is not possible, removal rate is quite low; surface tends to be rough for larger removal rates; not applicable to nonconducting materials

6.6 ELECTRON BEAM MACHINING (EBM)

Basically, electron beam machining is also a thermal process. Here, a stream of high speed electrons impinges on the work surface whereby the kinetic energy, transferred to the work material, produces intense heating. Depending on the intensity of the heat thus generated, the material can melt or vaporize. The process of heating by an electron beam can, depending on the intensity, be used for annealing, welding, or metal removal. Very high velocities can be obtained by using enough voltage; for example, an accelerating voltage of 150,000 V can produce an electron velocity of 228,478 km/sec. Since an electron beam can be focussed to a point with 10–200 μm diameter, the power density can go up to 6500 billion W/mm^2. Such a power density can vaporize any substance immediately. Thus, EBM is nothing but a very precisely controlled vaporization process. EBM is a suitable process for drilling fine holes and cutting narrow slots. Holes with 25–125 μm diameter can be drilled almost instantaneously in sheets with thicknesses up to 1.25 mm. The narrowest slot which can be cut by EBM has a width of 25 μm. Moreover, an electron beam can be manoeuvred by the magnetic deflection coils, making the machining of complex contours easy. However, to avoid a collision of the accelerating electrons with the air molecules, the process has to be conducted in vacuum (about 10^{-5} mm Hg); this makes the process unsuitable for very large workpieces. To indicate the wide range of applications of the electron beam, a plot of the power density versus the hot spot diameter is given in Fig. 6.69. It is obvious that the range of the electron beam is the largest. This is why the electron beam is used not only for machining but also for the other thermal processes.

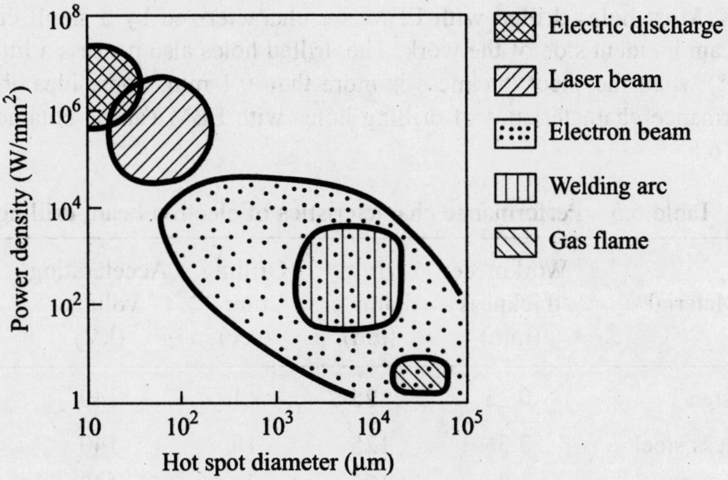

Fig. 6.69 Power density-hot spot diameter combination for various types of heat sources.

Figure 6.70 schematically shows the basic arrangement of an electron beam machine. The electrons are emitted from the cathode (a hot tungsten filament),

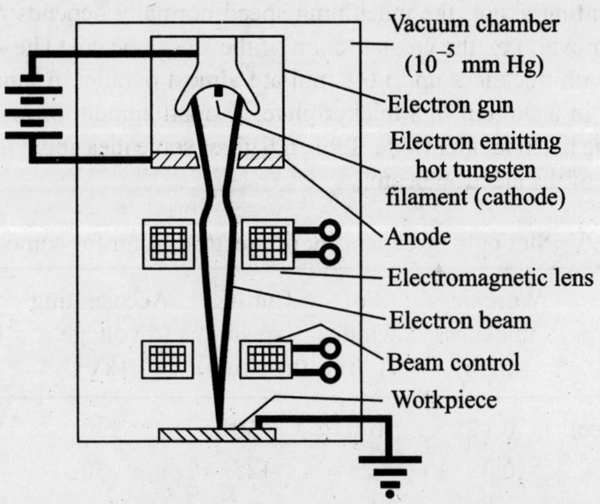

Fig. 6.70 Schematic view of electron beam machine.

the beam is shaped by the grid cup, and the electrons are accelerated due to a large potential difference between the cathode and the anode. The beam is focussed with the help of the electromagnetic lenses. The deflecting coils are used to control the beam movement in any required manner.

In case of drilling holes the hole diameter depends on the beam diameter and the energy density. When the diameter of the required hole is larger than the beam diameter, the beam is deflected in a circular path with proper radius. Most holes drilled with EBM are characterized by a small crater on the beam incident side of the work. The drilled holes also possess a little taper (2°–4°) when the sheet thickness is more than 0.1 mm. Some idea about the performance characteristics of drilling holes with EBM can be obtained from Table 6.5.

Table 6.5 Performance characteristics of electron beam drilling

Material	Workpiece thickness (mm)	Hole diameter (μm)	Drilling time (sec)	Accelerating voltage (kV)	Beam current (μA)
Tungsten	0.25	25	<1	140	50
Stainless steel	2.5	125	10	140	100
Stainless steel	1.0	125	<1	140	100
Aluminium	2.5	125	10	140	100
Alumina (Al_2O_3)	0.75	300	30	125	60
Quartz	3.0	25	<1	140	10

While cutting a slot, the machining speed normally depends on the rate of material removal, i.e., the cross-section of the slot to be cut. The sides of a slot in a sheet with thickness up to 0.1 mm are almost parallel. A taper of 1° to 2° is observed in a slot cut in a thicker plate. A small amount of material splatter occurs on the beam incident side. Table 6.6 gives some idea about the slot cutting capabilities of the electron beam.

Table 6.6 Slot cutting capability of electron beam for some materials

Material	Workpiece thickness (mm)	Slot width (μm)	Cutting speed (mm/min)	Accelerating voltage (kV)	Average beam current (μA)
Stainless steel	0.175	100	50	130	50
Tungsten	0.05	25	125	150	30
Brass	0.25	100	50	130	50
Alumina	0.75	100	600	150	200

The power requirement is found to be approximately proportional to the rate of metal removal. So, $P \approx CQ$, C being the constant of proportionality. Table 6.7 gives the approximate values of C for different work materials.

Table 6.7 Specific power consumption in EBM for various metals

Material	C (W/mm³/min)
Tungsten	12
Iron	7
Titanium	6
Aluminium	4

A very rough estimation of the machining speed for the given conditions is possible, using Table 6.7.

EXAMPLE 6.11 For cutting a 150-μm-wide slot in a 1-mm-thick tungsten sheet, an electron beam with 5 kW power is used. Determine the speed of cutting.

SOLUTION Let the speed of cutting be v mm/min. Then, the rate of material removal required is

$$Q = \frac{150}{1000} \times 1 \times v \text{ mm}^3\text{/min.}$$

The corresponding beam power is given by

$$P = C_{\text{tungsten}} Q = 12 \times \frac{150}{1000} v.$$

Since P is given to be 5000 W,

$$v = \frac{5000}{12 \times 0.15} \text{ mm/min} = 2778 \text{ mm/min} \equiv 4.6 \text{ cm/sec.}$$

We shall see in Example 6.13 that the speed arrived at in the foregoing example is much less than the actual speed. However, some idea about the order of magnitude of the power requirement for a given mrr is possible.

6.6.1 MECHANICS OF EBM

Electrons are the smallest stable elementary particles with a mass of 9.109 × 10^{-31} kg and a negative charge of 1.602 × 10^{-19} coulomb. When an electron is accelerated through a potential difference of V volts, the change in the kinetic energy can be expressed as $\frac{1}{2}m_e(u^2 - u_0^2)$ eV, where m_e is the electron mass, u is the final velocity, and u_0 is the initial velocity. If we assume the initial velocity of the emitting electrons to be negligible, the final expression for the electron velocity u in km/sec is

$$u \approx 600\sqrt{V}. \tag{6.67}$$

When a fast moving electron impinges on a material surface, it penetrates through a layer undisturbed. Then, it starts colliding with the molecules, and, ultimately, is brought to rest (Fig. 6.71). The layer through which the electron penetrates undisturbed is called the transparent layer. Only when the electron

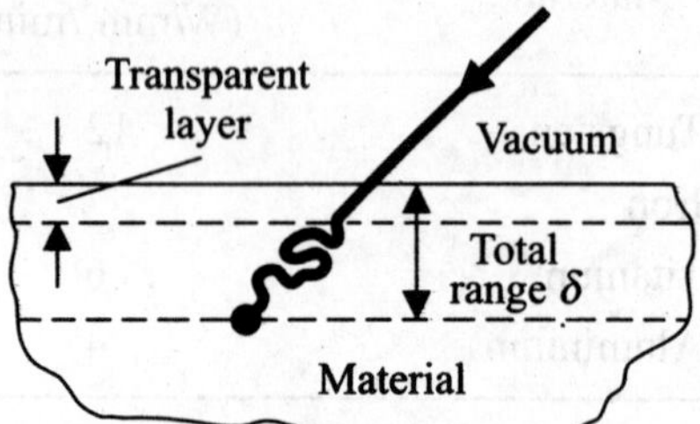

Fig. 6.71 Movement of electron below surface.

begins colliding with the lattice atoms does it start giving up its kinetic energy, and heat is generated. So, it is clear that the generation of heat takes place inside the material, i.e., below the transparent skin. The total range to which the electron can penetrate (δ) depends on the kinetic energy, i.e., on the accelerating voltage V. It has been found that

$$\delta = 2.6 \times 10^{-17} \frac{V^2}{\rho}, \tag{6.68}$$

where δ is the range in mm, V is the accelerating voltage in volts, and ρ is the density of the material in kg/mm^3.

EXAMPLE 6.12 During drilling of holes in a steel workpiece by EBM, an accelerating voltage of 150 kV is used. Determine the electron range.

SOLUTION The density of steel is about 76×10^{-7} kg/mm^3. So, directly using equation (6.68), we get

$$\delta = 2.6 \times 10^{-17} \frac{(150 \times 10^3)}{76 \times 10^{-7}} \text{ mm} = 7.7 \times 10^{-2} \text{ mm} = 77\ \mu\text{m}.$$

The temperature rise can be approximately estimated by solving the following one-dimensional heat conduction equation for the heat source placed inside the metal:

$$\frac{\partial \theta(z,t)}{\partial t} = \alpha \frac{\partial^2 \theta(z,t)}{\partial z^2} + \frac{1}{c\rho} H(z,t), \tag{6.69}$$

where θ is the temperature, α is the thermal diffusivity of the metal, z is the distance from the surface, t is time, c is the specific heat of the metal, ρ is the metal density, and $H(z, t)$ is the heat source intensity, i.e., heat generated per unit time per unit volume.

Assuming the beam to be steady, the heat source intensity depends only on z

according to the relation

$$H(z) = Ae^{-bz},$$

where A is a constant and b is a coefficient describing the energy absorption characteristics of the metal. Using this expression in equation (6.69), we obtain the final form of the equation describing surface heating due to an electron beam as

$$\frac{\partial\theta(z,t)}{\partial t} = \alpha\frac{\partial^2\theta(z,t)}{\partial z^2} + \frac{A}{\rho c}e^{-bz}. \tag{6.70}$$

We now assume that (i) the metal body is semi-infinite, (ii) the surface of the metal is insulated except for the hot spot, and (iii) the rate of heat input remains uniform with time during the pulse duration. The nature of variation of temperature, found by solving equation (6.70), plotted against z for different pulse durations τ is shown in Fig. 6.72. We see here that as the pulse duration increases, the peak temperature shifts towards the surface. For

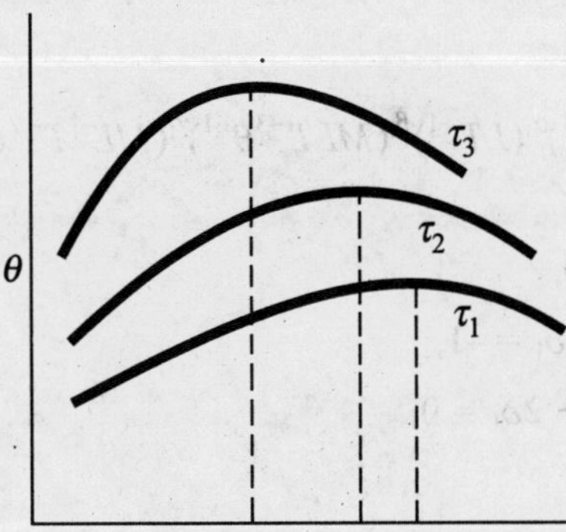

Fig. 6.72 Variation of temperature with distance from surface for various pulse durations.

a rough estimation of the functional characteristics during EBM, a dimensional analysis is found to be quite useful. The quantities of interest are the beam power P (= beam current × accelerating voltage), beam diameter d, velocity of beam with respect to work v, thermal conductivity of metal k, volume specific heat of metal ρc, melting temperature θ_m, and depth of penetration of melting temperature Z. The depth to which the melting temperature penetrates should depend on these quantities, and we expect that

$$Z = Z(P, v, d, k, \rho c, \theta_m). \tag{6.71}$$

If we select M (mass), L (length), T (time), and θ (temperature) as the basic dimensions, the dimensions of the foregoing quantities would then be

$$P = ML^2T^{-3}, \qquad v = LT^{-1}, \qquad d = L,$$

$$k = MLT^{-3}\theta^{-1}, \qquad \rho c = ML^{-1}T^{-2}\theta^{-1}, \qquad \theta_m = \theta, \qquad Z = L.$$

According to Buckingham's π-theorem, we can form $(n - m)$ number of dimensionless groups, where n is the number of parameters and m is the number of basic dimensional units. In this case, n is 7 and m is 4. So, we can have π_1, π_2, and π_3 as the three dimensionless groups. Taking Z, d, and θ_m as the quantities which directly go in π_1, π_2, and π_3, respectively, we obtain

$$\pi_1 = Z \cdot P^{\alpha_1} v^{\beta_1} k^{\gamma_1} (\rho c)^{\delta_1},$$

$$\pi_2 = d \cdot P^{\alpha_2} v^{\beta_2} k^{\gamma_2} (\rho c)^{\delta_2},$$

$$\pi_3 = \theta_m \cdot P^{\alpha_3} v^{\beta_3} k^{\gamma_3} (\rho c)^{\delta_3}.$$

Substituting the dimensions of each quantity, we equate to zero the ultimate exponent of each basic dimension since the π_is are dimensionless groups. Thus, α_i, β_i, γ_i, δ_i, where $i = 1, 2, 3$, can be solved. Since the dimensions of both Z and d are the same,

$$\alpha_1 = \alpha_2, \qquad \beta_1 = \beta_2, \qquad \gamma_1 = \gamma_2, \qquad \delta_1 = \delta_2.$$

Considering π_1, we get

$$\pi_1 = L \cdot (ML^2T^{-3})^{\alpha_1} (LT^{-1})^{\beta_1} (MLT^{-3}\theta^{-1})^{\gamma_1} (ML^{-1}T^{-2}\theta^{-1})^{\delta_1}.$$

Hence,

$$\alpha_1 + \gamma_1 + \delta_1 = 0,$$

$$2\alpha_1 + \beta_1 + \gamma_1 - \delta_1 = -1,$$

$$3\alpha_1 + \beta_1 + 3\gamma_1 + 2\delta_1 = 0,$$

$$\gamma_1 + \delta_1 = 0.$$

Solving, we get $\alpha_1 = 0$, $\beta_1 = 1$, $\gamma_1 = -1$, and $\delta_1 = 1$. Thus,

$$\pi_1 = \frac{Zv\rho c}{k}, \qquad \pi_2 = \frac{dv\rho c}{k}.$$

In a similar manner, α_3, β_3, γ_3, and δ_3 are found out, and finally we get

$$\pi_3 = \frac{k^2\theta_m}{P\rho c v}.$$

The ultimate relationship can be assumed to be of the form

$$\pi_i = f(\pi_j, \pi_k).$$

Let us assume $i = 1$, $j = 2$, and $k = 3$. Then, the functional relationship is of the form

$$\pi_1 = f(\pi_2, \pi_3)$$

or

$$\frac{Zv\rho c}{k} = f\left(\frac{dv\rho c}{k}, \frac{k^2\theta_m}{P\rho c v}\right). \tag{6.72}$$

It has been experimentally found that Z is directly proportional to P. Thus, equation (6.72) becomes

$$\frac{Zv\rho c}{k} = \frac{P\rho cv}{k^2\theta_{\mathrm{m}}} f_1\left(\frac{dv\rho c}{k}\right).$$

Simplifying, we get

$$\frac{Zk\theta_{\mathrm{m}}}{P} = f_1\left(\frac{dv\rho c}{k}\right).$$

It has, again, been experimentally observed that this dependence is of the form

$$\frac{Zk\theta_{\mathrm{m}}}{P} = 0.1\left(\frac{dv\rho c}{k}\right)^{-0.5}.$$

So, the final form of equation (6.71) is

$$Z = 0.1\frac{P}{k\theta_{\mathrm{m}}}\left(\frac{dv\rho c}{k}\right)^{-0.5}$$

or

$$Z = 0.1\frac{P}{\theta_{\mathrm{m}}\sqrt{kdv\rho c}}. \tag{6.73}$$

Any consistent system of units for the quantities should be used.

EXAMPLE 6.13 For cutting a 150-μm-wide slot in a 1-mm-thick tungsten sheet, an electron beam machine with 5 kW power is used. Determine the speed of cutting.

SOLUTION For tungsten, the value of volume specific heat ρc is 2.71 J/cm^3-°C, thermal conductivity is 2.15 W/cm-°C, and melting temperature is 3400°C.

Since a through slot has to be cut, the depth to which the melting temperature should reach must be equal to the plate thickness (actually, it is an underestimation). Thus, Z = 0.1 cm. The diameter of the beam at the focussed spot can be taken to be equal to the slot width, i.e., 0.015 cm. The power has been given to be 5×10^3 W. Substituting the data in equation (6.73), we get

$$0.1 = 0.1\frac{5\times 10^3}{3400(2.15\times 0.015\times v\times 2.71)^{1/2}}$$

or

$$2.15\times 0.015\times v\times 2.71 = \left(\frac{5\times 10^3}{3400}\right)^2$$

or

$$v = \left(\frac{50}{34}\right)^2 \times \frac{1}{2.15\times 0.015\times 2.71}\ \text{cm/sec} = 24.7\ \text{cm/sec}.$$

6.6.2. EFFECTS OF EBM ON MATERIALS

Since machining by an electron beam is achieved without raising the temperature of the surrounding material (except an extremely thin layer), there is no effect on the work material. Because of the extremely high energy density, the work material 25–50 μm away from the machining spot remains at the room temperature. Apart from this, the chance of contamination of the work is also less as the process is accomplished in vacuum.

6.6.3 SUMMARY OF EBM CHARACTERISTICS

Mechanics of material removal	Melting, vaporization
Medium	Vacuum
Tool	Beam of electrons moving at very high velocity
Maximum material removal rate	10 mm^3/min
Specific power consumption (typical)	450 W/mm^3-min
Critical parameters	Accelerating voltage, beam current, beam diameter, work speed, melting temperature
Materials application	All materials
Shape application	Drilling fine holes, cutting contours in sheets, cutting narrow slots
Limitations	Very high specific energy consumption, necessity of vacuum, expensive machine

6.7 LASER BEAM MACHINING (LBM)

Like a beam of high velocity electrons, a laser beam is also capable of producing very high power density. Laser is a highly coherent (in space and time) beam of electromagnetic radiation with wavelength varying from 0.1 μm to 70 μm. However, the power requirement for a machining operation restricts the effectively usable wavelength range to 0.4–0.6 μm. Because of the fact that the rays of a laser beam are perfectly parallel and monochromatic, it can be focussed to a very small diameter and can produce a power density as high as 10^7 W/mm^2. For developing a high power, normally a pulsed ruby laser is used. The continuous CO_2-N_2 laser has also been successfully used in machining operations.

Figure 6.73 shows a typical pulsed ruby laser. A coiled xenon flash tube is placed around the ruby rod and the internal surface of the container walls is made highly reflecting so that maximum light falls on the ruby rod for the pumping operation. The capacitor is charged and a very high voltage is applied to the triggering electrode for initiation of the flash. The emitted laser beam is focussed by a lens system and the focussed beam meets the work surface, removing a small portion of the material by vaporization and high speed ablation. A very small fraction of the molten metal is vaporized so quickly that a substantial mechanical impulse is generated, throwing out a large portion of the liquid metal. Since the energy released by the flash tube is much more than the energy emitted by the laser head in the form of a laser beam, the system must be properly cooled.

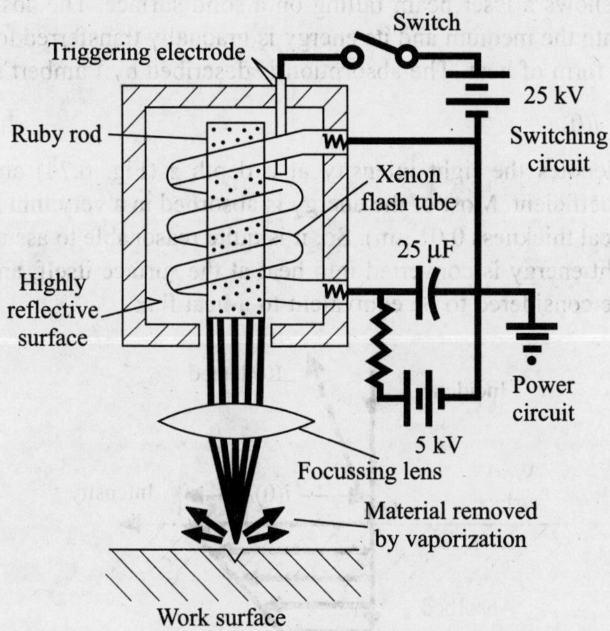

Fig. 6.73 Schematic diagram of laser beam machining.

The efficiency of the LBM process is very low—about 0.3–0.5%. The typical output energy of a laser is 20 J with a pulse duration of 1 millisecond. The peak power reaches a value 20,000 W. The divergence of the beam is around 2×10^{-3} rad, and, using a lens with a focal length of 25 mm, the spot diameter becomes about 50 μm.

Like the electron beam, the laser beam is also used for drilling microholes and cutting very narrow slots. Holes up to 250 μm diameter can be easily drilled by a laser. The dimensional accuracy is around ±0.025 mm. When the workpiece thickness is more than 0.25 mm, a taper of 0.05 mm per mm is noticed.

6.7.1 MECHANICS OF LBM

Machining by a laser beam is achieved through the following phases: (i) interaction of laser beam with work material, (ii) heat conduction and temperature rise, and (iii) melting, vaporization, and ablation. An accurate analysis of the whole process is difficult and beyond the scope of this text. We shall, however, discuss certain simple aspects of fundamental importance, considering only the increase in temperature of the work material up to the melting point; vaporization and ablation will not be taken into account in our analysis.

Interaction of Laser Beam with Work

The application of a laser beam in machining depends on the thermo-optic interaction between the beam and the solid work material. So, it is obvious that the work surface should not reflect back too much of the incident beam energy. Figure 6.74 shows a laser beam falling on a solid surface. The absorbed light propagates into the medium and its energy is gradually transferred to the lattice atoms in the form of heat. The absorption is described by Lambert's law as

$$I(z) = I(0)e^{-\mu z},$$

where $I(z)$ denotes the light intensity at a depth z (Fig. 6.74) and μ is the absorption coefficient. Most of the energy is absorbed in a very thin layer at the surface (typical thickness 0.01 μm). So, it is quite reasonable to assume that the absorbed light energy is converted into heat at the surface itself, and the laser beam may be considered to be equivalent to a heat flux.

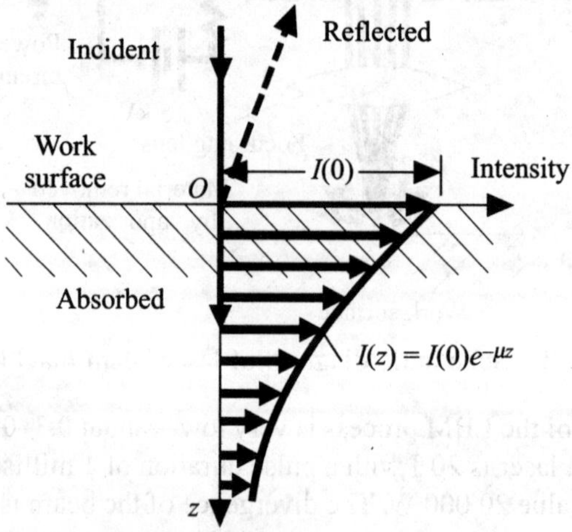

Fig. 6.74 Laser beam falling on work surface and variation of intensity below surface.

Heat Conduction and Temperature Rise

Reradiation from the surface at a temperature of 3000 K is of the order of only 600 W/cm^2 and it is negligible as compared with the input flux 10^5–10^7 W/cm^2. To make our analysis one-dimensional, the diameter of the beam spot is assumed to be larger than the depth of penetration. Also, the thermal properties, e.g., conductivity and specific heat, are considered to remain unaffected by the temperature change. So, the equivalent heat conduction problem is represented by a uniform heat flux $H(t)$ at the surface (Fig. 6.75) of a semi-infinite body. The equation for heat conduction for the region $z > 0$ is

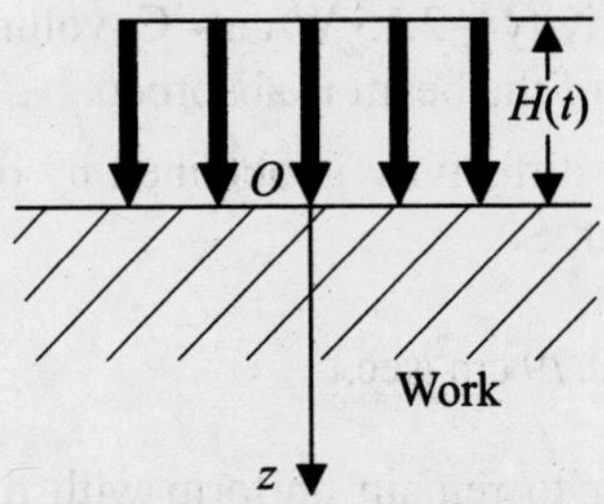

Fig. 6.75 Idealized heat flux for solving heat transfer problem.

$$\frac{\partial^2\theta(z,t)}{\partial z^2} - \frac{1}{\alpha}\cdot\frac{\partial\theta(z,t)}{\partial t} = 0,$$

where α is the thermal diffusivity and θ is the temperature. At the surface (i.e., $z = 0$),

$$\frac{\partial\theta}{\partial z} = -\frac{1}{k}H(t),$$

where k is the thermal conductivity. At $t = 0$, i.e., when the heat flux has just started, let the temperature of the body be zero. Then, the solution of the foregoing equation is given by

$$\theta(z,t) = \frac{1}{k}\sqrt{\frac{\alpha}{\pi}}\int_0^t [H(t-\tau)/\sqrt{\tau}]\exp\left(-\frac{z^2}{4\alpha\tau}d\tau\right). \tag{6.74}$$

When the laser pulse shape is known, the temperature $\theta(z, t)$ can be determined by numerical integration. Further, if the laser beam pulse is assumed to be a step function, i.e., the heat flux H remains constant, equation (6.74) can be written in the form

$$\theta(z,t) = \frac{2H}{k}\left[\sqrt{\frac{\alpha t}{\pi}}\exp\left(-\frac{z^2}{4\alpha\tau}\right) - \frac{z}{2}\operatorname{erfc}\left(z/2\sqrt{\alpha t}\right)\right]. \tag{6.75}$$

The surface temperature then obviously has the expression

$$\theta(0,t) = \frac{2H}{k}\left(\frac{\alpha t}{\pi}\right)^{1/2}. \tag{6.76}$$

If θ_m is the melting temperature of the material, the time required for the surface to reach this temperature (t_m) is then

$$t_m = \frac{\pi}{\alpha}\left(\frac{\theta_m k}{2H}\right)^2. \tag{6.77}$$

EXAMPLE 6.14 A laser beam with a power intensity of 10^5 W/mm^2 falls on a tungsten sheet. Find out the time required for the surface to reach the melting temperature. The given thermal properties of tungsten are melting temperature

= 3400°C, thermal conductivity = 2.15 W/cm-°C, volume specific heat = 2.71 J/cm^3-°C. Assume that 10% of the beam is absorbed.

SOLUTION The thermal diffusivity is obtained by dividing the conductivity by the volume specific heat. So,

$$\alpha = \frac{2.15}{2.71}\ \text{cm}^2/\text{sec} = 0.79\ \text{cm}^2/\text{sec}.$$

Assuming the beam power to remain uniform with time, the heat flux (H) is given by

$$H = 0.1 \times 10^5 \times 100\ \text{W/cm}^2 = 10^6\ \text{W/cm}^2.$$

So, substituting the values in equation (6.77), we get

$$t_m = \frac{\pi}{0.79}\left[\frac{3400 \times 2.15}{2 \times 10^6}\right]^2 \text{sec} = 0.000053\ \text{sec}.$$

A more realistic approach is to consider the heat flux to be on a circular spot with a diameter equal to that of the focussed beam. If the beam diameter is d and the heat flux is uniform, both in space and time (Fig. 6.76), then the solution

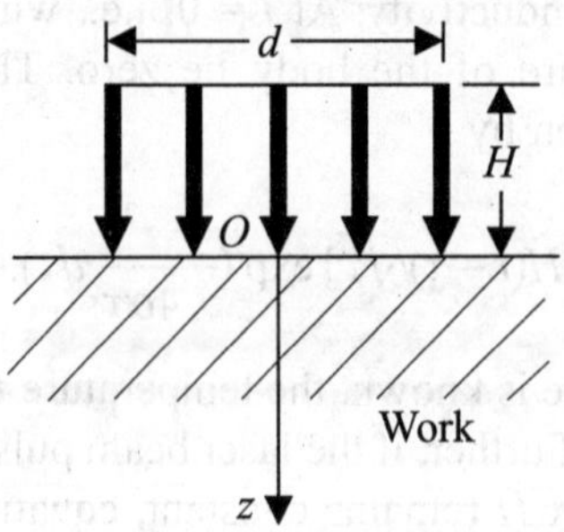

Fig. 6.76 Idealized heat flux, uniform both in space and time.

for the heat conduction equation with an appropriate boundary condition yields the following expression for the temperature along the z-axis:

$$\theta(z, t) = \frac{2H\sqrt{\alpha t}}{k}\left[\text{ierfc}\left(\frac{z}{2\sqrt{\alpha t}}\right) - \text{ierfc}\left(\frac{\sqrt{z^2 + d^2/4}}{2\sqrt{\alpha t}}\right)\right]. \tag{6.78}$$

So, at the point O, the expression for temperature is

$$\theta(0, t) = \frac{2H\sqrt{\alpha t}}{k}\left[\frac{1}{\sqrt{\pi}} - \text{ierfc}\left(\frac{d}{4\sqrt{\alpha t}}\right)\right]. \tag{6.79a}$$

EXAMPLE 6.15 A laser beam with a power intensity of 10^5 W/mm^2 falls on a tungsten sheet. The focussed diameter of the incident beam is 200 μm. How much time will it take for the centre of the circular spot to reach the melting temperature? Use the values of the thermal properties given in Example 6.15. Also, assume 10% absorption.

SOLUTION Substituting the appropriate values in equation (6.79a), we get

$$3400 = \frac{2 \times 0.1 \times 10^7 \sqrt{0.79 t_m}}{2.15} \left[\frac{1}{\sqrt{\pi}} - \text{ierfc} \frac{0.02}{4\sqrt{0.79 t_m}}\right],$$

where t_m indicates the time required for the centre of the circular spot to reach the melting temperature. Using the expression for ierfc (ζ) from Section 6.5.1, we find this equation becomes

$$3400 = 9.30 \times 10^5 \times \beta\left[\frac{1}{\sqrt{\pi}} - \frac{1}{\sqrt{\pi}} e^{-\zeta^2} + \zeta\{1 - \text{erf}\,(\zeta)\}\right],$$

where

$$\zeta = \frac{1}{200\beta}, \qquad \beta = \sqrt{0.79 t_m}.$$

The values of the error function (probability integral) erf (ζ) are available in standard tables. Solving the foregoing equation numerically, we get

$$\beta \approx 0.01 \quad \text{or} \quad t_m \approx 0.00013 \text{ sec.}$$

Comparing the results in Examples 6.14 and 6.15, we see that the approximate relation given by equation (6.77) does not yield an accurate result. However, if the power intensity is much higher, the second term within the brackets in equation (6.79a) will be very small as compared with $1/\sqrt{\pi}$ and the resulting relation will become the same as equation (6.77). As the power intensity of the beam decreases, the time required to reach the melting temperature increases. For a large value of time, equation (6.79a) for melting condition takes the form

$$\theta_m \approx \frac{2H}{k} \cdot \frac{d}{4}\left[1 - \text{erf}\left(\frac{d}{4\sqrt{\alpha t_m}}\right)\right]. \tag{6.79b}$$

Since the smallest possible value of erf (ζ) is zero, the minimum value of H with which it is possible to attain the melting temperature is given by

$$H_{cr} \approx \frac{2k\theta_m}{d}. \tag{6.80}$$

If the power intensity of the beam is such that H is below this critical value, the melting temperature will never be reached. Such an information, however, cannot be obtained from the simpler one-dimensional model.

EXAMPLE 6.16 If the diameter of the focussed laser beam incident on a tungsten work is 200 μm and 10% of the beam energy is absorbed, find out the minimum value of the beam power intensity to achieve the melting.

SOLUTION Using the values of the thermal properties of tungsten given in

Example 6.15, we see relation (6.80) takes the form

$$H_{cr} = \frac{2 \times 2.15 \times 3400}{0.02} \text{ W/cm}^2 = 7.31 \times 10^5 \text{ W/cm}^2.$$

Since only 10% of the energy is absorbed, the critical value of the beam power intensity is 7.31×10^6 W/cm^2.

Steady State Hole Penetration

The foregoing heat conduction analysis helped us get only a rough idea about the time required for the material to reach the melting temperature. But the determination of the dimensions of the molten portion of the material is quite complicated. However, if the molten pit (or hole) is deep and narrow, the major portion of heat conduction from the molten hole takes place through the side walls. When the heat input rate is equal to the rate of heat loss by the molten portion, it maintains its shape and size. In such a steady state condition, the rate of heat loss by the molten portion (Fig. 6.77) is given by

$$2\pi Zk \frac{\theta_m - \theta_0}{\ln (D/d)}.$$

From experience, it has been found that $D \approx 55d$. So, $\ln (D/d)$ may be

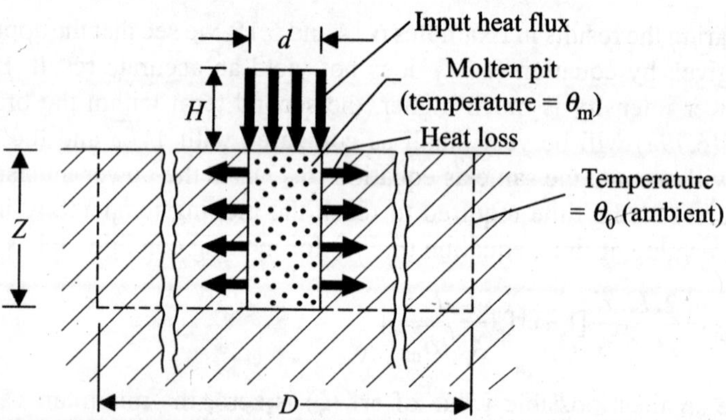

Fig. 6.77 Idealized model of melting process during LBM.

approximately taken to be 4, and equating the heat input rate to the heat loss rate, the relation we obtain is

$$\frac{H \cdot \pi d^2}{4} \approx \frac{\pi Zk(\theta_m - \theta_0)}{2}.$$

Finally, the depth of the molten pit, which can approximately represent the depth of the hole machined with a medium intensity beam, is given by

$$Z \approx \frac{Hd^2}{2k(\theta_m - \theta_0)}. \tag{6.81}$$

When the beam intensity is very high ($>10^7$ W/cm^2), the heating is very rapid, and the mechanism we have just given is not valid. The incident beam heats up the surface quickly and vaporizes it. Thus, the surface of the work where the beam falls recedes as the material vaporizes. So, if v is the velocity with which the surface recedes, the rate of heat input required to vaporize the material (equal to the rate of heat input from the incident beam) is

$$H \approx vL, \tag{6.82}$$

where L is the amount of energy to vaporize a unit volume of the material.

EXAMPLE 6.17 A laser beam with a power intensity of 10^5 W/mm^2 is used to drill holes in a tungsten sheet of 0.5 mm thickness. The drill diameter is 200 μm. If 3×10^4 joules/cm^3 are required to vaporize tungsten, estimate the time required to drill a through hole. The efficiency may be taken to be 10%.

SOLUTION From equation (6.82),

$$v = \frac{0.1 \times 10^7}{3 \times 10^4} \text{ mm/sec} = 3.3 \times 10^2 \text{ cm/sec.}$$

So, the time required to drill the hole is

$$t \approx \frac{0.5}{3.3 \times 10^2} \text{ sec} \approx 0.0015 \text{ sec.}$$

6.7.2 SUMMARY OF LBM CHARACTERISTICS

Mechanics of material removal	Melting, vaporization
Medium	Normal atmosphere
Tool	High power laser beam
Maximum material removal rate	5 mm^3/min
Specific power consumption (typical)	1000 W/mm^3/min
Critical parameters	Beam power intensity, beam diameter, melting temperature
Materials application	All materials
Shape application	Drilling fine holes
Limitations	Very large power consumption, cannot cut materials with high heat conductivity and high reflectivity

6.8 PLASMA ARC MACHINING (PAM)

A plasma is a high temperature ionized gas. The plasma arc machining is done with a high speed jet of a high temperature plasma. The plasma jet heats up the workpiece (where the jet impinges on it), causing a quick melting. PAM can be used on all materials which conduct electricity, including those which are resistant to oxy-fuel gas cutting. This process is extensively used for profile cutting of stainless steel, monel, and superalloy plates.

A plasma is generated by subjecting a flowing gas to the electron bombardment of an arc. For this, the arc is set up between the electrode and the anodic nozzle; the gas is forced to flow through this arc (Fig. 6.78).

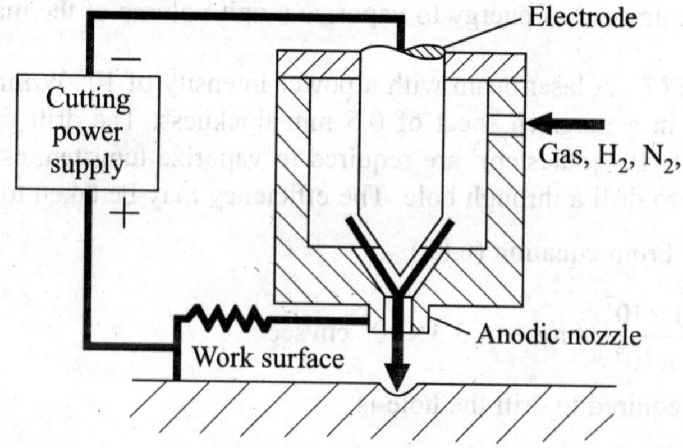

Fig. 6.78 Schematic diagram of plasma arc machining.

The high velocity electrons of the arc collide with the gas molecules, causing a dissociation of the diatomic molecules or atoms into ions and electrons resulting in a substantial increase in the conductivity of the gas which is now in plasma state. The free electrons, subsequently, accelerate and cause more ionization and heating. Afterwards, a further increase in temperature takes place when the ions and free electrons recombine into atoms or when the atoms recombine into molecules as these are exothermic processes. So, a high temperature plasma is generated which is forced through the nozzle in the form of a jet. The mechanics of material removal is based on (i) heating and melting, and (ii) removal of the molten metal by the blasting action of the plasma jet.

For more details, see the standard handbooks and reference books. Here, we shall list the basic characteristics to familiarize the reader with the process.

Summary of PAM characteristics

Mechanics of material removal	Melting
Medium	Plasma
Tool	Plasma jet
Maximum temperature	16,000°C
Maximum velocity of plasma jet	500 m/sec
Maximum material removal rate	150 cm^3/min

Specific energy	1000 W/cm^3-min
Power range	2–200 kW
Maximum plate thickness	Up to 200 mm (depends on material)
Cutting speed	0.1–7.5 m/min
Voltage	30–250 V
Current	Up to 600 amp
Critical parameters	Voltage, current, electrode gap, gas flow rate, nozzle dimensions, melting temperature
Materials application	All conducting materials
Shape application	Cutting plates
Limitation	Low accuracy

6.9 EXERCISE PROBLEMS

6.1 A cylindrical impression with a diameter of 10 mm and a depth of 1 mm has to be made on a tungsten carbide surface. The feed force is constant and equal to 5 N. The average diameter of the grains in the abrasive slurry is 0.01 mm. The tool oscillates with an amplitude of 30 μ at 20 kHz. The slurry contains 1 part of abrasive to about 1 part of water. The fracture hardness of tungsten carbide workpiece may be taken as 7000 N/mm^2. Estimate the machining time.

6.2 A square through hole of 5 mm × 5 mm has to be drilled in a 5-mm-thick tungsten carbide sheet. The slurry is made of 1 part of 10-μ-radius boron carbide grains mixed with $1\frac{1}{2}$ parts of water. The feed force is 4 N. The tool oscillates with an amplitude of 0.015 mm at 25 kHz. Assuming that only 20% of the pulses are effective, calculate the time required to complete the job.

6.3 In an ECM operation, a pure copper block is being machined. If a current of 5000 amp is used, determine the volume rate of material removal from the copper block.

6.4 The composition of a Nimonic alloy turbine blade is 18% cobalt, 62% Ni, and 20% chromium. It is being machined electrochemically with a current of 1500 amp. Find out the volume removal rate. The dissolution valency of chromium is 6, whereas that for both nickel and cobalt is 2.

6.5 The composition of a monel alloy workpiece undergoing electrochemical machining is as given here:

Ni	Cu	Fe	Mn	Si	C
63%	31.7%	2.5%	2%	0.5%	0.3%

If the machining current is 1000 amp, estimate the volume removal rate.

6.6 The equilibrium gap when machining (electrochemically) iron, using NaCl solution in water as the electrolyte, is found to be 0.2 mm with an operating voltage of 12 V. Iron dissolves at a valency 2, the density of iron is 7.8 g/cm^3, and the specific resistance of the electrolyte is 2.8 ohm-cm. Calculate the metal removal rate/unit work surface area. The overvoltage may be taken as 1.5 V.

6.7 In an electrochemical trepanning operation on a flat iron surface, an electrode in the form of a thin copper tube (with an outer diameter of 12 mm and a wall thickness of 1.5 mm) is being used. The properties of the NaCl-water electrolyte are

specific heat = 1.9 cal/g-°C, density = 1 g/cm^3,

specific resistance = 2.8 Ω-cm, boiling temperature = 95°C.

The gap between the electrode and the work surfaces is 0.2 mm. If the ambient temperature is 40°C and the operating voltage is 12 V, what is the minimum flow velocity of the electrolyte, taking the total overvoltage to be 1.5 V?

6.8 A semicylindrical impression has to be produced on an iron work surface. The radius of the cavity to be machined is 20 mm. The process data are

applied voltage = 12 V, total overvoltage = 1 V,

feed velocity = 0.5 mm/min, work material = Fe,

electrolyte conductivity = 0.3 Ω^{-1} cm^{-1}.

Find out the shape of the electrode required.

6.9 For spark machining of a 5 mm × 5 mm square through hole in a solid low carbon steel plate of 5 mm thickness, a brass tool is used with kerosene as the dielectric. The resistance and the capacitance in the relaxation circuit of the spark generator are 100 Ω and 15 μF, respectively. The supply dc voltage is 220 V and the gap is maintained at such a value that the discharge takes place at 120 V. Estimate the time required to complete the job.

6.10 A 10-mm-diameter hole has to be drilled in a 5-mm HSS sheet by EDM using a relaxation circuit. The required surface finish is 20 μ. Determine the capacitance to be used when the supply and discharge voltages are 220 V and 150 V, respectively, the resistance being 50 Ω. Also, estimate the time required to complete the job.

6.11 A 100-μm-wide slot is to be cut in a 1.5-mm-thick tungsten sheet, using an electron beam with a power of 7 kW. What will be the speed of cutting?

6.12 A laser beam with a power intensity of 2×10^5 W/mm^2 falls on a stainless steel sheet. Find out the time required for the stainless steel surface to reach the melting temperature, assuming that only 8% of the beam power is absorbed. Use suitable values for the thermal properties of stainless steel. Given $\alpha = 0.071$ cm^2/sec, $k = 0.27$ W/cm-°C, and $\theta_m = 1455$°C.

6.13 A laser beam with a power intensity of 2×10^5 W/mm^2 is used to drill a 0.2-mm diameter through hole in a tungsten sheet of 0.4 mm thickness. If the efficiency of the operation is only 10%, estimate the time required.

7 Manufacturing in the Twenty-First Century: Micromachining, Generative Manufacturing, and Self-Assembly

7.1 INTRODUCTION: WHITHER MANUFACTURING

The onset of industrial revolution (see also *Introduction*) with the invention of prime movers in the late eighteenth century was one of the most critical and decisive events in human history. Earlier, the source of mechanical power was either muscle power or strong wind or swift streams. Once man could convert thermal energy into mechanical energy, rapid industrialization became possible which changed the face of the world. Manufacturing became a major activity which started controlling the world economy.

The second industrial revolution was less noticed but has transformed our life during the last about three decades through revolutionary changes in communication, networking, control, and computation. This happened with the introduction of integrated circuits and microelectronics. The hallmark of this revolution has been the exponential increase in the number of transistors incorporated onto a single piece of silicon. This has led to a continuous increase in capability, performance, and even reliability. There are two basic aspects of these VLSI chips containing tens of millions of transistors, all virtually identical, integrated through an unbelievably complex circuit. These are (i) miniaturization, and (ii) low cost of production. Quite often, we tend to give more importance to the size but in reality what has resulted in the impact is the affordable cost of manufacturing. If the Pentium chip is 100 times bigger in size, not much will be different, but if it is 100 times costlier, the PC invasion into all walks of our life would not have been possible. So, miniaturization has definitely shrunk the size but at the same time has drastically cut down the cost, and that plays a very important role in converting our society from a manufacturing based one to a knowledge based one.

According to most technologists, we are on the verge of a third industrial revolution which may far surpass the second revolution in its impact. Though the silicon revolution has led to immense computing power, the interaction of these devices with the outside world is still through conventional devices which are bulky, slow, and expensive. Imagine a microscopic machine with components no bigger than a grain of fine dust going inside our fine arteries and repairing damaged nerves, or such a minuscule autonomous robot repairing tiny defects in integrated circuits of a complex system with a high degree of artificial intelligence and self-repairing capability. The possibilities are mind-boggling. Millions of such micromachines can be batch-produced making the cost very low helping their adoption in all walks of biology, medicine, sensors, actuators, etc.

These machines will have the ability to perform a variety of functions apart from computation including physical and chemical sensing, actuation, controlling light, and communication. It is obvious that such vanishingly small but smart machines can be developed through a complete integration of electronics and mechanical engineering. Both the design and fabrication have to be integrated. Such systems are being called microelectromechanical systems or MEMS in the abbreviated form. The functions of MEMS include sensing, computing, control, and mechanical actuation to interact with the outside world. The MEMS have the potential of becoming popular as they are made by exploiting the existing integrated circuit manufacturing infrastructure in many cases. At the same time, the conventional and unconventional manufacturing processes are undergoing revolutionary changes in their capability to fabricate microcomponents with extreme precision. Figure 7.1 shows how the tolerances in manufacturing and fabrication are changing with time. The technology for achieving such extreme tolerance values is closely associated with the technology used for producing extremely small dimensions. Fabricating objects with dimensions from submillimeter to micron and submicron levels is termed as micromanufacturing.

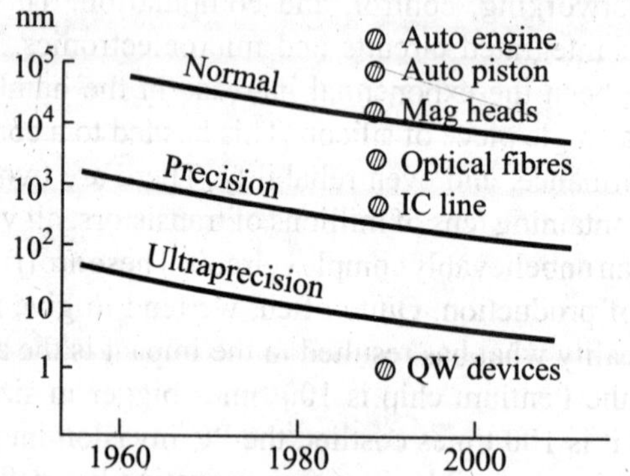

Fig. 7.1 Trends of tolerance in manufacturing.

Another important activity associated with the third industrial revolution is the development of many new exotic smart materials. Thus, the scope of this revolution so far as the material is concerned will be extremely wide in comparison to the second industrial revolution which was based only on silicon.

As mentioned at the very beginning of this book, manufacturing processes are primarily aimed at imparting a desired shape, size with a satisfactory degree of finish and accuracy. From the very beginning, the shaping methods were of two types—material conservation processes and material removal processes. In material conservation processes, the volume of the initial raw material is approximately the same as that of the final product. The casting, forming, and powder metallurgy (sintering) processes are of this type. Another method of achieving the desired shape is to remove the excess material from the original body. All types of machining operations belong to this class of removal processes. However, wisdom dawned on the technologists and towards the end of the twentieth century the concept of material addition processes emerged in the field of manufacturing. In nature, all living objects grow to definite shapes and sizes. But this shaping process is achieved through a gradual addition of material in small quanta (cells) in a well-controlled desired fashion. Towards the end of the twentieth century, the developments in both material science and computer technology reached the required levels for enabling man to imitate nature and develop shaping processes through gradual addition of material. Quite often, such processes are termed under a general heading—*rapid prototyping*—as, till now, such processes are used mainly for quick prototype development. But such a name is very restrictive as these processes are believed to become suitable for manufacturing also. Thus, technically, a better terminology can be *generative manufacturing*. Already a host of such processes have been developed and commercial machines are available. Figure 7.2 shows all the shaping processes.

In all shaping operations, the information on material and dimensions is not contained in the process itself. Such information comes from drawings, computer memory, etc. But in all natural growth processes, such information is encoded in the basic elements used in the building process. Thus, for achieving ultimate success in shaping operation, it is necessary to contain all information in the process itself. Though at a first glance the advantage may not be so apparent but a little thinking can reveal what a tremendous leap that is going to be. Most machines for manufacturing can then be abandoned. For colonizing other planets and satellites, that is essential as transporting all machines can make such projects impractical. Hence, in the ultimate stage, manufacturing process for shape generation is going to be self-generating where all information will be contained in the process itself. Objects will grow as babies grow in mother's womb. So, shaping operations can be classified as indicated in Fig. 7.3a. How the manufacturing processes will develop to this ultimate goal is shown in Fig. 7.3b indicating the important features at different stages of the development.

In this chapter, all the three major new emerging manufacturing processes—micromanufacturing, generative manufacturing, and self-assembly—will be discussed.

7.2 MICROMANUFACTURING—OVERVIEW

Though silicon technology and extension of that form an important component of micromanufacturing, there are other varieties of processes used to produce microcomponents. It should also be remembered that ultraprecision devices to measure and control dimensions at the microlevel need to be also developed

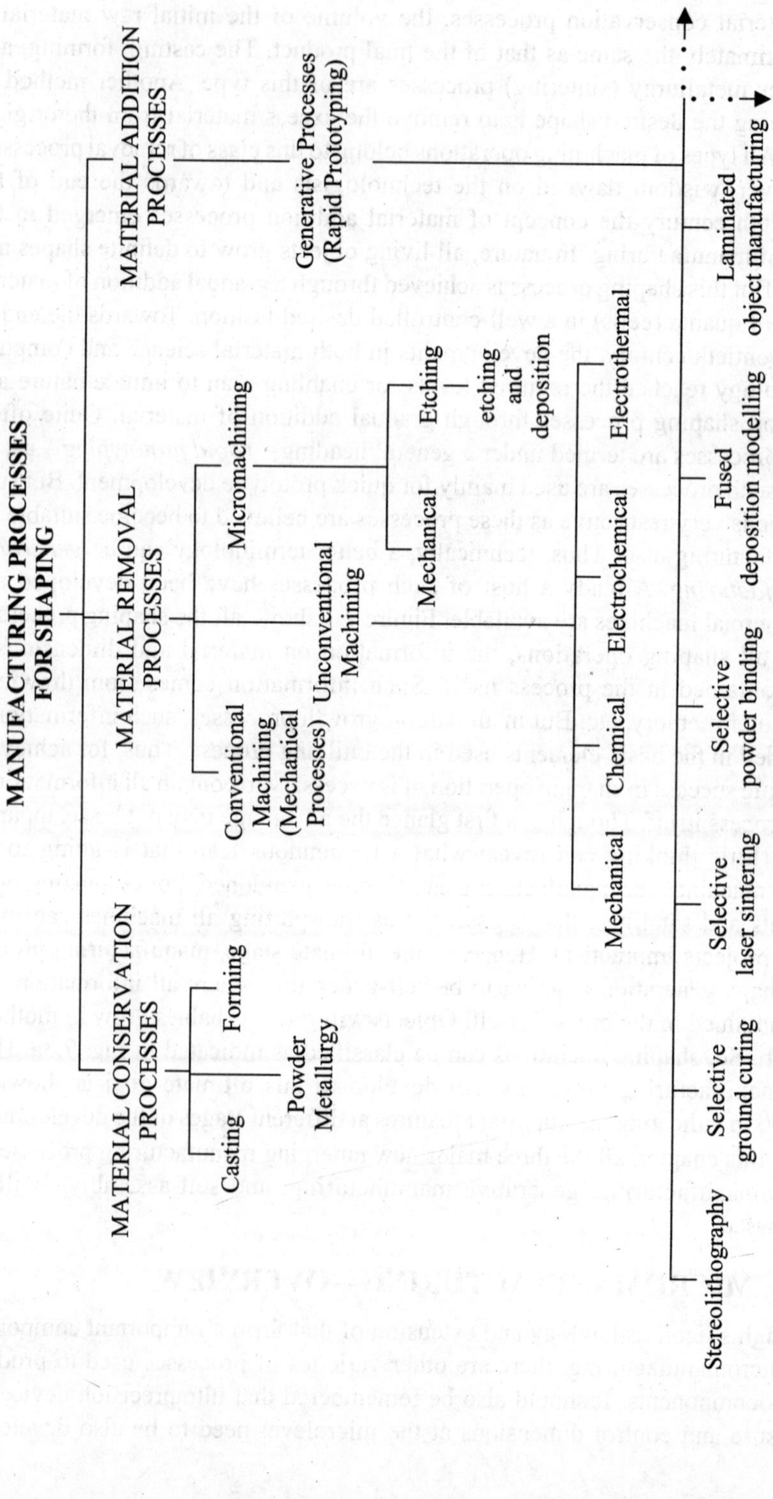

Fig. 7.2 Manufacturing processes for shaping.

SHAPING

NONSELF-GENERATING (OLD ERA PROCESSES)		SELF-GENERATING (NEW ERA PROCESSES)
Nongenerative	Generative	
Casting	Selective solidification	
Forming	Selective sintering	
Joining	Selective binding	
Machining	Selective deposition	

(a) Old era and new era shaping

Traditional processes	*Present generative processes*	*Future generative processes*	*Ultimate self-generative processes*
Material exists independent of shape	Shape and material grow simultaneously	Shape and material will grow simultaneously	Will be self-generating and self-controlled like biological growth
Individual components are single material objects	Multimaterial components possible	Multimaterial components will be common	
Material addition/ removal at macrolevel/ mesolevel	Material addition at microlevel	Material addition will go to nanolevel/ molecular level	
Information is from outside the process	Information is from outside the process	Information will be partially contained in the process itself	
Assembly is an essential step in manufacturing	Assembly can be eliminated in some cases	Most often subassemblies will be eliminated	

(b) Transformation of manufacturing processes in twenty-first century

Fig. 7.3 Shape generation and its transformation with time.

to make micromanufacturing possible. It is further important to note that in micromanufacturing machine tools/processing equipment and dimensional and surface quality measuring instruments must be integrated into a closed loop control system. Micromanufacturing and conventional precision manufacturing use many of the same techniques and equipment, but the objectives are different. In precision manufacturing, the objective is to produce parts with a size to tolerance ratio greater than 10,000. On the other hand, in micromanufacturing, the goal is to produce parts typically 1–100 μm in size with sufficient accuracy to achieve functional goals with required repeated satisfaction. Quite often, the accuracy and tolerances in micromanufacturing may be beyond the range of those prescribed in precision engineering.

Figure 7.4 indicates the dimensional magnitudes covered by micromanufacturing along with the wavelengths of different types of electromagnetic radiation. This is relevant as such radiations are used as tools for micromanufacturing of a few types as indicated. X-ray lithography can produce microcomponents

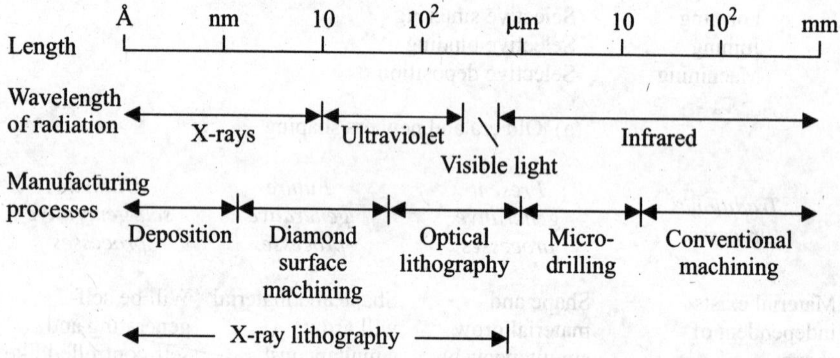

Fig. 7.4 Dimensional magnitudes of different micromanufacturing processes.

with ease because of its very small wavelength. It is also to be noted that in MEMS applications one of the key features is integration of components covering a range of a few orders of magnitude. Therefore, for MEMS, the required metrology has to also cover several orders of magnitude. So, the progress of MEMS and micromanufacturing has been closely associated with a parallel development in micrometrology.

7.2.1 DEVELOPMENTS IN METROLOGY

Features smaller than the wavelength of visible light cannot be seen directly. Scanning electron microscopes (SEM) can be considered to be the first development which made viewing of features smaller than 0.5 μm possible. In SEM, the wavelength of a high energy electron (accelerated by kilovolts) is very small. We know

$$E = h\nu = h \cdot \frac{c}{\lambda},$$

where E is the energy of a photon, c is the speed of light, ν and λ are the frequency and wavelength, and h is Planck's constant. So, if an electron is accelerated by 30 kV, the energy of the electron will be 30 keV which is equal to $30 \times 10^3 \times 1.6 \times 10^{-19}$ J $= 4.8 \times 10^{-15}$ J. Using

$$h = 6.63 \times 10^{-34} \text{ J-sec}, \qquad c = 3 \times 10^8 \text{ m/sec},$$

we get

$$\lambda = \frac{hc}{E} = \frac{6.63 \times 10^{-34} \times 3 \times 10^{8}}{4.8 \times 10^{-15}} \text{ m} = 4.1 \times 10^{-11} \text{ m}$$

$$\approx 0.4 \text{ Å}.$$

The short wavelength of an accelerated electron gives rise to a very high lateral resolution, typically 1 nm.

Optical interference techniques also yield good lateral resolution (in the micron range). Contact profilometers using pointed stylus with a few microns tip radius are also used for measurements with similar precision. High degree of vertical resolution can be obtained by interference microscopy method. Two beams are produced from one source; one of the two beams is considered as a reference wave while the other is the object wave which is modulated in phase by the object surface. When the two waves recombine, a fringe pattern results that depends on the modulation of the object wave. With this arrangement, it is possible to measure heights upto 500 μm with an accuracy of the order of ±5 μm However, for this technique to be successful, fairly reflective surfaces are required.

For mechanical type stylus instruments, the stylus movement is measured relative to a reference that conforms to the nominal shape of the surface under inspection. As mentioned earlier, a sharp conical or pyramidical diamond tip is used with about 2–10 μm radius. In scanning tunnelling microscopes (STM), noncontact type stylus is used. These instruments are capable of lateral resolution sufficient to resolve protruding atoms. The vertical resolution is as small as 0.02 μm. Figure 7.5 shows the basic principle of STM in which a very sharp stylus tip is brought very close to the surface of the sample and the gap is less

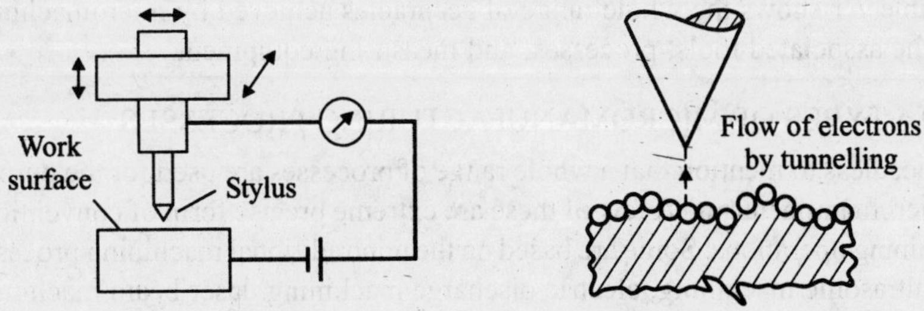

Fig. 7.5 Principle of STM.

than 1 nm. At such close proximity, the free electrons from work surface tunnel to the conductive sharp stylus tip, resulting in a very small current given by

$$I \propto e^{-2kd}, \tag{7.1}$$

where I is the tunnelling current, k is a constant, and d is the distance between the tip and the surface. The typical value of k is 1.0 Å^{-1}.

Another major development in micrometrology is the invention of atomic force microscopes (AFM). This is also a noncontact type instrument which uses a very fine stylus attached to the end of a thin cantilever. As the fine stylus tip is brought close to a surface (within 30–150 μm), an attractive force between the atoms is generated. This causes the cantilever to bend and a capacitive sensor measures its small deflection. Since the atomic force itself depends on the gap between the stylus tip and the work surface, profiling of the surface can be done.

Quite a few other types of micrometrology techniques like vibroscanning method, elastic transmission method, etc., have recently emerged. Of course, it should be remembered that MEMS do not necessarily require a very high resolution measurement. MEMS generally have relatively large vertical dimensions and measuring large vertical dimensions of features possessing smaller lateral dimensions is a typical challenge in MEMS micrometrology. The typical ranges of micrometrology tools are shown in Fig. 7.6.

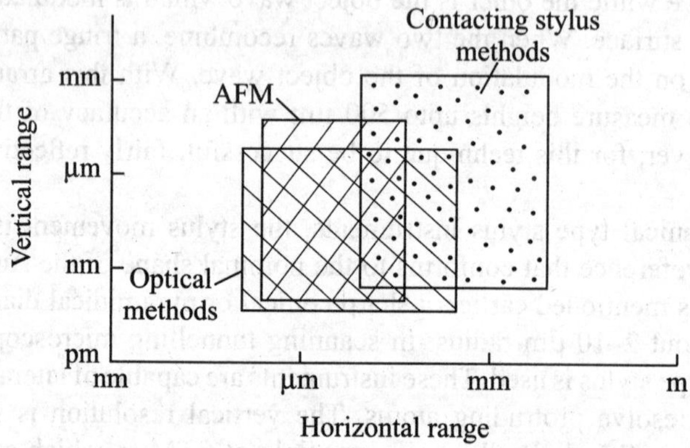

Fig. 7.6 Typical ranges of MEMS micrometrology.

Table 7.1 shows the whole range of accuracies achieved by micromachining and the associated tools, processes, and measuring equipment.

7.2.2 TYPES OF MICROMANUFACTURING PROCESSES

It is needless to mention that a whole range of processes are used for the purpose of micromanufacturing. A few of these are extreme precise form of conventional machining operations. Some are based on the nontraditional machining processes like ultrasonic machining, electric discharge machining, laser beam machining, etc., but with much higher degree of tolerance and more precise control. At the same time, two-dimensional (or even quasi two-dimensional) microsystems use marking and etching technology. A new concept in manufacturing has emerged recently which suggests development of shaped objects by adding material in small quanta. It is felt that at the highest level of sophistication, shaped objects will grow by itself utilizing the self-assembling capacity of smart materials. Before this becomes applicable to macro sized parts, for producing micro sized parts such techniques may prove to be very useful. Most of the available processes for micromanufacturing can be classified as shown in Fig. 7.7 (the term LIGA will be explained in Section 7.5).

Table 7.1 Aspects of microfabrication at various levels

Accuracy	Tools and materials	Processes	Related machine tool elements	Drive(s) and control	Measuring equipment
10 μm	Cutting tools Superhard alloys	EDM ECM WEDM	Ball and roller guideways Precision screws	Servomotors Stepper motors Relay logic controllers	Optical microscopes
1 μm	Abrasive grains Polycrystals Photoresists	Precision EDM Fine cutting and grinding Photolithography (visible light) EBM LBM	Dynamic hydrostatic bearings Ball/roller preloaded guides Electrostatic/ pneumatic bearings	dc servomotors Logic controllers Optimal control	UV ray microscopes
0.1 μm	Single point diamond cutting tools Photoresists Metallic oxides Abrasive grains	Ultraprecision grinding Vacuum deposition Chemical vapour deposition Photolithography (UV) Single point diamond tool cutting	Precision pneumostatic pressure bearings and guideways Elastic spring guideways Gem roller and ball bearings	Precision dc servomotors Adaptive control	Laser interferometer Electromagnetic comparators Precision differential transformer

Table 7.1 Aspects of microfabrication at various levels (cont.)

Accuracy	Tools and materials	Processes	Related machine tool elements	Drive(s) and control	Measuring equipment
0.01 μm	Reactive abrasive grains Lapping equipment Laser Electron X-ray	EEM Mechanochemical lapping Physical vapour deposition Electron beam exposure Synchrotron radiation exposure	Monostructure elastic spring guideways Electromagnetic or electrostatic fine movement guideways	High precision dc servomotors Electromagnetic servoactuators	Ultraprecision differential transformer Electromagnetic proximity sensors Laser optical sensors
0.001 μm = 1 nm	Atoms Molecules Ions Plasma	Noncontacting lapping Ion machining Sputter etching Reactive etching Ion deposition Ion implantation	Electrostatic and electromagnetic deflection Electrostrictive and magnetostrictive fine movements	Electrostriction Magnetostriction	Electron X-ray Scintillators Laser interferometers
Subnanometer	Atoms Molecules Neutrons	Substance synthesizing Self-assembly			

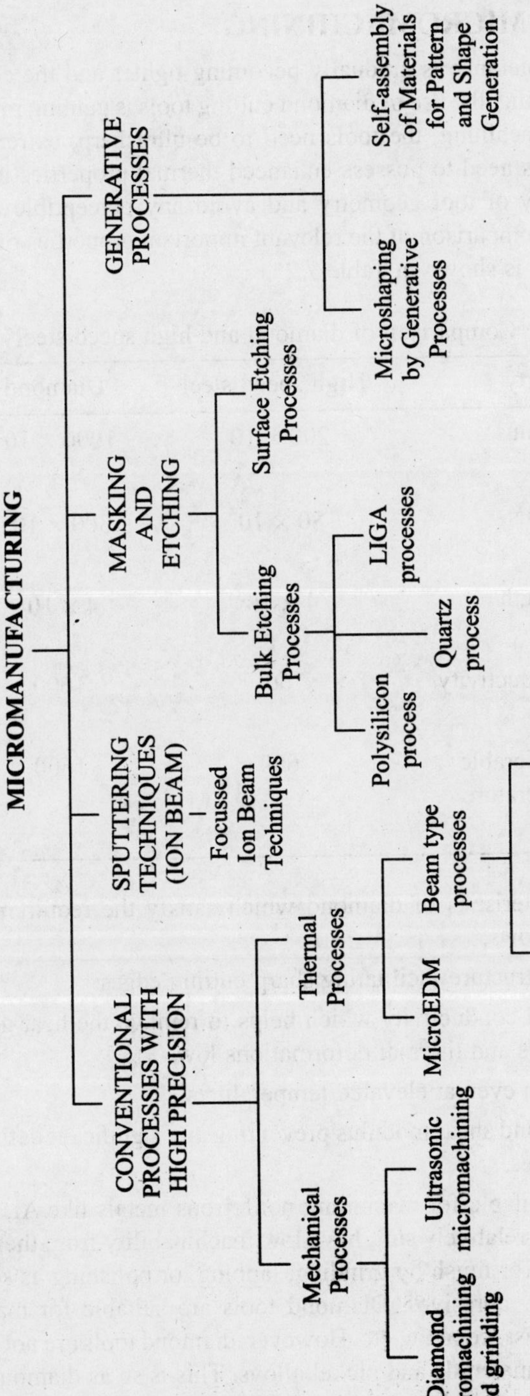

Fig. 7.7 Micromanufacturing processes.

7.3 DIAMOND MICROMACHINING

As the demand on tolerance is gradually becoming tighter and the component dimensions are reducing, the use of diamond cutting tools is gaining importance. For ultraprecision machining, the tools need to be ultrasharp, extremely hard and the tool materials need to possess enhanced thermal properties in order to maintain the integrity of tool geometry and avoid any perceptible deflection while machining. A comparison of the relevant important properties of diamond with high speed steel is shown in Table 7.2.

Table 7.2 Comparison of diamond and high speed steel

Property	High speed steel	Diamond
Elastic modulus (MPa)	200×10^3	1000×10^3
Shear modulus (MPa)	80×10^3	300×10^3
Tensile strength (MPa)	4×10^3	4×10^3
Thermal conductivity (W/m-°C)	60	1600
Maximum tolerable cutting temperature (°C)	650	1600

The special characteristics of diamond which satisfy the requirements we have stated are as follows:

(i) Crystalline structure facilitating sharp cutting edges.

(ii) High thermal conductivity which helps to remove the heat generated and keeps temperatures and thermal deformations low.

(iii) High strength even at elevated temperatures.

(iv) High elastic and shear modulus preventing any significant deformation due to machining forces.

Diamond is very suitable for machining nonferrous metals like Al, Cu, etc. These materials, being relatively soft, have low machinability from the point of view of producing mirror finish by grinding, lapping, or polishing as scratches are easily produced by abrasives. Diamond tools are suitable for machining brittle materials like glass, ceramic, etc. However, diamond tools are not suitable for machining ferrous materials and nickel alloys. This is so as diamond reacts chemically with ferrous materials resulting in high wear rate.

With CNC and very special machine tools, it is possible to precisely control the cutting tool position with respect to the job. Using diamond tools, one can produce very small components with high precision. Thus, diamond machining can produce very precise components with high degree of surface finish. This type of machining operation is of particular interest for optical and electronic industries.

Diamond tool machining can be used for producing either (i) small workpiece features with dimensions below 100 μm or (ii) macroscale components requiring submicrometric or nanometric tolerances or surface finish. Optical gratings with an accuracy of 1 nm and gratings with 1 nm resolution can be machined by diamond tools. Currently, it is possible to achieve 100 nm tolerance for components with dimensions of 250 mm.

Since extremely precise relative movements are required for diamond micromachining, the machine tools need to be specially designed. Hydrostatic oil or air bearings are used for spindles and guideways. The machine tools must possess very high loop stiffness between the tool and the workpiece. Special actuators are needed for providing extremely small movements required to ensure small feed and depth of cut—sometimes a few nanometers.

In diamond cutting, sometimes the depth of cut can only be a few layers of atoms as indicated in Fig. 7.8. It is obvious that extreme sharpness of the cutting tool edge is essential to achieve this. From this point of view, diamond has the

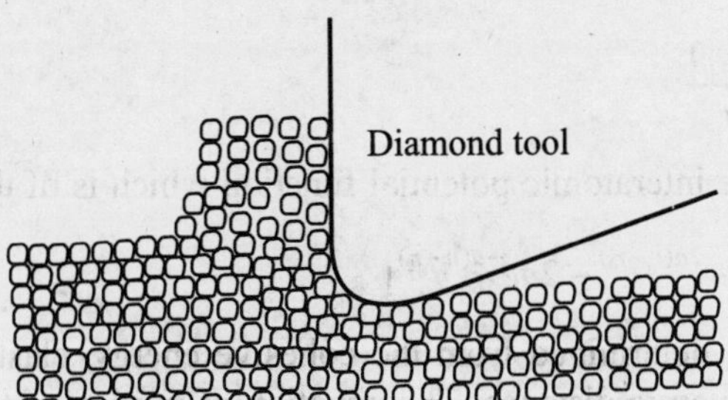

Fig. 7.8 Cutting by single point diamond tool.

real advantage because of its crystalline structure. Diamond tools can have theoretically an edge radius of a few nanometers. However, in practical cases, diamond tools with 15–20 nm edge radius are used. Since the cutting radius is comparable to the uncut thickness, the mechanics of the process is very complex and still not well understood. The effective rake angle varies continuously and there is substantial amount of rubbing at the flank side. The microcutting occurs in a small region containing only a few layers of molecules rendering the process discrete and atomistic. It is not appropriate to consider the work and tool materials to be continua. Theoretical models have been developed to analyze such cutting processes applying the principle of molecular dynamics. Figure 7.9 shows the machining region during the period when the tool advances by an amount Δs in a time interval Δt. The resultant interatomic force on an atom i of the workpiece at the location $\boldsymbol{r}_i$ is given by the sum of the forces from all other nearby atoms.

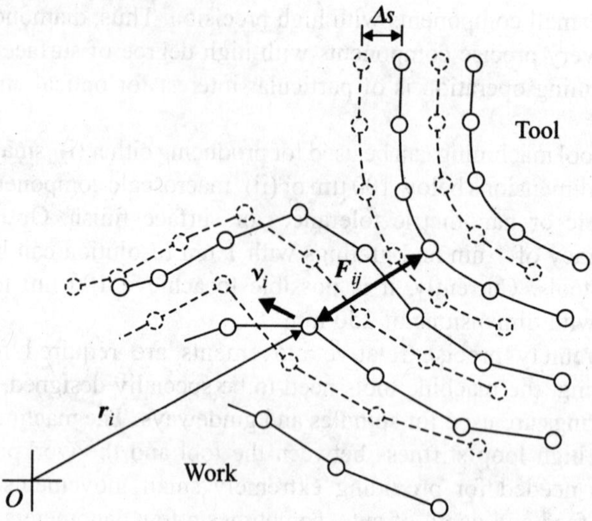

Fig. 7.9 Molecular dynamics simulation of diamond microcutting.

If $\boldsymbol{F}_i$ be the resultant force, then

$$F_i = \Sigma \frac{d\phi(r_{ij})}{dr}, \tag{7.2}$$

where $\phi(r_{ij})$ is the interatomic potential function which is of the form

$$\phi(r_{ij}) = D[e^{-2a(r_{ij}-r_0)} - 2ae^{-a(r_{ij}-r_0)}] \tag{7.3}$$

with D, a, and r_0 determined from the cohesive energy, elastic modulus, and equilibrium interatomic distance. The equation of motion of the i-th atom can be written as

$$\frac{d\mathbf{v}_i}{dt} = \frac{\boldsymbol{F}_i}{m}, \tag{7.4}$$

where m is the atomic mass and $\mathbf{v}_i$ is the atomic velocity. For the purpose of simulation, Δt is much smaller than the period of lattice vibration. The change in position and velocity of the i-th atom is then determined from the equations

$$\boldsymbol{r}_i(t+\Delta t) = \boldsymbol{r}_i(t) + \mathbf{v}_i(t) \cdot \Delta t + \frac{1}{2}\frac{\boldsymbol{F}_i}{m} \cdot \Delta t^2, \tag{7.5a}$$

$$\mathbf{v}_i(t+\Delta t) = \mathbf{v}_i(t) + \frac{\boldsymbol{F}_i}{m} \cdot \Delta t \tag{7.5b}$$

The degree of complexity of the simulation process is easily perceived and the interested reader can consult references. Figure 7.10 indicates the comparison of simulated and experimental results. Though the parameter values for the

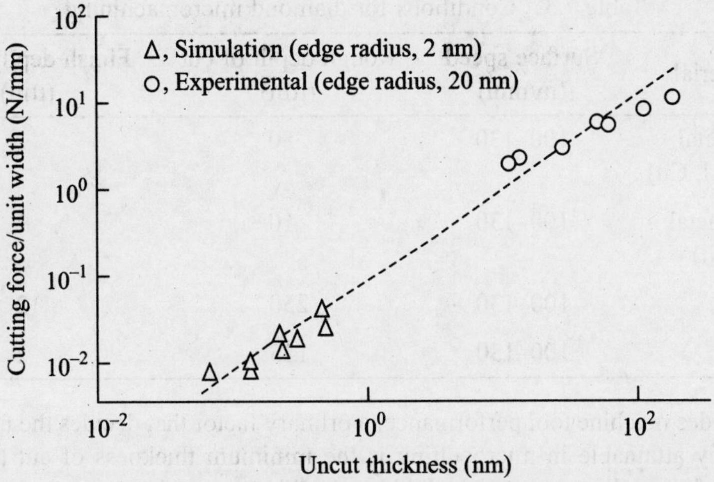

Fig. 7.10 Cutting force estimation by molecular dynamics simulation (work, copper; rake angle, 0°; clearance angle, 7°; cutting speed, 20 m/sec).

simulation and experiments are quite different, the trends match quite nicely. This is a field of active research and more results are expected. The cutting and the thrust components of the machining force suggest that a significant portion of mechanical energy is not effective in metal removal but is lost in rubbing and deformation at the machined surface. When results for a specific work are determined from simulation (for a small uncut thickness) and experiments, the general trend is as shown in Fig. 7.11. It is clear from this figure that specific work increases substantially with the decrease in uncut thickness.

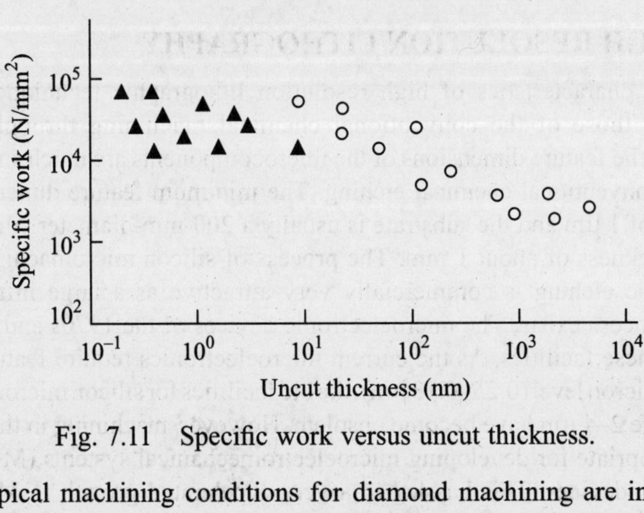

Fig. 7.11 Specific work versus uncut thickness.

The typical machining conditions for diamond machining are indicated in Table 7.3. Single crystal diamond tool turning can produce microgrooves 2.5 μm wide by 1.6 μm deep and diamond turning is being used increasingly as a high precision, high production rate process for a large number of applications.

Table 7.3 Conditions for diamond micromachining

Material	Surface speed (m/min)	Rough depth of cut (μm)	Finish depth of cut (μm)
Soft metal (e.g., Al, Cu)	100–130	50	2
Hard metal (e.g., Ni)	100–130	10	1
Plastic	100–130	250	12
Crystal	100–130	125	2

Besides machine tool performance, a primary factor that decides the ultimate accuracy attainable in microcutting is the minimum thickness of cut (MTC). This is the smallest uncut chip thickness stably removed from a work surface with specific machining parameters. Table 7.4 shows the molecular dynamics (MD) simulation results. In general, the MTC for copper and aluminium can be taken in the range (0.05–0.1)R and (0.1–0.2)R, respectively.

Table 7.4 Minimum thickness of cut (cutting speed, 200 m/sec)

Cutting edge radius (R)	Minimum thickness of cut	
	Copper	Aluminium
5 nm	0.3 nm	0.3 nm
10 nm	0.6 nm	1.2 nm

7.4 HIGH RESOLUTION LITHOGRAPHY

The basic characteristics of high resolution lithographic technique are very similar to those of the conventional chemical machining through etching. However, the feature dimensions of the microcomponents are much smaller than those in conventional chemical etching. The minimum feature dimension is of the order of 1 μm and the substrate is usually a 200-mm-diameter silicon wafer with a thickness of about 1 mm. The process of silicon micromachining using lithographic etching is commercially very attractive as a large infrastructure for this process exists. The microelectronic devices of the 1970s and the 1980s required these facilities. As the current microelectronics require features in the deep submicron level (0.25–0.13 μm), the old facilities for silicon micromachining in the range 2–4 μm have become obsolete. However, machining in this range is very appropriate for developing microelectromechanical systems (MEMS) and other microdevices. This has led to a very accelerated growth in MEMS. On the whole, the etching technique for micromachining plays the most important role in manufacturing micro-sized components and systems. The reasons are as follows:

- Already existing excellent infrastructure for microelectronic industry can be used.
- Batch manufacturing feature of this process makes the product very cheap.
- The process is capable of producing extremely small dimensions.

There are enormous variations in the detailed procedure of the etching techniques but all these lithography based processes can be classified as indicated in Fig. 7.12. In all such processes, the basic elements are lithography and etching

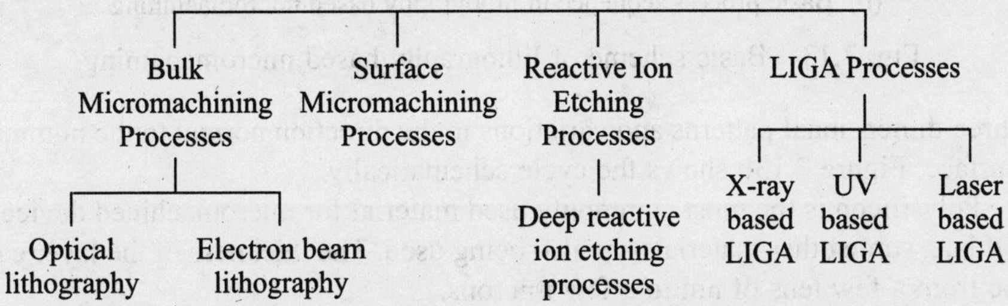

Fig. 7.12 Types of lithography based microfabrication processes.

(and sometimes forming as in LIGA). Figure 7.13a shows the typical steps and sequence. The basic process sequence in lithography based micromachining involves addition of a material layer on a silicon substrate and covering it with a suitable photoresist layer to be patterned by lithography followed by etching. In some cases, the cycle needs to be repeated for generating complex

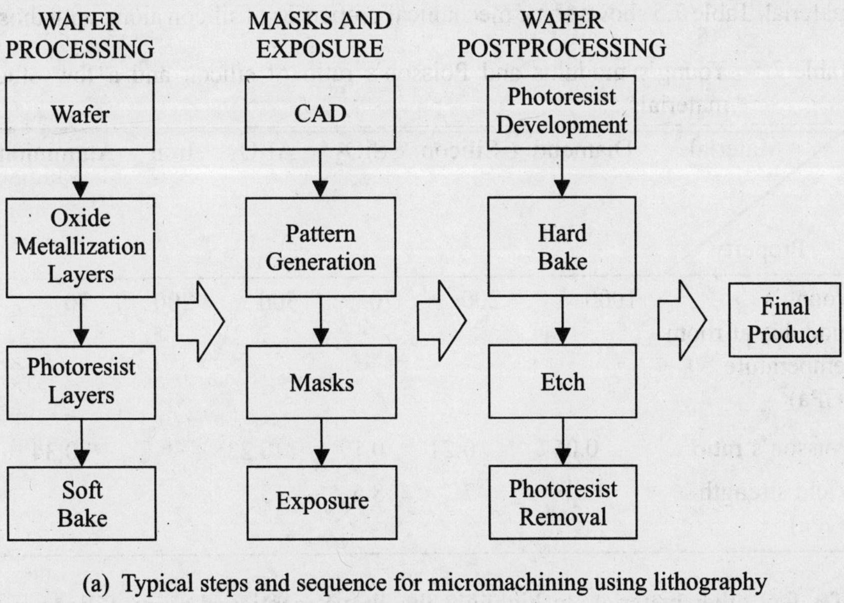

(a) Typical steps and sequence for micromachining using lithography

Fig. 7.13 Basic scheme of lithography based micromachining (cont.).

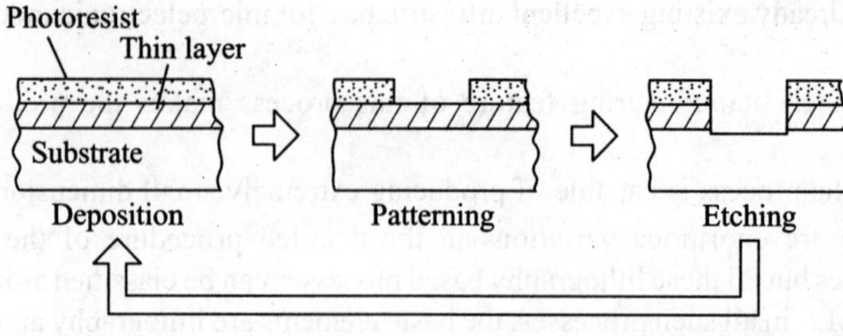

(b) Basic process sequence in lithography based micromachining

Fig. 7.13 Basic scheme of lithography based micromachining.

three-dimensional patterns and variations in the direction normal to the nominal surface. Figure 7.13b shows the cycle schematically.

Polysilicon is the most commonly used material for micromachined devices. Of late, some other materials are also being used. The thickness of the layer can be from a few tens of nm to a few microns.

7.4.1 SILICON AS A MATERIAL FOR MICROELECTROMECHANICAL DEVICES

There are a number of reasons for using silicon as a material for microdevices. Unlike most other materials it can be economically manufactured in single crystal substrates with a high degree of purity and perfection. This provides a number of electrical and mechanical advantages. Silicon has been at the root of the present day microelectronics and, at the same time, it is mechanically an elastic and strong material. Table 7.5 shows some mechanical properties of silicon along with those

Table 7.5 Young's modulus and Poisson's ratio of silicon and a few other materials

Property \ Material	Diamond	Silicon	SiO_2	Al_2O_3	Iron	Aluminium
Young's modulus at room temperature (GPa)	1000	200	70	500	200	70
Poisson's ratio	0.067	0.21	0.17	0.23		0.34
Yield strength (GPa)		7	8.5	15		

of a few other materials to highlight the above-mentioned point. Silicon is a suitable material platform for electronic, mechanical, thermal, optical, and fluidic functions. This wide range of functional integration results in convenient batch manufacturing which, in turn, lowers the cost of production. Ultrapure electronic-grade crystalline silicon wafers are available at relatively low cost (≈\$10 for a 100-mm-diameter and 500-μm-thick wafer). Silicon exists in three forms—crystalline, polycrystalline, and amorphous. Polycrystalline silicon (usually called polysilicon) is also a very suitable material for microelectromechanical devices. Silicon retains its mechanical integrity at temperatures upto 500°C and it is reasonably stable while in contact with commonly-used chemicals.

Silicon oxide is stable and electrically and thermally insulating. Furthermore, SiO_2 is also used as sacrificial layers in fabricating microdevices as SiO_2 can be preferentially etched by HF acid. SiO_2 is thermally grown by oxidizing silicon at temperatures above 800°C.

The crystal structure of crystalline silicon and polysilicon plays an important role when machining is done by chemical etching. This is because the etching rate depends on the orientation of the crystallographic plane on which the etching reagent acts. Si has a valence 4 structure, i.e., each atom is bonded to four neighbours in a tetragonal configuration as shown in Fig. 7.14a. The crystal lattice is face-centred cubic (FCC) as shown in Fig. 7.14b. The lattice constant

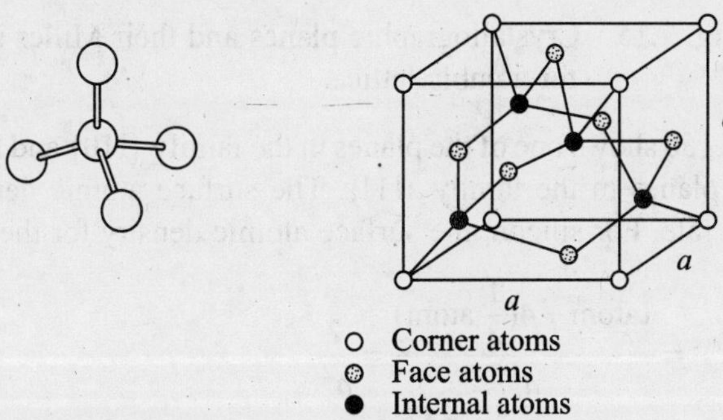

(a) Silicon atom bond structure (b) Unit cell

Fig. 7.14 Crystal structure of silicon.

a, defining the size of the cubic lattice, is equal to 5.43 Å at room temperature and the nearest neighbour's distance is equal to $\sqrt{3}(a/4)$. The atomic density can be determined as

$$N = N_i + \frac{1}{2}N_f + \frac{1}{8}N_c,$$

where N_i is the number of atoms inside the cube, N_f is the number of atoms on the six faces, and N_c is the number of corner atoms. For silicon (Fig. 7.14b), we get $N_i = 4$, $N_f = 6$, and $N_c = 8$. Thus,

$$N = 4 + \frac{1}{2} \times 6 + \frac{1}{8} \times 8 = 8.$$

The atomic density is, therefore, 8 atoms per cube and this gives the density as

$8/a^3$ (≡ 5 × 10^{22} atoms/cm^3).

The three principal axes are [100], [010], and [001][1] along the three edges of the cubic lattice as shown in Fig. 7.15. The third bracket is used to indicate a direction; thus, the +x-direction is indicated by [100] and the +y-direction by [010]. The plane is indicated by the direction that is normal to the plane. So, a plane normal to the y-axis is indicated (see Fig. 7.15a) by (010). Furthermore, all planes (100), (010), and (001) are crystallographically equivalent. This is indicated by the second bracket. So, the (100), (010), and (001) planes can be represented by {100}.

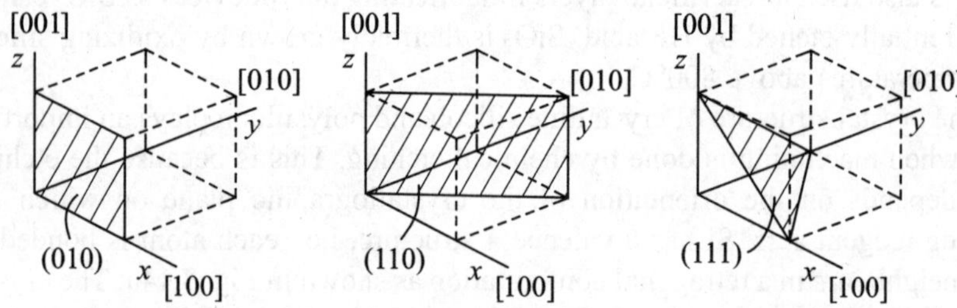

(a) 010 plane (b) 110 plane (c) 111 plane

Fig. 7.15 Crystallographic planes and their Miller indices for a cubic lattice.

Figure 7.15b shows one of the planes in the family {110} and Fig. 7.15c shows one of the planes in the family {111}. The surface atomic density determines the etching rate. For silicon, the surface atomic density for the {100} planes is

$$\rho_{100}^{\text{surface}} = \frac{1\text{atom} + 4(\frac{1}{4}\text{ atom})}{a^2} = \frac{2}{a^2}$$

$$= 6.8 \times 10^{14} \text{ atoms/cm}^2.$$

Similarly, for the {110} planes (see Fig. 7.16),

$$\rho_{110}^{\text{surface}} = \frac{2\text{atoms} + 4(\frac{1}{4}\text{ atom}) + 2(\frac{1}{2}\text{ atom})}{\sqrt{2}(a \cdot a)} = \frac{4}{\sqrt{2}(a^2)}$$

$$= 9.6 \times 10^{14} \text{ atoms/cm}^2.$$

The shape of an etched object depends on the orientation of the various planes and the angles between them. The angle between the {100} and {110} planes is equal to 45°. The planes {100} and {111} are at an angle 54.74°. The planes {111} and {110} can intersect each other at 35.26°, 90°, or 144.74°. The angle between {100} and {111} is particularly important as KOH etches the {100} planes but not the {111} plane. Thus, a cavity etched will be bounded by the {111} planes.

[1]Miller indices.

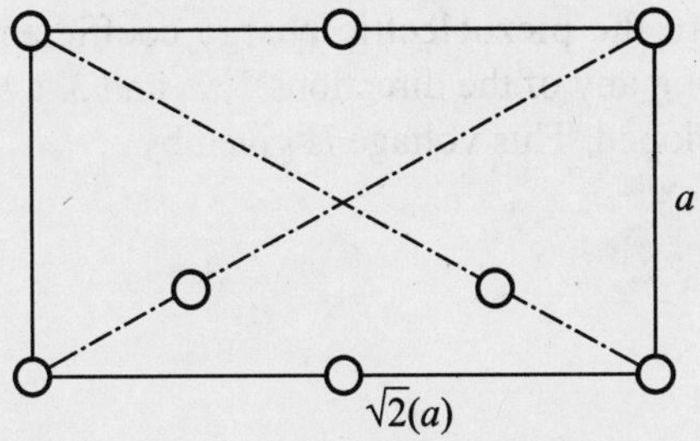

Fig. 7.16 Atoms in a {110} plane.

The commercially available silicon is in the form of circular wafers cut from a larger body along specific crystal planes. The top surface of a wafer, i.e., the cut plane, characterizes the wafers and is encoded on the wafer. Thus, for a {100} orientation wafer, the top surface is a {100} plane (that could be any of the six equivalent (100) planes). The commercially available wafers are mostly of {100} orientation.

7.4.2 OTHER MATERIALS FOR MICROMACHINED DEVICES

Though silicon is the most commonly used material for producing microelectromechanical devices, some other materials are also used for developing microdevices. Among these, the technologies for quartz and glass are well developed. Glass and quartz are suitable for applications requiring optical transparency and electrical insulation. Apart from this, quartz has piezoelectric property and, in many instances, microdevices use piezoelectric actuation. Sometimes, the piezoelectric effect is also utilized for sensing purpose. Figure 7.17 shows a piezoelectric object of rectangular shape subjected to a voltage V_a applied across the

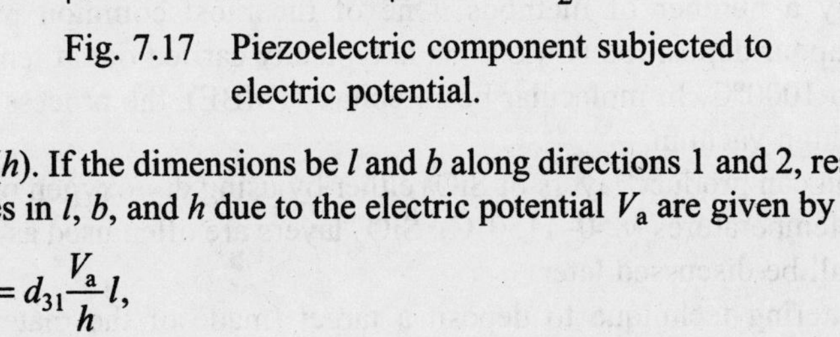

Fig. 7.17 Piezoelectric component subjected to electric potential.

thickness (h). If the dimensions be l and b along directions 1 and 2, respectively, the changes in l, b, and h due to the electric potential V_a are given by

$$\Delta l = d_{31}\frac{V_a}{h}l, \tag{7.6a}$$

$$\Delta b = d_{31}\frac{V_a}{h}b, \tag{7.6b}$$

$$\Delta h = d_{33}V_a, \tag{7.6c}$$

where d_{31} and d_{33} are the piezoelectric charge coefficients. Conversely, if a force F is applied along any of the directions 1, 2, and 3, a voltage V_d across the thickness h gets developed. This voltage is given by

$$V_d = d_{31}\frac{F}{\varepsilon l}, \tag{7.7a}$$

$$V_d = d_{31}\frac{F}{\varepsilon b}, \tag{7.7b}$$

$$V_d = d_{33}\frac{F \cdot h}{\varepsilon lb}, \tag{7.7c}$$

where ε is the dielectric permittivity of the material. For quartz, the following properties can be used for designing microdevices:

$d_{33} = 2.3 \times 10^{-12}$ coulomb/newton,
$\varepsilon = 4.5$,
density = 2.65 g/cm^3,
Young's modulus (E) = 107 GPa.

Gallium arsenide is another material that can be used for micromachined devices. Silicon carbide and diamond have also been used occasionally.

7.4.3 DEPOSITION

As indicated in Fig. 7.13, a thin layer has to be deposited on the substrate. There are a number of methods for deposition of a crystalline layer on the substrate. The common methods are epitaxy, sputtering, oxidation, chemical vapour deposition, anodic bonding, etc.

Epitaxy is a popular method of growing a thin layer of a crystal on a single crystal substrate; it happens in such a way that the crystallographic characteristics of the substrate are reproduced in the deposited layer. The epitaxial layers are typically 1–20 µm thick and possess the same crystallographic orientation as the underlying crystalline substrate. When the layer is grown over an amorphous substrate, the grown layer becomes polycrystalline. Epitaxial growth can be achieved by a number of methods. One of the most common methods is chemical vapour deposition (CVD) that is typically carried out at temperatures greater than 1000°C. In molecular beam epitaxy (MBE), the process is carried out in ultrahigh vacuum.

Oxidation can produce layers of SiO_2 either by using dry oxygen or in steam at elevated temperatures (850–1150°C). SiO_2 layers are often used as sacrificial layers as will be discussed later.

The sputtering technique to deposit a target (made of the material to be deposited) is bombarded by a stream of inert ions (e.g., argon, helium) in a vacuum chamber. Materials in atomic level are ejected and get deposited on the wafer. The deposition by sputtering can take place at relatively low temperatures (below 150°C).

Chemical vapour deposition is based on chemical reaction in a vacuum chamber resulting in deposition on a heated substrate. Typically, the deposition takes place at temperatures above 300°C.

7.4.4 LITHOGRAPHY

Once the material to be micromachined is deposited on the substrate, the lithographic process can be initiated. This involves three sequential steps, namely:

(i) Application of a photoresist on the surface. This is a photosensitive emulsion layer.

(ii) Exposure of this resist to print an image of the mask (developed according to the pattern to be machined).

(iii) Dissolution of the exposed resist in a developer solution so that the appropriate portions dissolve, exposing the surface of the layer to be machined.

The mask is generated using a computer aided design (CAD) tool and transferred to the thin chromium layer on a clear glass substrate. Complicated micromachining and all cases of surface micromachining involve a number of lithographic operations. For simpler and cheaper operations, Fe_2O_3 on soda lime glass or a photographic emulsion on soda lime glass can be used. Chromium on quartz glass is most expensive and is generally used in case of deep UV lithography. For 3″ wafers, the mask plate dimensions are 4″ × 4″ × 0.06″. In "light-field" masks, the drawn feature is opaque and the rest is clear. For "dark-field" masks, it is mostly opaque and the drawn feature is clear. Alignment of the mask with the wafer is an important requirement.

Positive photoresists are organic resin materials. A wafer is coated with an etch mask layer like SiO_2 or SiN_4 after cleaning. The liquid photoresist is poured over the SiO_2 layer and the wafer is spun, thereby creating a uniform photoresist layer. The whole scheme is shown in Fig. 7.18. In spin coating, the wafer holding chuck is rotated at 3000–6000 rpm for about 15–30 seconds.

The thickness of a resist layer is determined by the viscosity of the photoresist and the speed of spinning. Empirically, the resist layer thickness h depends on the parameters according to the equation

$$h = kC^2/(N/1000)^{1/2}, \tag{7.8}$$

where k is the spinner constant with a numerical value between 80 and 100, C is the resist solid content in per cent, and N is the spinner rotational speed in rpm. For commercial silicon processes, the resist thickness lies in the range 0.5–10 μm. Through a prebaking operation (at 90–100°C for 20 minutes in a convection oven) the coating solvent is allowed to evaporate. Microwave heating or IR lamps are also used in production lines. The thickness of the resist usually decreases by about 25% during prebake for both positive and negative resists. The prebake temperature affects the dissolution rate during development.

After prebaking, the mask is aligned with the wafer and the assembly is subjected to optical exposure. The sensitizer contained in the resist breaks down when exposed to light in the UV (200 nm) to blue (450 nm) range. The exposed

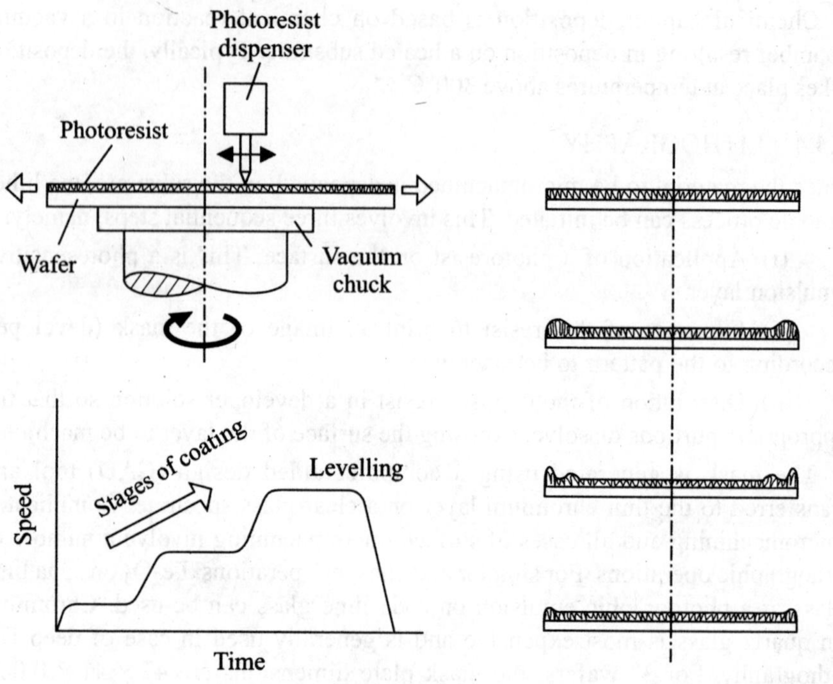

Fig. 7.18 Photoresist spin coating.

portions dissolve when the wafer is put in developer solution in case of positive photoresists. For a negative photoresist, the unexposed portions dissolve during development. Figure 7.19a indicates the difference.

The optical exposure can be accomplished in three different modes, viz., contact, proximity, and projection. Figures 7.19b, 7.19c, and 7.19d schematically show these three modes of exposure. In contact exposure, the mask touches the resist and shortens the life of the mask and leaves undesirable residue on the wafer. In proximity exposure, the mask is brought very near to the resist surface but is not allowed to touch the resist. A gap of about 25–30 μm is maintained between the mask and the resist. In projection mode, an imaging optics is placed between the mask and the resist for forming a sharp image on the resist surface. Though the life of the mask is more in proximity mode, the image quality is poorer. The resolution in proximity mode can be expressed as

$$\text{resolution} \approx [\lambda(\text{gap} + \text{photoresist thickness})]^{1/2}, \tag{7.9}$$

where λ is the wavelength of the exposing radiation. In projection printing, an image is formed on the photoresist surface and the achieved resolution by this method can be expressed as

$$\text{resolution} \approx K\lambda/A, \tag{7.10}$$

where A is the numerical aperture of the optical system and K is a constant

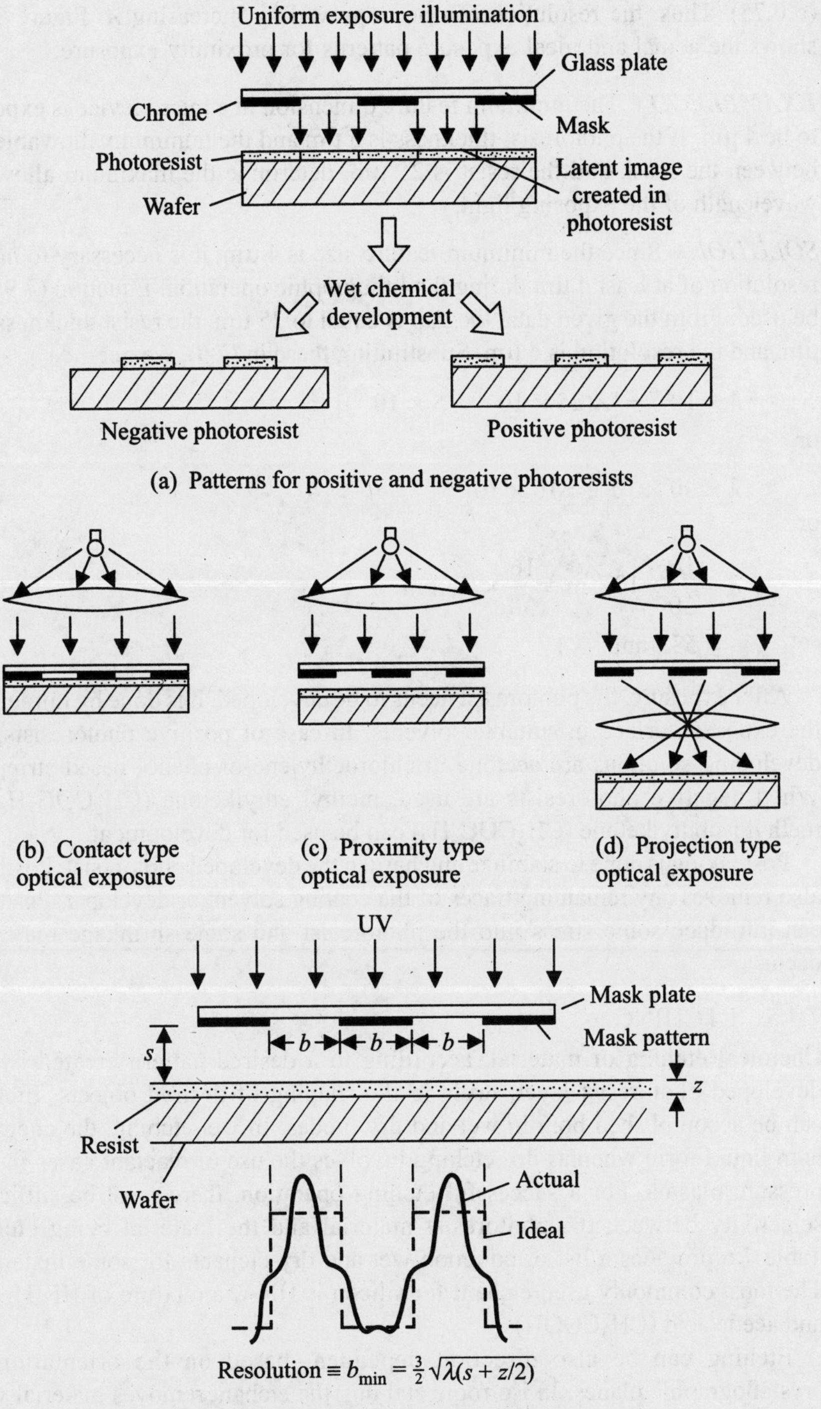

(a) Patterns for positive and negative photoresists

(b) Contact type optical exposure

(c) Proximity type optical exposure

(d) Projection type optical exposure

(e) Real and ideal exposure patterns

Fig. 7.19 Photoresist patterning and exposure techniques.

(≈0.75). Thus, the resolution can be improved by increasing *A*. Figure 7.19e shows the actual and ideal exposure patterns for proximity exposure.

EXAMPLE 7.1 The minimum feature dimension in a microdevice is expected to be 4 μm. If the photoresist thickness is 5 μm and the minimum allowable gap between the mask and the resist is 25 μm, determine the maximum allowable wavelength of the exposing light.

SOLUTION Since the minimum feature size is 4 μm, it is necessary to have a resolution of at least 4 μm during the lithographic operation. Equation (7.9) can be used. From the given data, the gap is equal to 25 μm, the resist thickness is 5 μm, and the resolution is 4 μm. Substituting these in (7.9), we get

$$4 \times 10^{-6} = [\lambda(25 \times 10^{-6} + 5 \times 10^{-6})]^{1/2}$$

or

$$\lambda \times 30 \times 10^{-6} = 16 \times 10^{-12}$$

or

$$\lambda = \frac{16}{30} \times 10^{-6}\ \text{m} \equiv \frac{16}{30} \times 10^{3}\ \text{nm}$$

$$= 533\ \text{nm}.$$

After exposure, the photoresist needs to be developed. It is done by immersing the exposed surface in suitable solvents. In case of positive photoresists, the developing solutions are acetone, trichloroethylene or phenol based strippers. When negative photoresists are used, methyl ethylketone ($CH_3COC_2H_5$) or methyl isobutylketone ($CH_3COC_4H_9$) can be used for development.

Postbaking is done to stabilize and harden the developed photoresist. Postbake also removes any remaining traces of the coating solvent or developer. Postbake can introduce some stress into the photoresist and some shrinkage may also occur.

7.4.5 ETCHING

Chemical etching of material, according to a desired pattern created by the developed photoresist layer, leads to the shaping of desired objects. Etching can be accomplished both in wet and dry modes. In wet etching, the chemical is in liquid form whereas dry etching involves the use of reactant gases in low pressure plasma. For a successful etching operation, there must be sufficient selectivity between the photoresist material and the material being etched. Table 7.6 provides a list of common wet and dry etchants for some materials. The most commonly used etchant for silicon is HNA, a mixture of HF, HNO_3, and acetic acid (CH_3COOH).

Etching can be also direction dependent based on the orientation of crystallographic planes. In isotropic etching, the etchant removes material with equal speed in all directions. So, when a film undergoes isotropic etching, the shape of the cavity produced will be as shown in Fig. 7.20. In the case of isotropic

Table 7.6 Some commonly-used etchants

Material	Characteristic parameter(s)	Wet			Dry (Plasma)	
		HNA	KOH	EDP	SF_6	SF_6/C_4F_8 (DRIE)
Si	Anisotropic	No	Yes	Yes	Varies	Yes
	Temperature	25°C	70–90°C	115°C	0–100°C	20–80°C
	ER (μm/min)	1–20	0.5–2 {100} and {110}	0.75	0.1–0.5	1–3
	ER {111} : ER {100}	1 : 1	1 : 300	1 : 100		
	p^{++} doping stop	No	Yes	Yes	No	No
SiN_4	ER (nm/min)	Low	<1	≈0.1	≈200	≈200
SiO_2	ER (nm/min)	30–70	≈10	≈0.2	≈10	≈10

etching, the shape etched is shown in Fig. 7.20a. When the etchant attacks the material preferably in a direction normal to the exposed top surface, then one can realise straight side walls as shown in Fig. 7.20b. It is, thus, obvious that much better aspect ratio of the micromachined components can be obtained by

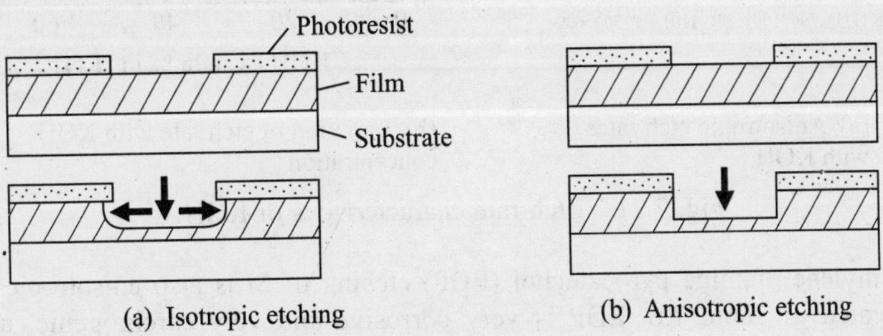

Fig. 7.20 Types of etching.

anisotropic etching. The aqueous solution of KOH is the most commonly used etchant for anisotropic etching. Ethylene diamine pyrochatechol (EDP) is also used for anisotropic wet etching. As can be seen from Table 7.6 that etch rate (ER) can be made dependent on doping and this phenomenon is often used as etch stop. In more advanced processes, electrochemical action is also employed to control and stop etching.

In the case of isotropic etching of Si with HNA, the mechanism of material removal is as follows. HNO_3 oxidizes Si which is removed by HF. Acetic acid stabilizes the pH value.

For anisotropic wet etching, typically the aqueous solutions of alkali metal hydroxides like KOH, NaOH are used. Generally, 15–40% (by weight) concentration at 70–90°C is employed. The etch mechanism is

$$\underset{\substack{\uparrow \\ \text{Material}}}{Si} + 2OH^- + 2H_2O \rightarrow \underset{\substack{\uparrow \\ \text{Soluble complex}}}{SiO_2(OH)_2^{-2}} + \underset{\substack{\uparrow \\ \text{Bubbles}}}{2H_2}$$

The ratios of the etch rates for different directions in Si are indicated in Fig. 7.21a when KOH etchant is used. The dependence of etch rates using KOH on Si for various concentrations is indicated in Fig. 7.21b.

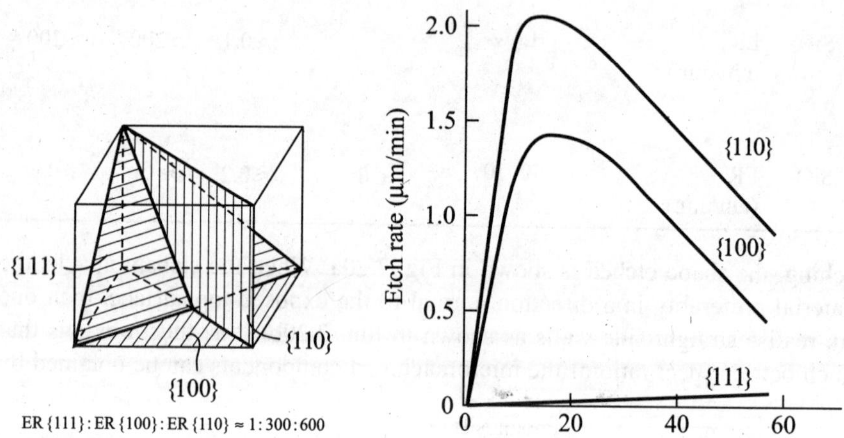

(a) Anisotropic etch rates with KOH

(b) Variation of etch rate with KOH concentration

Fig. 7.21 Etch rate characteristics of KOH.

Ethylene diamine pyrocatechol (EDP) etching of Si is also anisotropic as indicated in Table 7.6. EDP is very corrosive and very carcinogenic, and therefore generally not allowed in the mainstream fabrication. There must be fume collecting arrangements.

It is often necessary to stop etching at a particular level. To achieve this, etch stop layers can be used; these layers drastically reduce the etch rate and a reasonable control over etch depth is feasible. In the case of Si etching, stop layers are created by boron doping. Table 7.7 indicates the magnitude by which the etch rate is slowed down for different etchants. Etch stop can also be achieved through electrochemical effects.

Table 7.7 Effect of boron doping on etch rate of Si

Etchant	Boron doping (atoms/cm^3)	ER reduction factor
HNA		ER actually gets speeded up
KOH	$>10^{20}$	20
NaOH	$>3 \times 10^{20}$	10
EDP	$>7 \times 10^{29}$	50

Anisotropic Etching

Since the etch rates are drastically different depending on the direction in the case of anisotropic etching (the most common example is KOH etching of Si as already mentioned), such processes can be effectively employed for creating high aspect ratio components and produce definite shapes. Once the dependence of dissolution rate with respect to the orientation is known, it is possible to guess the profile that will be produced. The Wulff-Jaccodine diagram is often used for this purpose as now explained.

In this procedure, the polar plots of etch rates are drawn with respect to the edges of the photoresist mask. Figure 7.22a shows O_1 and O_2 as the edges of the mask where the portion O_1O_2 is free for the etchant to interact with the material. Let us consider a two-dimensional situation. The surfaces, those will be generated, will be normal to the smallest etch rate vectors corresponding to the cusps formed by the etch rate vector polar plots. In Fig. 7.22a, the cusps are C_1, C_2, C_3, and C_4, the corresponding etch rate vectors being ER_1, ER_2, ER_3, and ER_4. It should be remembered that etched surfaces are produced in the material. Thus, the shape of the cavity produced will be similar to *PQRST*. The size will grow with time. Figure 7.22b shows the etch rate vectors and the corresponding shape produced.

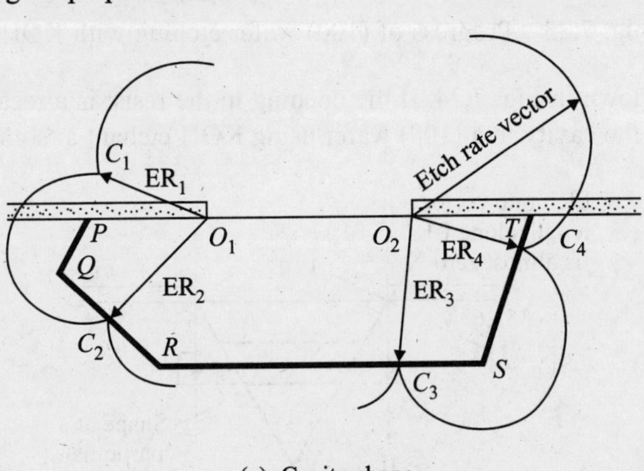

(a) Cavity shape

Fig. 7.22 Wulff-Jaccodine diagram (cont.).

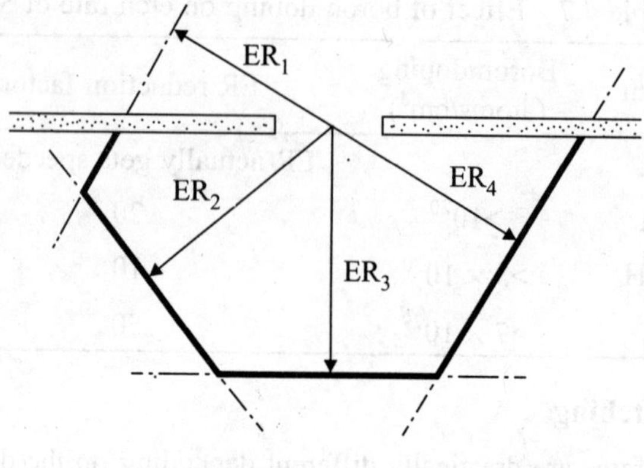

(b) Etch rate vectors

Fig. 7.22 Wulff-Jaccodine diagram.

Etching of {100} Si Wafer

In the case of Si etching by KOH, the {100} planes are etched quickly whereas the {111} planes are left almost intact. So, in the case of a (100) wafer, KOH etching will etch the (100) plane (parallel to the top surface), leaving the (111) planes making 54.74° with the (100) planes intact. Figure 7.23 shows how etching progresses and finally stops. The corresponding Wulff-Jaccodine

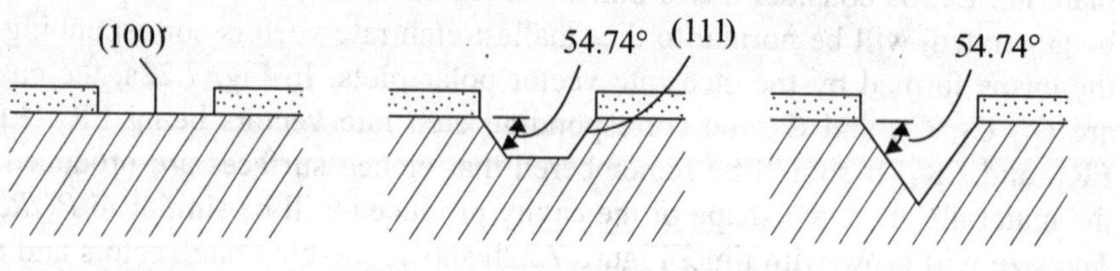

Fig. 7.23 Progress of (100) wafer etching with KOH.

diagram is shown in Fig. 7.24. If the opening in the resist is a rectangular one, the shape of the cavity in a (100) wafer using KOH etchant at some instants of

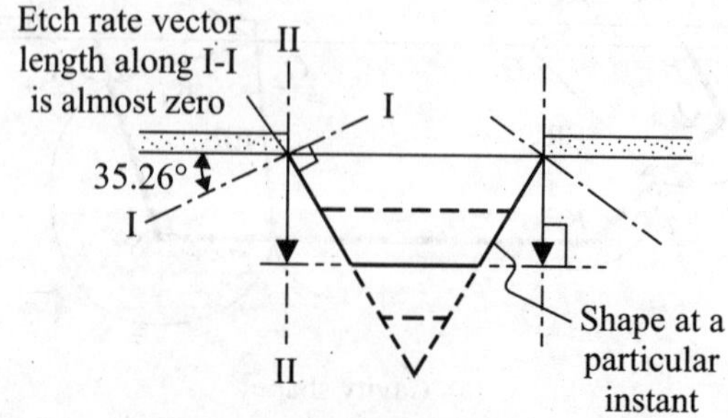

Fig. 7.24 Wulff-Jaccodine diagram for (100) Si wafer.

time will be as shown in Fig. 7.25. In Fig. 7.25a, the cavity shape is shown at some intermediate instant of time whereas the configuration at which the etching process stops is shown in Fig. 7.25b. This kind of situation in which the passive planes meet and the etching comes to an end is called self-limiting etch.

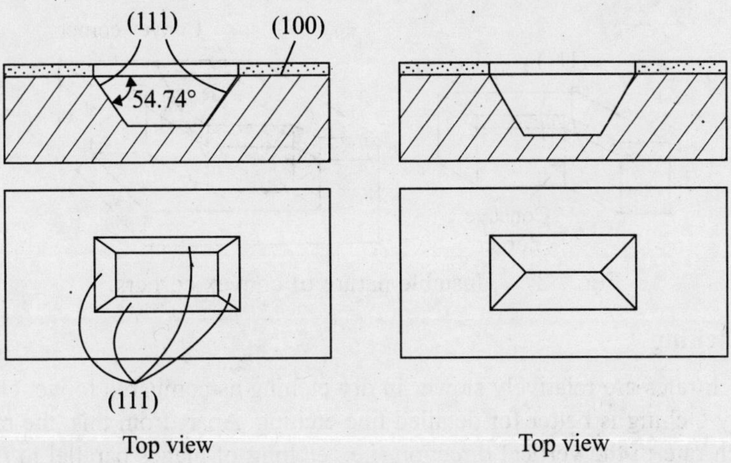

(a) Etching in progress (b) Etched cavity after etching stops

Fig. 7.25 Cavity shape by anisotropic etching of (100) Si with KOH.

Etching of {110} Si Wafers

The shapes of the cavities produced by anisotropic etching of {110} Si wafers are very different from those obtained with {100} wafers. In {110} wafers, four out of eight {111} planes are normal to the (110) wafer surface. The other four {111} planes are inclined to the wafer surface at 35.26°. The four vertical {111} planes intersect to form a parallelogram with an inside angle of 70.5°. Figure 7.26 shows the shape of a cavity produced by anisotropic etching of {110} wafer.

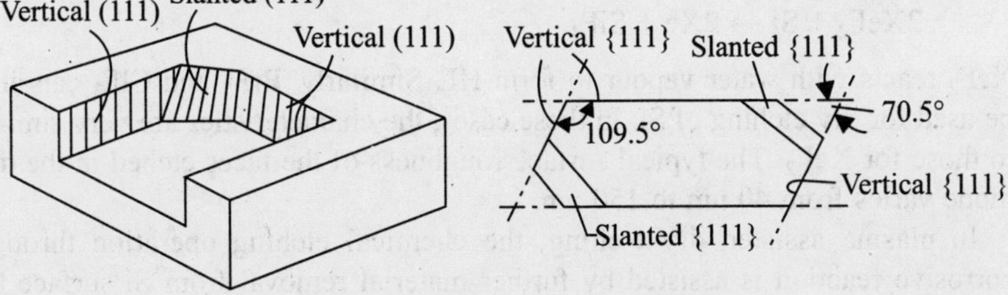

Fig. 7.26 Etched cavity in {110} wafer.

During anisotropic etching of Si with KOH, the corners need to be taken care of. While the concave corners bounded by the {111} planes remain intact, the convex corners are immediately attacked by the etchant. It is because of the fact that a slight erosion of such convex corners exposes the planes other than the {111} planes which are prone to etching action. Figure 7.27 shows this effect. This effect is sometimes effectively employed to create suspended structures.

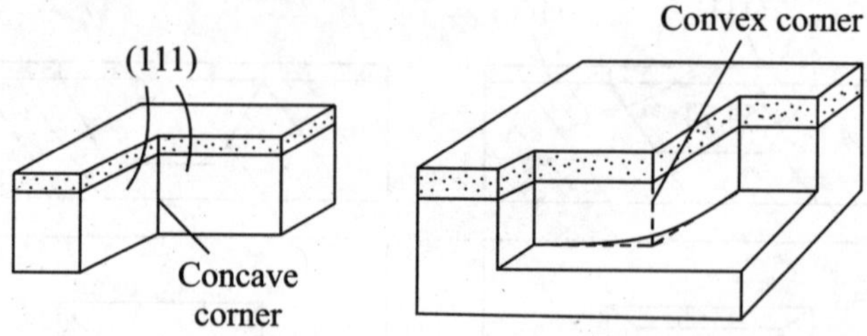

Fig. 7.27 Unstable nature of convex corners.

Dry Etching

The etch rates are relatively slower in dry etching as compared to wet etching. But dry etching is better for detailed fine etching. Apart from this, the ratio of the etch rate in the vertical direction (i.e., etching of planes parallel to the top surface) is about six times greater than that in the lateral direction. This inherent anisotropy is helpful in achieving higher aspect ratios without depending too much on the crystal orientation. It not only uses lesser amount of chemicals but eliminates the hazard of handling unsafe chemicals also. Of course, some gases used in dry etching can be toxic and corrosive. Redeposition of nonvolatile compounds is also another disadvantage.

Dry etching can be either nonplasma based using an appropriate reactive gas mixture or it can be plasma based using radio frequency (RF) power to drive chemical reaction. Typically, the fluorine based gases are used in nonplasma based dry etching because such gas mixtures readily etch Si but do not attack metals or SiO_2. The process is very conveniently controlled by varying the temperature and pressure. The etching operation is isotropic for Si and the typical etch rates are 1–3 μm/min. The chemical reaction for the etching operation (with XeF_2 gas) is

$$2XeF_2 + Si \rightarrow 2Xe + SiF_4$$

XeF_2 reacts with water vapour to form HF. Similarly, BrF_3 and ClF_3 can also be used for dry etching of Si. In these cases, the characteristics are very similar to those for XeF_2. The typical surface roughness of the faces etched in the dry mode varies from 40 nm to 150 nm.

In plasma assisted dry etching, the chemical etching operation through corrosive reaction is assisted by further material removal from Si surface by bombardment of ions. It is the directionality of the accelerated ions that gives reactive ion etching (RIE) its characteristic anisotropy. Energetic electrons, excited by a high frequency (radio frequency, RF) electromagnetic field, collide with gas molecules, resulting in ions (e.g., SF_x^+) and reactive but electrically neutral species (e.g., F, Cl). This mixture of ions, electrons, and neutrals is called plasma. The creation of ion-electron pairs is a continuous process and an equilibrium is maintained by creation and spontaneous destruction through recombination. The typical kinetic energy (KE) of an electron in a plasma is 2–8 eV. Since the KE can be expressed either as $\frac{1}{2}mv^2$ or $\frac{3}{2}kT$, where m is the mass, v is the velocity, T is the absolute temperature, and k is the Boltzmann constant; for a 2 eV electron, $T \approx 15{,}000$ K and $v = 8.4 \times 10^5$ m/sec. The most basic form of a plasma etching system is shown in Fig. 7.28. In Fig. 7.28a, the wafer is not connected to power and the etching is by reaction of plasma with the wafer. This is normally called plasma etching (PE). In the case shown in Fig. 7.28b, the wafer is connected to power. As a result, the ions hit the wafer with extra momentum caused by the acceleration of ions towards the wafer. This process is called reactive ion etching (RIE) and possesses both anisotropy and higher etch rate. In plasma etching, plasma is used to produce chemically reactive atoms, ions, etc., from an inert molecular gas (usually argon). These reactive species diffuse into the wafer surface, resulting in adsorption. Through chemical reaction, the wafer atoms produce a byproduct which leaves the body through desorption and diffuses into the bulk gas. It should be remembered that the production of gaseous byproduct is essential for successful plasma etching. Figure 7.28c shows the various steps in the process of etching.

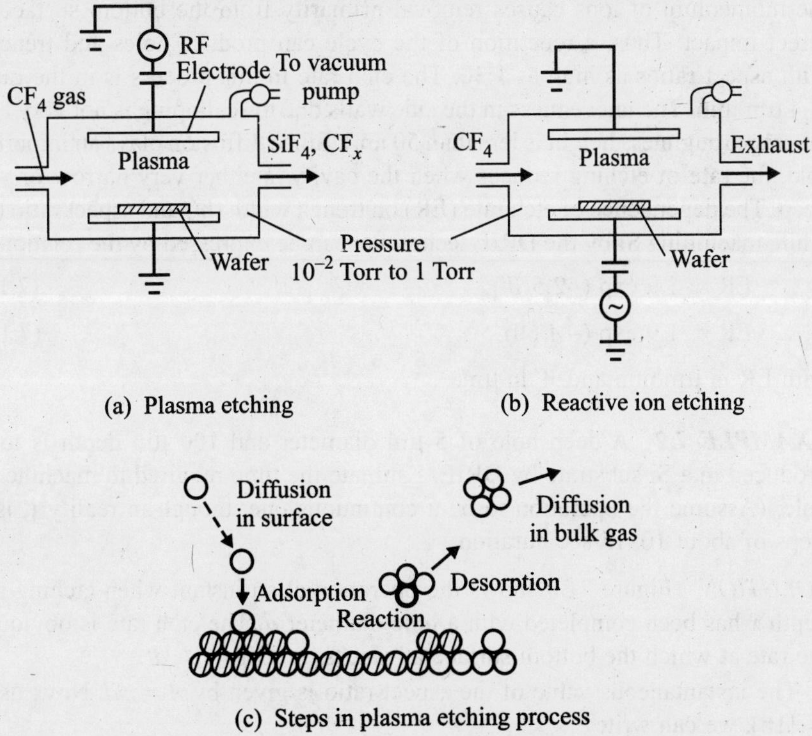

Fig. 7.28 Etching devices and etching mechanics.

Deep Reactive Ion Etching (DRIE)

This process evolved from reactive ion etching for the purpose of producing microcomponents with high aspect ratios. There are quite a few variations in the procedures adopted by different companies. In the procedure followed by Bosch, plasma is produced by glow discharge at low pressure and the plasma is used to alternately etch silicon and deposit an etch-resistant thin polymer layer on side walls and bottom surface. During a typical etch cycle (usually of 5–15 sec duration), in the first half the SF_6 plasma is used to etch Si; in the next half, the C_4F_8 plasma deposits a thin ($\approx$10 nm) layer of flurocarbon polymer on the surfaces (see Figs. 7.29a and 7.29b). During etching in the next cycle (Fig. 7.29c), the passivating layer on the bottom surface breaks and Si is etched to a lower depth. The side wall polymer layer remains relatively intact as

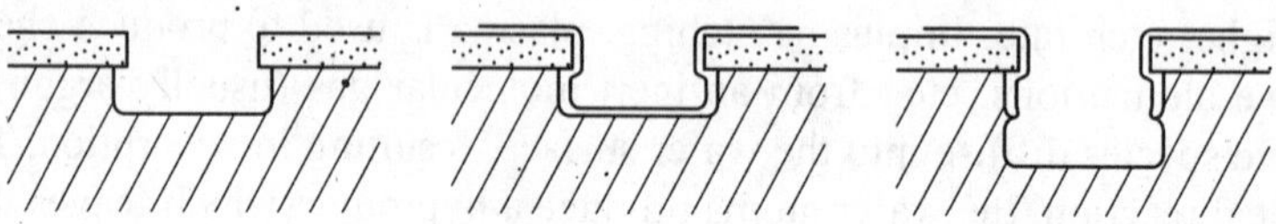

(a) Etch with SF_6 (b) Passivate with C_4F_8 (c) Etch with SF_6

Fig. 7.29 Operating cycle in DRIE (Bosch process).

the momentum of ions causes removal primarily from the bottom surface by direct impact. Thus, a repetition of the cycle can produce holes and trenches with aspect ratios as high as 1:30. The etch rate in this process is in the range 1–4 μm/min. The unevenness in the side walls due to scalloping is not very high (average roughness height is less than 50 nm). Since diffusion plays an important role, the rate of etching reduces when the cavity is either very narrow or very deep. The dependence of etch rate (ER) on trench width (W) and aspect ratio (A), while machining Si by the DRIE technique, can be expressed by the relations

$$\text{ER} \approx 1.9 \exp(-2.5/W), \tag{7.11a}$$

$$\text{ER} \approx 1.9 \exp(-A/60) \tag{7.11b}$$

with ER in μm/min and W in μm.

EXAMPLE 7.2 A deep hole of 5 μm diameter and 100 μm depth is to be produced in a Si substrate by DRIE. Estimate the time required to machine the hole. (Assume the operation to be a continuous one, though in reality it is in steps of about 10–15 sec duration.)

SOLUTION Figure 7.30 shows the operation at an instant when etching to a depth x has been completed with a hole diameter d. The etch rate is obviously the rate at which the bottom surface advances. So, ER = dx/dt.

The instantaneous value of the aspect ratio is given by $A = x/d$. Now, using (7.11b), we can write

$$dx/dt = 1.9 \exp\{-x/(60d)\}$$

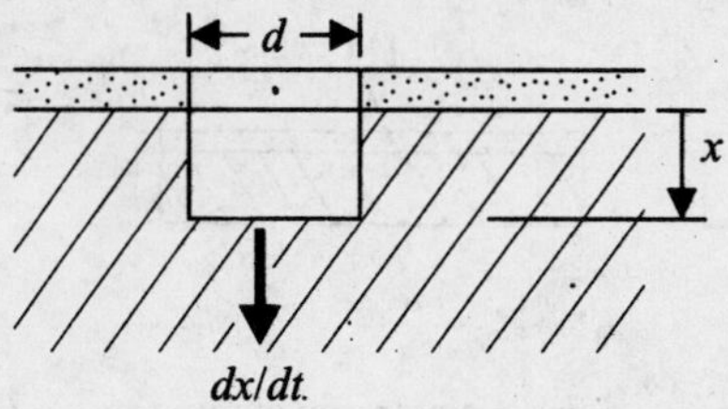

Fig. 7.30 Deep hole etching.

or

$$\int \exp\{x/(60d)\}\,dx = 1.9\int dt + C,$$

where C is a constant to be determined by the initial condition. Completing the integration this relation becomes

$$60d \exp\{x/(60d)\} = 1.9t + C,$$

but at $t = 0$, $x = 0$. Hence, $C = 60d$ and the final form of the relation between the depth and the time of etching comes out as

$$t = 31.58d[\exp\{x/(60d)\} - 1],$$

where x is the depth in microns, d is the diameter of the hole in microns, and t is the time in minutes. So, when $x = 100$ μm, for $d = 5$ μm, the time is

$$t = 31.58 \times 5[\exp\{100/(60 \times 5)\} - 1]$$

$$\approx 62.5 \text{ min.}$$

7.4.6 BULK AND SURFACE MICROMACHINING

Bulk micromachining is basically a subtractive process wherein micromechanical structures are realized through selective etching of the material from the bulk of a single silicon wafer. This technique is most commonly used for manufacturing most of the commercially available microsensors and -devices. In the thickness direction, such microstructures may cover a range from a submicron level to a full wafer thickness (≈200 μm). In the lateral direction, the feature size may cover a range from a few microns to the dimensions of a full wafer. Both isotropic and anisotropic etchings are used. Inkjet printer nozzles, pressure sensors, accelerometers, gyrosensors, microthrusters can be conveniently produced by bulk micromachining. Figure 7.31 shows the various stages of production of a micronozzie.

In surface micromachining, a micromechanical structure is produced not by machining the bulk silicon. In this technique, a stack of polysilicon thin films with alternating layers of sacrificial SiO_2 is built. Eventually, the composite structure, made of sacrificial layers (SiO_2) and structural layers (Si), is treated by HF to remove SiO_2 and only the features produced by the structural layers

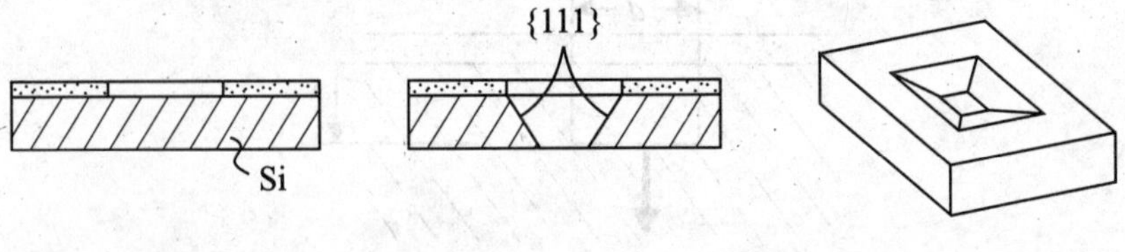

(a) Si wafer with patterned photoresist (b) Anisotropic etching of micronozzle (c) Pictorial view of etched micronozzle

Fig. 7.31 Bulk micromachining of a micronozzle.

remain. Typically, a stack may contain 4–5 layers though in more complex cases the stack may contain upto ten layers. Thus, surface micromachining is characterized by depositing alternating layers of structural material and SiO_2 and subjecting the surfaces to repeated photolithographic operation and eventual removal of all sacrificial layers. Figure 7.32 shows the procedure of building a tuning fork by surface micromachining.

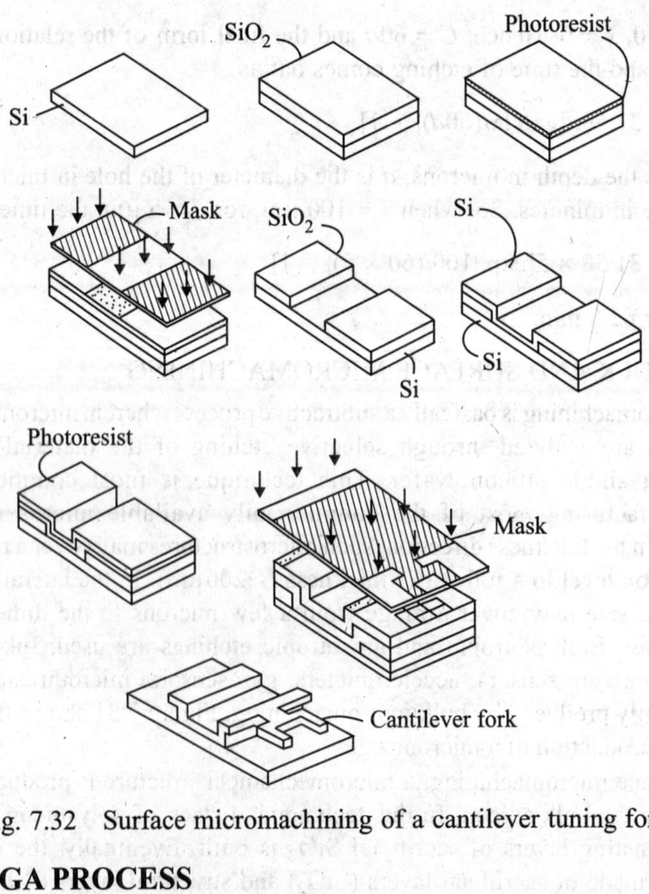

Fig. 7.32 Surface micromachining of a cantilever tuning fork.

7.5 LIGA PROCESS

In the case of conventional silicon micromachining, the aspect ratio cannot be large because of limitations inherent in etching. Generally, the depth to which machining can be done is limited to a few tens of microns only. This imposes severe constraints for developing three-dimensional microsystems. To produce three-dimensional microdevices with high aspect ratio components, deep etch lithography is necessary. Such a process using high energy X-ray (wavelength 0.2–0.6 nm) lithography followed by electroplating was developed towards the end of the 1970s at Karlsruhe in Germany. Later, similar processes were developed using deep etch UV lithography.

7.5.1 BASIC LIGA PROCESS

The acronym LIGA stands for LIthographie (lithography), Galvanoformung (electroplating), and Abformung (moulding). The comparison of the characteristic lateral and vertical depth dimensions in Si microlithography with those in LIGA is presented in Fig. 7.33. It is clear from the figure that for LIGA the key consideration is high aspect ratio. It can be as high as 50. The LIGA process

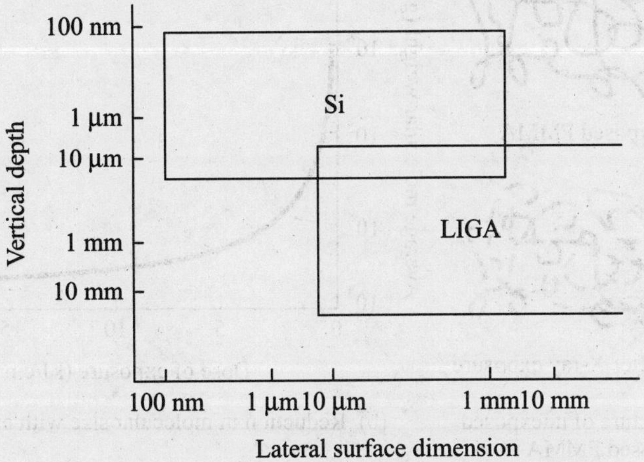

Fig. 7.33 Comparison of dimension ranges in LIGA and surface micromachining.

is based on deep X-ray lithography to achieve such high aspect ratios. Since the basic technique depends on shadow printing of the mask pattern and X-ray beams cannot be collimated by optics, high energy X-ray synchrotron radiation is used. The most commonly used X-ray sensitive photoresist is polymethyl methacrylate (PMMA)[1]. In general, PMMA is attached to a substrate that can be used as an electroplating base. The steps in LIGA are shown in Fig. 7.34. After the PMMA layer is attached to the substrate and the X-ray mask is ready, the photoresist is exposed to synchrotron radiation through the mask. The regions of PMMA irradiated by X-ray experience bond scission as indicated in Fig. 7.35a. The X-ray induced main chain scission in PMMA results in a reduction

[1]PMMA is a transparent vinyl polymer sensitive to X-rays.

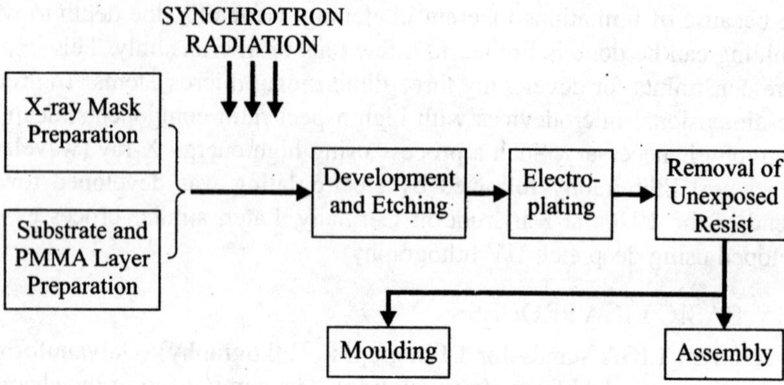

Fig. 7.34 Steps in a typical LIGA process.

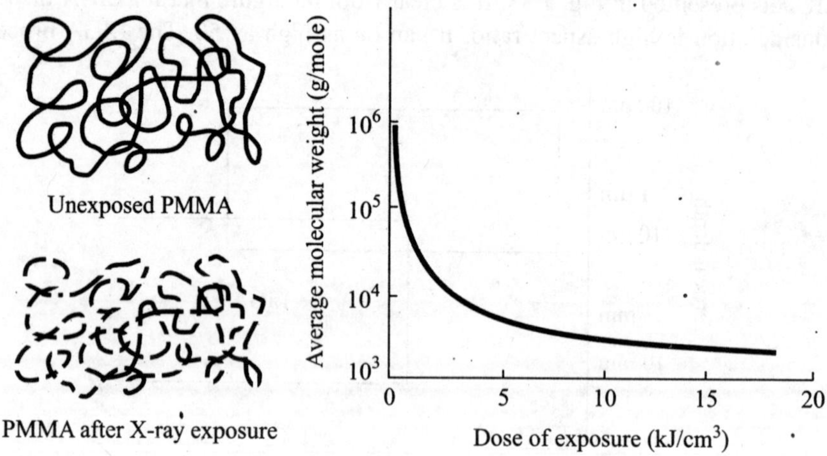

(a) Structure of unexposed and exposed PMMA

(b) Reduction in molecular size with exposure

Fig. 7.35 Effect of radiation on PMMA characteristics.

in the molecular weight as the molecules become smaller. This phenomenon is indicated in Fig. 7.35b as a reduction in molecular weight with irradiation dose. When subjected to chemical development, the exposed portions selectively dissolve. After the development of PMMA, the cavities are filled up by some suitable metal through electroplating from the conductive base. Next, the unexposed PMMA is dissolved and the remaining patterned metal part can be used either as a mould for further mass production of microparts (of some polymer) or the resulting metallic object can be directly used as a micropart. Figure 7.34 shows these two end usages of the micropart obtained by the LIGA process. Figure 7.36 shows the process through simple diagrams.

7.5.2 LIGA MASKS

The deep X-ray lithography for the LIGA process requires special X-ray masks.

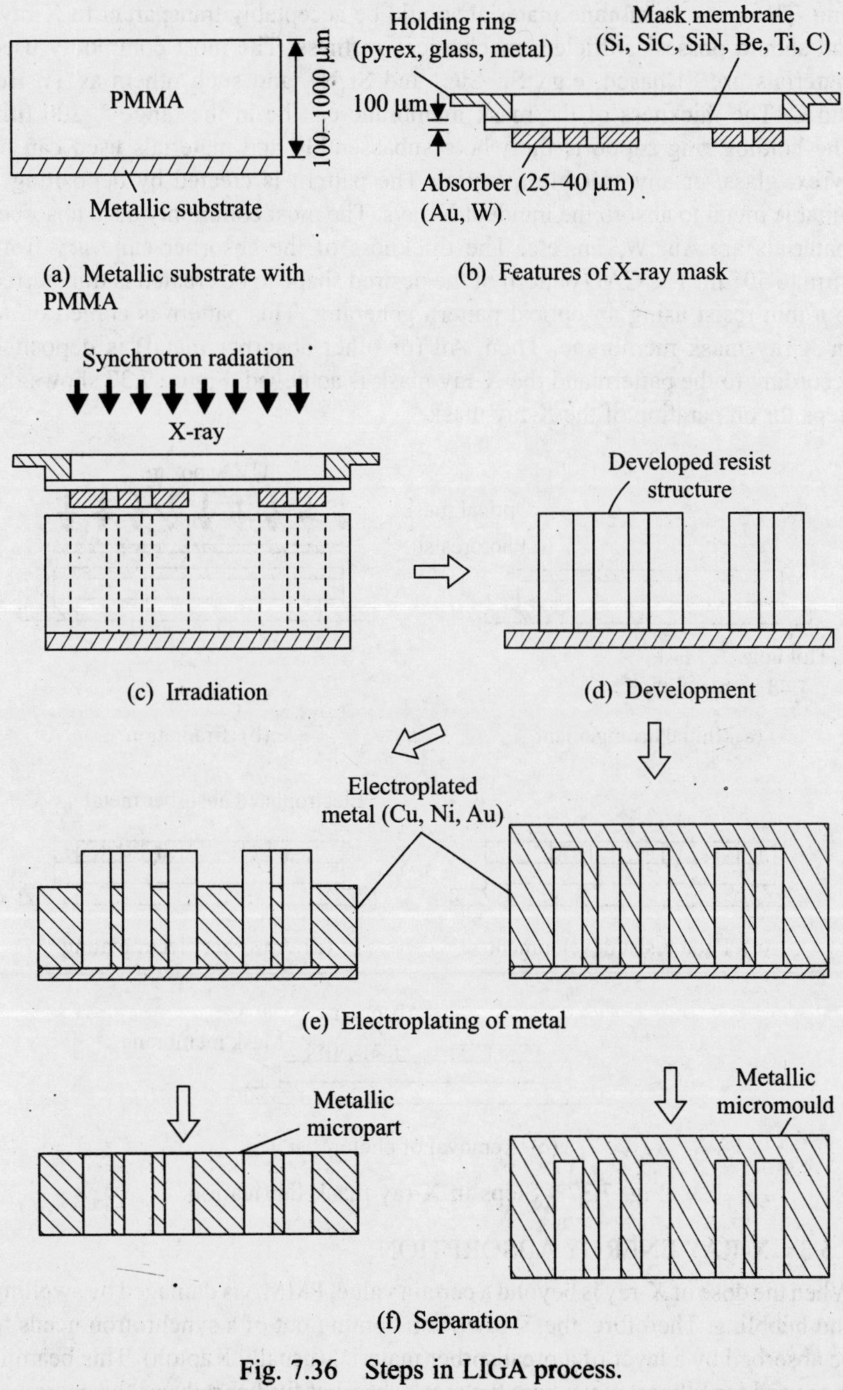

Fig. 7.36 Steps in LIGA process.

These masks have primarily three components, viz. (see Fig. 7.36b), (i) mask membrane for supporting the patterned absorber, (ii) absorber, and (iii) holding ring. The mask membrane material should be acceptably transparent to X-rays and should possess sufficient mechanical stiffness. The most commonly used materials are Si based, e.g., Si, SiC, and Si_3N_4, and such others as Ti, Be, and C. The thickness of the mask membrane can be in the range 5–200 μm. The holding ring supports the whole subassembly and materials used can be pyrex, glass, or any suitable material. The pattern is created by depositing a suitable metal to absorb the incident X-rays. The most commonly used absorber materials are Au, W, Tn, etc. The thickness of the absorber can vary from 1 μm to 50 μm. The CAD pattern of the desired shape to be created is transferred to a thin resist using an optical pattern generator. This pattern is copied on to an X-ray mask membrane. Then, Au (or other absorber metal) is deposited according to the pattern and the X-ray mask is achieved. Figure 7.37 shows the steps for preparation of the X-ray mask.

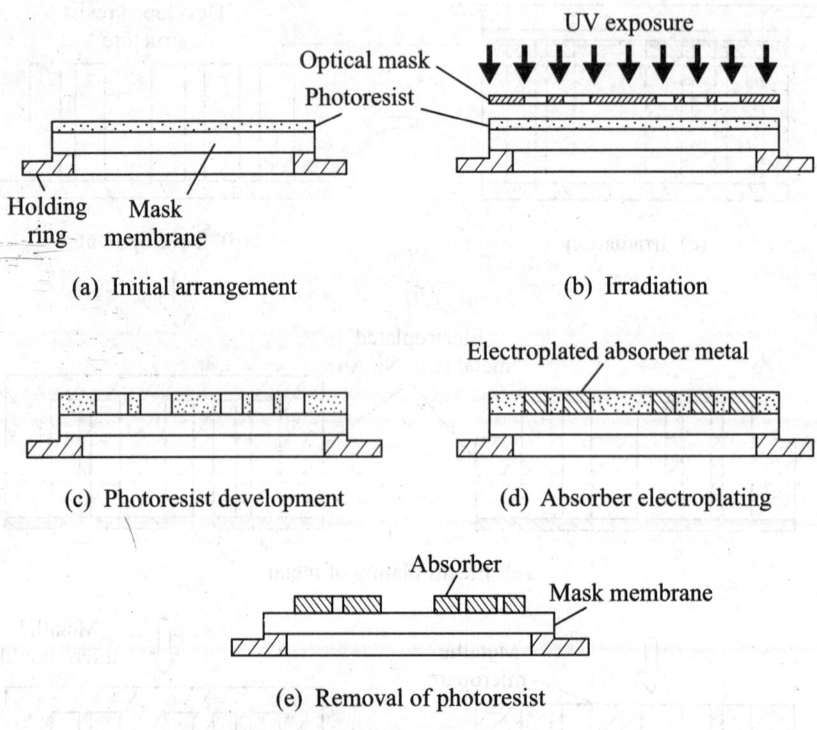

Fig. 7.37 Steps in X-ray mask fabrication.

7.5.3 X-RAY ENERGY ABSORPTION

When the dose of X-ray is beyond a certain value, PMMA is damaged by swelling and bubbling. Therefore, the X-ray beam coming out of a synchrotron needs to be absorbed by a layer of a preabsorber material (usually Kapton). This beam is then used for lithography where the mask absorber further reduces the energy of the X-ray to bring it below the threshold level. At this exposure, PMMA remains intact while developing. Figure 7.38 shows how the X-ray energy reduces at the

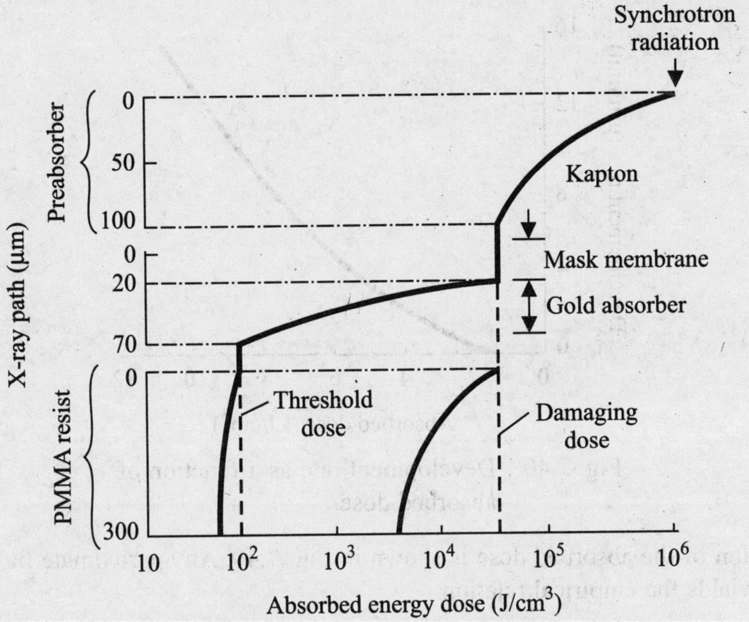

Fig. 7.38 Level of X-ray dose at different locations during irradiation.

time of irradiation. The absorbing characteristics of various materials commonly used in X-ray mask preparation are indicated in Fig. 7.39.

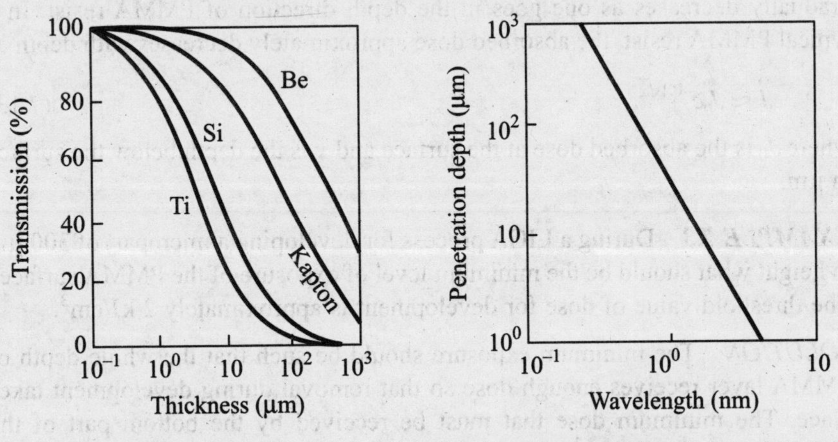

(a) Absorption characteristics of materials

(b) Penetration characteristics of X-rays as a function of energy

Fig. 7.39 Material characteristics for X-ray penetration.

The development rate of the polymer resist is usually specified in micron of layer thickness removed per unit time (i.e., μm/min). This rate depends on the type of the polymer (primarily the length of PMMA molecule chains) and the absorbed X-ray dose. For a typical case of PMMA, the development rate as a

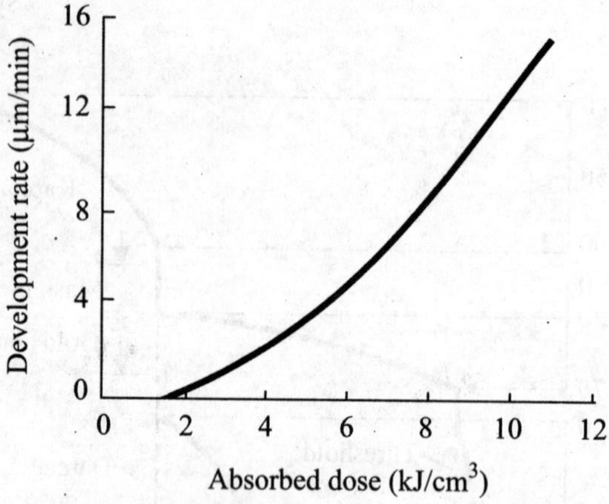

Fig. 7.40 Development rate as a function of absorbed dose.

function of the absorbed dose is shown in Fig. 7.40. An approximate fit of the data yields the empirical relation

$$\text{ER} \approx \frac{1}{3}(J^{1.6} - 3)\ \mu\text{m/min}, \tag{7.12}$$

where ER is the development rate (i.e., the etching rate while developing) and J is the absorbed dose in kJ/cm^3. As shown in Fig. 7.38, the absorbed dose also gradually decreases as one goes in the depth direction of PMMA resist. In a typical PMMA resist, the absorbed dose approximately decreases with depth as

$$J \approx J_0 e^{-0.1\sqrt{x}}, \tag{7.13}$$

where J_0 is the absorbed dose at the surface and x is the depth below the surface in μm.

EXAMPLE 7.3 During a LIGA process for developing a micropart of 300 μm in height what should be the minimum level of exposure of the PMMA surface? The threshold value of dose for development is approximately 2 kJ/cm^3.

SOLUTION The minimum exposure should be such that the whole depth of PMMA layer receives enough dose so that removal during development takes place. The minimum dose that must be received by the bottom part of the PMMA layer is 2 kJ/cm^3. So, using (7.13), the corresponding dose at the top surface can be determined as

$$2 \approx J_0 e^{-0.1\sqrt{300}}\ \text{kJ/cm}^3$$

or

$$J_0 \approx 2e^{0.1\sqrt{300}}\ \text{kJ/cm}^3$$
$$\approx 11.3\ \text{kJ/cm}^3.$$

EXAMPLE 7.4 A 300-μm PMMA layer is irradiated by a synchrotron radiation. The dose at the top surface is 15 kJ/cm^3, the threshold dose being 2 kJ/cm^3. Determine the time required to develop the PMMA resist.

SOLUTION The etch rate as a function of local dose is given by (7.12). Again, the level of local dose as a function of the depth below the surface can be obtained from (7.13). Combining the two, we can express the etch rate as a function of depth. So, at a depth x below the surface, the rate of development can be expressed as

$$\text{ER} = \frac{dx}{dt} \approx \frac{1}{3}[\{J_0 e^{-0.1\sqrt{x}}\}^{1.6} - 3] \text{ μm/min},$$

where x is in μm, J_0 is the surface dose in kJ/cm^3, and t is in minutes. This equation becomes (after putting $J_0 = 15$)

$$\frac{dx}{dt} \approx \frac{1}{3}(76.16e^{-0.16\sqrt{x}} - 3).$$

Using the initial condition at $t = 0$, $x = 0$, this nonlinear equation can be solved numerically. The value of t when $x = 300$ is found out to be about 157 minutes. In fact, the solution to the foregoing equation yields t versus x as shown in Fig. 7.41.

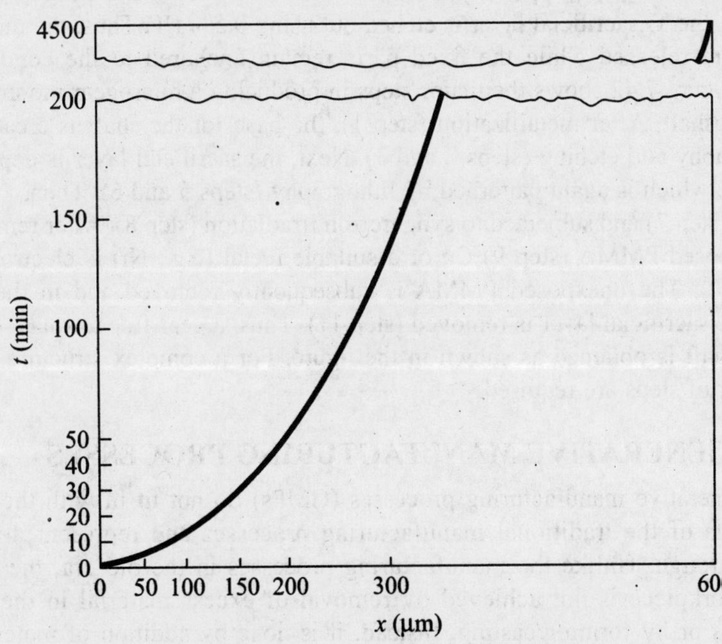

Fig. 7.41 Dependence of time on PMMA thickness for resist development.

7.5.4 EXTENDED LIGA PROCESS

The parts of a microsystem can be produced by the conventional LIGA technology. When microstructures are produced by the LIGA process, these are generally fixed on a base plate. To obtain microsystems with parts possessing the capability of relative motion, the sacrificial layer technique along with LIGA is necessary.

To start with, a thin metallization layer is deposited onto the substrate; this consists of an adhesion layer of a good electrical conductor like Ag. This layer serves as a plating layer for the electroforming process. This layer is patterned as required by standard optical lithography and wet etching technique. Next, the sacrificial layer is deposited onto this prepared substrate. In the case of extended LIGA process, Ti is a very suitable material for the sacrificial layer as it provides excellent adhesion to the PMMA resist, and a conductive layer; at the same time, it etches selectively to Ni, Cu, Ag, etc., used for electroplating. This sacrificial layer of Ti is next patterned by optical lithography and etching. The typical thickness of this sacrificial layer is about 3–5 μm. The X-ray mask is so designed (and aligned) that the movable parts of the microsystem are patterned onto the Ti covered areas and the fixed parts are positioned directly onto the conductive layer. Al is also used as sacrificial layers in extended LIGA process. After the PMMA resist is irradiated and developed, electroforming of parts/components of the microsystem is done. Next, the PMMA resist (not removed during development) is removed as in the standard LIGA process. Finally, the Ti sacrificial layer is etched out using 0.5% HF. Thus, the movable parts are released while the fixed parts remain anchored to the conductive base. Figure 7.42 shows the major steps in producing a microgear mounted on a microshaft. After metallization (step 1), the base for the shaft is created by lithography and etching (steps 2 and 3). Next, the sacrificial layer is deposited (step 4) which is again patterned by lithography (steps 5 and 6). Then, PMMA is laid (step 7) and subjected to synchrotron irradiation (step 8). After removing the exposed PMMA (step 9) Cu or a suitable metal (e.g., Ni) is electroplated (step 10). The unexposed PMMA is subsequently removed and in the final step the sacrificial layer is removed (step 11). Thus, a gear free to rotate on the microshaft is obtained as shown in the figure. For a complex structure, more number of steps are required.

7.6 GENERATIVE MANUFACTURING PROCESSES

The generative manufacturing processes (GMPs) do not fit in with the basic concepts of the traditional manufacturing processes and represent a major breakthrough. Unlike the manufacturing processes in the old era, the shape of a workpiece is not achieved by removal of excess material in the form of chips or by forming/casting. Instead, it is done by addition of material in small quanta without any particular form. One important aspect which makes these processes so eminently suitable for the future is its basic nature being so amenable to computer control. In GMPs, material is added or created (by solidification/bonding) where it is needed. The first commercial GMP was based on solidification of a liquid by a laser beam (called stereolithography) and was

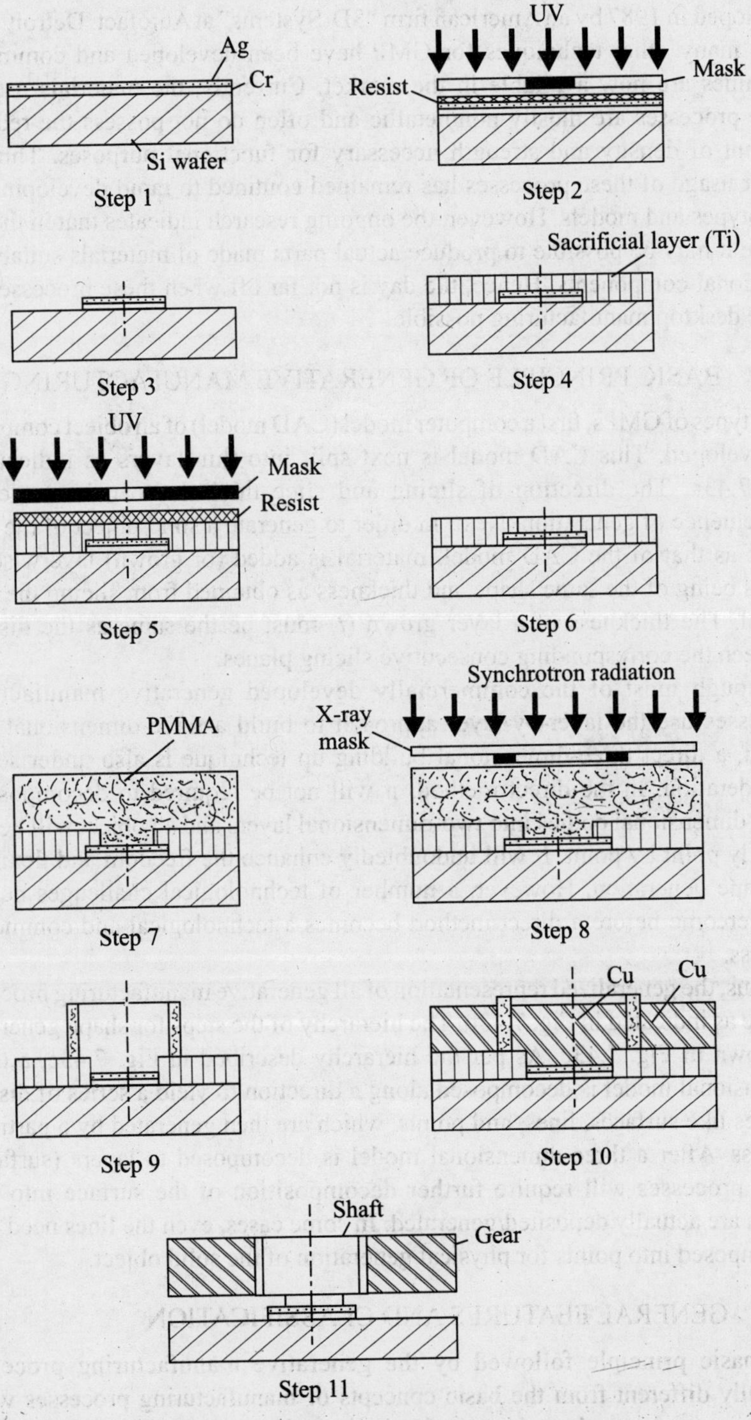

Fig. 7.42 Steps in producing a microgear by extended LIGA process.

developed in 1987 by an American firm "3D-Systems" at Autofact, Detroit. Since then, many other techniques for GMP have been developed and commercial machines are now available in the market. Currently, the materials used for these processes are mostly nonmetallic and often do not possess the requisite amount of density and strength necessary for functional purposes. Thus, the major usage of these processes has remained confined to rapid development of prototypes and models. However, the ongoing research indicates that in the near future it may be possible to produce actual parts made of materials suitable for functional components. Hence, the day is not far off when these processes will make desktop manufacturing possible.

7.6.1 BASIC PRINCIPLE OF GENERATIVE MANUFACTURING

In all types of GMPs, first a computer model (CAD model) of an object component is developed. This CAD model is next split into thin layers as indicated in Fig. 7.43a. The direction of slicing and slice thickness can be varied for convenience of generation. Next, in order to generate a solid object of the same shape as that of the CAD model, material is added (or grown) layerwise, the layers being of the same shape and thickness as obtained from slicing the CAD model. The thickness of a layer grown (t) must be the same as the distance between the corresponding consecutive slicing planes.

Though most of the commercially developed generative manufacturing processes use the layer-by-layer approach to build a three-dimensional solid object, a direct three-dimensional building up technique is also under serious consideration. In the direct method, it will not be required to decompose the three-dimensional bodies into two-dimensional layers and an object will be built directly point by point. It will undoubtedly enhance the freedom and flexibility in shape generation. However, a number of technological challenges need to be overcome before a direct method becomes a technological and commercial success.

Thus, the generalized representation of all generative manufacturing processes can be as indicated in Fig. 7.43b. The hierarchy of the steps for shape generation is shown in Fig. 7.43c. As per the hierarchy described in Fig. 7.43c, a three-dimensional model is decomposed along a direction to yield a series of discrete entities like surfaces, lines, and points, which are then generated by a particular process. After a three-dimensional model is decomposed to layers (surfaces), some processes will require further decomposition of the surface into lines which are actually deposited/generated. In some cases, even the lines need to be decomposed into points for physical generation of the solid object.

7.6.2 GENERAL FEATURES AND CLASSIFICATION

The basic principle followed by the generative manufacturing process is radically different from the basic concepts of manufacturing processes which have remained prevalent since the beginning of human civilization. The world had to wait for the necessary achievements in computer technology for the

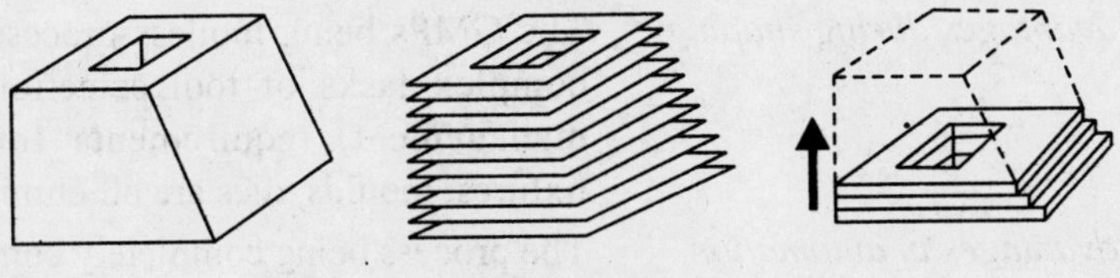

(a) Basic principle of layered GMP

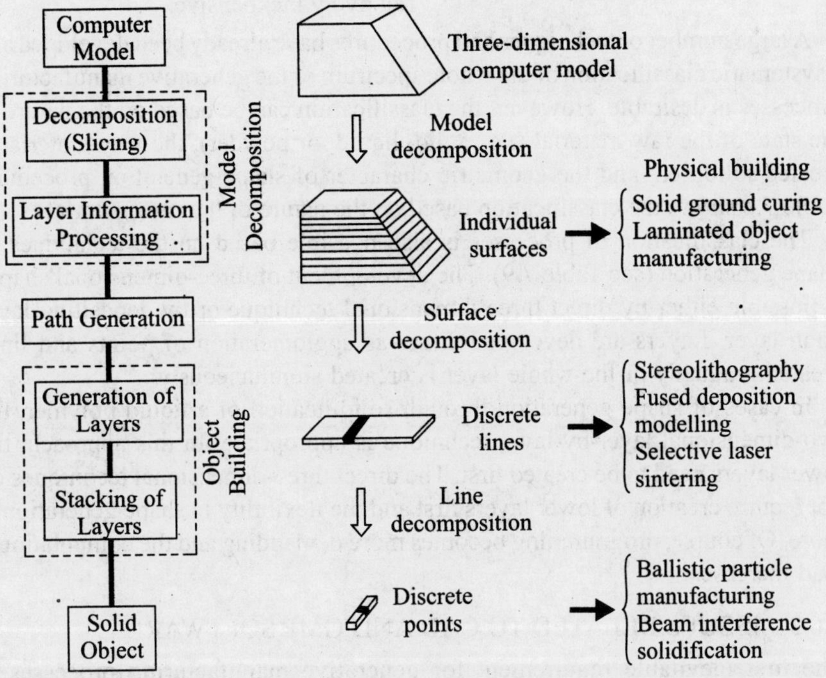

(b) Basic processes (c) Decomposition and building process hierarchy

Fig. 7.43 General features of generative manufacturing processes.

GMPs to become commercially viable. These processes are also free from the traditional problems of manufacturing as the material is created wherever needed during a GMP. The major advantages of the generative manufacturing processes are as now given:

- *Advantages during design* Manufacturing process is quite independent of the part features and there is no need for feature based design.
- *Advantages in planning* No blanks are required and no planning for the blanks is required. The GMPs are based on single operation only and no complicated scheduling and routing problems are faced.

- *Advantages during shaping* The GMPs being toolless processes the complex tasks of tool selection, tool management, requirements for jigs, fixtures, moulds, dies are all eliminated.
- *Advantages in automation* The process being completely computer-oriented, integration and automation of the manufacturing process is easy and relatively inexpensive.

A large number of techniques and procedures have already been developed and a systematic classification of the whole spectrum of the generative manufacturing processes is desirable. However, the classification can be based on the nature of the state of the raw material (i.e., solid, liquid, or powder), the way material is created/solidified, and the geometric character of shape generation procedure. Table 7.8 shows the classification based on the nature of the raw material.

The classification of processes is also possible based on the techniques of shape generation (see Table 7.9). The development of three-dimensional shapes is possible either by direct three-dimensional technique or by depositing layer upon layer. Layers are developed either as agglomeration of points and lines created gradually or the whole layer is created simultaneously.

In cases of shape generation through solidification of a liquid polymer, the two-dimensional layer-by-layer technique is appropriate. In this approach, the lower layers need to be created first. The direct three-dimensional techniques do not require creation of lower layers first and the flexibility in shape generation is more. Of course, programming becomes more demanding and the computational load increases.

7.6.3 ISSUES RELATED TO CAD AND GMP SOFTWARE

The first inevitable requirement for generative manufacturing processes is representation of the desired three-dimensional object in the form of a computer generated model. Certain issues related to CAD and subsequent processing of CAD model are discussed in brief.

In most cases, the GMP systems receive their data from CAD systems in either three-dimensional surface models or three-dimensional solid models. 3D Systems Inc., who first marketed a GMP machine based on stereolithography (STL), developed an STL file format. Since such machines outnumber all other types of machines, the STL format has become the de facto standard for almost all types of GMP machines. The technique is based on creating a mesh of interconnected triangles oriented three-dimensionally. Figure 7.44 shows the representation of three-dimensional surfaces by triangular mesh. The vertices of the triangles are ordered to indicate which side of the triangle contains material, and thus need to be generated during the process. It is not difficult to realize that an increase in the number of triangles results in improvement of accuracy when curved surfaces are involved.

The computational slicing of the CAD model is done by using a ray tracing

Table 7.8 Classification of GMPs based on state of raw material

State of raw material	Type of material	Mechanism(s) involved	Type of energy employed	Source of energy	Process
Liquid	Photopolymer	Liquid photo-polymerization	Monochromatic light	Lamp Laser beam Holography	Solid ground curing Stereolithography Holographic interference solidification
			Light (two frequencies)	Two laser beams	Beam interference solidification
	Thermosetting polymer	Liquid thermal polymerization	Heat	Laser beam	Thermal stereolithography
	Nonmetal	Melting and solidification	Heat	Heated nozzle	Fused deposition modelling Ballistic particle manufacturing
	Metal	Melting and solidification	Heat	Electric arc Laser beam Electrochemical discharge	Shape melting Fused deposition modelling Fused deposition modelling

Table 7.8 Classification of GMPs based on state of raw material (cont.)

State of raw material	Type of material	Mechanism(s) involved	Type of energy employed	Source of energy	Process
Solid	Thin sheet or foil	Selective gluing and cutting	Adhesive bonds, cutting energy	Glue and laser beam	Laminated object manufacturing
	Semipolymerized plastic foil	Foil polymerization	Light	Lamp	Solid foil polymerization
Powder	Single component	Selective sintering	Heat	Laser beam	Selective laser sintering
	Coated powder	Selective sintering	Heat	Laser beam	Selective laser sintering
	One component and one binder	Selective powder binding	Chemical bond	Fine droplet beam of binder liquid	Three-dimensional printing or selective powder binding

Table 7.9 Classification of GMPs based on mode of shape generation

Development of solid object	Basic element of creation	Nature of connectivity	Process
Two-dimensional layer-by-layer technique	Point	Discrete	Stereolithography Thermal polymerization Foil polymerization Selective laser sintering Selective powder binding Ballistic particle manufacturing
		Continuous	Stereolithography Fused deposition modelling Shape melting
	Layer		Laminated object manufacturing Solid ground curing Repetitive masking and deposition
Direct three-dimensional technique	Point	Discrete	Beam interference solidification Ballistic particle manufacturing
		Continuous	Fused deposition modelling Shape melting
	Surface		Holographic interference solidification
	Volume		Programmable moulding

algorithm which scans through a particular z-level of the model. The resulting cross-section can be one or more closed paths and a complete representation of the area to be filled with material. Such areas are to be generated by suitable cross-hatching algorithms to generate the trajectory of the creating element (the laser beam or material ejecting nozzle, etc.). Cross-hatching is important and should be optimal so as to generate the object in the shortest possible time,

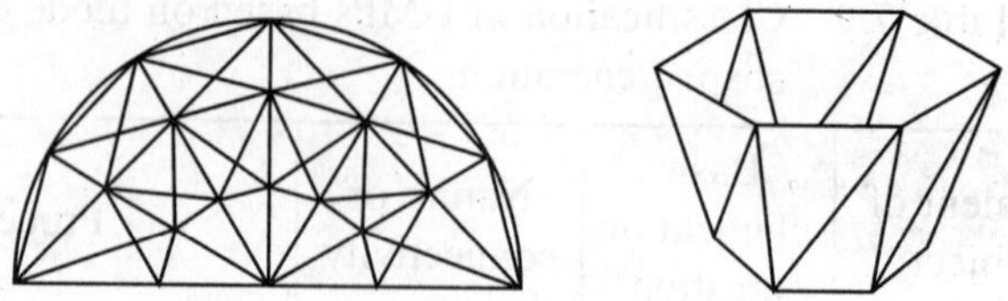

(a) Hemispherical surface (b) Truncated cone

Fig. 7.44 Representation of surfaces by connected triangles.

maintaining the required density and strength. Distortion also depends to some extent on the hatching pattern and should be taken note of. The orientation of the object needs to be chosen judiciously to optimize time and accuracy. The software needed for slicing and generation of data to control the GMP system movements is not a general one and depends on the specific system. The effect of orientation on build time and accuracy is indicated by the example shown in Fig. 7.45. It is seen that the orientation of the part as shown in Fig. 7.45a results in stair-step appearance and to generate a smooth curved surface the thickness of the layers has to be very small increasing the time required to generate the object. If the orientation of the object is changed to that shown in Fig. 7.45b, a smoother curved surface can be generated without making the layer thickness too small.

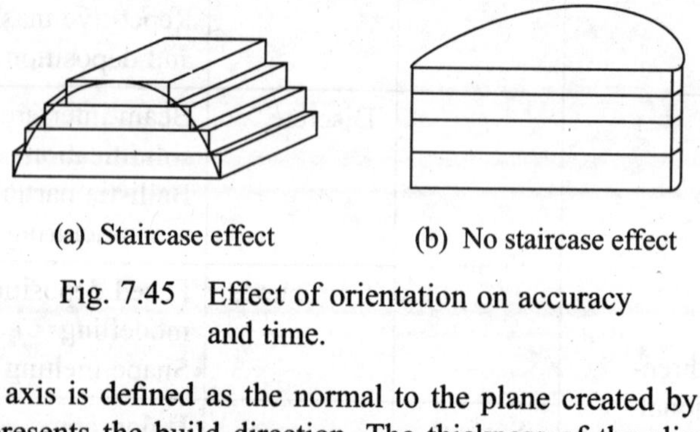

(a) Staircase effect (b) No staircase effect

Fig. 7.45 Effect of orientation on accuracy and time.

The slice axis is defined as the normal to the plane created by slicing and this also represents the build direction. The thickness of the slice ultimately becomes the thickness of the corresponding layer created and therefore dictates the texture, accuracy, and build time. Normally, layer thickness is in the range 0.0625–0.75 mm. However, it should be noted that the use of thicker layers does not necessarily reduce the build time though the number of layers to be created reduces. This is so as the scanning speed, while creating the material, depends on the thickness. Figure 7.46 shows the dependence of build time on layer thickness. As can be seen, the build time reduces with the beam power for a given layer thickness. This is so because the scanning can be done at higher speeds with increased beam power. However, the dependence of build time on layer thickness suggests the existence of an optimum thickness with which the build time is minimum. The range 0.125 mm to 0.25 mm is recommended for such operations.

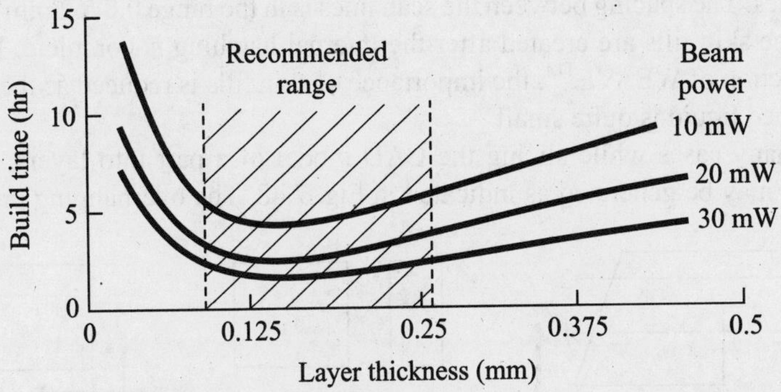

Fig. 7.46 Dependence of build time on layer thickness.

As mentioned earlier, internal hatching is used to solidify (or create) the area of a layer inside the outside boundaries of the material object. A properly chosen hatching pattern can generate a part in minimum time with the right kind of properties like strength, density, etc., and reduces distortion to a minimum. Initially, the boundary lines are created and then the interior is criss-crossed with lines, giving the part adequate stiffness.

Figure 7.47 shows a commonly-used hatching pattern that is called Tri-Hatch. It consists of parallel lines making 0°, 60°, and 120° with the x-axis, resulting in an internal structure that consists of equilateral triangles. The spacing between the consecutive lines is about 0.625 mm. When liquid photopolymers are used in the process, the material trapped inside the triangles remains liquid. It is solidified when the part is subjected to a curing operation.

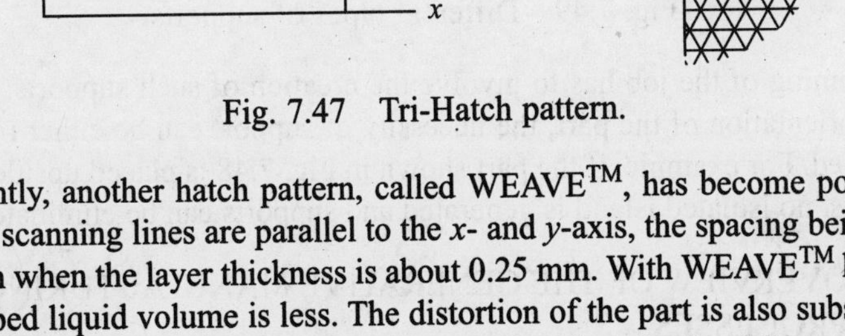

Fig. 7.47 Tri-Hatch pattern.

Recently, another hatch pattern, called WEAVE™, has become popular. In this, the scanning lines are parallel to the x- and y-axis, the spacing being about 0.28 mm when the layer thickness is about 0.25 mm. With WEAVE™ hatching, the trapped liquid volume is less. The distortion of the part is also substantially reduced. However, a much better control of curing depth while scanning is necessary.

It is obvious that the outer surface of the generated part cannot end up being porous. Therefore, skins are created by skin fills which consist of closely spaced scan lines. The spacing between the scan lines is in the range 0.0762 mm to 0.127 mm. The skin fills are created after the internal hatching is complete. With the introduction of WEAVE™, the importance of skin fills is reduced as the volume of trapped liquid is quite small.

In many cases while slicing the CAD model of a part into layers, isolated regions may be generated as indicated in Fig. 7.48. The overhanging projection

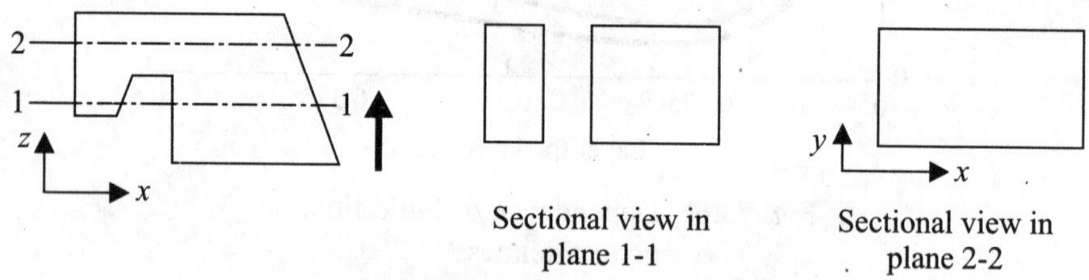

Fig. 7.48 Formation of isolated regions.

is connected to the main body of the part from the top. Since the building up is from the bottom, the connection is generated at a later time. Hence, it becomes essential to design a support for the overhanging part to prevent it from falling down under the action of gravity. Of course, such situations arise when a liquid based process, e.g., stereolithography, is used. Sometimes, supports are necessary even though no isolated region is created. For example, parts having cantilever beams or simple beams require a support structure. Otherwise, when the overhanging beam is just started, the initial few layers will not be able to hold itself under the action of gravity. Figure 7.49 shows various types of supports. After the generation of the complete part is over, such supports are removed. The

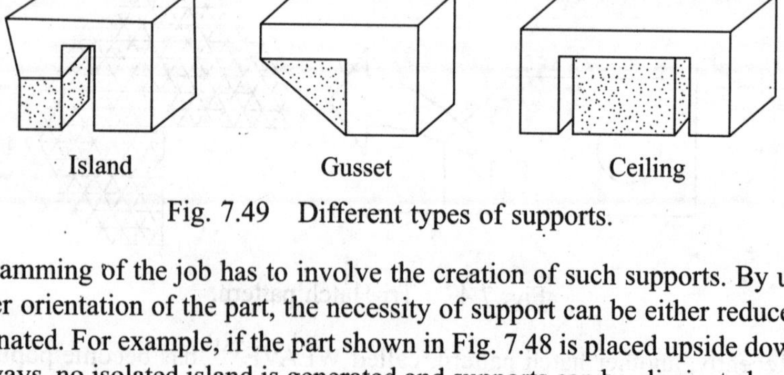

Fig. 7.49 Different types of supports.

programming of the job has to involve the creation of such supports. By using proper orientation of the part, the necessity of support can be either reduced or eliminated. For example, if the part shown in Fig. 7.48 is placed upside down or sideways, no isolated island is generated and supports can be eliminated.

7.6.4 OVERVIEW OF THE GENERATIVE MANUFACTURING PROCESSES

In this section, it will be attempted to present a quick overview of the different generative processes. Since this area is growing and many processes are still not fully matured, only brief descriptions can be presented. However, the process stereolithography was the first GMP to be developed and commercialized by 3D Systems Inc. and will be discussed in more details. The presentation of the various processes will be attempted according to the scheme of shape generation. Thus, first the processes following two-dimensional layer-by-layer technique will be discussed. This will be followed by a brief discussion of the direct three-dimensional shape generation processes.

Stereolithography (STL) with Photopolymerization

This process is based on curing of a liquid photopolymer (or monomer) by an ultraviolet laser beam. The basic principle of this process is shown in Fig. 7.50.

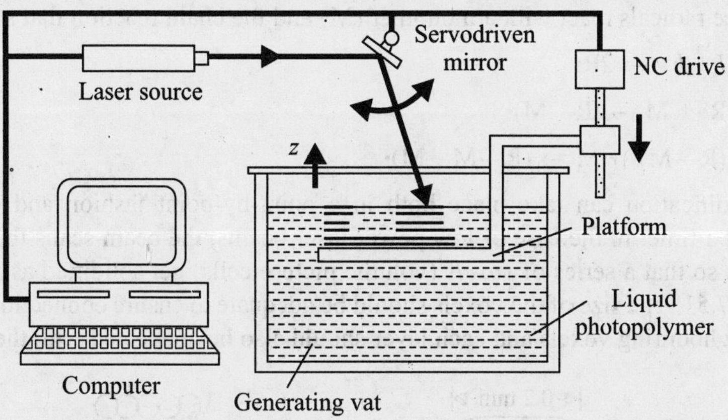

Fig. 7.50 Basic principle of stereolithography (STL).

The generating vat contains a UV sensitive liquid photopolymer. An NC drive controls the height of the platform on which the work is generated. Initially, it is positioned along the top surface of the liquid. A UV laser beam is reflected on to the surface of the liquid with the help of a mirror mounted on a pair of orthogonally scanning galvanometers whose positions are controlled by a computer. The computer controls the movement of the mirror so that the beam traces the desired path on the liquid surface while generating a particular cross-section of the part being produced. The beam cures the liquid to produce the solid layer with a depth of a few tenths of a mm which is equal to the thickness of the particular cross-section (i.e., the thickness of a layer between two consecutive slices). Once the first layer is cured, the platform is lowered by a distance equal to the thickness of a layer. Then, the laser beam scans the liquid surface again to produce the next layer. The process is repeated till the topmost cross-section of the object is generated. Next, the part is removed from the vat and the excess material is removed from the crevices and openings using ultrasonic cleaning technique. An alcohol bath is used to clean any unused polymer. To solidify the trapped liquid polymer inside the hatched pattern, the part is subjected to postcuring operation that is carried out by applying intense long wave UV radiation.

In this process, solidification of the liquid resin is accomplished through a process called *photopolymerization* which links small molecules, called *monomers*, into large molecules, called *polymers*. This also results in major changes in the bulk properties. Since the loose van der Waals interaction among neighbouring monomers is replaced by a network of covalent bonds, the shear strength increases significantly, changing the liquid to solid. The polymerization process is very energy efficient as the process is exothermic. This allows low power UV lasers in stereolithography process. In comparison to powder sintering type of operation, stereolithography requires about 1000 times less power! Upon exposure to UV radiation, free radicals (R•) are produced from the photo-initiator (I) when the photons of certain frequencies are absorbed. The free radicals react with a monomer (M) and the chain reaction that starts is

$$I + h\nu \rightarrow 2\mathrm{R}\bullet$$

$$\mathrm{R}\bullet + \mathrm{M} \rightarrow (\mathrm{R—M})\bullet$$

$$(\mathrm{R—M})\bullet + \mathrm{M} \rightarrow (\mathrm{R—M—M})\bullet$$

Solidification can take place both in a point-by-point fashion and curing lines at a time. In the case of low power laser beams, the beam scans the resin surface so that a series of *voxels* (volume picture cells) get solidified as shown in Fig. 7.51. The size of the voxels should be adequate to ensure connection with the neighbouring voxels and each layer should also be connected with the layer

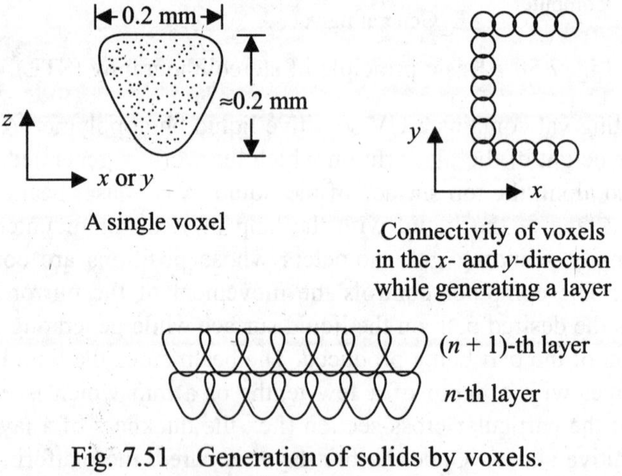

Fig. 7.51 Generation of solids by voxels.

solidified prior to the current one as indicated in Fig. 7.51. When the power of the laser beam is low, voxel formation is achieved by a point-to-point NC control of the mirror that causes the laser beam to stop at each voxel point. The beam need not be switched off in between voxels and the traversing speed being high, polymerization does not take place during this period. When high power lasers are used, continuous lines can be cured, forming a solid parabolic cylinder as indicated in Fig. 7.52. The degree of solidification through photopolymerization

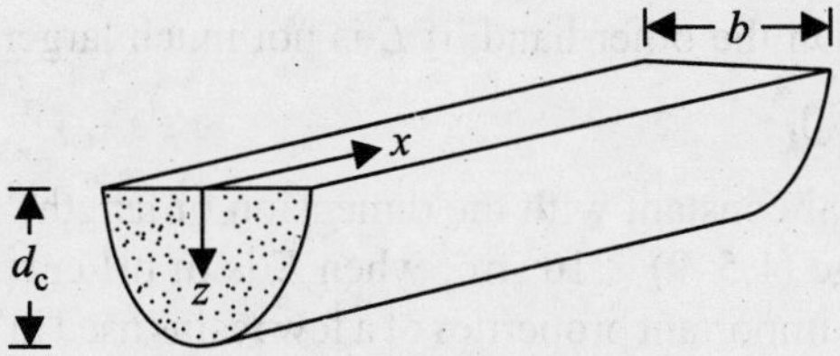

Fig. 7.52 A parabolic prism cured by a laser beam.

depends on the number of photons impinging on to a particular target volume, i.e., the dose of radiation received. Solidification requires a minimum critical dose. The surface energy received while traversing a line can be expressed as[1]

$$E_0 = \frac{P}{vb},$$

where P is the laser's average performance (intensity), v is the traversing speed, and b is the line width. The variation of the beam intensity along the depth (z-direction) follows the Beer-Lambert relation. Thus,

$$E(z) = E_0 \exp(-z/d_p),$$

where d_p is defined as the optical penetration depth (i.e., the depth at which the intensity drops by a factor e; it is a material property). If E_c be the critical dose for solidification, then the depth of a cured line can be expressed as

$$d_c = d_p \ln(E_0/E_c)$$

$$= d_p \ln\left(\frac{P}{vbE_c}\right).$$

This is called the *working curve equation*. However, the beam intensity is not uniform across the whole spot area. It is almost Gaussian and the effective spot radius, R, is typically 0.275 mm. If the beam traversing speed be v, the exposure time is $2R/v$. Typically, the characteristic exposure time, t_c, is in the range 0.07–2 millisecond. If P be the power of the laser beam, the surface energy for a typical stereolithography system can be approximately expressed as

$$E_0 \approx \frac{64P}{v},$$

where P is the power in watts and v is in cm/sec.

The strength of a photopolymerized resin depends on the total exposure. Young's modulus Y, which can be an indicator of the mechanical strength of the solidified resin, is a function of the energy E. However, it is obvious that when $E < E_c$, there will be no solidification and $Y = 0$. When E exceeds E_c, $Y(E)$ increases sharply and approaches a limiting value Y_{max} asymptotically. When

[1]Actually, the maximum value of E, i.e., E_0, for a Gaussian beam is given by $E_0 = \sqrt{2/\pi}[P/(wv)]$, where w is the $1/e^2$ Gaussian half width of the laser spot on the resin surface.

$E \gg E_c$, $Y = Y_{max}$. On the other hand, if E is not much larger than E_c, then

$$Y = k(E - E_c),$$

where k is a material constant with the dimension of length^{-1} and its numerical value is in the range $(1.5–9) \times 10^5$ m^{-1} when E is in mJ/cm^2 and Y is in N/m^2. Table 7.10 gives the important properties of a few resins used in stereolithography.

Table 7.10 Data for a few resins used in STL

Resin	d_p (mm)	E_c (mJ/cm^2)	k (10^5 m^{-1})
XB 5081–1	0.177	5.6	5.7
XB 5131	0.175	5.0	8.7
XB 5134–1	0.168	7.6	1.6
XB 5139	0.160	5.1	5.9
XB 5143	0.145	4.3	4.0

It is possible to develop a process of shape generation, somewhat similar to the STL process we have described, in which a resin is not solidified through photopolymerization but a thermosetting liquid polymer is selectively solidified by heat. Quadrax has marketed a system of this type in which a 5-W Ar-Ion laser is used. However, this company ultimately did not succeed in capturing the market for a number of reasons, patent infringement and litigation being one of the main reasons.

Selective Laser Sintering (SLS)

This process was first conceptualized and developed in 1987 at the University of Texas at Austin; the process was commercialized and marketed by DTM Corporation, USA. In this process, a part is created following a layer-by-layer approach by fusing powdered thermoplastic materials with a high power laser beam. This process of getting powder grains bonded together by localized partial melting is called sintering. In general, sintering occurs when a particle's viscosity drops due to higher temperature and the surface tension effect overpowers the viscosity. SLS can be of two basic types, viz., indirect SLS and direct SLS. In the case of indirect SLS, the metal or ceramic powders get bonded together through a polymeric material which softens (or melts) and forms necks between adjacent grains. The binding takes place through polymer-polymer bonds. In the indirect SLS, either metal (or ceramic) powders are mixed with a small amount of polymer or metallic (or ceramic) powder grains coated with polymer. In the direct SLS process, laser beams of much higher energy are used to partially melt the metallic powder and get sintered. Figure 7.53 shows the nature of indirect and direct SLS products. Because of the very nature of the solidification process, the products are generally porous and of low density. (In the case of indirect SLS, the density is lower than that achieved through direct SLS which can be

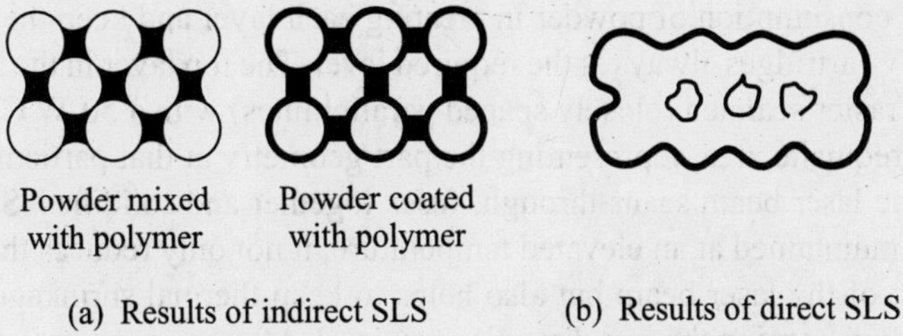

Fig. 7.53 Product character with indirect and direct SLS processes.

of the order of 80–90%.) The density produced by indirect SLS is 80–90% of that achieved through direct SLS.

The layers formed by the sintering process reside within the powder and no separate support structure is required. This, to an extent, simplifies the planning and designing task. When all the layers are formed, the product, embedded in the loose powder, is removed and cleaned.

The machine marketed by DTM Corporation consists of a cylindrical container in which the parts are generated. There are two powder supply cartridges on both sides, a counter-rotating roller for laying fresh layer of powder on the top of the newly created layer of the part, a 50-W CO_2 laser and its associated optics along with a computer controlled mirror to guide the laser beam. The basic scheme of the machine is shown in Fig. 7.54. The part building chamber

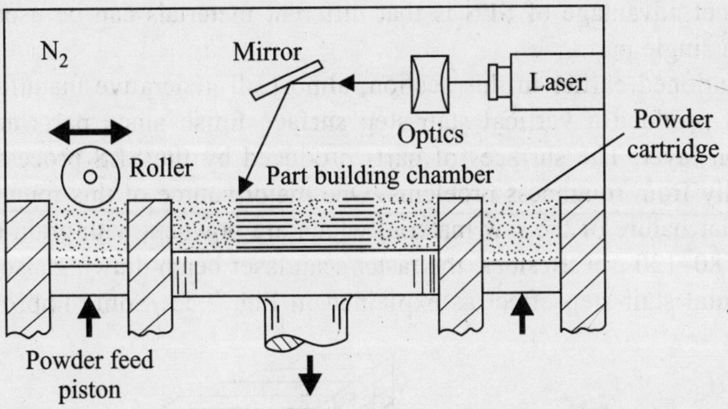

Fig. 7.54 Scheme of selective laser sintering.

is purged with an inert gas, generally nitrogen. The chamber is also maintained at an elevated temperature so that the powder in the top layer is just below the sintering temperature. This helps to minimize the power requirement of the laser beam, and also increases the speed of sintering process.

In the SLS process, a thin layer (a few tenths of a millimeter thick) of powder is spread into a cylindrical part building chamber with the help of a counter-rotating roller which moves from one side to another of the cylindrical chamber flanked by two powder feed cartridges. The powder feed pistons move up with the consumption of powder in creating each layer and keep the powder in the supply cartridges always at the required level. The top layer in the building chamber is raster scanned (closely spaced parallel lines) with a 50-W CO_2 laser beam. The requisite area, representing the part geometry at that particular layer on which the laser beam scans through, fuses together and solidifies. Since the chamber is maintained at an elevated temperature, it not only reduces the power requirement of the laser beam but also helps to keep thermal shrinkage during fabrication low keeping the part distortion minimal. After every layer is sintered, the part supporting platform is lowered by the equivalent of one layer thickness and a fresh coat of powder is laid. The process is then repeated. The unsintered powder remains to support the developing part as a form-fitted 'cake'. Thus, the part is well stabilized during the building process as already mentioned. After all layers are built, the supporting piston is raised; the unsintered powder mostly falls off the part and a spatula is used to remove the additional powder. The unsintered powder can be reused. A layer of unsintered powder, about 25 mm thick, is left covering the part that serves as an insulation which helps to reduce distortion as the part cools. The coating is removed by brushes and air jets after the cooling is over.

The parts produced by SLS are of low density as no compaction pressure is applied unlike that in the conventional powder metallurgy based processes. The parts produced by sintering polyvinyl chloride have a relative density of only 60% (i.e., 40% of the part volume is air). The materials used for SLS include PVC, polycarbonate, investment wax, nylon, and ceramic and metal powders. One distinct advantage of SLS is that different materials can be used while building a single part.

As mentioned earlier in this section, almost all generative manufacturing processes produce a vertical stair-step surface finish since parts are built layer upon layer. The surfaces of parts produced by the SLS process suffer additionally from roughness problems. One major source of this roughness is the granular nature of the raw material which are powders with grain sizes in the range 80–120 μm. Besides, the raster scan laser beam drawing also results in horizontal stair-step effect as explained in Fig. 7.55. Some improvement

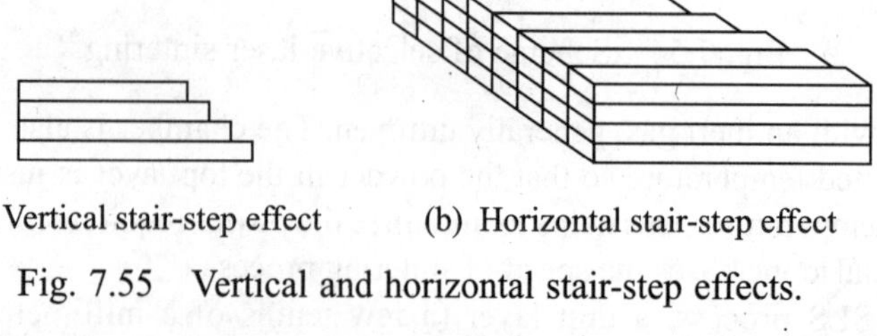

(a) Vertical stair-step effect (b) Horizontal stair-step effect

Fig. 7.55 Vertical and horizontal stair-step effects.

of surface finish may be possible by rotating the orientation of raster by 90° at every alternate layer. This will distribute the roughness evenly on all surfaces. Further improvement of surface finish is possible by outlining each cross-section prior to the drawing of rasters, though this will increase the build time.

Selective Powder Binding (SPB) or Three-Dimensional Printing

Selective powder binding (also called three-dimensional printing) was originally developed at MIT in the early 1990s. The process is based on creating a solid object from a refractory powder by selective binding through the application of a colloidal liquid silica binder. The liquid binder is applied selectively to thin layers of powder, in the form of droplets, using the inkjet technology causing the particles of powder to stick together wherever the binder droplets are applied. Thus, it has certain similarities with SLS. Soligen licensed the three-dimensional printing technology patents from MIT for direct shell production casting. Z Corp also procured licence of the three-dimensional printing technology for building models.

The inkjet technique has also been used to develop solid objects by directly injecting material droplets at required locations. Either one or two inkjets can be employed for depositing tiny drops of hot liquid thermoplastic materials. More than one material can be used for making the part if two inkjets are used. Usually, one jet is employed for creating the support structure and the other forms the main part. With a different material, the support structure can be easily removed after the part formation is complete. Solidoscape commercialized such inkjet machines.

In the SPB type three-dimensional printing process, a thin layer of ceramic powder is laid on a flat bed (Fig. 7.56). Next, a fine jet of ceramic binder is ejected

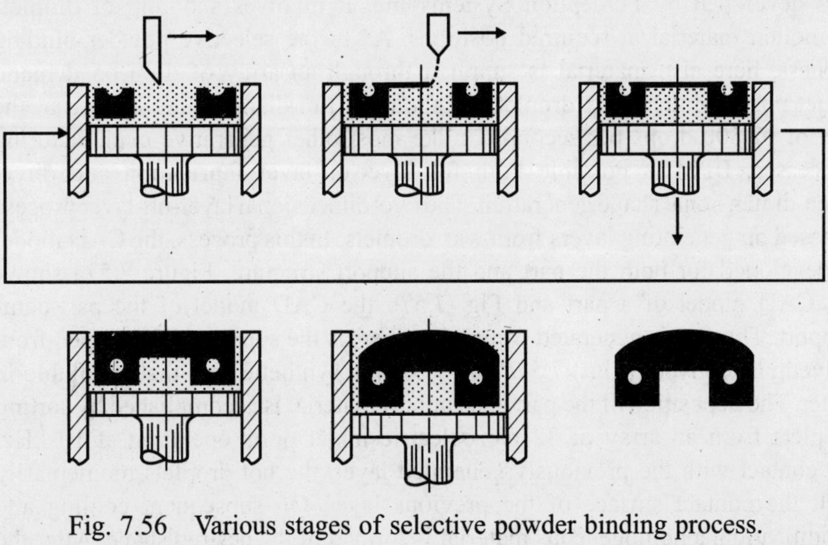

Fig. 7.56 Various stages of selective powder binding process.

onto the powder at locations where solidification is desired. The application of binder droplets can be done in two different ways. In the drop on demand technique, a droplet is ejected by the inkjet mechanism when a drop is needed while the inkjet mechanism traverses the layer of powder. In continuous jet systems, the droplets are ejected continuously while the nozzle traverses the layer surfaces. But at locations which are not to be solidified the inkjet is diverted by an electric field. The droplets are electrically charged while leaving the nozzle tip. The nozzle is moved across the powder surface in a raster scan while the computer generated electrical signals control the deposit of the binder. After the completion of the selective binding operation of the powder layer, the platform supporting the part is lowered by one layer thickness and the cycle is repeated. Figure 7.56 shows the various stages of the process, including the completed part. The inkjet based printhead consists of an array of a large number of jet nozzles, each one capable of operating at few tens of kHz. Typically, for a layer of 0.5 m × 0.5 m, it takes about 4 sec to complete the cycle using the drop on demand technique. When continuous jets are used, the time for solidification of the layer can be much smaller and even as low as a fraction of a second.

After all the layers are completed, the part is cured at 120°C for about two hours. The unbound powder is then removed. For ceramic parts, a final firing at 1000–1500°C is required to impart the object its full strength. The typically used powders are aluminium oxide, zirconia, zircon, and silicon carbide. The minimum feature size is about a fraction of a mm. Difficulty in achieving good surface finish is one of the major problems of the process. The removal of unbound powder from narrow passages and enclosed cavities also poses difficulties.

Ballistic Particle Manufacturing (BPM): Layered Process

This technique for creating three-dimensional solid objects from the CAD model was developed by Perception Systems Inc. It involves shooting of droplets of molten material at required positions. As in the selective powder binding process, here also material is supplied through an array of drop on demand inkjet ports. Molten wax droplets of about 50 μm diameter are ejected at the rate of 12,500 drops per second. Unlike most other generative manufacturing processes, BPM is possible both for layer-on-layer fabrication and direct three-dimensional shape generation. The two-dimensional layer-on-layer process is based on generating layers from wax droplets. In this process, the CAD model is developed for both the part and the support structure. Figure 7.57a shows the CAD model of a part and Fig. 7.57b the CAD model of the part-cum-support. The part is generated from wax whereas the support is developed from polyethylene glycol (Figs. 7.57c and 7.57d), a synthetic wax that is soluble in water. The deposition of the part and support material is accomplished by sorting droplets from an array of 32 piezoelectric inkjet ports operating at 10 kHz. On contact with the previously generated layer, the hot droplets momentarily melt the contact surface of the previous layer. On subsequent cooling and solidification, a homogeneous material is formed of the desired shape. After the completion of deposition of all layers, the object is placed in warm water bath to dissolve the support material, leaving the desired part. The accuracy of the process depends on the accuracy of the position of droplets, which in turn is

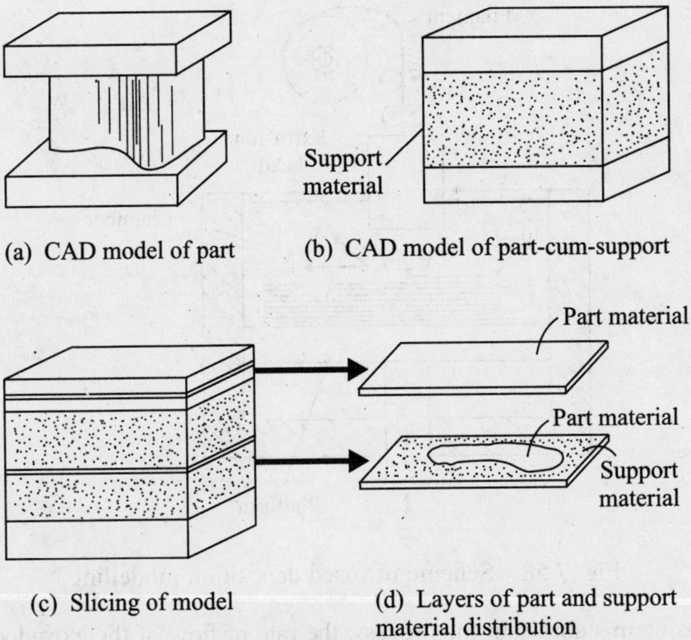

Fig. 7.57 Support material configuration in GMPs.

dependent on the accuracy of the location of the piezoelectric jet system and the ballistic paths of the individual droplets. Thus, it is desirable that the jet ports be as close to the substrate as possible. The layer thickness is monitored by a feedback loop using proximity sensors for measuring the distance between the jet ports and the substrate. In the commercially available systems, parts have been generated with 90 μm layer thickness.

Fused Deposition Modelling (FDM)

This is one of the more popular methods for generative manufacturing. In this process, three-dimensional objects are produced by depositing a molten thermoplastic material layer by layer. Stratasys Inc. is commercially manufacturing machines for FDM. A solid filament of thermoplastic material with 1.25 mm diameter is fed into an *x-y* controlled extrusion head. The material is melted by a resistance heater at a temperature of 180°F (1°F above its melting temperature). As the head is moved along the required trajectory using computer control, the thermoplastic material is deposited by extruding it through a nozzle by a precision volumetric pump. As the extruded material, deposited as a fine layer, comes out with a temperature just above the melting point, it resolidifies within 0.1 second by natural cooling. To ensure proper adhesion of the deposited fused material to the previously deposited layer, the object temperature is maintained just below the solidification temperature. After one layer is deposited, the platform, supporting the object, is lowered by one layer thickness. Figure 7.58 shows the process schematically.

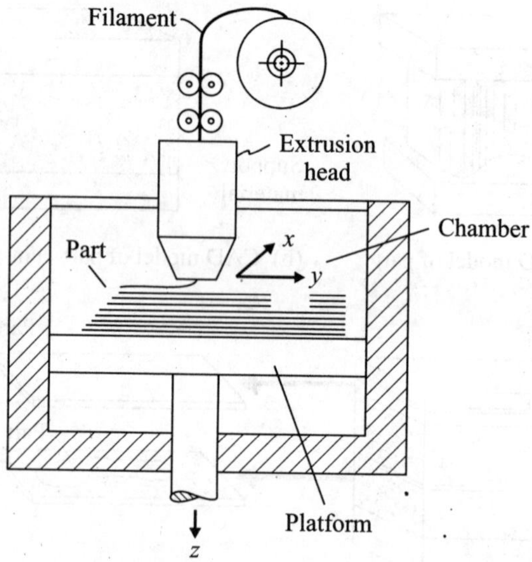

Fig. 7.58 Scheme of fused deposition modelling.

To maintain stability in the process, the rate of flow of the extruded molten filament is controlled to match (i) the travelling speed of the depositing head (which can go upto 380 mm/sec), (ii) the desired thickness of the layer (that varies from 0.025 mm to 1.25 mm), and (iii) the width of the deposited line (which varies from 0.23 mm to 6.25 mm). The repeatability and positional accuracy of this process are claimed to be about ±0.025 mm with an overall tolerance of 0.125 mm over a cube with 305 mm sides. The FDM process is still not very suitable for parts with very small features. The typically-used materials for the process include investment casting wax, wax filled adhesive material, and tough nylon-like material. Polymer type thermoplastics can also be used.

Other Layered Type Processes

There are quite a few other layered type generative manufacturing processes which have been developed with lesser degree of commercial success. Of these, two are briefly described here.

Laminated object manufacturing (LOM) In this process, parts are produced by successive bonding of layers of sheet materials (mostly paper type material) and laser cutting of each cross-section. Thin sheets of plastics and composites are also used. The part acquires a wood-like structure and quality. Helisys Inc. released the first commercial model of LOM system. Figure 7.59 shows the basic features of LOM schematically. Sheet material is supplied from a roll and the unused portion of the material is wound up at the take-up roll. The sheet is coated with a heat sensitive adhesive. When the fresh material comes over the work table, a heated roller presses it down to the uppermost layer of the object being fabricated. Then, a CO_2 laser beam cuts the outline of respective

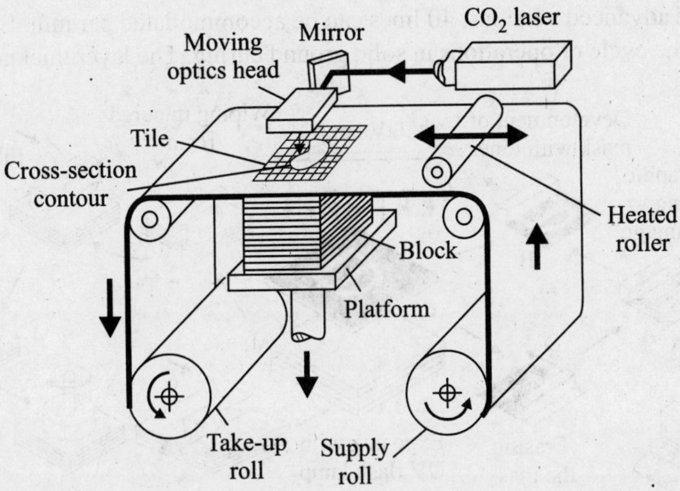

Fig. 7.59 Scheme of laminated object manufacturing (LOM).

cross-sections. The beam intensity and speed are so adjusted that only one sheet is cut. The sheet material outside the desired cross-section is cross-hatched by the beam into squares (called *tiles*) so as to separate the part easily after it is generated. After the cutting of one layer is completed, the platform is lowered by one sheet thickness and the rollers supply a fresh area of the sheet material. Interlayer adhesion near the boundaries is a problem for LOM. This problem is partially reduced by a method called *burn out*. The area on the previously laid layer where gluing is undesirable is cut with a tightly spaced cross-hatched pattern. A hollow part cannot be generated by LOM as the excess material remains trapped inside.

Solid ground curing (SGC) In this process, whole layers are simultaneously cured according to the required cross-section. Cubital marketed an SGC system called Solider. A thin layer of liquid photosensitive resin is applied and then exposed to a strong UV radiation through a mask. The transparent areas of the mask correspond to the desired cross-section of a particular layer. The UV radiation solidifies the exposed areas of the resin and the uncured liquid resin is removed from the unexposed areas. These areas are replaced by wax to build up the support structure. Finally, the cured resin and the deposited wax are both machined to a predetermined thickness using an end mill cutter. The cycle is then repeated till the complete object is formed. Concurrently, another cyclic operation is performed to prepare the mask for each layer. A glass plate is ionographically charged to create a pattern according to the required cross-section, which is then developed by using a toner, as done in photocopying units. Once the curing of the required cross-section is over, the pattern from the glass mask is erased and it is used for creating the mask for the next layer. A computer containing the part geometry controls the charge pattern. Typically, the charge is deposited on the glass plate in raster lines with 11.8 lines per mm.

In more advanced versions, 40 lines can be accommodated per mm. Figure 7.60 shows the cycle of operations in solid ground curing. The layer thickness ranges

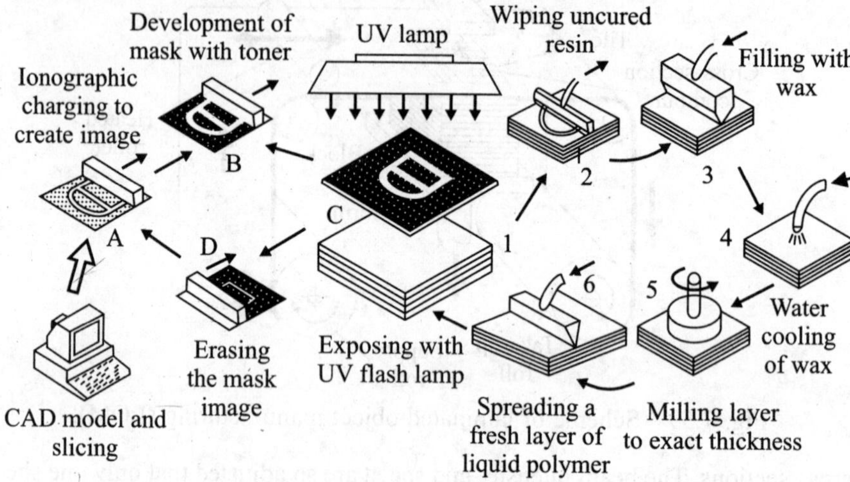

Fig. 7.60 Cycle of operations in solid ground curing (SGC).

from 0.03 mm to 1.27 mm. High viscosity low shrinkage resins can be used for the SGC process. Since curing is done of a whole layer, the build time per layer is independent of part geometry, and it is about 50 seconds. An accuracy of 0.03 mm can be achieved in a 25 mm part dimension.

Direct Three-Dimensional Generative Manufacturing Processes

Though most of the commercially available generative manufacturing process units fabricate parts layer by layer, it is possible to build objects directly in three-dimensional space. In all layer-by-layer methods, the lower layers have to be created before the next layer is deposited. But direct three-dimensional techniques do not require to create the lower portions first. Direct three-dimensional techniques can be of three types as follows:

- Shape generation through point by point.
- Shape generation through surface by surface.
- Shape generation through simultaneous creation of the whole object.

Some techniques discussed now are still at the developing stage.

Beam interference solidification (BIS) This method was patented by Formiographic Engine. The material which is used in the process is a photosensitive transparent liquid plastic (monomer). When the monomer is subjected to a laser beam of a particular frequency, it reaches a reversible metastable state and no bonding reaction takes place. But when a part of the liquid that is already in a metastable state is hit by another laser beam of a specific (but different) frequency, polymerization of the metastable state takes place, resulting in the solidification of a voxel (volume picture cell) at the intersection of the two beams. The basic scheme of the process is shown in Fig. 7.61. By moving the

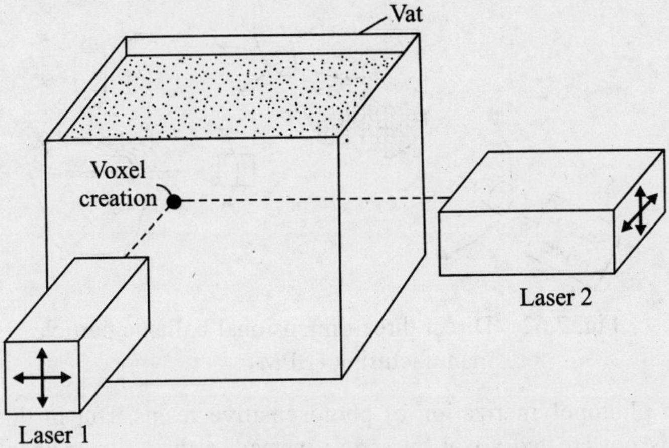

Fig. 7.61 Principle of beam interference solidification.

two laser guns in a particular way, the volume of the desired object can be generated voxel by voxel. In spite of the elegance of the concept, there are a number of serious difficulties for its practical application. The intensity of the beam decreases continuously during its passage through the resin because of absorption. This poses a serious difficulty in programming the laser beams to maintain uniformity in the characteristics of all voxels. The problem is further compounded because of shadow effects produced by the portions of the object already solidified.

Ballistic particle manufacturing (BPM) The process has already been discussed in its layer-by-layer application. In the direct three-dimensional approach, the part building is achieved by shooting of molten droplets on the top of each other. Figure 7.62 shows the technique schematically. Two piezoelectric inkjet printing nozzles are guided by manipulators to deposit the molten droplets according to the need from any direction instead of using an array of such inkjet nozzles as done in the layer-by-layer approach. Six-axis robots can be used for the purpose. One important advantage of this approach is the elimination of support structures.

Direct metal deposition (DMD) The University of Michigan, Ann Arbor, has commercialized a laser based machine for generating three-dimensional metallic parts. Metal powder is supplied which is melted by a high power laser beam and functional metallic part can be generated. The deposition head is guided by a manipulator.

Holographic interference solidification (HIS) This exotic process is also

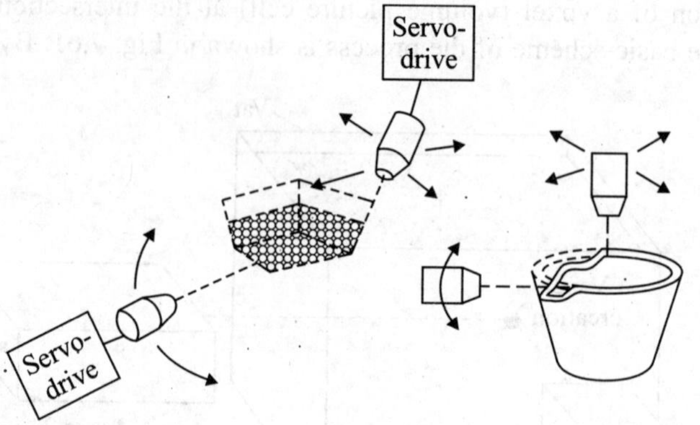

Fig. 7.62 Direct three-dimensional ballistic particle manufacturing (BPM).

based on photopolymerization of photosensitive resins. But in this process, the part is not created voxel by voxel; instead, a three-dimensional image is projected in a vat containing a photosensitive liquid monomer and a whole three-dimensional surface gets solidified as a whole. The holographic film for projecting the image is created with a CAD system. A system based on this principle has been developed by Quadtec Pty., Melbourne.

The long-standing desire of manufacturing engineers to produce solid parts directly from the design data stored in a computer seems to be on the verge of realization. Though the current methods of generative manufacturing are used for rapid prototyping, primarily, the trends of development indicate that soon these processes will be used for production of functional parts. Some indirect benefits will also be there once functional parts are produced by these techniques. At present, products have a limited life mainly because of the cost of spare parts and their nonavailability after some time. Keeping a large stock of spares for an indefinite period of time means blocking of capital. But when generative processes will mature enough to produce spare parts, spares can be stored electronically eliminating the blockage of capital.

The major advantages of the generative manufacturing processes can be summarized as follows:

- No tools and fixtures are required.
- No restriction on geometry of the part shape exists.
- Composite parts and assemblies can be produced in one go.
- The process is very suitable for computer integration.
- Small batch production of complex parts is economically viable.

The generative processes will also play a very major role in manufacturing microparts. These processes are also responsible for bringing *bottom-up* approach in the manufacturing scenario and can be considered to be the forerunner of the ultimate process—manufacturing by self-assembly of material. The next section offers a glimpse of the possibilities.

7.7 SELF-ASSEMBLY

7.7.1 WHY SELF-ASSEMBLY

The ultimate in manufacturing is the process through which three-dimensional devices and objects will be manufactured by self-assembly of material(s) without any continued intervention of human action. It has been emphasized earlier in this chapter that designing and developing autonomous, intelligent micron-sized (even nano-sized) machines will be a major development in future engineering activities. However, to successfully exploit the potential of such devices, it is essential that the manufacturing cost be low. It has also been mentioned that the top-down manufacturing processes are limited in their capacity to manufacture micron-sized (or nano-sized) three-dimensional features. The currently practised photolithographic techniques are capable of producing submicron level features (as required in VLSI chips) but the capability of lithographic processes is limited to two-dimensional systems primarily. For producing truly three-dimensional objects, the scope of lithography is very doubtful. There have been some recent developments like focussed ion beam (FIB) technology, using which even nano-sized three-dimensional features can be produced. But, unfortunately, this technique is not very suitable for batch production. Thus, the cost of the objects produced using FIB is prohibitive. The other top-down approaches for micromanufacturing have limited capabilities like microdrilling, microslot cutting, etc. Besides, the cost of manufacturing is generally high.

From the previous sections, it is clear that the bottom-up approach can be very suitable for developing very small-size three-dimensional objects. However, the usual philosophy of adding (or generating) micro-sized elements in succession to develop the whole object is not only time-consuming but also not amenable to batch production. If, on the other hand, the process of material addition takes place spontaneously without human intervention, not only large scale batch production will be possible but the time taken for manufacturing will also be within reasonable limits. The process will be somewhat like producing a huge number of bacteria in a vat. The addition of material will be spontaneous as in the case of the growth of a living object.

7.7.2 WHAT IS SELF-ASSEMBLY

Self-assembly (or self-organization) is a process in which components automatically come together to form aggregates. All living objects and biological systems as well as a large number of nonliving physical systems exhibit self-organizing behaviour. Examples of nonbiological systems formed by self-assembly are many, the most common being crystals, micelle, colloids, and self-assembled monolayers. The key point in this mode of ordered structure formation is that all the information required to specify the desired shape and size is available within the process itself. Self-assembly can involve components from the molecular to the planetary (weather systems) scales and can be based on many different kinds of interactions. Traditionally, self-assembly has been associated with atoms and molecules only and its study has been a common subject among the chemists. However, the subject has gained importance recently with the advent of nanotechnology. There are mainly three scales in which the process operates—molecular, nanoscale (colloids, nanowires, etc.), and meso to macroscopic (objects with dimensions from microns to centimeters). Of course, the subject is at its infancy and neither our understanding nor our control of self-assembly is anywhere near the stage when it will be possible to map out the road to the final destination—self-assembly of microdevices and -machines. It is hoped that the ultimate self-assembled structures will be multifunctional, self-correcting and, perhaps, self-replicating.

Basic Principles of Self-Assembly Process

Like any spontaneously occurring process (like rolling of a ball down an inclined plane resulting in lower potential energy), minimization of energy is the key motivating factor in self-assembly and self-organization. The process is governed by different types of interaction forces acting between assembling components and the resulting structure formation is associated with minimization of energy. The self-assembly processes can be classified into two main groups as now explained:

- *Uncoded self-assembly* In these processes, minimization of energy is the only motivation while forming aggregates. No very specific geometric features are involved.
- *Coded* (*directed*) *self-assembly* With uncoded self-assembly, formation of three-dimensional features is generally not possible. In coded self-assembly, information on the geometric features of the desired object is encoded into the basic components. A suitable combination of the motivation for energy minimization and the geometric features produces complex three-dimensional objects. The best example is the recombination of two DNA strands. There is only a unique way two strands can join.

Successful self-assembly depends on five important requirements, viz.:

- *Components/Elements* Groups of molecules, segments of macro-molecules, nanoparticles/microparticles that interact with one another are the components of a self-assembly process. The primary objective is to create a more ordered structure from the randomly-oriented large number of these basic building blocks.
- *Intercomponent interactions* The creation of an ordered agglomeration of the components requires a suitable interaction among the components which brings them together and keeps them assembled. These interactions should lead to weak bonding so that readjustment of the orientations becomes possible. These interactions motivate the components to come together, overcoming the energy barrier due to thermal fluctuations.
- *Ability to readjust positions* If the interactive forces are too strong, the components get attached to one another in an irreversible manner, i.e., once attached, no further spontaneous change of orientation of their relative positions

is possible. Such a situation leads to the formation of amorphous and disordered structures. Thus, to provide the ability to readjust positions, the strength of the bonds must be comparable to those which try to readjust the positions. Figure 7.63a shows a situation where adjustability is not possible. The assembly process leads to disordered amorphous states. On the other hand, Fig. 7.63b

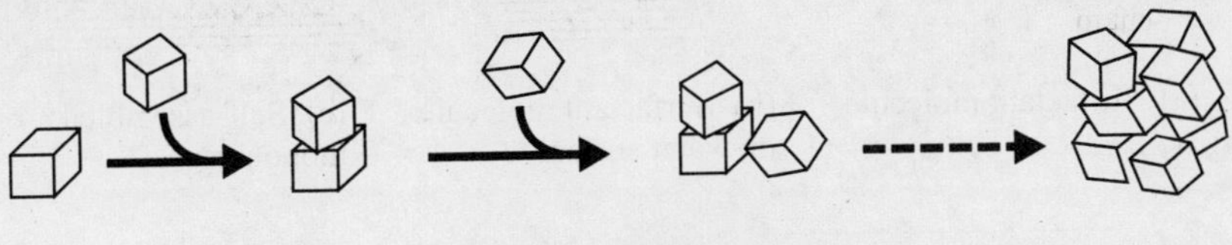

(a) Without readjustability

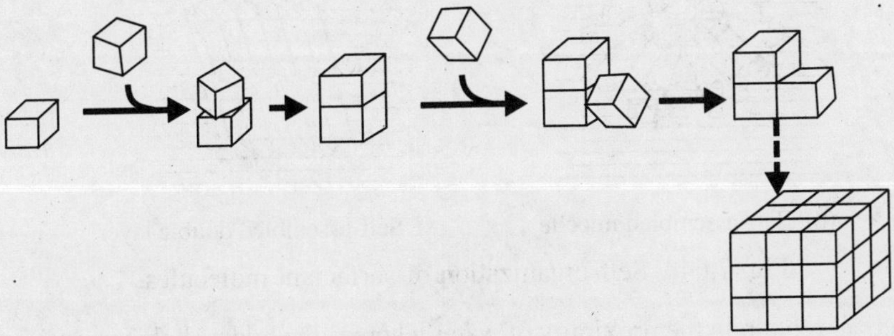

(b) With readjustability

Fig. 7.63 Self-assembly.

shows a self-assembly process with reversibility which leads to an ordered regular shaped object.

- *Environment* All self-assembly processes require mobility of the components and therefore are facilitated in solutions instead of dry environment.
- *Mobility and transportation* The components in a self-assembly process need to be mobile for obvious reason.
- *Equilibration* Reaching a state of equilibrium is necessary in self-assembly processes. Otherwise, the assembled object may not possess structural stability.

As in the case of any spontaneously occurring process, e.g., rolling down of a ball on an inclined plane, resulting in lower potential energy, minimization of total energy is the key motivation for any self-organizing or self-assembly process. The process is governed by different types of interactions among the components with a desire to minimize the overall energy. In appropriate cases, this is achieved through the realization of a particular pattern or shape. The concept of self-organization based on energy minimization can be explained with the classic example of monolayer and micelle formation with surfactant molecules in aqueous solutions. Surfactant molecules have spatially different domains within the structure, and, in the simplest form, consist of a hydrophilic head group and a hydrophobic tail as shown in Fig. 7.64a. The hydrophilic head

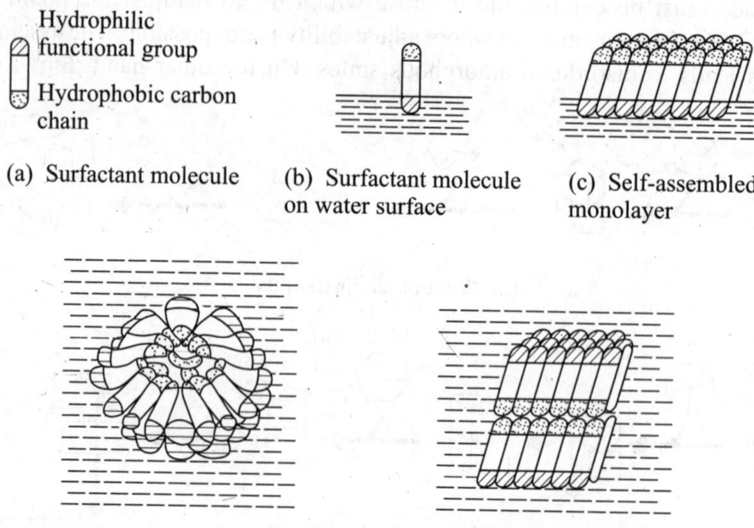

(a) Surfactant molecule (b) Surfactant molecule on water surface (c) Self-assembled monolayer

(d) Self-assembled micelle (e) Self-assembled double layer

Fig. 7.64 Self-organization of surfactant molecules.

prefers to remain in the proximity of water whereas the hydrophobic end tries to avoid contact with water. Such hydrophilic and hydrophobic characteristics can be attributed to their surface tensions. In general, the surface energy (or surface tension) of a hydrophilic material is higher than that of water and is preferentially wetted by water. In contrast, a hydrophobic material has surface energy less than that of water and is not wetted by water. When such a surfactant molecule is put in water, the hydrophilic end is wetted and prefers to remain surrounded by water. Conversely, the hydrophobic end prefers to avoid contact with water. Under the influence of the resulting antagonistic force field originating from the interfacial tensions of different domains of the molecule, it is most suited if the molecule migrates to the water-air interface and orients itself so that the hydrophilic head remains in water and the tail is in the air as shown in Fig. 7.64b. This is the most basic example of self-organization as the orientation of the molecule is predetermined by the interfacial tensions. As more such molecules are added, all of them migrate to the water surface and organize themselves in a head-down position, forming an ordered monolayer (Fig. 7.64c). When the concentration of these molecules is increased, the water-air interface gets fully covered by the monolayer (at this stage, the distance between the neighbouring molecules is governed by the steric repulsion between the head groups—i.e., the heads nearly touch each other as indicated in Fig. 7.64c). The newly added molecules are forced to stay within water, but instead of getting distributed in a random fashion (which is thermodynamically not favourable), the surfactant molecules form spherical aggregates called micelles. The structure of a micelle is shown in

Fig. 7.64d which hides all the hydrophobic ends from water. Depending on the situation, a bilayer can also form as shown in Fig. 7.64e. There exists a critical concentration [known as critical micellar concentration (cmc)] above which all the surfactant molecules added to the solution will form micelle. If an organic solvent is used instead of water, the scenario becomes just the opposite as the hydrocarbon tail is energetically favoured and the hydrophilic head is repelled. Thus, depending on the concentration, either a monolayer or reverse micelle is formed automatically. However, many other kinds of other structural aggregates like rods, lamella, etc., can also form depending on the situation.

Many different types of patterns and structures can be self-assembled by designing the basic components. Figure 7.65 shows a particular case where a

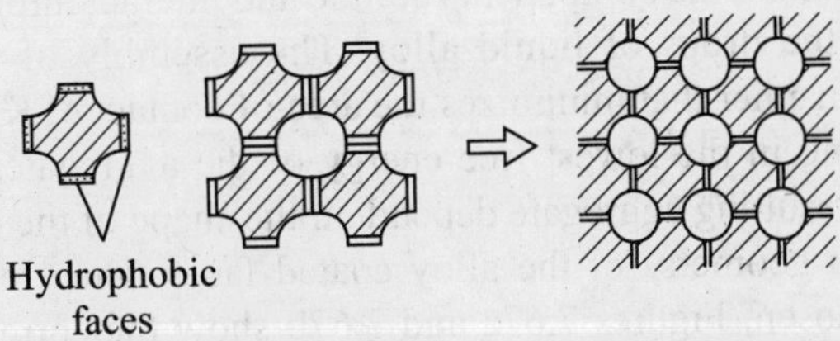

Fig. 7.65 Self-assembly of nanoparticles with designed characteristics.

porous structure is developed. When hexagonal components with alternate faces as hydrophobic and hydrophilic are used, again a porous nanostructured pattern is generated as shown in Fig. 7.66a. On the other hand, if all the faces are made hydrophobic, a nonporous structure is formed as shown in Fig. 7.66b. Thus, it

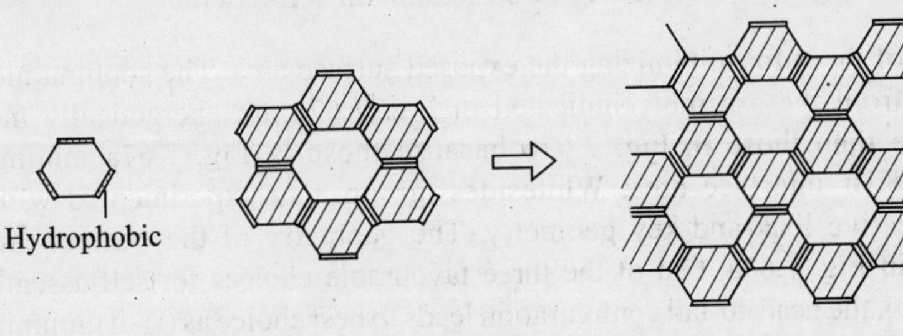

(a) Alternate faces hydrophobic

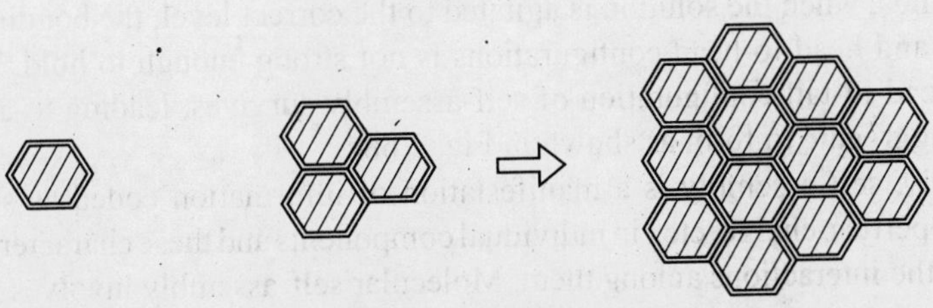

(b) All faces hydrophobic

Fig. 7.66 Self-assembly using hexagonal components.

is seen that more complex shape generation can be achieved when the basic components are designed suitably. Such type of self-organization is termed as templated self-assembly. In such cases, the interactions between the components with a pre-existing regular pattern determine the formation of the final structures. In essence, such type of templated self-assembly combines both the top-down and bottom-up concepts to some extent. We now give examples of templated self-assembly.

Whitesides[1, 2] and his group have attempted such templated self-assembly. They prepared millimeter-size components of different regular shapes and coated the chosen faces with a low melting metallic alloy, and suspended the resulting particles in aqueous KBr solution kept at the melting temperature of the alloy. On agitation, the components collide and interact through the capillary forces between the drops of liquid alloy. The assembly of the components takes place in a manner that minimizes the area of contact of KBr solution with the alloy, resulting in the lowest free energy of the aggregate. The shape and character of the resulting aggregate depend on the shape of the components and the locations and geometry of the alloy coated faces when regular polyhedra and cubes are chosen. Figures 7.67a and 7.67b show the correct and incorrect

(a) Correct matching (b) Incorrect matching

Fig. 7.67 Matching of basic units in self-assembly.

matching of faces for minimizing the exposed surface area. The configurations resulting from the assembly indicated in Fig. 7.67a are energetically more favourable than those in Fig. 7.67b because those in Fig. 7.67a minimize the alloy-KBr interface area. Whitesides' group also experimented with a shape-selective lock-and-key geometry. The geometry of the component is indicated in Fig. 7.68a. Out of the three favourable choices for self-assembly (Fig. 7.68b), the head-to-tail configuration leads to best choice as (i) it minimizes the area of exposed hydrophobic faces, and (ii) it leads to better kinematic stability. Thus, when the solution is agitated to the correct level, the bonding for tail-to-tail and head-to-head configurations is not strong enough to hold. Thus, only the head-to-tail configuration of self-assembly survives, leading to a long strip-like aggregate to form as shown in Fig. 7.68c.

Generally, self-assembly is a manifestation of information coded as shape, surface properties, charge, etc., in individual components and these characteristics determine the interactions among them. Molecular self-assembly involves weak

[1]Breen, T.L., Tien, J., Oliver, S.R.J., Hddzic, T. and Whitesides, G.M., *Science*, **284**, 1999.
[2]Bowden, N., Terfort, A., Carbeck, J. and Whitesides, G.M., *Science*, **276**, 1997.

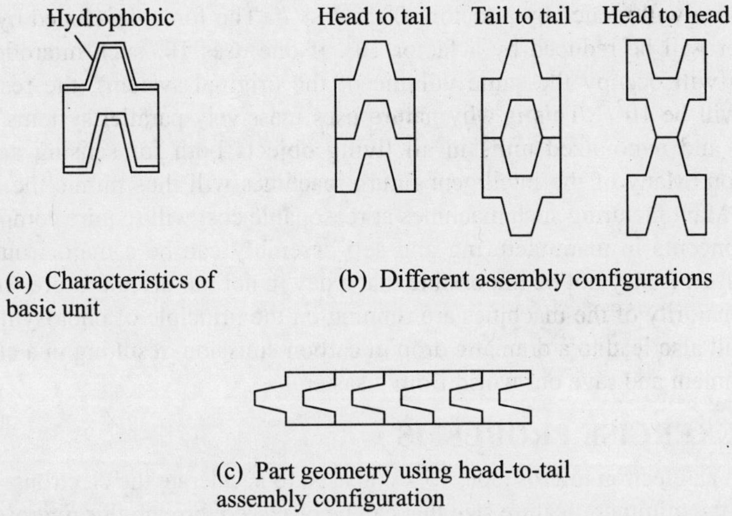

Fig. 7.68 Templated self-assembly with lock-and-key geometry.

covalent interactions like van der Waals, electrostatic, acid-base interactions, interactions based on hydrophobicity and hydrophilicity, hydrogen bonds, etc. In contrast, in the case of components of larger sizes (tens of nanometers to hundreds of microns), interactions such as electric and magnetic fields, capillary and entropic interactions are of interest. For self-assembly to generate structures more complex than crystals, it is important that the components must come together only in some predetermined unique way.

7.8 FUTURE POSSIBILITIES

Manufacturing processes are primarily based on the needs of making things and, as a result, they depend on the nature of the ultimate products. Thus, in future, the new concepts in manufacturing science will depend on the nature of the future machines and devices. Although the conventional machines and devices will continue to exist, a completely new brand of machines and devices will emerge which will be closely linked with the principles of life science. A totally new subject, *synthetic biology*, has just started appearing on the scene. It is aimed at developing inanimate devices operating on the principles of biology and using biochemical energy. The main objective of developing a new class of machines will be to make them truly intelligent and autonomous. Apart from the need to use a huge number and enormous range of sensing devices, the actuation of the moving members will follow biological principles. One example can explain the basic idea. Let a device of volume V develop a force F for actuation following the conventional engineering principle which uses the electromagnetic phenomena for generating the forces. It is known that the electrostatic force depends on l^2, where l represents the length dimension of the actuator/motor. On the other hand, the electromagnetic force varies as l^4. Now, if we reduce the size of an electrostatic actuator by a factor 100, the volume of the device will reduce by a factor 10^6 as $V \propto l^3$. The force developed by each actuator will be reduced by a factor 10^4. If one uses 10^6 such microdevices (which will occupy the same volume of the original system), the resultant force will be 10^2F. This is why nature uses massively parallel systems using micro- and nano-sized units in all living objects both for sensing and for actuation. Many of the intelligent future machines will thus mimic the living world. Manufacturing such machines at reasonable cost will require completely new concepts in manufacturing and self-assembly can be a major route for such cases. It may not be impossible that a day in not too distant future will see that a majority of the machines are running on the principle of photosynthesis. This will also lead to a dramatic drop in carbon emission, resulting in a cleaner environment and save our world from disaster.

7.9 EXERCISE PROBLEMS

7.1 In an electron microscope, 75 kV is used to accelerate the electrons. What will be the minimum feature size that can be observed through this microscope?

7.2 Find out the required rotational speed for spin coating a resist layer of thickness 5 μm. The spinner constant is 80 and the resist solid content is 75%.

7.3 In a lithographic operation, the wavelength of the UV exposing radiation is 600 nm. The thickness of the photoresist is 2 μm and the minimum allowable gap between the mask and the resist is 15 μm. Estimate the minimum feature dimension that can be achieved in the process.

7.4 A through hole of 4 μm diameter has to be machined in a 60-μm Si wafer using the DRIE process. Assuming the operation to be continuous, estimate the time required for machining.

7.5 Draw the Wulff-Jaccodine diagram for cutting a slot (measuring 25 μm at the top surface) in a (100) Si wafer using KOH.

7.6 Determine the minimum dose of exposure for developing a 250-μm PMMA layer during a LIGA process if the threshold dose is 2 kJ/cm^3 and the time allowed is 2 hr.

Bibliography

General

Campbell, J.S., Principles of Manufacturing Materials and Processes, McGraw-Hill, New York, 1961.

Cook, N.H., Manufacturing Analysis, Addison-Wesley, Reading, Massachusetts, 1966.

DeGarmo, E.P., Materials and Processes in Manufacturing, Collier Macmillan, New York, 1979.

Doyle, L.E., Keyser, C.A., Leach, J.L., Scharder, G.F., and Singer, M.B., Manufacturing Processes and Materials for Engineers, Prentice-Hall, Englewood Cliffs, New Jersey, 1969.

Lindberg, R.A., Processes and Materials of Manufacture, Allyn and Bacon, Boston, 1977.

Pollack, H.W., Manufacturing and Machine Tool Operations, Prentice-Hall, Englewood Cliffs, New Jersey, 1979.

Radford, J.D. and Richardson, D.B., Production Engineering Technology, Macmillan, London, 1969.

Schey, J.A., Introduction to Manufacturing Processes, McGraw-Hill, New York, 1977.

Chapter 1

Anderson, J.C. and Leaver, K.D., Materials Science, Thomas Nelson, London, 1971.

Dieter, G.E., Mechanical Metallurgy, McGraw-Hill, New York, 1976.

Smith, C.O., The Science of Engineering Materials, Prentice-Hall, Englewood Cliffs, New Jersey, 1977.

Wulff, J. (ed.), The Structure and Properties of Materials, Vols. 1–4, Wiley, New York, 1964–66.

Chapter 2

Flinn, R.A., Fundamentals of Metal Casting, Addison-Wesley, Reading, Massachusetts, 1963.

Geiger, G.H. and Poirier, D.R., Transport Phenomena in Metallurgy, Addison-Wesley, Reading, Massachusetts, 1973.

Veynik, A.I., Thermodynamics for the Foundryman, MacLaren, London, 1968.

Chapter 3

American Society of Tool and Manufacturing Engineers, High Velocity Forming of Metals, Prentice-Hall, Englewood Cliffs, New Jersey, 1964.

Avitzur, B., Metal Forming: Processes and Analysis, McGraw-Hill, New York, 1968.

Dieter, G.E., Mechanical Metallurgy, McGraw-Hill, New York, 1976.

Eary, D.F. and Reed, E.A., Techniques of Pressworking Sheet Metal, Prentice-Hall, Englewood Cliffs, New Jersey, 1974.

Masterov, V. and Berkovsky, V., Theory of Plastic Deformation and Metal Working, MIR Publishers, Moscow, 1975.

Rowe, G.W., Elements of Metalworking Theory, Edward Arnold, London, 1979.

Chapter 4

Armarego, E.J.A. and Brown, R.H., The Machining of Metals, Prentice-Hall, Englewood Cliffs, New Jersey, 1969.

Boothroyd, G., Fundamentals of Metal Machining and Machine Tools, Scripta Book Company, Washington, 1975.

Sen, G.C. and Bhattacharyya, A., Principles of Metal Cutting, New Central Book Agency, Calcutta, 1969.

Shaw, M.C., Metal Cutting Principles, MIT Press, Cambridge, Massachusetts, 1957.

Trent, E.M., Metal Cutting, Butterworths, London, 1977.

Chapter 5

Houldcroft, P.T., Welding Process Technology, Cambridge University Press, London, 1977.

Lancaster, J.F., Metallurgy of Welding, Brazing and Soldering, Allen and Unwin, London, 1980.

Little, R.L., Welding and Welding Technology, McGraw-Hill, New York, 1973.

Nikolaev, G. and Olshansky, N., Advanced Welding Processes, MIR Publishers, Moscow, 1977.

Chapter 6

Bhattacharyya, A., New Technology, Institution of Engineers (India), Calcutta, 1973.

de Barr, A.E. and Oliver, D.A., Electrochemical Machining, Elsevier, Amsterdam, 1968.

Esibyan, E., Plasma Arc Equipment, MIR Publishers, Moscow, 1973.

Lazarenko, B.R., Electrospark Machining of Metals, Consultants Bureau, New York, 1964.

McGough, J.A., Principles of Electrochemical Machining, Chapman and Hall, London, 1974.

Pande, P.C. and Shan, H.S., Modern Machining Processes, Tata McGraw-Hill, New Delhi, 1980.

Rosenberg, L.D., Kazantsev, V.F., Makarov, L.O., and Yakhimovich, D.F., Ultrasonic Cutting, Consultants Bureau, New York, 1964.

Rykalin, N., Uglov, A., and Kakora, A., Laser Machining and Welding, MIR Publishers, Moscow, 1978.

Springborn, R.K., Nontraditional Machining Processes, American Society of Tool and Manufacturing Engineers, Dearborn, Michigan, 1967.

***Chapter** 7*

Bucknall, David G., Nanolithography and Patterning Techniques in Microelectronics, Woodhead, Cambridge (U.K.), 2005.

Dai, Liming, Intelligent Macromolecules for Smart Devices, Springer-Verlag, New York, 2003.

Ghosh, Amitabha, Rapid Prototyping: A Brief Introduction, East-West Press, New Delhi, 1997.

H Su, Tai-Ran, MEMS & Microsystems: Design and Manufacture, Tata McGraw-Hill, New Delhi, 2002.

Jacobs, Paul F., Rapid Prototyping and Manufacturing, Society of Manufacturing Engineers, Dearborn, Michigan, 1992.

Madou, Marc J., Fundamentals of Microfabrication, 2nd edn., CRC Press, Boca Raton, 2002.

Maluf, Nadim, An Introduction to Microelectromechanical Systems Engineering, Artech House, Boston, Massachusetts, 2000.

Answers to Selected Problems

Chapter 2

2.3 8538 cm^3 **2.4** 4.45 cm^2
2.6 $[2(0.5+2e_{fs})^{1/2}/g][(p/\rho-gh_2)^{1/2}-\{p/\rho-g(h_2+h_3)\}^{1/2}]$ **2.7** 11.08 sec
2.8 26.23 sec **2.9** 4.07 cm, 1.55 cm **2.10** No **2.11** 1527°C
2.13 65.81 mm **2.14** $t_{cyl}=1.084t_{par}$
2.15 (i) 10.89 cm, (ii) 9.98 cm, (iii) 5.64 cm **2.16** 1016 cm^3

Chapter 3

3.1 (i) 0.1, (ii) 0.1 rad, (iii) $\theta_n = 1.379°$
3.2 (i)(a) 680 kN (b) 34.8 kW, (ii) 1.3 mm **3.3** 36.08 kW
3.4 0.79 MN **3.5** 1.13 MN **3.6** 5.23 MN **3.7** 31.6 kW
3.8 12.7 mm → 11.34 mm → 10.2 mm
3.11 0.14 MN, 85.6% (first approach); 0.17 MN, 80.3% (second approach)
3.12 45.161 kN **3.13** (i) 10.055 kN, (ii) 6.14 mm, plastic buckling
3.14 (i) 4.5 mm, (ii) 89.86°, (iii) 283.4 N
3.16 (i) 0.257 mm, (ii) 309, (iii) 0.72 W/punch
3.18 Reduce half-cone angle to 20°

Chapter 4

4.1 25.9°, 22.95°, 18.17°, 10.2°; 2.26, 2.59, 3.28, 5.74
4.2 (i) 0.86, (ii) 480.7 N, 413.5 N **4.3** 56.4° **4.5** 233.4 MPa
4.6 280 N **4.8** 306°C **4.9** 710°C
4.10 $F_C = 730.8$ N, $F_f = 423.5$ N, $F_n = 154.1$ N
4.12 $F_C = 541.9$ N, $F_f = 246.3$ N, $F_R = 114.8$ N, 638.4 W
4.13 29.9 N·m **4.14** 4.1 N·m, 1004.6 N
4.15 (i) 4398 W, (ii) 4070 N, 717.7 N **4.16** 3331.4 W
4.17 4.93 sec **4.18** +24.6% **4.19** 2828 rpm
4.20 94.86 W, 189.72 W **4.21** C 30 Q·12 V **4.22** 19.12 m/min
4.23 20.6 m/min

Chapter 5

5.1 80 V, 1000 A **5.2** 590.6 A ≈ 631.5 A **5.4** $\sqrt{2}$
5.5 0.027 m/sec **5.6** 0.544

Chapter 6

6.1 13.36 min **6.2** 63.13 min **6.3** 0.37 cm^3/sec
6.4 0.0363 cm^3/sec **6.5** 2.5 cm^3/min **6.6** 0.4 cm/min
6.7 0.62 cm/sec **6.9** 3 hr 2.9 min **6.10** 3 hr **6.11** 32.33 cm/sec
6.12 0.67 μsec **6.13** 0.597 msec

Index